Biomedical Science and Engineering

Biomedical Science and Engineering

Edited by **Mark Walters**

C LANRYE
INTERNATIONAL

New Jersey

Published by Clanrye International,
55 Van Reypen Street,
Jersey City, NJ 07306, USA
www.clanryeinternational.com

Biomedical Science and Engineering
Edited by Mark Walters

International Standard Book Number: 978-1-63240-085-7 (Hardback)

Printed in the United States of America.

Contents

Preface

This book was inspired by the evolution of our times; to answer the curiosity of inquisitive minds. Many developments have occurred across the globe in the recent past which has transformed the progress in the field.

A comprehensive account of biomedical science and engineering is available in this book. Biomedical science is a vast field dealing with disease progression, paradigms and therapeutic measures. For instance, biomaterial implants are analyzed under biomedical sciences. Biomedical engineering encompasses biological materials, clinical test assessments, medical devices like contact lenses etc. Mathematical drafting of physiological systems and their assessment comes under physiological engineering. Even in hospital management, both biomedical science and engineering are required in order to run hospitals efficiently. This book will prove to be a valuable source for clinicians, scientists and students.

This book was developed from a mere concept to drafts to chapters and finally compiled together as a complete text to benefit the readers across all nations. To ensure the quality of the content we instilled two significant steps in our procedure. The first was to appoint an editorial team that would verify the data and statistics provided in the book and also select the most appropriate and valuable contributions from the plentiful contributions we received from authors worldwide. The next step was to appoint an expert of the topic as the Editor-in-Chief, who would head the project and finally make the necessary amendments and modifications to make the text reader-friendly. I was then commissioned to examine all the material to present the topics in the most comprehensible and productive format.

I would like to take this opportunity to thank all the contributing authors who were supportive enough to contribute their time and knowledge to this project. I also wish to convey my regards to my family who have been extremely supportive during the entire project.

Editor

Part 1

Biomaterials and Implants

Multifunctional Magnetic Hybrid Nanoparticles as a Nanomedical Platform for Cancer-Targeted Imaging and Therapy

Husheng Yan[1], Miao Guo[1] and Keliang Liu[2]
*[1]Key Laboratory of Functional Polymer Materials, Ministry of Education,
Institute of Polymer Chemistry, Nankai University, Tianjin,
[2]Beijing Institute of Pharmacology and Toxicology, Beijing,
P. R. China*

1. Introduction

Nanotechnology offers tremendous potential for use in biomedical applications, including imaging, disease diagnosis, and drug delivery. The development of nanosystems has improved the molecular understanding of many diseases and permitted the controlled nanoscale manipulation of materials (Couvreur & Vauthier, 2006). Nanomedical platforms offer many advantages as delivery, sensing, and image-enhancing agents. In recent years, many studies have focused on multifunctional nanomedical platforms that incorporate therapeutic and diagnostic agents with molecular targeting capabilities. Gregoriadis et al. first proposed liposomes as drug carriers in cancer chemotherapy in 1974 (Gregoria et al., 1974). Today, drug delivery systems made of lipids or polymers frequently are exploited for the controlled delivery of therapeutic drugs in the body (Jain, 2005; Vasir et al., 2005).

Nanosized particles for biomedical platforms can be made from a variety of materials, including lipids (liposomes, nanoemulsions, and solid-lipid nanoparticles), self-assembling amphiphilic molecules, nondegradable and degradable polymers, dendrimers, metals, and inorganic semiconductor nanocrystals. The selection of the platform material is determined by the desired diagnostic or therapeutic goal, payload type, material safety profile, and administration route. Among the various types of functional nanostructures, nanomedical platforms based on magnetic nanoparticles (MNPs) are of particular interest in biomedical applications. Most frequently, MNPs are constructed of superparamagnetic iron oxides (SPIOs) (e.g., Fe_3O_4 or γ-Fe_2O_3), although metals such as cobalt and nickel are also employed. The characteristics of MNPs, including their composition, size, morphology, and surface chemistry, are tailored by various processes for their wide application in the detection, diagnosis, and treatment of illnesses. The most popular MNPs for biomedical applications are comprised of a magnetic inorganic nanoparticle core and a biocompatible surface coating that provides stabilization under physiological conditions. The additional application of a suitable surface chemistry allows the integration of functional ligands, such that MNPs can perform multiple functions. The modification and functionalization of MNPs improve their magnetic properties and affect their behavior in vivo (Tartaj et al., 2003; A.K. Gupta & M. Gupta, 2005).

Multifunctional MNPs (MFMNPs) are a major class of nanoscale materials with the potential to revolutionize clinical diagnostic and therapeutic techniques. Due to their unique magnetic properties and ability to function at the cellular and molecular levels of biological interactions, MFMNPs have been investigated as an attractive nanomedical platform. MFMNPs in the form of SPIOs have been actively investigated as contrast enhancement agents for magnetic resonance imaging (MRI) and hyperthermia in response to an external alternating magnetic field, due to their ability to enhance the proton relaxation of specific tissues. MFMNPs have been evaluated extensively as a nanomedical platform for the targeted delivery of pharmaceuticals through magnetic drug targeting (Neuberger et al., 2005) and through the attachment of high-affinity ligands (Zhang et al., 2002; Torchilin, 2006).

2. Surface coatings and functionalization of MNPs

2.1 Core–shell structure

Iron oxides with a core/shell structure are widely used as sources of MFMNP platforms. Iron oxides have several crystalline polymorphs, but only γ-Fe_2O_3 (maghemite) or Fe_3O_4 (magnetite) can be used for biomedical applications. These particles, which range in diameter from about 5–20 nm, have unique advantages, including (1) superparamagnetic behavior, with no magnetism after removal of the magnetic field; (2) high saturation magnetization values and high magnetic susceptibility, for effective magnetic enrichment; (3) biocompatibility and rapid removal through extravasation and renal clearance; and (4) easily tailored surface chemistry and functionalization.

Iron oxide nanoparticles have a significant tendency to agglomerate as a result of their high surface energy. Massart (1981) first prepared stable aqueous dispersions of Fe_3O_4 nanoparticles (ferrofluids) that were stabilized by electrical double layers. However, the colloidal electrostatic stabilization arising from surface charge repulsion on the nanoparticles typically is inadequate to prevent aggregation in biological solutions, due to the presence of salts or other electrolytes that can neutralize the charges. Furthermore, the iron oxide surfaces may be subjected to plasma protein adsorption or opsonization, leading to their rapid clearance by the reticuloendothelial system (RES) (Berry & Curtis, 2003).

To solve the above problems, proper surface coatings have been exploited as an integral component of the MFMNP platform for biomedical applications. The iron oxide core can be coated by organic materials [e.g., polymers such as dextrant (Thorek et al., 2006) and polyethylene glycol (PEG) (Gref et al., 1994)], inorganic metallic materials [e.g., gold (Ji et al., 2007)], or oxides [e.g., silica or alumina (Bumb et al., 2008)]. Polymer coatings will be introduced in detail in the next section. Silica shells are attractive as protective coatings on the iron oxide core, due to their stability under aqueous conditions and ease of synthesis. Recently, Ma et al. (2006) described one such core–shell MFMNP, composed of an iron oxide core (approximately 10 nm diameter) surrounded by a SiO_2 shell (10–15 nm thick). They doped an organic dye, tris(2,2'-bipyridine) ruthenium, inside a second silica shell to provide luminescence and prevent quenching by interaction with the magnetic core. As a core–shell structure exhibiting superparamagnetic and luminescent properties, this MFMNP platform can be used as a multifunctional imaging agent for biomedical applications.

Gold offers several advantages as a coating material for iron oxide cores, due to its low chemical reactivity and unique ability to form self-assembled monolayers on the core surface using alkanethiols (Prime & Whitesides, 1991). A variety of methods (reversed

microemulsion, combined wet chemical, and laser irradiation) can be used to synthesize gold-coated iron oxides (A. H. Lu et al., 2007).
The core/shell structure of MFMNPs offers several advantages, including good dispersibility and high stability against oxidation. In addition, an appreciable amount of therapeutic agent can be loaded on the MFMNP shell. Functionalization chemistries generally are better established when a coating material is used.

2.2 Polymer coatings

Polymers comprise some of the most important materials used as shells. Polymer coatings not only provide a steric barrier to prevent nanoparticle agglomeration, but also allow MNPs to evade uptake by the RES and thereby to maintain a long plasma half-life. Polymer coatings provide a means to tailor the surface properties of MNPs, such as the surface charge and chemical functionality. An ideal polymer coating will have a high affinity for the iron oxide core, as well as nonimmunogenic and nonantigenic properties. It also will prevent opsonization by plasma proteins. Polymer materials comprised of lipids, proteins, dendrimers, gelatin, dextran, chitosan, pullulan, PEG, poly(ethylene-co-vinyl acetate), poly(vinylpyrrolidone), poly(vinyl alcohol) (PVA), or poly(glycerol monoacrylate) (PGA) are often chosen as the surface coatings for MNPs.
PEG is the most widely used polymer for nanoparticle coating in biomedical applications. PEG provides a very attractive combination of properties: excellent solubility in aqueous solutions; high flexibility of its polymer chain; very low toxicity, immunogenicity, and antigenicity; lack of accumulation in the RES cells; and minimal influence on the specific biological properties of modified pharmaceuticals (Yamaoka et al., 1994). As a so-called "stealth" surface, PEG prevents the nanomedical platform from being recognized by RES, and thereby extends its blood circulation time in vivo. On the biological level, coating nanoparticles with PEG sterically hinders the interaction of blood components with the nanoparticle surface and reduces the binding of plasma proteins. Mechanisms of preventing opsonization by PEG include the shielding of the surface charges, increased surface hydrophilicity (Gabizon & Papahadjopoulos, 1992), and enhanced repulsive interaction between polymer-coated nanoparticles and blood components (Needham et al., 1992). Various methods have been utilized to attach PEG to the MNP surface, including silane grafting to the oxide surface (Butterworth et al., 2001), alkaline coprecipitation of ferric and ferrous ions in the presence of PEG-containing block copolymers (Wan et al., 2005), direct attachment of PEG-containing block copolymers (Guo et al., 2010), polymerization at the MNP surface (Flesch et al., 2005), and modification through sol-gel approaches (Y. Lu et al., 2002).
Polysaccharide dextran is another polymer coating that has been used widely and successfully in vivo. Dextran-coated iron oxide nanoparticles have become an important part of clinical cancer imaging, and have been shown to increase the accuracy of cancer nodal staging (Harisinghani & Weissleder, 2004; Ferrari, 2005). Because the dextran coating is not strongly associated with the iron oxide core, the polymer is susceptible to detachment. Accordingly, cross-linked iron oxide nanoparticles have been developed by cross-linking the dextran shell with epichlorohydrin (Josephson et al., 1999). The resulting particle offers superb stability under harsh conditions, without causing any change in size or blood half-life or loss of the dextran coat. Chemical functionality can be established by treating cross-linked iron oxide nanoparticles with ammonia to provide primary amino groups for the

attachment of biomolecules such as proteins or peptides (Wunderbaldinger et al., 2002; Schellenberger et al., 2002). These formulations of dextran-coated iron oxide nanoparticles have been evaluated extensively for a variety of MRI applications (Josephson et al., 1999).

In addition to the traditional polymer coatings, a new kind of biocompatible polymer material has been reported by Wan et al. (2005): namely, homopolymers of glycerol monoacrylate or glycerol monomethacrylate, or their block copolymers. Highly stable aqueous magnetic fluids were prepared by coating Fe_3O_4 nanoparticles with poly(glycerol monoacrylate) (PGA), poly(glycerol monomethacrylate) (PGMA), or diblock copolymers with PGA or PGMA segments. As shown in Fig. 1, the proposed mechanism of stabilization was the multidentate interactions of 1,2-diols on the polymer chain with iron atoms at the surface of the iron oxide nanoparticles (Wan et al., 2005). This process was a good choice for the preparation of stable magnetic fluids with tailored surfaces; PGA or PGMA binds very tightly to the iron oxide surface, is highly hydrophilic, and does not introduce charges on the surface. Moreover, various block copolymers containing PGA or PGMA can be used to modify the iron oxides and to introduce tailored functional groups for further functionalization.

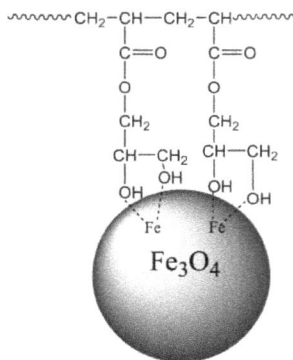

Fig. 1. Proposed structure in the interaction between the iron oxide surface and PGA (Wan et al., 2005).

2.3 Functional ligands

As discussed in the above sections, the core–shell structure of MFMNPs provides a means to tailor the nanoparticle surface properties, such as surface charge and chemical functionality. Various functional ligands, including targeting agents, permeation enhancers, optical dyes, and therapeutic agents, can be conjugated on the surface or incorporated within the nanostructure. The modification of the nanoparticle surface with targeting ligands was described recently as a promising biotargeting strategy. To generate target-specific nanoparticles, various biological molecules, such as antibodies, proteins, small molecular targeting agents, etc., can be bound to the coating surfaces of the MFMNPs by chemical coupling. Tumor cells are rapidly proliferating and overexpress certain receptors that lead to the enhanced uptake of nutrients, including folic acid, vitamins, sugars, and proteins. MFMNPs conjugated with these molecules can be targeted to tumor cells that overexpress the corresponding receptors. Table 1 summarizes a number of different ligands and their corresponding functions that have been investigated for the in vivo targeting of MFMNPs.

Targeting ligand	Functional activity	References
Folic acid	Preferentially targets cancer cells that overexpress folate receptors and facilitates internalization	Zhang et al., 2002
CREKA peptide or F3 peptide	Targets an antigen associated with colorectal carcinoma cells	Reddy et al., 2006; Simberg et al., 2007
Pullulan	Increases receptor-mediated hepatic uptake	Kaneo et al., 2001
Elastin	Cross-linked protein; Provides elasticity for many tissues	Debelle & Tamburro, 1999
RGD peptide	Enhances cell spreading, differentiation, and DNA synthesis	Bhadriraju & Hansen, 2000
Tat-peptide	Membrane-permeating peptide; Enhances intracellular delivery of nanoparticles	Josephson et al., 1999; Lewin et al., 2000
Transferrin	Targets primary proliferating cells by transferrin receptors	Weissleder et al., 2000; Moore et al., 2001
Insulin	Hormone; Regulates blood glucose levels	Gupta et al., 2003
Monoclonal antibody A7	Targets an antigen associated with colorectal carcinoma cells	Toma et al., 2005

Table 1. Selected functional ligands used for MFMNPs in biomedical applications

Organic dyes or fluorophores have been loaded on MNPs as optical imaging agents to allow detection by multiple imaging modalities. In addition to their use as contrast enhancement agents, FITC- (Zhang et al., 2002), rhodamine- (Bertorelle et al., 2006), or other fluorophore-labeled MNPs can be used for the in vitro fluorescent imaging of cells. Since both MRI and optical signals come from the same nanoparticles, the MR image can serve as a roadmap to the fluorescently labeled tumor cells. The conjugation of near-infrared fluorescent (NIRF) dyes to MNPs has received recent attention due to the deep penetration of NIRF light in the tissues (Weissleder & Ntziachristos, 2003). The integration of NIRF detectability allows for these nanoparticles to be used for presurgical planning by MRI and intraoperative resection of malignant tissues by optical imaging.

3. Biomedical applications of MFMNPs

3.1 Targeted drug delivery

One promising biomedical application of MFMNPs is as carriers for site-specific drug delivery. Many therapeutic agents, while pharmacologically effective, also exhibit side-effects because of their toxicities. For example, cytotoxic compounds used in cancer therapy kill not only target cells but also normal cells in the body, resulting in undesired side-effects. Meanwhile, many barriers to the delivery of therapeutic agents are presented, including renal clearance of small molecular therapeutic agents and overexpressed membrane-associated multi-drug resistance developed by tumor cells. Therefore, many therapeutic agents are limited in their clinical application. As widely used nanocarriers, MFMNPs have been considered as alternatives for the target-specific delivery of drugs to different sites in the body. These engineered nanoparticulate carriers offer some advantages, including passive targeting due to the enhanced permeability and retention (EPR) effect and functionalized surface features for target-specific localization. It also may be possible to develop nanocarriers that respond to physiological stimuli, or to combine drugs with energy (heat, light, and sound) delivery for synergistic therapeutic effects.

3.1.1 Passive targeting

Passive targeting relies on the properties of the delivery system and the disease pathology to accumulate the drug preferentially at the site of interest and avoid nonspecific distribution. Long-circulating nanoparticles of 20-200 nm in diameter containing surface PEG or poly(ethylene oxide) (PEO) blocks can accumulate at sites of disease such as tumors, infection, or inflammation through passive targeting via the EPR effect. Maeda and colleagues (2001) first described the EPR effect in their study of vascular abnormalities of solid tumors. Blood vessels in most solid tumors possess unique characteristics that are not usually observed in normal blood vessels, including: active angiogenesis and high vascular density; extensive production of vascular mediators that facilitate extravasation; defective vascular architecture (lack of smooth muscle layer cells, lack of receptors for angiotensin, large gap in endothelial cell-cell junctions, and anomalous conformations); and impaired lymphatic clearance of macromolecules and lipids from interstitial tissue.

Due to the EPR effect, nanopharmaceuticals (macromolecular drugs and drug-loaded nanoparticles) accumulate in tumor tissues with remarkable selectivity as schematically illustrated in Fig. 2. For example, the administration of polymer-drug conjugates results in 10-100 fold higher drug concentrations in the tumor compared to the administration of free drug (van Vlerken et al., 2007). This selective drug targeting to solid tumors results in substantial therapeutic benefits due to the higher drug accumulation in the tumor tissue, as well as fewer side effects. The EPR effect also has been observed in inflammatory and infectious tissues. Thus, the application of nanocarriers is expected to have therapeutic benefits for treating these diseases as well (Allen & Cullis, 2004).

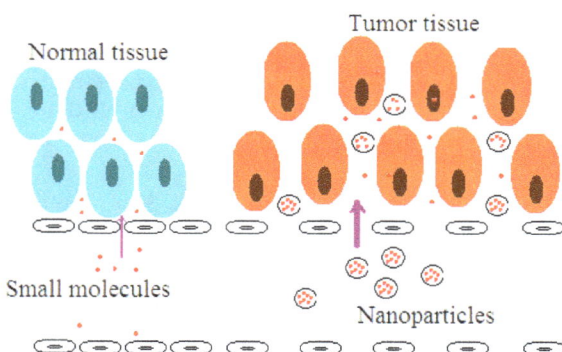

Fig. 2. Schematic illustration for passive targeting using the EPR effect.

Another approach for passive targeting involves the tendency of nanoparticles to localize in the RES. This phenomenon provides an opportunity for nanoparticles to accumulate at high concentrations in the liver or spleen, where many macrophages are present. Overall, the passive targeting strategy provides a means of delivering MFMNPs (as contrast agents or drug carriers) or other nanoparticles to the targeted organs or tissues.

3.1.2 Active targeting

Another promising approach towards increasing the local accumulation of nanoparticles in diseased tissue is known as active (or specific) targeting. Active targeting involves the conjugation of targeting molecules that possess high affinity toward unique molecular signatures found on malignant cells. Targeting ligands, such as proteins, peptides, aptamers and small molecules, have been investigated to increase the site-specific accumulation of MFMNPs. For example, there are certain receptors that are overexpressed on the surface of solid tumor cells, such as antigens, integrin receptors, and folate receptors (Table 1). By bonding with these targeting molecules, MFMNPs can be targeted to the corresponding tumor cells and internalized by receptor- or antigen-mediated endocytosis.

Monoclonal antibodies (mAbs) were the first targeting agents to exploit molecular recognition to deliver MNPs; mAbs continue to be used widely, due to their high specificity. For instance, Herceptin®, an FDA-approved mAb to the HER2/neu (erbB2) receptor, has been used to modify DMSA-coated magnetite nanoparticles. When these MFMNPs were used as contrast enhancement agents, the MR imaging of mice bearing xenograft tumors showed a T2 decrease of ~20% due to the specific accumulation of the nanoprobe in the tumor (Huh et al., 2005). Nanoparticles modified with an HER2-specific antibody (Trastuzumab® or Herceptin®) also are able to localize and deliver the therapeutic payload specifically in HER2-expressing tumor cells (Kirpotin et al., 2006). Certain tumor cells express specific integrin receptors, such as $\alpha_v\beta_5$ or $\alpha_v\beta_3$ that can bind to the arginine-glycine-aspartic acid (RGD) peptide sequence. The RGD peptide has been utilized for the delivery of MNPs to a variety of neoplastic tissues, including breast tumors, malignant melanomas, and squamous cell carcinomas (Montet et al., 2006).

Among the small targeting molecules, folate has been used to modify nanoparticles for targeted delivery to tumor cells that overexpress folate receptors. Recently, our group reported multilayer MFMNPs with a folate-modified surface and doxorubicin (an anticancer chemotherapeutic agent) loaded in the inner shell (Fig. 3) (Guo et al., 2011). The folate-conjugated MFMNPs displayed a much greater cellular uptake than nonfolate-conjugated MFMNPs by a folate receptor–mediated endocytosis process (Fig. 4). Folate conjugation significantly increased nanoparticle cytotoxicity against human cervical carcinoma HeLa cells (Guo et al., 2011).

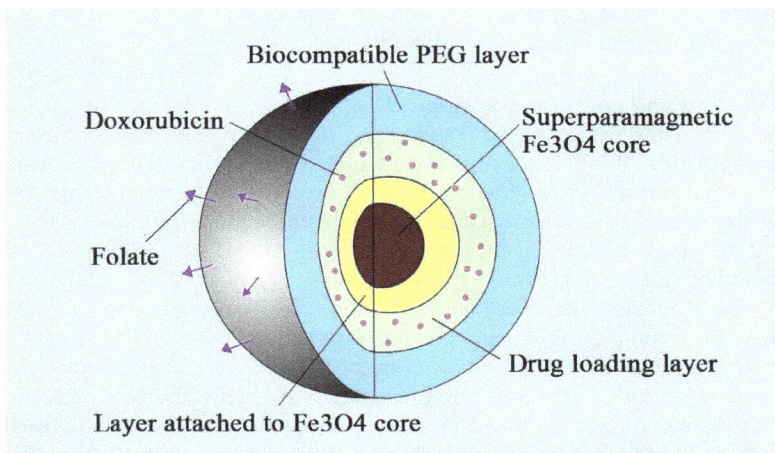

Fig. 3. Schematic illustration of multilayer MFMNPs with folate as the targeting ligand and loaded doxorubicin as the anticancer chemotherapeutic agent in the inner shell.

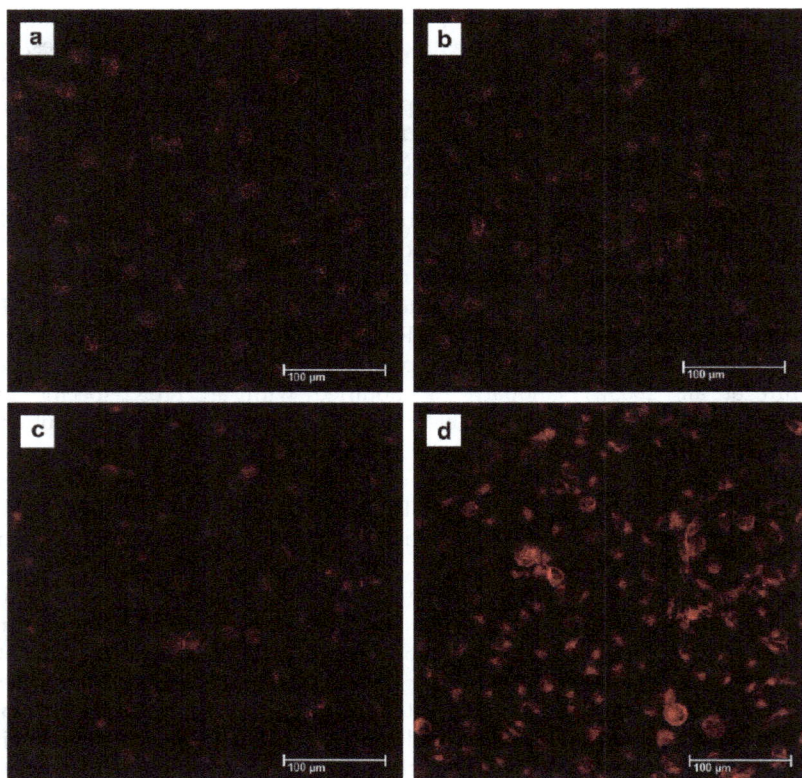

Fig. 4. Confocal microscopic images of HeLa cells incubated with (a, b) doxorubicin-loaded MFMNPs without folate conjugation or (c, d) folate–conjugated and doxorubicin-loaded MFMNPs in (a, c) folate-containing or (b, d) folate-free media (Guo et al., 2011).

3.1.3 Drug loading and controlled release

An essential step in the use of MFMNPs for drug delivery is the controlled release of the therapeutic payload in the desired tumor cells or tissues. The drug-loading capacity and release rate are correlated with the binding affinity of the drug. Strong carrier–drug interactions may enhance the loading capacity and decrease the release rate of the drug from the carrier; therefore, the choice of proper carrier–drug interactions is critical in the design and preparation of nanocarriers for drug delivery.

Successful MNP delivery devices with a prolonged circulation time can carry a chemotherapeutic payload and can be engineered to release its drugs after cell internalization. To successfully integrate a drug into a NP system, several design strategies can be explored, including physical complexation with hydrophobic drugs, or covalent bonding with cleavable linkages for intracellular release. Currently, several chemical drug formulations have been combined with MNPs, including paclitaxel, doxorubicin, and methotrexate, all specifically developed for cancer therapy. For example, methotrexate, an anticancer drug, has an affinity to the target cells, and after grafting the drug to the surface,

MNPs can be internalized more rapidly. Kohler et al. (2006) first demonstrated this utility in a study where methotrexate was covalently attached to the surface of PEG-coated MNPs via a cleavable amide linkage. Recently, Sun et al. (2008b) further modified the same MFMNP system with chlorotoxin to enhance the NP's targeting abilities against brain tumor cells.

Ideal drug delivery systems should be stable with a long circulation time, and should keep the loaded drugs unreleased during circulation in the bloodstream or in normal tissues. Upon reaching the tumor tissues and being taken up by cancer cells, the systems should release the drugs rapidly to kill cancer cells.

To achieve this purpose, stimuli-triggered drug delivery systems have been used that respond to characteristics of the local microenvironment, such as pH, temperature, redox potential, etc. [reviewed by Danhier et al. (2010) and Muthu et al. (2009)]. In particular, pH gradients have been used widely to design responsive nanoparticle delivery systems. Various nanocarriers with pH-responsive delivery behaviors have been developed on the basis of the differential pH values of blood plasma (pH 7.4), extracellular tumor matrix (pH 5.8–7.2), and endocytic compartments such as endosomes (pH 5–6) and lysosomes (pH 4–5). Drugs have been loaded into polymeric nanocarriers by acidic pH-induced cleavable covalent bonds, creating smart drug delivery systems that respond to the endosomal/lysosomal pH (Yoo et al., 2002; Bae et al., 2003; Gillies et al., 2004; Hruby et al., 2005). The pH-induced cleavage of such bonds can accelerate drug release from the nanocarriers.

Drugs are otherwise loaded into the core of polymeric micelles by noncovalent (e.g., hydrophobic) interactions. Compared to chemical attachment, noncovalent entrapment is convenient and easy to achieve. Nasongkla et al. (2006) described the preparation of micelles of PEG-b-poly(D, L-lactide) that encapsulated doxorubicin and a cluster of SPIO nanoparticles by noncovalent hydrophobic interactions. The protonation of doxorubicin under acidic conditions increased its water-solubility and induced its release.

Recently, our group reported a MFMNP platform that can load drugs with ionizable groups and hydrophobic moieties by the combined action of ionic bonding and hydrophobic interactions (Guo et al., 2008, 2010) (Fig. 5). The use of double noncovalent interactions resulted in a high loading affinity at a neutral pH (7.4), preventing premature release into the bloodstream. At an endosomal/lysosomal pH (<5.5), protonation of polycarboxylate anions in the polymer chains led to ionic bond breakage and drug release. The release process was controlled, responded well to pH, and displayed good kinetics.

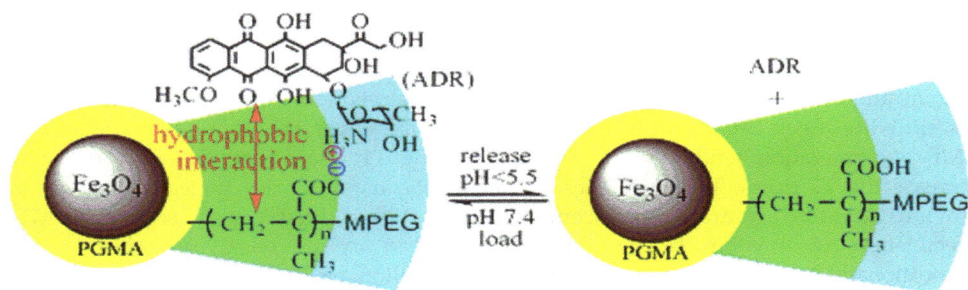

Fig. 5. Schematic illustration of the MFMNP structure, and the load and release of model drug adriamycin (ADR) (Guo et al., 2008).

3.2 MRI

MRI is a powerful noninvasive imaging modality that is utilized widely in clinical medicine. MRI is based on the property that hydrogen protons will align and process around an external alternating magnetic field. The subsequent process through which these protons return to their original state is referred to as the relaxation phenomenon. Two independent processes, longitudinal relaxation (T1-recovery) and transverse relaxation (T2-decay), are monitored to generate the MR image. Local variations in relaxation, corresponding to image contrast, arise from the proton density and the chemical and physical natures of the different tissues. Due to their ability to enhance proton relaxation and accumulation in specific tissues, MFMNPs have been actively investigated as contrast enhancement agents for MRI.

3.2.1 Magnetic properties of iron oxide nanoparticles

Particles whose unpaired electron spins align themselves spontaneously so that the material can exhibit magnetization without being in a magnetic field are called ferromagnetic particles. Materials such as iron oxide nanoparticles that exhibit ferromagnetism can be permanently magnetized. The magnetic properties of iron oxide nanoparticles can be described by the dependence of the magnetic induction B on the magnetic field H. For most materials, the relationship between B and H is linear: $B = \mu H$, where μ is the magnetic permeability of the particles. Iron oxide particles exhibit paramagnetism if $\mu > 1$, and diamagnetism if $\mu < 1$.

Usually, ferromagnetic properties arise only when a certain number of atoms are bound together in solid form; single atoms cannot exhibit ferromagnetism. When the size of particles is smaller than the ferromagnetic domain, they are no longer ferromagnetic but exhibit superparamagnetism (Bansil et al., 1998). Magnetic nanoparticles smaller than ~ 30 nm show superparamagnetic behavior without any magnetic remanence (i.e., restoration of the induced magnetization to zero upon removal of the external magnetic field), but the particles still exhibit very strong paramagnetic properties with a very large susceptibility. This is one important advantage for magnetic nanoparticles: it enables their stability and dispersion upon removal of the magnetic field, as no residual magnetic force exists between the particles.

In MRI, superparamagnetic nanoparticles made of iron oxide act as contrast enhancement agents by shortening both the T1 and T2 relaxations of the surrounding protons. The influence on the T1 relaxation depends strongly on the local MNP concentration, and the shortening processes can be hindered by the coating thickness. The effect of MNPs on T2 shortening is caused by the large susceptibility difference between the particles and surrounding medium, which results in microscopic magnetic field gradients. At low concentrations, a T1-positive contrast can be observed; at high concentrations, the susceptibility effects cause irreversible destruction of the MR signal around the particles.

NMP agglomeration tends to slightly decrease the T1 relaxation times but markedly decrease the T2 times. Therefore, superparamagnetic nanoparticles typically are used to provide negative contrast enhancement using T2-weighted pulse sequences. The effectiveness of a contrast agent can be described by its relaxivity, which is the proportionality constant of the measured rate of relaxation, or R1 (1/T1) and R2 (1/T2). The relaxivity depends on not only the composition, size, and magnetic properties of the MNP, but also depends on experimental variables such as the field strength, temperature, and medium in which the measurements are made.

3.2.2 Molecular imaging in cancer

The generation of new molecular targets that are closely related to pathophysiology will open the way for the development of new treatment paradigms for currently untreatable diseases. In recent years, many new diagnostic technologies have been developed, including molecular diagnostic compounds and new imaging technologies such as MR molecular imaging. Molecular imaging is the noninvasive imaging of targeted macromolecules, cells, and biological or cellular processes in living organisms. Due to their ability to act as molecularly targeted imaging agents, MNPs play an integral role in the applications of early disease detection, individualized treatment, and drug development. In the clinical imaging of tumors, MNPs can be used as contrast enhancement agents to improve the detection, diagnosis, and therapeutic management of solid tumors by exploiting the unique molecular signatures of the diseases. MNPs also have been investigated to improve the delineation of the tumor position, boundaries, and volume.

The first clinical indication for iron oxide nanoparticles was the imaging of liver tumors and metastases. After intravenous injection, MNPs are taken up rapidly by hepatic specialized macrophages. This process causes a drop in MR signal intensity and generates hypointense images, mostly because of a susceptibility effect. However, tumors lack a permanent decrease in signal intensity after MNP administration; tumors are almost devoid of macrophages, which are located exclusively in the healthy hepatic parenchyma. Therefore, MNPs can markedly increase the contrast between healthy and diseased tissue. The clinical imaging of liver tumors and metastases through RES-mediated uptake of MNPs has allowed the detection of lesions as small as 2–3 mm (Semelka & Helmberger, 2001). In combination with MRI, MNPs also have been shown to be effective in the identification of lymph node metastases of 5–10 mm in diameter (Harisinghani et al., 2003). The use of MFMNPs as contrast enhancement agents provides increased lesion conspicuousness and lesion detection compared to nonenhanced imaging.

MFMNPs are currently under evaluation for use in improving the delineation of brain tumor boundaries and quantifying tumor volumes (Enochs et al., 1999; Neuwelt et al., 2004). Some recent approaches have explored utilizing iron oxide nanoparticles as drug delivery vehicles for the MRI-monitored magnetic targeting of brain tumors (Chertok et al., 2008). The accumulation of iron oxide nanoparticles in gliosarcomas is enhanced by magnetic targeting and successfully quantified by MRI (Chertok et al., 2008) (Fig. 6). Such noninvasive approaches for cancer diagnosis and therapy also have been adopted in the treatment of prostate, breast, and colon cancers.

As both drug delivery devices and MRI contrast enhancement agents, MNPs retain the ability to track the movement of drug through the body. This is significant because it allow clinicians to monitor the effectivity of injected therapeutics to reach their target sites. There remains significant flexibility in the contrast agents implemented in these constructs and the manner in which drugs are delivered. Medarova et al. (2007) recently developed cross-linked iron oxide nanoparticles modified with a NIR fluorophore, therapeutic siRNA sequences, and a cell penetrating peptide. The MNPs used passive targeting by the EPR effect to direct tumor localization. In vivo, these MNPs demonstrated therapeutic efficacy against target tissue, as determined by real time PCR and histological evaluation, while simultaneously demonstrating image contrast in both MR and optical imaging. In a study by Sun et al. (2008a) active cell targeting was shown by PEG-coated MNPs to which the chemotherapeutic, methotrexate, and targeting molecule, chlorotoxin, were attached. The selective contrast enhancement of the 9L brain tumor by these MNPs indicates preferential accumulation compared with the same MNP construct without the chlorotoxin peptide in a 3-day study.

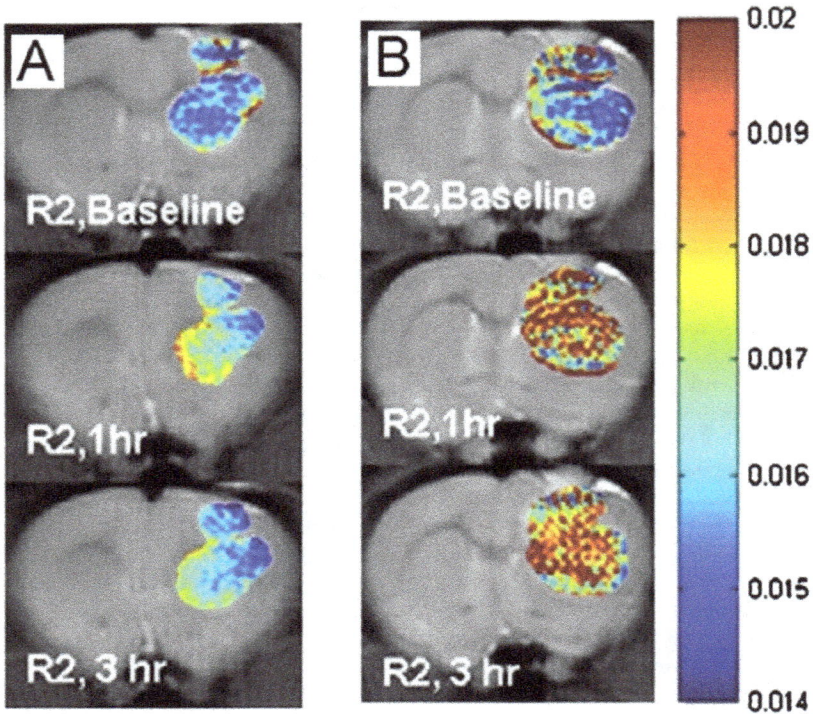

Fig. 6. MR images of brain tumor. Change in the R2 relaxation for the tumor regions before (baseline) and 1–3 h after MNPs administration in (A) control and (B) targeted rats (Chertok et al., 2008).

In another recent study by Yang et al. (2007) simultaneous targeted drug delivery and MR imaging of breast cancer tumors were demonstrated through the use multifunctional magneto-polymeric nanohybrids composed of magnetic nanocrystals and doxorubicin which were simultaneously encapsulated within an amphiphilic block copolymer shell. The surfaces of these micelles were additionally functionalized with the breast cancer targeting/therapeutic ligand, anti-Herceptin antibody. In vivo evaluations of this nanoparticle system were performed in nude mice bearing NIH3T6.7 breast cancer tumors. The quantitative evaluation of MR images revealed preferential accumulation of the targeted MNPs compared to the control MNPs. The therapeutic functionality of the MNPs developed in this study were additionally evaluated and it was determined that the HER-MMPNs which were decorated with targeting ligands and loaded with doxorubicin were most effective in inhibiting tumor growth. Combined, these findings illustrate the functionality and efficacy of targeted multifunctional MNPs for simultaneous MR imaging and drug delivery.

3.3 Hyperthermia
Hyperthermia is the method of using heat as a treatment for malignant tumors. It is based on the observation that tumor cells are more susceptible to heat than normal cells, due to the

higher rates of metabolism of cancer cells. Cancer cells typically show signs of apoptosis and necrosis when heated to 41-47 °C, whereas normal cells can survive at higher temperatures (Milleron & Bratton, 2007). Hyperthermia with targeted nanoscale heaters is recognized as a useful therapeutic modality to kill cancer by essentially "cooking" malignant cells from the inside out.

Magnetic nanoparticle hyperthermia is actualized by the exposure of cancer tissues to an alternating magnetic field. The magnetic field cannot be absorbed by the living tissues and can be applied to deep regions in the living body. When MNPs are injected into an organ with a tumor, they tend to accumulate in the tumor due to passive and active targeting strategies (as described above). Subsequent exposure to an alternating magnetic field causes heat to be generated in the tumor tissue due to magnetic hysteresis loss. This process effectively destroys the tumor but not the surrounding healthy tissue. The amount of heat generated depends on the nature of the MNPs and magnetic field parameters used.

The use of MFMNPs for targeted hyperthermia has shown a therapeutic effect in several types of tumors. Using dextran-coated MNPs conjugated to breast cancer–targeting chimeric L6 mAb, DeNardo et al. (2005) demonstrated the feasibility of this method for treating breast cancer cells. Kobayashi et al. constructed a novel therapeutic tool of magnetite nanoparticle-loaded anti-HER2 immunoliposomes that was applicable to the treatment of HER2-overexpressing cancer (Ito et al., 2004).

A clinical breakthrough in MNP use was made in 2007, when Maier-Hauff et al. (2007) reported the results of using heated implanted MNPs for therapeutic hyperthermia in humans. In that study, 14 patients with recurrent glioblastoma multiforme, a type of severe brain cancer, received an intratumoral injection of aminosilane-coated MNPs. The tumor sites were located by several comprehensive MRI scans, and the patients were exposed to an alternating magnetic field to induce particle heating. The nanoparticle deposits were stable for several weeks, and all patients tolerated the nanoparticles without any complications. These findings indicate that MNP hyperthermia may be an effective therapeutic method to cure human brain cancer.

As a potential approach for the treatment of malignant tumors, MNP hyperthermia has the following advantages: it provides a noninvasive way to raise cell temperatures to a therapeutic level; MNPs can be visualized using MRI, thus combining diagnostic and therapeutic approaches in one type of particle; and the particles can be functionalized and combined with other types of treatment, such as chemotherapy or radiotherapy.

4. Conclusions

Multifunctional nanomaterials have been widely used as nanoplatforms for multimodal imaging or simultaneous imaging and therapy. As a multifunctional nanoplatform with many biomedical applications, MFMNPs of various formulations have been developed to diagnose and treat diseases for which conventional therapy has shown limited efficacy, such as cancer. The use of MFMNPs as MRI contrast enhancement agents and anticancer drug carriers has drawn enormous attention, and may provide new opportunities for early cancer detection and targeted therapies. This technology not only will minimize the need for invasive procedures, but also will reduce the side effects to healthy tissues, which are primary concerns in conventional cancer therapies.

5. Acknowledgment

This work was supported by the National Natural Science Foundation of China (20974052), the National Key Technologies R & D Program for New Drugs of China (2009ZX09301-002) and the Natural Science Foundation of Tianjin Municipality (09JCZDJC22900).

6. References

Allen, T. M. & Cullis, P. R. (2004). Drug Delivery Systems: Entering the Mainstream. *Science,* Vol. 303, No. 5665, pp. 1818-1822.

Bae, Y.; Fukushima, S.; Harada, A. & Kataoka, K. (2003). Design of Environment-Sensitive Supramolecular Assemblies for Intracellular Drug Delivery: Polymeric Micelles that Are Responsive to Intracellular pH Change. *Angewandte Chemie International Edition,* Vol. 42, No. 38, pp. 4640-4643.

Bansil, A.; Sawatzky, G.; Fujii, Y.; Shelton, R. & Prosser, D. (1998). Journal of Physics and Chemistry of Solids 40th Anniversary – Preface. *Journal of Physics and Chemistry of Solids,* Vol. 59, No. 4, pp. III-IIV.

Berry, C. C. & Curtis, A. S. G. (2003). Functionalisation of Magnetic Nanoparticles for Applications in Biomedicine. *Journal of Physics D-Applied Physics,* Vol. 36, No. 13, pp. R198-R206.

Bertorelle, F.; Wilhelm, C.; Roger, J.; Gazeau, F.; Menager, C. & Cabuil, V. (2006). Fluorescence-Modified Superparamagnetic Nanoparticles: Intracellular Uptake and Use in Cellular Imaging. *Langmuir,* Vol. 22, No. 12, pp. 5385-5391.

Bhadriraju, K. & Hansen, L. K. (2000). Hepatocyte Adhesion, Growth and Differentiated Function on RGD-Containing Proteins. *Biomaterials,* Vol. 21, No. 3, pp. 267-272.

Bumb, A.; Brechbiel, M. W.; Choyke, P. L.; Fugger, L.; Eggeman, A.; Prabhakaran, D.; Hutchinson, J. & Dobson, P. J. (2008). Synthesis and Characterization of Ultra-Small Superparamagnetic Iron Oxide Nanoparticles Thinly Coated with Silica. *Nanotechnology,* Vol. 19, No. 33, pp. 335601-335606.

Butterworth, M. D.; Illum, L. & Davis, S. S. (2001). Preparation of Ultrafine Silica- and PEG-Coated Magnetite Particles. *Colloids and Surfaces A-Physicochemical and Engineering Aspects,* Vol. 179, No. 1, pp. 93-102.

Chertok, B.; Moffat, B. A.; David, A. E.; Yu, F. Q.; Bergemann, C.; Ross, B. D. & Yang, V. C. (2008). Iron Oxide Nanoparticles as a Drug Delivery Vehicle for MRI Monitored Magnetic Targeting of Brain Tumors. *Biomaterials,* Vol. 29, No. 4, pp. 487-496.

Couvreur, P. & Vauthier, C. (2006). Nanotechnology: Intelligent Design to Treat Complex Disease. *Pharmaceutical Research,* Vol. 23, No. 7, pp. 1417-1450.

Danhier, F.; Feron, O. & Preat V. (2010). To Exploit the Tumor Microenvironment: Passive and Active Tumor Targeting of Nanocarriers for Anti-Cancer Drug Delivery. *Journal of Controlled Release,* Vol. 148, No. 2, pp. 135–146.

Debelle, L. & Tamburro, A. M. (1999). Elastin: Molecular Description and Function. *International Journal of Biochemistry & Cell Biology,* Vol. 31, No. 2, pp. 261-272.

DeNardo, S. J.; DeNardo, G. L.; Miers, L. A.; Natarajan, A.; Foreman, A. R.; Gruettner, C.; Adamson, G. N. & Ivkov, R. (2005). Development of Tumor Targeting Bioprobes (In-111-Chimeric L6 Monoclonal Antibody Nanoparticles) for Alternating Magnetic Field Cancer Therapy. *Clinical Cancer Research,* Vol. 11, No. 19, pp. 7087S-7092S.

Enochs, W. S.; Harsh, G.; Hochberg, F. & Weissleder, R. (1999). Improved Delineation of Human Brain Tumors on MR Images Using a Long-Circulating,

Superparamagnetic Iron Oxide Agent. *Journal of Magnetic Resonance Imaging*, Vol. 9, No. 2, pp. 228-232.

Ferrari, M. (2005). Cancer Nanotechnology: Opportunities and Challenges. *Nature Reviews Cancer*, Vol. 5, No. 3, pp. 161-171.

Flesch, C.; Unterfinger, Y.; Bourgeat-Lami, E.; Duguet, E.; Delaite, C. & Dumas, P. (2005). Poly(ethylene glycol) Surface Coated Magnetic Particles. *Macromolecular Rapid Communications*, Vol. 26, No. 18, pp. 1494-1498.

Gabizon, A. & Papahadjopoulos, D. (1992). The Role of Surface-Charge and Hydrophilic Groups on Liposome Clearance In Vivo. *Biochimica et Biophysica Acta*, Vol. 1103, No. 1, pp. 94-100.

Gillies, E. R.; Goodwin, A. P. & Frechet, J. M. J. (2004). Acetals as pH-Sensitive Linkages for Drug Delivery. *Bioconjugate Chemistry*, Vol. 15, No. 6, pp. 1254-1263.

Gref, R.; Minamitake, Y.; Peracchia, M. T.; Trubetskoy, V.; Torchilin, V. & Langer, R. (1994). Biodegradable Long-Circulating Polymeric Nanospheres. *Science*, Vol. 263, No. 5153, pp. 1600-1603.

Gregoria, G.; Wills, E. J.; Swain, C. P. & Tavill, A. S. (1974). Drug-Carrier Potential of Liposomes in Cancer Chemotherapy. *Lancet*, Vol. 1, No. 7870, pp. 1313-1316.

Guo, M.; Que, C. L.; Wang, C. H.; Liu, X. Z.; Yan, H. S. & Liu, K. L. (2011). Multifunctional Superparamagnetic Nanocarriers with Folate-Mediated and pH-Responsive Targeting Properties for Anticancer Drug Delivery. *Biomaterials*, Vol. 32, No. 1, pp. 185-194.

Guo, M.; Yan, Y.; Liu, X. Z.; Yan, H. S.; Liu, K. L.; Zhang, H. K.; & Cao, Y. J. (2010). Multilayer Nanoparticles with a Magnetite Core and a Polycation Inner Shell as pH-Responsive Carriers for Drug Delivery. *Nanoscale*, Vol. 2, No. 3, pp. 434-441.

Guo, M.; Yan, Y.; Zhang, H. K.; Yan, H. S.; Cao, Y. J.; Liu, K. L.; Wan, S. R.; Huang, J. S. & Yue, W. (2008). Magnetic and pH-responsive Nanocarriers with Multilayer Core-Shell Architecture for Anticancer Drug Delivery. *Journal of Materials Chemistry*, Vol. 18, No. 42, pp. 5104-5112.

Gupta, A. K.; Berry, C.; Gupta, M. & Curtis, A. S. G. (2003). Receptor-Mediated Targeting of Magnetic Nanoparticles Using Insulin as a Surface Ligand to Prevent Endocytosis. *IEEE Transactions on Nanobioscience*, Vol. 2, No. 4, pp. 255-261.

Gupta, A. K. & Gupta, M. (2005). Synthesis and Surface Engineering of Iron Oxide Nanoparticles for Biomedical Applications. *Biomaterials*, Vol. 26, No. 18, pp. 3995-4021.

Harisinghani, M. G. & Weissleder, R. (2004). Sensitive, Noninvasive Detection of Lymph Node Metastases. *Plos Medicine*, Vol. 1, No. 3, pp. 202-209.

Hruby, M.; Konak, C. & Ulbrich, K. (2005). Polymeric Micellar pH-Sensitive Drug Delivery System for Doxorubicin. *Journal of Controlled Release*, Vol. 103, No. 1, pp. 137-148.

Huh, Y. M.; Jun, Y. W.; Song, H. T.; Kim, S.; Choi, J. S.; Lee, J. H.; Yoon, S.; Kim, K. S.; Shin, J. S.; Suh, J. S. & Cheon, J. (2005). In Vivo Magnetic Resonance Detection of Cancer by Using Multifunctional Magnetic Nanocrystals, *Journal of the American Chemical Society*, Vol. 127, No. 35, pp. 12387-12391.

Ito, A.; Kuga, Y.; Honda, H.; Kikkawa, H.; Horiuchi, A.; Watanabe, Y. & Kobayashi, T. (2004). Magnetite Nanoparticle-Loaded Anti-Her2 Immunoliposomes for Combination of Antibody Therapy with Hyperthermia. *Cancer Letters*, Vol. 212, No. 2, pp. 167-175.

Jain, K. K. (2005). The Role of Nanobiotechnology in Drug Discovery. *Drug Discovery Today*, Vol. 10, No. 21-24, pp. 1435-1442.

Ji, X. J.; Shao, R. P.; Elliott, A. M.; Stafford, R. J.; Esparza-Coss, E.; Bankson, J. A.; Liang, G.; Luo, Z. P.; Park, K.; Markert, J. T. & Li, C. (2007). Bifunctional Gold Nanoshells with a Superparamagnetic Iron Oxide-Silica Core Suitable for both MR Imaging and Photothermal Therapy. *Journal of Physical Chemistry*, Vol. 111, No. 17, pp. 6245-6251.

Josephson, L.; Tung, C. H.; Moore, A. & Weissleder, R. (1999). High-Efficiency Intracellular Magnetic Labeling with Novel Superparamagnetic-Tat Peptide Conjugates. *Bioconjugate Chemistry*, Vol. 10, No. 2, pp. 186-191.

Kaneo, Y.; Tanaka, T.; Nakano, T. & Yamaguchi, Y. (2001). Evidence for Receptor-Mediated Hepatic Uptake of Pullulan in Rats. *Journal of Controlled Release*, Vol. 70, No. 3, pp. 365-373.

Kirpotin, D. B.; Drummond, D. C.; Shao, Y.; Shalaby, M. R.; Hong, K. L.; Nielsen, U. B.; Marks, J. D.; Benz, C. C. & Park, J. W. (2006). Antibody Targeting of Long-Circulating Lipidic Nanoparticles does not Increase Tumor Localization but does Increase Internalization in Animal Models. *Cancer Research*, Vol. 66, No. 13, pp. 6732-6740.

Kohler, N.; Sun, C.; Fichtenholtz, A.; Gunn, J.; Fang, C. & Zhang, M. Q. (2006). Methotrexateimmobilized poly(ethylene glycol) magnetic nanoparticles for MR imaging and drug delivery. *Small*, Vol. 2, No. 6, pp. 785-792.

Lewin, M.; Carlesso, N.; Tung, C. H.; Tang, X. W.; Cory, D.; Scadden, D. T. & Weissleder, R. (2000). Tat Peptide-Derivatized Magnetic Nanoparticles Allow In Vivo Tracking and Recovery of Progenitor Cells. *Nature Biotechnology*, Vol. 18, No. 4, pp. 410-414.

Lu, A. H.; Salabas, E. L. & Schuth, F. (2007). Magnetic Nanoparticles: Synthesis, Protection, Functionalization, and Application. *Angewandte Chemie International Edition*, Vol. 46, No. 8, pp. 1222-1244.

Lu, Y.; Yin, Y. D.; Mayers, B. T. & Xia, Y. N. (2002). Modifying the Surface Properties of Superparamagnetic Iron Oxide Nanoparticles through a Sol-Gel Approach. *Nano Letters*, Vol. 2, No. 3, pp. 183-186.

Ma, D. L.; Guan, J. W.; Normandin, F.; Denommee, S.; Enright, G.; Veres, T. & Simard, B. (2006). Multifunctional Nano-Architecture for Biomedical Applications. *Chemistry of Materials*, Vol. 18, No. 7, pp. 1920-1927.

Maeda, H. (2001). The Enhanced Permeability and Retention (EPR) Effect in Tumor Vasculature: The Key Role of Tumor-Selective Macromolecular Drug Targeting. *Advances in Enzyme Regulation*, Vol 41, v. 41, pp. 189-207.

Maier-Hauff, K.; Rothe, R.; Scholz, R.; Gneveckow, U.; Wust, P.; Thiesen, B.; Feussner, A.; von Deimling, A.; Waldoefner, N.; Felix, R. & Jordan, A. (2007). Intracranial Thermotherapy Using Magnetic Nanoparticles Combined with External Beam Radiotherapy: Results of a Feasibility Study on Patients with Glioblastoma Multiforme. *Journal of Neuro-Oncology*, Vol. 81, No. 1, pp. 53-60.

Massart, R. (1981). Preparation of Aqueous Magnetic Liquids in Alkaline and Acidic Media. *IEEE Transactions on Magnetics*, Vol. 17, No. 2, pp. 1247-1248.

Medarova, Z.; Pham, W.; Farrar, C.; Petkova, V. & Moore, A. (2007). In Vivo Imaging of siRNA Delivery and Silencing in Tumors. *Nature Medicine*, Vol. 13, No. 3, pp. 372-377.

Milleron, R. S. & Bratton, S. B. (2007). 'Heated' Debates in Apoptosis. *Cellular and Molecular Life Sciences*, Vol. 64, pp. 2329-2333.

Montet, X.; Montet-Abou, K.; Reynolds, F.; Weissleder, R. & Josephson, L. (2006). Nanoparticle Imaging of Integrins on Tumor Cells. *Neoplasia*, Vol. 8, No. 3, pp. 214-222.

Moore, A.; Josephson, L.; Bhorade, R. M.; Basilion, J. P. & Weissleder, R. (2001). Human Transferrin Receptor Gene as a Marker Gene for MR Imaging. *Radiology*, Vol. 221, No. 1, pp. 244-250.

Muthu, M. S.; Rajesh, C. V.; Mishra A. & Singh S. (2009). Stimulus-Responsive Targeted Nanomicelles for Effective Cancer Therapy. Nanomedicine, Vol. 4, No. 6, 657-667.

Nasongkla, N.; Bey, E.; Ren, J. M.; Ai, H.; Khemtong, C.; Guthi, J. S.; Chin, S. F.; Sherry, A. D.; Boothman, D. A. & Gao, J. M. (2006). Multifunctional Polymeric Micelles as Cancer-Targeted, MRI-Ultrasensitive Drug Delivery Systems. *Nano Letters*, Vol. 6, No. 11, pp. 2427-2430.

Needham, D.; Mcintosh, T. J. & Lasic, D. D. (1992). Repulsive Interactions and Mechanical Stability of Polymer-Grafted Lipid-Membranes. *Biochimica et Biophysica Acta*, Vol. 1108, No. 1, pp. 40-48.

Neuberger, T.; Schopf, B.; Hofmann, H.; Hofmann, M. & von Rechenberg, B. (2005). Superparamagnetic Nanoparticles for Biomedical Applications: Possibilities and Limitations of a New Drug Delivery System. *Journal of Magnetism and Magnetic Materials*, Vol. 293, No. 1, pp. 483-496.

Neuwelt, E. A.; Varallyay, P.; Bago, A. G.; Muldoon, L. L.; Nesbit, G. & Nixon, R. (2004). Imaging of Iron Oxide Nanoparticles by MR and Light Microscopy in Patients with Malignant Brain Tumours. *Neuropathology and Applied Neurobiology*, Vol. 30, No. 5, pp. 456-471.

Prime, K. L. & Whitesides, G. M. (1991). Self-Assembled Organic Monolayers - Model Systems for Studying Adsorption of Proteins at Surfaces. *Science*, Vol. 252, No. 5009, pp. 1164-1167.

Reddy, G. R.; Bhojani, M. S.; McConville, P.; Moody, J.; Moffat, B. A.; Hall, D. E.; Kim, G.; Koo, Y. E. L.; Woolliscroft, M. J.; Sugai, J. V.; Johnson, T. D.; Philbert, M. A.; Kopelman, R.; Rehemtulla, A. & Ross, B. D. (2006). Vascular Targeted Nanoparticles for Imaging and Treatment of Brain Tumors. *Clinical Cancer Research*, Vol. 12, No. 22, pp. 6677-6686.

Schellenberger, E. A.; Bogdanov, A.; Hogemann, D.; Tait, J.; Weissleder, R. & Josephson, L. (2002). Annexin V-CLIO: a Nanoparticle for Detecting Apoptosis by MRI. *Molecular Imaging*, Vol. 1, No. 2, pp. 102-107.

Semelka, R. C. & Helmberger, T. K. G. (2001). Contrast Agents for MR Imaging of the Liver. *Radiology*, Vol. 218, No. 1, pp. 27-38.

Simberg, D.; Duza, T.; Park, J. H.; Essler, M.; Pilch, J.; Zhang, L.; Derfus, A. M.; Yang, M.; Hoffman, R. M.; Bhatia, S.; Sailor, M. J. & Ruoslahti, E. (2007). Biomimetic Amplification of Nanoparticle Homing to Tumors. *Proceedings of the National Academy of Sciences of the United States of America*, Vol. 104, No. 3, pp. 932-936.

Sun, C.; Fang, C.; Stephen, Z.; Veiseh, O.; Hansen, S.; Lee, D.; Ellenbogen, R. G.; Olson, J. & Zhang, M. Q. (2008a). Tumor-Targeted Drug Delivery and MRI Contrast Enhancement by Chlorotoxin-Conjugated Iron Oxide Nanoparticles. *Nanomedicine*, Vol. 3, No. 4, pp. 495-505.

Sun, C.; Veiseh, O.; Gunn, J.; Fang, C.; Hansen, S.; Lee, D.; Sze, R.; Ellenbogen, R. G.; Olson, J. & Zhang, M. (2008b). In Vivo MRI Detection of Gliomas by Chlorotoxin-Conjugated Superparamagnetic Nanoprobes. *Small*, Vol. 4, No. 3, pp. 372-379.

Tartaj, P.; Morales, M. D.; Veintemillas-Verdaguer, S.; Gonzalez-Carreno, T. & Serna, C. J. (2003). The Preparation of Magnetic Nanoparticles for Applications in Biomedicine. *Journal of Physics D-Applied Physics,* Vol. 36, No. 13, pp. R182-R197.

Thorek, D. L. J.; Chen, A.; Czupryna, J. & Tsourkas, A. (2006). Superparamagnetic Iron Oxide Nanoparticle Probes for Molecular Imaging. *Annals of Biomedical Engineering,* Vol. 34, No. 1, p. 23-38.

Toma, A.; Otsuji, E.; Kuriu, Y.; Okamoto, K.; Ichikawa, D.; Hagiwara, A.; Ito, H.; Nishimura, T. & Yamagishi, H. (2005). Monoclonal Antibody A7-Superparamagnetic Iron Oxide as Contrast Agent of MR Imaging of Rectal Carcinoma. *British Journal of Cancer,* Vol. 93, No. 1, pp. 131-136.

Torchilin, V. P. (2006). Multifunctional nanocarriers. *Advanced Drug Delivery Reviews,* Vol. 58, No. 14, pp. 1532-1555.

van Vlerken, L. E.; Duan, Z. F.; Seiden, M. V. & Amiji, M. M. (2007). Modulation of Intracellular Ceramide Using Polymeric Nanoparticles to Overcome Multidrug Resistance in Cancer. *Cancer Research,* Vol. 67, No. 10, pp. 4843-4850.

Vasir, J. K.; Reddy, M. K. & Labhasetwar, V. D. (2005). Nanosystems in Drug Targeting: Opportunities and Challenges. *Current Nanoscience,* Vol. 1, No. 1, pp. 47-64.

Wan, S. R.; Zheng, Y.; Liu, Y. Q.; Yan, H. S. & Liu, K. L. (2005). Fe_3O_4 Nanoparticles Coated with Homopolymers Of Glycerol Mono(meth)acrylate and Their Block Copolymers. *Journal of Materials Chemistry,* Vol. 15, No. 33, pp. 3424-3430.

Weissleder, R.; Moore, A.; Mahmood, U.; Bhorade, R.; Benveniste, H.; Chiocca, E. A. & Basilion, J. P. (2000). In Vivo Magnetic Resonance Imaging of Transgene Expression. *Nature Medicine,* Vol. 6, No. 3, pp. 351-354.

Weissleder, R. & Ntziachristos, V. (2003). Shedding Light onto Live Molecular Targets. *Nature Medicine,* Vol. 9, No. 1, pp. 123-128.

Wunderbaldinger, P.; Josephson, L. & Weissleder, R. (2002). Tat Peptide Directs Enhanced Clearance and Hepatic Permeability of Magnetic Nanoparticles. *Bioconjugate Chemistry,* Vol. 13, No. 2, pp. 264-268.

Yamaoka, T.; Tabata, Y. & Ikada, Y. (1994). Distribution and Tissue Uptake of Poly(ethylene glycol) with Different Molecular-Weights after Intravenous Administration to Mice. *Journal of Pharmaceutical Sciences,* Vol. 83, No. 4, pp. 601-606.

Yang, J.; Lee, C. H.; Ko, H. J.; Suh, J. S.; Yoon, H. G.; Lee, K.; Huh, Y. M. & Haam, S. (2007). Multifunctional Magneto-Polymeric Nanohybrids for Targeted Detection and Synergistic Therapeutic Effects on Breast Cancer. *Angewandte Chemie International Edition,* Vol. 46, No. 46, pp. 8836-8839.

Yoo, H. S.; Lee, E. A. & Park, T. G. (2002). Doxorubicin-Conjugated Biodegradable Polymeric Micelles Having Acid-Cleavable Linkages. *Journal of Controlled Release,* Vol. 82, No. 1, pp. 17-27.

Zhang, Y.; Kohler, N. & Zhang, M. Q. (2002). Surface Modification of Superparamagnetic Magnetite Nanoparticles and Their Intracellular Uptake. *Biomaterials,* Vol. 23, No. 7, pp. 1553-1561.

2

Non-Thermal Plasma Surface Modification of Biodegradable Polymers

N. De Geyter and R. Morent
Research Unit Plasma Technology – Department of Applied Physics,
Faculty of Engineering and Architecture – Ghent University,
Belgium

1. Introduction

Biodegradable polymers are often used as packaging materials. However, these polymers can also play an important role in tissue engineering as so called scaffolds (i.e. three-dimensional porous structures). The success of these biodegradable scaffolds strongly depends on the reaction of them with their surrounding biological environment. This reaction is mainly governed by the surface features of the scaffold and different approaches have already been tested to change the surface properties of biodegradable polymers. In particular, the research field on the use of non-thermal plasmas for a selective surface modification has known a steep rise. Therefore, this chapter will give an introductory and critical overview on recent achievements in plasma-assisted surface modification of biodegradable polymers. Firstly, we will discuss in short the most commonly biodegradable polymers. Secondly, we will go into more detail about surface modification by a non-thermal plasma and finally we will focus on some examples of plasma-treated biodegradable polymers.

2. Biodegradable polymers

2.1 Biomedical applications

A biodegradable polymer is defined as a polymer that preserves its mechanical strength and other material performances during its practical application, but that is finally degraded to low molecular weight compounds such as H_2O, CO_2 and other non-toxic by-products (Ikada & Tsuji, 2000). Next to their use as packaging material, which is not in the scope of this chapter, biodegradable polymers could play a key role in biomedical engineering for a variety of reasons (Ikada & Tsuji, 2000). Firstly, since the polymer degrades, it is clear that a device made of such a polymer can be implanted in the human body without necessitating a second surgery to remove the device (Athanasiou et al., 1998, Middleton & Tipton, 2000). Moreover, this prevented second operation makes the use of biodegradable polymers even more beneficial in other ways. For example, a fractured bone, fixated with a rigid, non-biodegradable stainless steel implant, has an inclination to fracture again when the implant is taken away because during the healing process the bone does not carry sufficient load, since the load is entirely intercepted by the rigid steel implant. This is in contrast to a biodegradable implant which degrades little by little and transfers by degrees the load from the implant to the fractured bone. This gradually movement of the load results in less bone re-fracture (Middleton & Tipton, 2000, Athanasiou et al., 1998).

Secondly, another interesting application field for biodegradable polymers is tissue engineering. This research branch aims to produce completely biocompatible tissues which could be employed to replace damaged or diseased tissues in reconstructive surgery (Djordjevic et al., 2008). Today's focus in this field is the use of so called scaffolds. These scaffolds are 3D artificial matrices that guarantee optimal support and conditions for growth of tissue (Djordjevic et al., 2008). Optimally, these scaffolds should fulfil the following two requirements (Ryu et al., 2005):

- being capable of supporting initial cell growth and further proliferation
- having the ability to degrade over time while leaving behind a reproduced functional tissue.

Finally, biodegradable polymers can also contribute in controlled drug delivery (Amass et al., 1998). A gradual delivery of antibiotics can be beneficial for the treatment of deep skeletal infections after a surgery, while the healing process of a fractured bone can be enhanced by a delivery in stages of bone morphogenetic proteins (Agrawal et al., 1995, Wang et al., 1990, Ramchandani & Robinson, 1998).

2.2 Overview of the most commonly used biodegradable polymers

Both natural and synthetic polymers have been extensively studied as biodegradable biomaterials (Nair & Laurencin, 2007). At first, natural polymers were considered as promising candidates. Nevertheless, several studies quickly demonstrated that some of these materials involve different drawbacks: possibility of disease transmission, strong immunogenic reactions and major difficulties with the purification process (Nair & Laurencin, 2007). In a later stage, synthetic biodegradable polymers were developed by the synthesis of polymers with hydrostatically unstable linkages in their backbones (Middleton & Tipton, 2000). These hydrostatically unstable groups are esters, ortho-esters, anhydrides and amines (Middleton & Tipton, 2000). The major advantages of synthetic over natural polymers are (Nair & Laurencin, 2007):

- synthetic biodegradable polymers are biologically inert
- they have more predictable properties
- their characteristics can be tailored with a specific application in mind

Various synthetic biodegradable polymers were invented of which the aliphatic polyesters appear to be the most attractive for biomedical applications. For this reason, this chapter will only discuss this type of polymers. An aliphatic polyester is a thermoplastic polymer which contains hydrolysable aliphatic ester linkage in its backbone (Donglu, 2010). In theory all polyesters are degradable, however only aliphatic polyesters with reasonably short aliphatic chains between the ester bonds will degrade within a time interval suitable for biomedical applications (Nair & Laurencin, 2007). Such aliphatic polyesters can be derived from a wide range of monomers through ring-opening and condensation polymerisation routes, but in some cases bacterial processes can also be used. In what follows we will briefly discuss the most common biodegradable aliphatic polyesters. The structural chemical formula of each discussed polymer is given in Table 1.

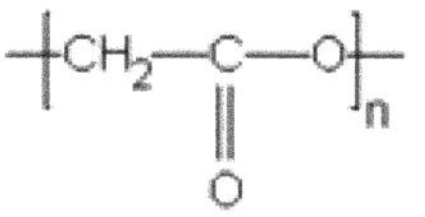 polyglycolic acid (PGA)	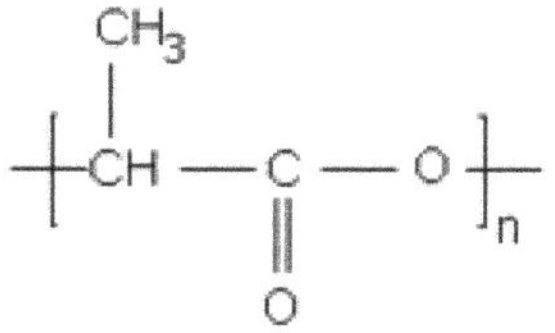 polylactic acid (PLA)

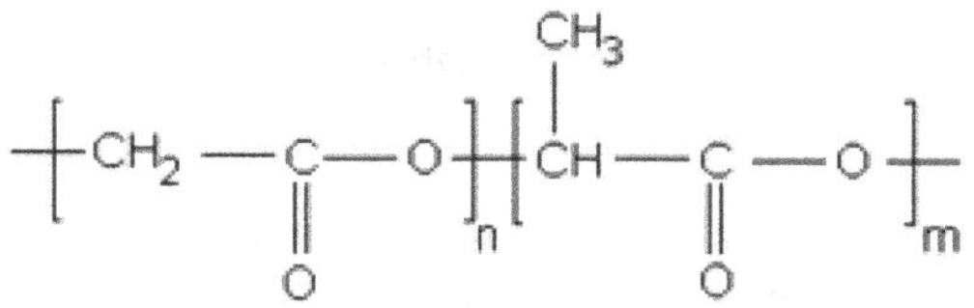

poly(lactic-co-glycolic acid)

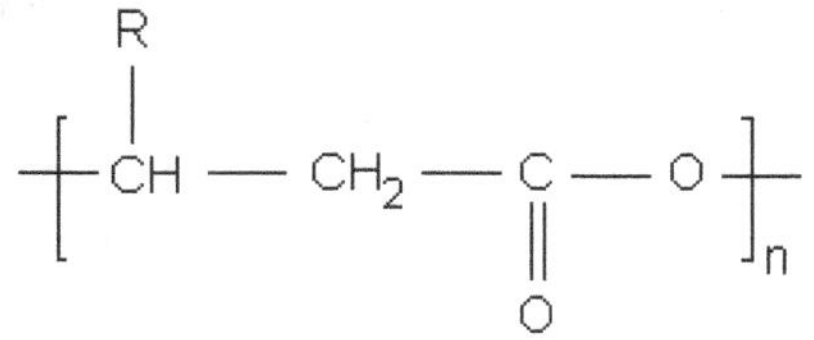

polyhydroxyalkanoates (PHA) R = CH₃: poly-3-hydroxybutyrate (PHB) R = CH₂CH₃: poly-3-hydroxyvalerate (PHV)	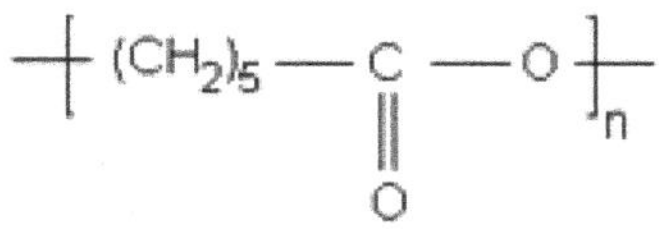 polycaprolactone (PCL)

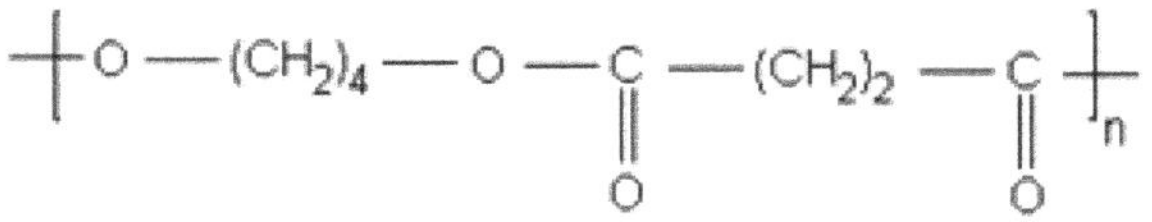

polybutylene succinate (PBS)

Table 1. Structural chemical formula of the most common biodegradable polymers.

2.2.1 Polyglycolic acid (PGA)

Polyglycolic acid (PGA) or polygycolide is the most simple linear biodegradable polymer and known as a rigid thermoplastic polymer (Vroman & Tighzert, 2009). Due to its high crystallinity (44-55 %) PGA provides excellent mechanical properties and it exhibits a low solubility in most organic solvents. Alow solubility (Nair & Laurencin, 2007). Nevertheless, it is soluble in highly-fluorinated solvents, such as hexafluoroisopropanol (Donglu, 2010). PGA has initially been studied for use as resorbable synthetic sutures and was for the first time commercialised in 1962 under the name DEXON® (Gilding & Reed, 1979). Today, due to their excellent biodegradability, good cell viability and good initial properties, PGA non-woven fabrics are also extensively utilized as scaffolds for tissue engineering (Nair & Laurencin, 2007).

PGA loses its strength in 1 to 2 months when hydrolyzed and its mass within 6 to 12 months (Nair & Laurencin, 2007). When inserted in the body, PGA breaks down into glycolic acid. Glycolic acid is not toxic and can be excreted in the urine or converted into H_2O and CO_2 and subsequently removed from the body via the respiratory system (Maurus & Kaeding, 2004). Despite the above-mentioned non-toxicity of glycolic acid, it may result in an increased and localized acid concentration leading to tissue dammage (Gunatillake & Adhikari, 2003, Taylor et al., 1994). This presents in particular problems for orthopaedic applications where implants with substantial dimensions are needed (Gunatillake & Adhikari, 2003). Together with the high degradation rate and low solubility, these acidic degradation products have hampered the use of PGA for biomedical engineering applications.

2.2.2 Polylactic acid (PLA)

The monomer building block of polylactic acid, lactic acid, is formed by converting sugar or starch from vegetable origin (e.g. wheat, corn, rice, etc.) via either bacterial fermentation or via a petrochemical process (Rasal et al., 2010). If PLA is implanted, it hydrolyses to its building block lactic acid which is a normal human metabolic by-product (Gunatillake & Adhikari, 2003). Lactic acid is degraded into H_2O and CO_2 which can be further removed by the respiratory system (Nair & Laurencin, 2007, Maurus & Kaeding, 2004). As can be seen in Table 1, lactic acid is a chiral molecule and therefore different forms of PLA occur. The two most important forms are poly(L-lactic acid) (PLLA) and poly(DL-lactic acid) (PDLLA).

Similar to PGA, PLLA has a high degree of crystallinity (± 37% depending on molecular weight and production processes) (Nair & Laurencin, 2007). Compared to PGA, PLLA slowly degrades: when PLLA is hydrolyzed, it loses its strength in circa 6 months. However, no mass loss is observed for a very long time and total degradation amount up to several years (Middleton & Tipton, 2000, Nair & Laurencin, 2007, Bergsma et al., 1995). Next to this slow degradation, PLLA offers good tensile strength, a high tensile modulus and low extension and can therefore be applied in load bearing applications like in orthopaedic fixation devices (Nair & Laurencin, 2007). PLLA fibres are also often used as surgical sutures, while PLLA composites, porous membranes or sponges can be employed as scaffolding matrices for tissue regeneration (Hu & Huang, 2010, Heino et al., 1996, Lam et al., 1995, Vaquette et al., 2008, Ma et al., 2006, Chen & Ma, 2004). For some other applications, the long degradation time of PLLA however presents a major concern.

PDLLA has an amorphous nature resulting in a substantial lower strength compared to PLLA (Nair & Laurencin, 2007). Moreover, PDLLA loses its strength in 1 to 2 months and its mass within 12 to 16 months (Maurus & Kaeding, 2004). Taking into account this low strength and its fast degradation rate, PDLLA can be employed as drug delivery system or as low strength scaffolding matrix for tissue engineering (Nair & Laurencin, 2007, Xie & Buschle-Diller, 2010).

2.2.3 Poly(lactic-co-glycolic acid) (PLGA)

A lot of research has been carried out on the development of a full range of poly(lactic-co-glycolic acid) (PLGA) polymers. This research has indicated that the degradation rate of PLGA strongly depends on the lactic acid/glycolic acid ratio (Gilding & Reed, 1979, Reed & Gilding, 1981, Miller et al., 1977). It is common knowledge that the intermediate co-polymers are much more unstable than the homo-polymers: a 50/50 PLGA and an 85/15 PLGA degrade in 1-2 months and 5-6 months respectively (Middleton & Tipton, 2000). This opportunity to tune the degradation rate of the polymer by varying the monomer ratio has made PLGA an ideal candidate for biomedical applications in the drug delivery and tissue engineering domain. However, the first commercial use of the co-polymer PLGA was as suture material under the name Vicryl ®(Nair & Laurencin, 2007, Gunatillake & Adhikari, 2003).

2.2.4 Polycaprolactone (PCL)

Polycaprolactone (PCL) is of great interest since it can be obtained from the relatively cheap monomer unit ε-caprolactone (Storey & Taylor, 1998). PCL degrades very slowly and complete degradation can take several years. Due to this slow degradation, its non-toxicity and its high permeability to small drug molecules, PCL has in the beginning been studied as a polymer for long-term drug delivery systems. PCL also offers excellent biocompatibility. Therefore, recently extensive research has been done on the use of PCL as scaffold matrices in tissue regeneration (Chiari et al., 2006, Mondrinos et al., 2006). Also several co-polymers have been developed to increase the degradation rate compared to pure PCL (Li et al., 2002, Li et al., 2003, Qian et al., 2000, Wang et al., 2001). For co-polymers synthesized from L-lactide and ε-caprolactone, the degradation rate was again strongly influenced by the L-lactide/ε-caprolactone ratio.

2.2.5 Polyhydroxyalkanoates (PHA)

Polyhydroxyalkanoates (PHA) are structurally related to PLA and are a polyester class derived from hydroxyalkanoic acids which can vary in chain length and in the hydroxyl group positions (Breulmann et al., 2009). As is the case for PLA, PHA can be obtained from renewable resources like starch, sugars or fatty acids, however, chemical transformation is not needed. The most widespread PHA is poly-3-hydroxybutyrate (PHB) which was discovered in 1920 as produced by the bacteria "Bacillus megaterium" (Nair & Laurencin, 2007). Subsequent research showed that PHB could also be synthesized via other bacterial strains and via chemical routes. **Subsequent research showed that PHB could also be synthesized via other bacterial strains and via chemical routes** (Shelton et al., 1971).

PHB degrades into D-3-hydroxybutyrate which is a normal element of human blood (Wang et al., 2001). To be used directly as biopolymer, PHB has the disadvantage of a very low degradation rate in the body compared with other biodegradable polyesters and is often considered too brittle for many applications (Nair & Laurencin, 2007, Pompe et al., 2007). Therefore, co-polymers of 3-hydroxybutyrate with other monomers such as 3-hydroxyvalerate have been synthesized (Nair & Laurencin, 2007). This poly(3-hydroxybutyrate-co-3-hydroxyvalerate) (PHBV) is far less brittle and thus offers more potential as biomaterial (Nair & Laurencin, 2007, Ojumu et al., 2004). Moreover, PHBV is piezoelectric which enables electrical stimulation – known for promoting bone healing – of the implant (Nair & Laurencin, 2007). Although the faster degradation rate of PHBV compared to PHB, it has been observed that the in vivo degradation of both polymers remains slow. Therefore, these polymers may be potential candidates for long term implants.

2.2.6 Polybutylene succinate (PBS)

Polybutylene succinate (PBS) was discovered in 1990 and commercialized under the trade name Bionolle® (Fujimaki, 1998). PBS degrades via naturally occurring enzymes and micro-organisms into H_2O and CO_2 (Tserki et al., 2006). PBS can be easily produced in a wide variety of forms and structures, such as yarns, non-wovens, films, mono-filaments and it offers excellent mechanical properties comparable with polyethylene or polypropylene (Li et al., 2005, Vroman & Tighzert, 2009). These characteristics makes PBS an excellent choice for use as scaffolds in tissue regeneration.

3. Plasma-assisted surface modification of biodegradable polyesters

3.1 Introduction

Biodegradable polymers are non-toxic, possess low immunogenicity and good mechanical properties. Moreover, their degradation rate can be adjusted and therefore recently they have been extensively studied as scaffold matrices for tissue engineering (Shen et al., 2007). This research indicated that due to their hydrophobicity and their low surface energy cells only poorly attach, spread and proliferate on these biodegradable polyesters. Therefore, the surface of these polyesters should usually be modified and already several approaches have been presented to increase their cell affinity (Desmet et al., 2009). Typically the polyesters are chemically modified by introducing specific functional groups on their surface. Two possible wet-chemical routes are surface aminolysis and surface hydrolysis. Surface aminolysis in for example 1,6-hexanediamine leads to the production of free amino groups on the surface of the polyester which improves cell adhesion (Zhu et al., 2002, Zhu et al., 2004). By applying surface hydrolysis with the use of a NaOH solution, the ester group is hydrolyzed by the hydroxide anion leading to a rupture of the polymer chain and the formation of carboxylic acid and hydroxyl groups on the tail ends of the two new chains.

The presence of these groups results in an enhanced hydrophilicity and in improved cell-material interactions (Zhu et al., 2002, Zhu et al., 2004). Although these wet-chemical processes have their merit, some disadvantages cannot be neglected. Such surface modifications are quite rough and can thus possibly lead to unwanted side-effects such as a faster degradation rate and a reduction of mechanical performance (Chong et al., 2007, Desmet et al., 2009). In addition, research indicated that these techniques can lead to irregular surface etching and that the degree of modification strongly depends on molecular weight, crystallinity or tacticity and may therefore not be reproducible (Goddard & Hotchkiss, 2007, Desai & Singh, 2004). Moreover, it is clear that these wet-chemical techniques use substantial amounts of water or other liquids and consequentially generate hazardous chemical waste. Other approaches like peroxide oxidation, ozone oxidation, UV- and γ-radiaton can also introduce reactive chemical groups on the polyester surface, however, most of these techniques also lead to degradation of the polyesters (Ho et al., 2007, Koo & Jang, 2008, Loo et al., 2004, Place et al., 2009, Montanari et al., 1998).

Opposite those above-mentioned techniques, plasma-assisted surface modification offers a very suitable strategy to incorporate reactive functional groups on the polyester surface. Without the use of a solvent, these groups are efficiently introduced on the surface without altering the bulk properties of the polymer (Ho et al., 2006, Desmet et al., 2009). In addition, complex shaped scaffolds can be uniformly treated (Shen et al., 2007). Next to the incorporation of functional groups, plasma treatment can also be employed for the deposition of polymer coatings or for the immobilization of proteins or other biomolecules (Yang et al., 2002, Cheng & Teoh, 2004, Barry et al., 2005, Barry et al., 2006, Guerrouani et al., 2007, Zelzer et al., 2009).

Due to these numerous advantages, surface modification of biodegradable polymers by plasma treatment offers several excellent prospects. Therefore, in this section 3, we will give a more general introduction on plasma-surface interactions, while in section 4 we will focus on some successful examples of plasma modification of biodegradable aliphatic polyesters.

3.2 Plasma-surface interactions and surface modification strategies

Plasma is sometimes referred to as the fourth state of matter as introduced by Langmuir (Langmuir, 1928). Plasma is a partly ionized, but quasi-neutral gas in the form of gaseous or fluid-like mixtures of free electrons, ions and radicals, generally also containing neutral particles (atoms, molecules) (Denes & Manolache, 2004). Some of these particles may be excited and can return to their ground state by emission of a photon. The latter process is at least partially responsible for the luminosity of a typical plasma. In plasma several electrons are not bound to molecules or atoms, but free. Therefore, positive and negative charges can move somewhat independently from each other.

Plasmas are frequently subdivided into equilibrium (or non-thermal/low-temperature/cold) and non-equilibrium (or thermal/high-temperature/hot) plasmas (Denes & Manolache, 2004, Bogaerts et al., 2002, Fridman et al., 2008). Thermal equilibrium implies that the temperature of all particles (electrons, ions, neutrals and excited species) is the same. This is, for example true for stars, as well as for fusion plasmas. High temperatures are required to form these type of plasmas (Bogaerts et al., 2002, Lieberman & Lichtenberg, 2005). In contrast, plasmas with strong deflection from kinetic equilibrium have electron temperatures that are a lot more elevated than the temperature of the ions and neutrals. Such plasmas are classified as non-equilibrium or non-thermal plasmas. It is clear that the high temperatures used in thermal plasmas are destructive for heat-sensitive polymers and most applications for surface modification of polymers will make use of non-thermal or cold plasmas. Since a non-thermal plasma contains a mixture of reactive species, different interactions between the plasma and a surface are possible, including plasma treatment, plasma polymerization and plasma etching (Denes & Manolache, 2004, Rausher et al., 2010, Gomathi et al., 2008). These different interactions between a plasma and the surface can be divided into 4 different approaches to modify the biodegradable polymer. These 4 approaches will briefly be introduced in the following paragraphs, while section 4 will give some practical examples.

3.2.1 Plasma treatment

Plasma treatment is mostly used to enhance the surface energy of a polymer. Figure 1 shows the decrease in contact angle of a PLA surface after treatment in different discharge atmospheres. Oxygen or nitrogen containing groups are introduced on the surface of a (biodegradable) polymer when the material is exposed to a cold plasma generated in O_2, N_2, air or NH_3 (Morent et al., 2008a, Morent et al., 2008b). These functionalities are polar hydrophilic groups which are formed during the interaction of the plasma active species with the polymer molecules. Next to oxygen- and nitrogen-containing discharges, plasmas generated in pure helium or argon will lead to the creation of free radicals that can be used for cross-linking or grafting of oxygen-containing groups when the surface is exposed to oxygen or air after the treatment (Desmet et al., 2009, De Geyter et al., 2007, Ding et al., 2004). Finally, it should be mentioned that the induced surface characteristics are not permanent. The treated surfaces will tend to partially recover to their untreated state during storage in e.g. air (so-called hydrophobic recovery) and they will also undergo post-plasma oxidation reactions (De Geyter et al., 2008, Morent et al., 2010, Siow et al., 2006).

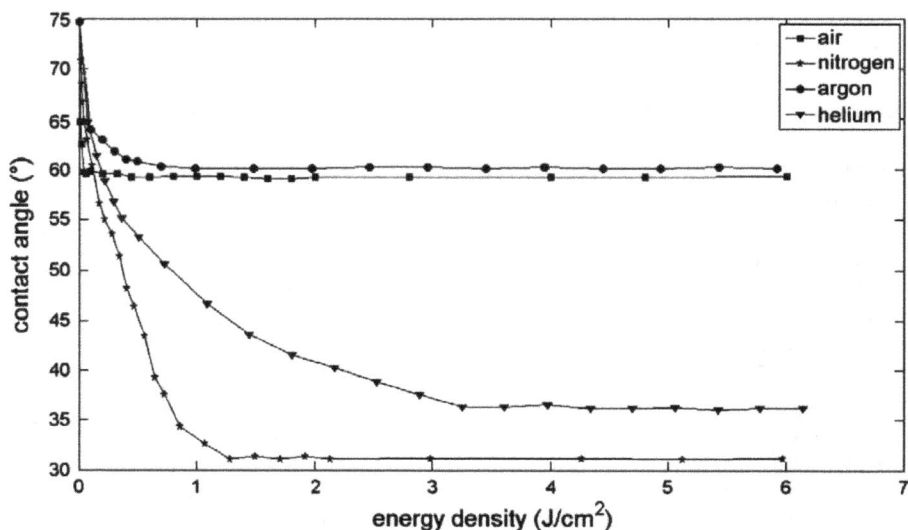

Fig. 1. Water contact angles as a function of energy density for air, nitrogen, argon and helium plasma-treated PLA samples. (Reprinted from (De Geyter et al., 2010) with permission of Elsevier).

3.2.2 Plasma post-irradiation grafting

Plasma post-irradiation grafting is a two-step process of which the first step is a plasma treatment as described in the previous paragraph. The induced functionalities can then be applied to initiate polymerization reactions (Desmet et al., 2009). In contrast to plasma treatment, this technique results in a permanent effect. In the second step, the activated polymer surface is brought into direct contact with a monomer. The monomer can be in the gas phase or the substrate can be immersed into a monomer solution (Vasilets et al., 1997, Zhu et al., 2007). It is important to notice that in both cases the monomer is not subject to the reactive plasma environment. Therefore, the grafted polymers will have similar characteristics as polymers synthesized by conventional polymerization processes (Desmet et al., 2009).

3.2.3 Plasma syn-irradiation

Firstly, a monomer is adsorbed to a material, after which the substrate is exposed to a plasma (Ding et al., 2004). This plasma will generate radicals in the adsorbed monomer layer and the surface of the substrate. This approach will lead to a cross-linked polymer top-layer (Desmet et al., 2009). Opposite to the plasma post-irradiation grafting described in the previous paragraph, the monomer is in plasma syn-irradiation directly subjected to the plasma.

3.2.4 Plasma polymerization

Thin films with unique chemical and physical properties can be developed by plasma polymerization and are called plasma polymers (Gomathi et al., 2008). During plasma polymerization, gaseous or liquid monomers are typically via a carrier gas inserted into the discharge zone in which they are converted into reactive fragments (Morent et al., 2009). These reactive fragments recombine to polymers and a polymer film is deposited on the substrate exposed to the plasma. The formed plasma polymers will not necessarily have the same chemical structure and composition as polymers obtained via conventional polymerization processes (Desmet et al., 2009). In general, plasma polymers are pinhole-free and highly cross-linked and are therefore insoluble, thermally stable, chemically inert and mechanically tough. Furthermore, such films are often highly coherent and adherent to a variety of substrates including conventional polymer, glass and metal surfaces (Morent et al., 2011, Morent et al., 2009).

4. Examples of plasma-assisted surface modification of biodegradable polyesters

In this last section, we will give some examples of the above-mentioned approaches of plasma-assisted surface modification of biodegradable aliphatic polyesters. Due to the introductory nature of this chapter, not all literature will be discussed in detail. For a more complete overview of literature, the reader could consult the review paper on the same subject of our research group (Morent et al., 2011). Plasma treatment as described in section 3.2.1 is by far the most occurring approach used to modify the surface of biodegradable polyesters and numerous examples can be found in literature. In this chapter, we will try to give some examples for all the biodegradable polyesters discussed in section 2. Therefore, section 4.1 will be much more extensive than the other sections (4.2 up to 4.4) since the availability of literature on these latter approaches is less pronounced.

4.1 Plasma treatment of biodegradable polymers

Hirotsu et al. published in 1997 one of the first studies on plasma modification of biodegradable polymers and treated PLA fabrics with a low pressure radio frequent (RF) discharge generated in pure oxygen and nitrogen (Hirotsu et al., 1997). The same group reported in 2002 about an enhancement of the wettability of PLLA sheets and showed a strong decrease in water contact angle from 80° to approximately 55° after 30 seconds of oxygen and helium plasma treatment (Hirotsu et al., 2002). They suggested that this increased wettability was only due to chemical changes of the surface, since pronounced etching is not likely to happen after such short treatment times. However, they were not able to determine the groups incorporated at the surface.

	C (at%)	O (at%)	N (at%)
no treatment	68	32	0
air plasma treatment	62	38	0

Table 2. Atomic composition of untreated and air plasma-treated PLA films (De Geyter et al., 2010).

To identify the functionalities incorporated at the surface, De Geyter et al. did detailed XPS studies on PLA sheets plasma-treated with a medium pressure dielectric barrier discharge (DBD) sustained in air. Table 2 shows the atomic composition of the PLA films plasma-treated in air. This table suggests that air plasma mainly adds oxygen atoms to the PLA surfaces. From high-resolution XPS scans, the authors have concluded that after plasma treatment in air, the concentration of C-O and O-C=O groups increases, while the C-C and C-H functional groups decrease. Hirtosu et al. observed a gradual increase in water contact angle when PLA samples were kept in dry air (Hirotsu et al., 2002). A similar hydrophobic recovery has recently been examined in detail by Morent et al., who employed a medium pressure DBD in different atmospheres for the surface modification of PLA. They concluded that during storage in air, the induced polar chemical groups reorientate or migrate to the bulk of the material (Morent et al., 2010). The introduction of specific functional groups on the surface of PLA samples and the accompanying increase in wettability often have the aim to improve the cell-material interactions. These interactions between B65 nervous tissue cells and oxygen plasma-treated PLLA films were studied by Khorasani et al. (Khorasani et al., 2008). Figure 2 shows optical photomicrographs of B65 cell attachment and growth on untreated and plasma-modified PLLA surfaces and it can clearly be observed that this oxygen plasma treatment substantially improves cell attachment and growth. The authors concluded that plasma-modified PLLA surfaces are very suitable for nervous tissue engineering purposes.

The majority of the above-discussed research is on flat 2D PLLA surfaces. However, from biomedical point of view, 3D porous polymer scaffolds are needed in the field of tissue engineering in order to offer sufficient support for tissue growth (Djordjevic et al., 2008). Only few authors have worked with 3D structures because of two reasons: (1) the insufficient knowledge on the penetration of plasma into porous structures and (2) the difficulty of characterisation of the interior surface with classical surface analytical tools. Wan et al. have modified 4 mm thick PLLA scaffolds with an ammonia plasma (Wan et al., 2006). To examine the plasma effect, they have immersed the scaffolds in blue ink after treatment to demonstrate the influence of treatment time on the modifying depth. Figure 3 shows that due to the poor hydrophilicity of the internal surface of the untreated PLLA sample only the most outside layer is dyed. However, with increasing treatment time, the ink increasingly penetrates the PLLA scaffold and after a treatment of half an hour the interior part of the scaffold is fully dyed.

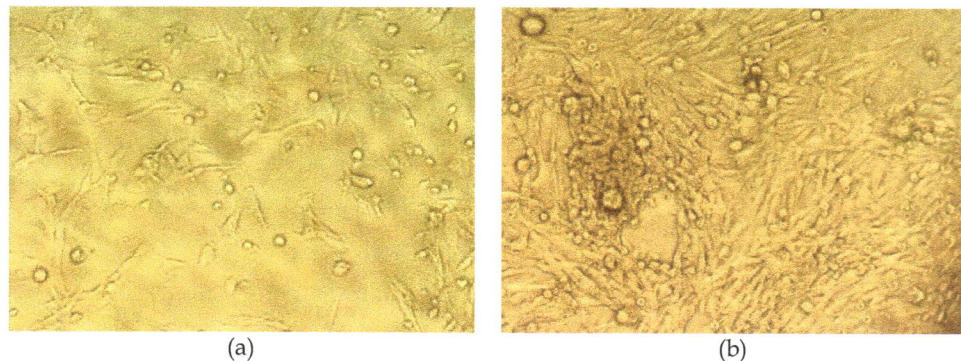

(a)   (b)

Fig. 2. B65 attachment on (a) untreated PLLA and (b) oxygen plasma-treated PLLA (magnification 400x) (Reprinted from (Khorasani et al., 2008) with permission of Elsevier).

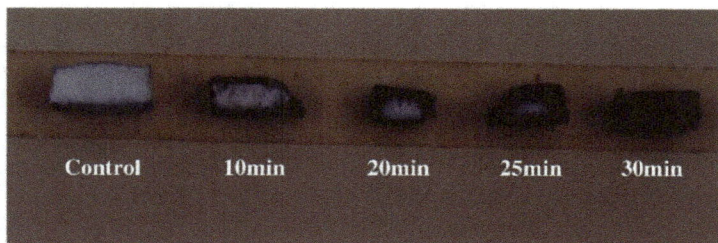

Fig. 3. Effect of plasma treatment time on the modifying depth of PLLA scaffolds (Reprinted from (Wan et al., 2006) with permission of Elsevier).

To our knowledge, no literature on plasma treatment of polyglycolic acid has been published so far. However, quite a few research articles deal with plasma modification of the copolymer PLGA (Khorasani et al., 2008, Hasirci et al., 2010, Khang et al., 2002, Park et al., 2007, Park et al., 2010, Safinia et al., 2007, Safinia et al., 2008, Shen et al., 2008, Wang et al., 2004). 50/50 PLGA films were modified in an oxygen plasma at low pressure and a decrease of the contact angle from 67° to below 40° after plasma treatment was observed. XPS revealed that oxygen containing functionalities are introduced and cell culture tests (3T3 fibroblasts) showed a higher cell attachment and proliferation on oxygen plasma-treated PLGA surfaces. As discussed in the previous paragraph on PLLA treated surfaces, Khorasani et al. also investigated in the same paper the interaction between nervous tissue cells and plasma modified PLGA samples (Khorasani et al., 2008). Figure 4 shows that oxygen plasma treatment clearly improves attachment and growth of B65 cells, however, the effect of oxygen plasma treatment seems less pronounced as was the case for PLLA surfaces (see Figure 2). Khang et al. studied and compared several modification methods including chemical methods (sulphuric acid, chloric acid, sodium hydroxide) as well as physical methods (atmospheric pressure air discharge) for the surface treatment of PLGA (Khang et al., 2002). Their results clearly evidenced that both chemical methods and plasma treatment could enhance cell attachment and growth. The high potential of non-thermal plasma for the surface modification of biodegradable polymers was clearly demonstrated since plasma treatment showed to be almost as efficient in increasing cell-material interactions as a chloric acid treatment and more efficient than sulphuric acid and sodium hydroxide treatments.

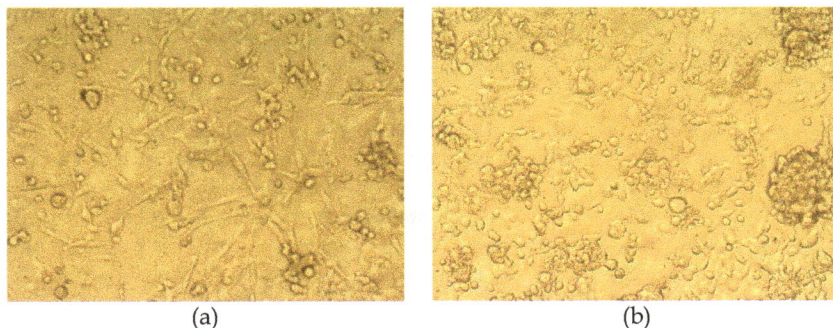

(a)   (b)

Fig. 4. B65 cell attachment on (a) untreated PLGA and (b) oxygen plasma-treated PLGA (magnification 400x) (Reprinted from (Khorasani et al., 2008) with permission of Elsevier).

Surface modification of PCL with oxygen, helium and air plasmas has resulted into similar effects as on PLA and PLGA: an increased hydrophilicity, a higher oxygen amount and consequently an enhanced cell attachment and proliferation (Yildirim et al., 2008, Hirotsu et al., 2000a, Lee et al., 2009, Prabhakaran et al., 2008, Little et al., 2009). Lee and co-workers treated PCL with atmospheric pressure plasmas with different discharge gases (Lee et al., 2008). Figure 5 shows the enhancement in hydrophilicity (Figure 5 (a)), the increased cell attachment (Figure 5 (b)) and the increased cell proliferation (Figure 5 (c)).

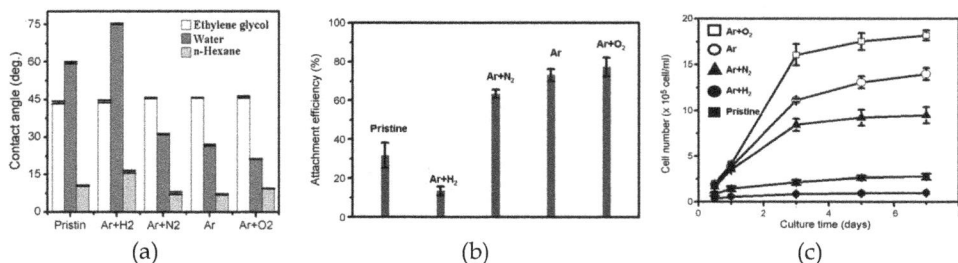

(a)   (b)   (c)

Fig. 5. Plasma treatment of PCL: (a) contact angle (b) cell attachment and (c) cell proliferation of human epithelial cells (Reprinted from (Lee et al., 2008) with permission of Elsevier).

The most common polyhydroxyalkanoate subjected to plasma treatments is the co-polymer poly(3-hydroxybutyrate-co-3-hydroxyvalerate) (PHBV) and oxygen plasmas have been widely employed to modify this co-polymer (Wang et al., 2006, Hasirci et al., 2003, Tezcaner et al., 2003, Kose et al., 2003b, Kose et al., 2003a, Ferreira et al., 2009). A low pressure oxygen plasma was employed to PHBV films containing 8% hydroxyvalerate in its structure by Hasirci et al. (Hasirci et al., 2003). A decrease in water contact angle upon oxygen plasma treatment was observed which was attributed to the incorporation of oxygen-containing functional groups on the PHBV surface. A subsequent study showed that O_2 plasma treatment significantly enhanced the interaction between retinal pigment epithelium (RPE) cells and PHBV (Tezcaner et al., 2003).

Hirotsu et al. published an interesting article on the plasma modification of self-made PBS sheets in different discharge atmospheres (O_2, N_2 and helium) (Hirotsu et al., 2000b). Contact angle measurements on the plasma-modified samples clearly showed that plasmas are able to increase the hydrophilicity, however, it was not stated which chemical groups contributed to this increased wettability.

4.2 Plasma post-irradiation grafting of biodegradable polyesters

Section 4.1 clearly focussed on the observation that plasma treatment can easily induce desired functionalities onto the surface of biodegradable polymers resulting in an improved cell affinity. However, hydrophobic recovery acts as a brake on practical applications of plasma-treated polyesters. Nevertheless, this drawback can be solved by covalently immobilizing bioactive molecules on plasma-treated surfaces (Gupta et al., 2002). Typically extracellular matrix (ECM) proteins such as gelatine, collagen or fibrin have been grafted on the surface of biodegradable polyesters since these proteins are known to enhance cell adhesion and proliferation (Ma et al., 2007). Different authors have studied the immobilization of collagen on PCL films (Ma et al., 2007, Chong et al., 2007,

Cheng & Teoh, 2004). Firstly, an argon plasma is applied to a PCL film to generate radicals on the polyester surface. Exposure to the atmosphere for several minutes leads to the formation of functionalities such as surface peroxides and hydroperoxides that will be employed as initiator sites for UV-induced graft polymerization of acrylic acid. To pre-activate the carboxyl groups, the grafted substrates are immersed into a carbodiimide solution. In a final step, the material is immersed into a collagen solution leading to the production of a collagen-immobilized biodegradable polyester. These collagen-modified PCL surfaces have been tested with a diversity of cells including human dermal fibroblasts, human myoblasts, human endothelial cells and human smooth muscle cells and all demonstrated favourable response from these cells (Ma et al., 2007, Chong et al., 2007, Cheng & Teoh, 2004). Next to collagen, Kang et al. also immobilized insulin on the surface of a PHBV co-polymer and observed that the proliferation of human fibroblasts was significantly accelerated on these films compared to the untreated samples (Kang et al., 2001).

4.3 Plasma syn-irradiation of biodegradable polyesters
As described in section 3.2.3, a polymer can also be grafted on the surface of a biodegradable polyester by pre-adsorption of the monomer followed by a plasma treatment. However for the specific case of biodegradable polymers, we were able to track only one research paper using this plasma approach (Ding et al., 2004). In this paper, Ding et al. tried to modify the surface of PLLA films with a chitosan layer. However, results indicated poor cell adhesion, but acceptable cell proliferation.

4.4 Plasma polymerization on biodegradable polyesters
Plasma polymerization differs from plasma grafting in that respect that it coats the substrate rather than covalently binds species to a plasma-modified polymer surface (Barry et al., 2005). Allylamine is one of the most frequently used monomers to plasma polymerize on biodegradable polymers such as PLLA, PCL and PHBV (Barry et al., 2005, Guerrouani et al., 2007, Carlisle et al., 2000). Plasma polymerized allylamine films on biodegradable polymers resulted in highly hydrophilic surfaces with contact angles of 20° or lower due to the amine groups on the surface.

As there is a great interest in the surface modification of 3D implants, plasma polymerization has also been performed on the surface of 3D PLA scaffolds. Plasma grafting has been compared with plasma polymerization using allylamine (Barry et al., 2005). In the case of plasma grafting, the scaffolds were first pre-treated with an oxygen plasma and afterwards exposed to allylamine vapour, while plasma polymerization was carried out by exposing the scaffolds to an allylamine vapour plasma after an oxygen plasma pre-treatment. XPS measurements of the scaffolds at different points across the scaffold diameter demonstrated that the grafting process resulted in a more homogeneous nitrogen concentration through the scaffold while the concentration of nitrogen on the internal surface of the scaffold on which the plasma deposit was formed decreased from the edge to the core of the scaffold, as can be seen in Figure 6. However, at the lowest nitrogen concentration, the nitrogen concentration on the internal surface of the plasma-polymerized scaffold was still greater than that of the grafted surface. The plasma-coated scaffolds also showed a higher metabolic activity than the plasma-grafted samples. Moreover, fibroblasts were detected in the centre of the plasma-coated samples, which was not the case for the grafted scaffolds (Barry et al., 2005).

5. Conclusion

The growing research fields of tissue engineering and regenerative medicine are a leverage for surface engineering of biodegradable polymers. Next to chemical surface modification techniques which encounter problems with the use of hazardous organic solvents in relation to cell viability, non-thermal plasma technology knows a steep growth as solvent-free technique. Plasma treatments are already commonly performed on biodegradable polymers such as PLA and PLGA, while treatment of more advanced biodegradable polymers (such as PCL, PHBV, PBS and composites) and other plasma-based techniques (such as plasma grafting and plasma polymerization) are only at the verge of breaking through. Non-thermal plasma technology can greatly enhance cell-material interactions, however, a better understanding of these interactions is of crucial importance. This knowledge can provide us information on which plasma-based strategies should exactly be pursued.

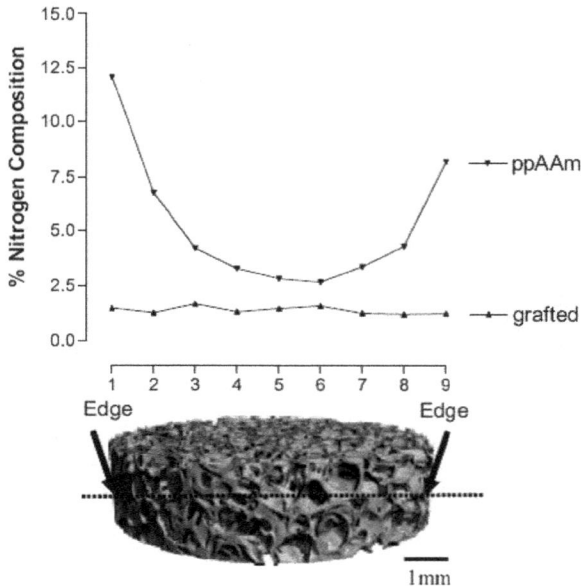

Fig. 6. Nitrogen concentration as determined by XPS at set points across the internal diameter of grafted and plasma-polymerized allylamine (ppAAm) scaffolds (Reprinted from (Barry et al., 2005) with permission from Wiley-VCH Verlag GmbH & Co. KGaA).

6. References

Agrawal, C. M., Best, J., Heckman, J. D. & Boyan, B. D. 1995. Protein Release Kinetics of A Biodegradable Implant for Fracture Non-Unions. *Biomaterials,* 16, 1255-1260.

Amass, W., Amass, A. & Tighe, B. 1998. A review of biodegradable polymers: Uses, current developments in the synthesis and characterization of biodegradable polyesters, blends of biodegradable polymers and recent advances in biodegradation studies. *Polymer International*, 47, 89-144.

Athanasiou, K. A., Agrawal, C. M., Barber, F. A. & Burkhart, S. S. 1998. Orthopaedic applications for PLA-PGA biodegradable polymers. *Arthroscopy-the Journal of Arthroscopic and Related Surgery*, 14, 726-737.

Barry, J. J. A., Howard, D., Shakesheff, K. M., Howdle, S. M. & Alexander, M. R. 2006. Using a core-sheath distribution of surface chemistry through 3D tissue engineering scaffolds to control cell ingress. *Advanced Materials*, 18, 1406-+.

Barry, J. J. A., Silva, M. M. C. G., Shakesheff, K. M., Howdle, S. M. & Alexander, M. R. 2005. Using plasma deposits to promote cell population of the porous interior of three-dimensional poly(D,L-lactic acid) tissue-engineering scaffolds. *Advanced Functional Materials*, 15, 1134-1140.

Bergsma, J. E., Rozema, F. R., Bos, R. R. M., Boering, G., Debruijn, W. C. & Pennings, A. J. 1995. In-Vivo Degradation and Biocompatibility Study of In-Vitro Pre-Degraded As-Polymerized Polylactide Particles. *Biomaterials*, 16, 267-274.

Bogaerts, A., Neyts, E., Gijbels, R. & van der Mullen, J. 2002. Gas discharge plasmas and their applications. *Spectrochimica Acta Part B-Atomic Spectroscopy*, 57, 609-658.

Breulmann, M., Künkel, A., Philipp, S., Reimer, V., Siegenthaler, K. O., Skupin, G. & Yamamoto, M. 2009. Biodegradable polymers. *Ullmann's Encyclopedia of Industrial Chemistry*.

Carlisle, E. S., Mariappan, M. R., Nelson, K. D., Thomes, B. E., Timmons, R. B., Constantinescu, A., Eberhart, R. C. & Bankey, P. E. 2000. Enhancing hepatocyte adhesion by pulsed plasma deposition and polyethylene glycol coupling. *Tissue Engineering*, 6, 45-52.

Chen, V. J. & Ma, P. X. 2004. Nano-fibrous poly(L-lactic acid) scaffolds with interconnected spherical macropores. *Biomaterials*, 25, 2065-2073.

Cheng, Z. Y. & Teoh, S. H. 2004. Surface modification of ultra thin poly (epsilon-caprolactone) films using acrylic acid and collagen. *Biomaterials*, 25, 1991-2001.

Chiari, C., Koller, U., Dorotka, R., Eder, C., Plasenzotti, R., Lang, S., Ambrosio, L., Tognana, E., Kon, E., Salter, D. & Nehrer, S. 2006. A tissue engineering approach to meniscus regeneration in a sheep model. *Osteoarthritis and Cartilage*, 14, 1056-1065.

Chong, M. S. K., Lee, C. N. & Teoh, S. H. 2007. Characterization of smooth muscle cells on poly(epsilon-caprolactone) films. *Materials Science & Engineering C-Biomimetic and Supramolecular Systems*, 27, 309-312.

De Geyter, N., Morent, R., Desmet, T., Trentesaux, M., Gengembre, L., Dubruel, P., Leys, C. & Payen, E. 2010. Plasma modification of polylactic acid in a medium pressure DBD. *Surface & Coatings Technology*, 204, 3272-3279.

De Geyter, N., Morent, R. & Leys, C. 2008. Influence of ambient conditions on the ageing behaviour of plasma-treated PET surfaces. *Nuclear Instruments & Methods in Physics Research Section B-Beam Interactions with Materials and Atoms*, 266, 3086-3090.

De Geyter, N., Morent, R., Leys, C., Gengembre, L. & Payen, E. 2007. Treatment of polymer films with a dielectric barrier discharge in air, helium and argon at medium pressure. *Surface & Coatings Technology*, 201, 7066-7075.

Denes, F. S. & Manolache, S. 2004. Macromolecular plasma-chemistry: an emerging field of polymer science. *Progress in Polymer Science*, 29, 815-885.

Desai, S. M. & Singh, R. P. 2004. *Surface modification of polyethylene*.

Desmet, T., Morent, R., De Geyter, N., Leys, C., Schacht, E. & Dubruel, P. 2009. Nonthermal Plasma Technology as a Versatile Strategy for Polymeric Biomaterials Surface Modification: A Review. *Biomacromolecules*, 10, 2351-2378.

Ding, Z., Chen, J. N., Gao, S. Y., Chang, J. B., Zhang, J. F. & Kang, E. T. 2004. Immobilization of chitosan onto poly-L-lactic acid film surface by plasma graft polymerization to control the morphology of fibroblast and liver cells. *Biomaterials*, 25, 1059-1067.

Djordjevic, I., Britcher, L. G. & Kumar, S. 2008. Morphological and surface compositional changes in poly(lactide-co-glycolide) tissue engineering scaffolds upon radio frequency glow discharge plasma treatment. *Applied Surface Science*, 254, 1929-1935.

Donglu, S. 2010. Synthetic biodegradable polymers. *Introduction to biomaterials*. World Scientific Publishing Co.

Ferreira, B. M. P., Pinheiro, L. M. P., Nascente, P. A. P., Ferreira, M. J. & Duek, E. A. R. 2009. Plasma surface treatments of poly(L-lactic acid) (PLLA) and poly(hydroxybutyrate-co-hydroxyvalerate) (PHBV). *Materials Science & Engineering C-Biomimetic and Supramolecular Systems*, 29, 806-813.

Fridman, G., Friedman, G., Gutsol, A., Shekhter, A. B., Vasilets, V. N. & Fridman, A. 2008. Applied plasma medicine. *Plasma Processes and Polymers*, 5, 503-533.

Fujimaki, T. 1998. Processability and properties of aliphatic polyesters, 'BIONOLLE', synthesized by polycondensation reaction. *Polymer Degradation and Stability*, 59, 209-214.

Gilding, D. K. & Reed, A. M. 1979. Biodegradable Polymers for Use in Surgery - Polyglycolic-Poly(Actic Acid) Homopolymers and Copolymers .1. *Polymer*, 20, 1459-1464.

Goddard, J. M. & Hotchkiss, J. H. 2007. Polymer surface modification for the attachment of bioactive compounds. *Progress in Polymer Science*, 32, 698-725.

Gomathi, N., Sureshkumar, A. & Neogi, S. 2008. RF plasma-treated polymers for biomedical applications. *Current Science*, 94, 1478-1486.

Guerrouani, N., Baldo, A., Bouffin, A., Drakides, C., Guimon, M. F. & Mas, A. 2007. Allylamine plasma-polymerization on PLLA surface evaluation of the biodegradation. *Journal of Applied Polymer Science*, 105, 1978-1986.

Gunatillake, P. A. & Adhikari, R. 2003. Biodegradable synthetic polymers for tissue engineering. *European Cells and Materials*, 5, 1-16.

Gupta, B., Hilborn, J., Plummer, C., Bisson, I. & Frey, P. 2002. Thermal crosslinking of collagen immobilized on poly(acrylic acid) grafted poly(ethylene terephthalate) films. *Journal of Applied Polymer Science*, 85, 1874-1880.

Hasirci, N., Endogan, T., Vardar, E., Kiziltay, A. & Hasirci, V. 2010. Effect of oxygen plasma on surface properties and biocompatibility of PLGA films. *Surface and Interface Analysis,* 42, 486-491.

Hasirci, V., Tezcaner, A., Hasirci, N. & Suzer, S. 2003. Oxygen plasma modification of poly(3-hydroxybutyrate-co-3-hydroxyvalerate) film surfaces for tissue engineering purposes. *Journal of Applied Polymer Science,* 87, 1285-1289.

Heino, A., Naukkarinen, A., Kulju, T., Tormala, P., Pohjonen, T. & Makela, E. A. 1996. Characteristics of poly(L-)lactic acid suture applied to fascial closure in rats. *Journal of Biomedical Materials Research,* 30, 187-192.

Hirotsu, T., Ketelaars, A. A. J. & Nakayama, K. 2000a. Plasma surface treatment of PCL/PC blend sheets. *Polymer Engineering and Science,* 40, 2324-2331.

Hirotsu, T., Masuda, T., Matumura, Y. & Takahashi, M. 1997. Surface effects of plasma treatments on some biodegradable polyesters. *Journal of Photopolymer Science and Technology,* 10, 123-128.

Hirotsu, T., Nakayama, K., Tsujisaka, T., Mas, A. & Schue, F. 2002. Plasma surface treatments of melt-extruded sheets of poly(L-lactic acid). *Polymer Engineering and Science,* 42, 299-306.

Hirotsu, T., Tsujisaka, T., Masuda, T. & Nakayama, K. 2000b. Plasma surface treatments and biodegradation of poly(butylene succinate) sheets. *Journal of Applied Polymer Science,* 78, 1121-1129.

Ho, M. H., Hou, L. T., Tu, C. Y., Hsieh, H. J., Lai, J. Y., Chen, W. J. & Wang, D. M. 2006. Promotion of cell affinity of porous PLLA scaffolds by immobilization of RGD peptides via plasma treatment. *Macromolecular Bioscience,* 6, 90-98.

Ho, M. H., Lee, J. J., Fan, S. C., Wang, D. M., Hou, L. T., Hsieh, H. J. & Lai, J. Y. 2007. Efficient modification on PLLA by ozone treatment for biomedical applications. *Macromolecular Bioscience,* 7, 467-474.

Hu, W. & Huang, Z. M. 2010. Biocompatibility of braided poly(L-lactic acid) nanofiber wires applied as tissue sutures. *Polymer International,* 59, 92-99.

Ikada, Y. & Tsuji, H. 2000. Biodegradable polyesters for medical and ecological applications. *Macromolecular Rapid Communications,* 21, 117-132.

Kang, I. K., Choi, S. H., Shin, D. S. & Yoon, S. C. 2001. Surface modification of polyhydroxyalkanoate films and their interaction with human fibroblasts. *International Journal of Biological Macromolecules,* 28, 205-212.

Khang, G., Choee, J. H., Rhee, J. M. & Lee, H. B. 2002. Interaction of different types of cells on physicochemically treated poly(L-lactide-co-glycolide) surfaces. *Journal of Applied Polymer Science,* 85, 1253-1262.

Khorasani, M. T., Mirzadeh, H. & Irani, S. 2008. Plasma surface modification of poly (L-lactic acid) and poly (lactic-co-glycolic acid) films for improvement of nerve cells adhesion. *Radiation Physics and Chemistry,* 77, 280-287.

Koo, G. H. & Jang, J. 2008. Surface modification of poly(lactic acid) by UV/Ozone irradiation. *Fibers and Polymers,* 9, 674-678.

Kose, G. T., Ber, S., Korkusuz, F. & Hasirci, V. 2003a. Poly(3-hydroxybutyric acid-co-3-hydroxyvaleric acid) based tissue engineering matrices. *Journal of Materials Science-Materials in Medicine,* 14, 121-126.

Kose, G. T., Kenar, H., Hasirci, N. & Hasirci, V. 2003b. Macroporous poly(3-hydroxybutyrate-co-3-hydroxyvalerate) matrices for bone tissue engineering. *Biomaterials,* 24, 1949-1958.

Lam, K. H., Nijenhuis, A. J., Bartels, H., Postema, A. R., Jonkman, M. F., Pennings, A. J. & Nieuwenhuis, P. 1995. Reinforced Poly(L-Lactic Acid) Fibers As Suture Material. *Journal of Applied Biomaterials,* 6, 191-197.

Langmuir, I. 1928. Oscillations in ionized gases. *Proceedings of the National Academy of Sciences of the United States of America,* 14, 627-637.

Lee, H. U., Jeong, Y. S., Jeong, S. Y., Park, S. Y., Bae, J. S., Kim, H. G. & Cho, C. R. 2008. Role of reactive gas in atmospheric plasma for cell attachment and proliferation on biocompatible poly epsilon-caprolactone film. *Applied Surface Science,* 254, 5700-5705.

Lee, H. U., Jeong, Y. S., Koh, K. N., Jeong, S. Y., Kim, H. G., Bae, J. S. & Cho, C. R. 2009. Contribution of power on cell adhesion using atmospheric dielectric barrier discharge (DBD) plasma system. *Current Applied Physics,* 9, 219-223.

Li, G. M., Cai, Q., Bei, J. Z. & Wang, S. G. 2002. Relationship between morphology structure and composition of polycaprolactone/poly (ethylene oxide)/polylactide copolymeric microspheres. *Polymers for Advanced Technologies,* 13, 636-643.

Li, G. M., Cai, Q., Bei, J. Z. & Wang, S. G. 2003. Morphology and levonorgestrel release behavior of polycaprolactone/poly(ethylene oxide)/polylactide tri-component copolymeric microspheres. *Polymers for Advanced Technologies,* 14, 239-244.

Li, H. Y., Chang, J., Cao, A. M. & Wang, J. Y. 2005. In vitro evaluation of biodegradable poly(butylene succinate) as a novel biomaterial. *Macromolecular Bioscience,* 5, 433-440.

Lieberman, M. A. & Lichtenberg, A. J. 2005. *Principles of plasma discharges and materials processing - Second edition,* Hoboken, New Jersey, John Wiley & Sons, Inc.

Little, U., Buchanan, F., Harkin-Jones, E., Graham, B., Fox, B., Boyd, A., Meenan, B. & Dickson, G. 2009. Surface modification of poly(epsilon-caprolactone) using a dielectric barrier discharge in atmospheric pressure glow discharge mode. *Acta Biomaterialia,* 5, 2025-2032.

Loo, S. C. J., Ooi, C. P. & Boey, Y. C. F. 2004. Radiation effects on poly(lactide-co-glycolide) (PLGA) and poly(L-lactide) (PLLA). *Polymer Degradation and Stability,* 83, 259-265.

Ma, P. X., Xiaohua, L. & Youngjun, W. 2006. Porogen-induced surface modification of nano-fibrous poly(l-lactic acid) scaffolds for tissue engineering. *Biomaterials,* vol.27, no.21, 3980-3987.

Ma, Z. W., Mao, Z. W. & Gao, C. Y. 2007. Surface modification and property analysis of biomedical polymers used for tissue engineering. *Colloids and Surfaces B-Biointerfaces,* 60, 137-157.

Maurus, P. B. & Kaeding, C. C. 2004. Bioabsorbable implant material review. *Operative Techniques in Sports Medicine,* 12, 158-160.

Middleton, J. C. & Tipton, A. J. 2000. Synthetic biodegradable polymers as orthopedic devices. *Biomaterials,* 21, 2335-2346.

Miller, R. A., Brady, J. M. & Cutright, D. E. 1977. Degradation Rates of Oral Resorbable Implants (Polylactates and Polyglycolates) - Rate Modification with Changes in Pla-Pga Copolymer Ratios. *Journal of Biomedical Materials Research*, 11, 711-719.

Mondrinos, M. J., Dembzynski, R., Lu, L., Byrapogu, V. K. C., Wootton, D. M., Lelkes, P. I. & Zhou, J. 2006. Porogen-based solid freeform fabrication of polycaprolactone-calcium phosphate scaffolds for tissue engineering. *Biomaterials*, 27, 4399-4408.

Montanari, L., Costantini, M., Signoretti, E. C., Valvo, L., Santucci, M., Bartolomei, M., Fattibene, P., Onori, S., Faucitano, A., Conti, B. & Genta, I. 1998. Gamma irradiation effects on poly(DL-lactictide-co-glycolide) microspheres. *Journal of Controlled Release*, 56, 219-229.

Morent, R., De Geyter, N., Desmet, T., Dubruel, P. & Leys, C. 2011. Plasma Surface Modification of Biodegradable Polymers: A Review. *Plasma Processes and Polymers*, 8, 171-190.

Morent, R., De Geyter, N., Gengembre, L., Leys, C., Payen, E., Van Vlierberghe, S. & Schacht, E. 2008a. Surface treatment of a polypropylene film with a nitrogen DBD at medium pressure. *European Physical Journal-Applied Physics*, 43, 289-294.

Morent, R., De Geyter, N. & Leys, C. 2008b. Effects of operating parameters on plasma-induced PET surface treatment. *Nuclear Instruments & Methods in Physics Research Section B-Beam Interactions with Materials and Atoms*, 266, 3081-3085.

Morent, R., De Geyter, N., Trentesaux, M., Gengembre, L., Dubruel, P., Leys, C. & Payen, E. 2010. Influence of Discharge Atmosphere on the Ageing Behaviour of Plasma-Treated Polylactic Acid. *Plasma Chemistry and Plasma Processing*, 30, 525-536.

Morent, R., De Geyter, N., Van Vlierberghe, S., Dubruel, P., Leys, C., Gengembre, L., Schacht, E. & Payen, E. 2009. Deposition of HMDSO-based coatings on PET substrates using an atmospheric pressure dielectric barrier discharge. *Progress in Organic Coatings*, 64, 304-310.

Nair, L. S. & Laurencin, C. T. 2007. Biodegradable polymers as biomaterials. *Progress in Polymer Science*, 32, 762-798.

Ojumu, T. V., Yu, J. & Solomon, B. O. 2004. Production of Polyhydroxyalkanoates, a bacterial biodegradable polymer. *African journal of Biotechnology*, 3, 18-24.

Park, H., Lee, J. W., Park, K. E., Park, W. H. & Lee, K. Y. 2010. Stress response of fibroblasts adherent to the surface of plasma-treated poly(lactic-co-glycolic acid) nanofiber matrices. *Colloids and Surfaces B-Biointerfaces*, 77, 90-95.

Park, K. E., Lee, K. Y., Lee, S. J. & Park, W. H. 2007. Surface characteristics of plasma-treated PLGA nanofibers. *Macromolecular Symposia*, 249, 103-108.

Place, E. S., George, J. H., Williams, C. K. & Stevens, M. M. 2009. Synthetic polymer scaffolds for tissue engineering. *Chemical Society Reviews*, 38, 1139-1151.

Pompe, T., Keller, K., Mothes, G., Nitschke, M., Teese, M., Zimmermann, R. & Werner, C. 2007. Surface modification of poly(hydroxybutyrate) films to control cell-matrix adhesion. *Biomaterials*, 28, 28-37.

Prabhakaran, M. P., Venugopal, J., Chan, C. K. & Ramakrishna, S. 2008. Surface modified electrospun nanofibrous scaffolds for nerve tissue engineering. *Nanotechnology*, 19.

Qian, H. T., Bei, J. Z. & Wang, S. G. 2000. Synthesis, characterization and degradation of ABA block copolymer of L-lactide and epsilon-caprolactone. *Polymer Degradation and Stability*, 68, 423-429.

Ramchandani, M. & Robinson, D. 1998. In vitro and in vivo release of ciprofloxacin from PLGA 50 : 50 implants. *Journal of Controlled Release*, 54, 167-175.

Rasal, R. M., Janorkar, A. V. & Hirt, D. E. 2010. Poly(lactic acid) modifications. *Progress in Polymer Science*, 35, 338-356.

Rausher, H., Perucca, M. & Buyle, G. 2010. *Plasma Technology for hyperfunctional surfaces*, Weinheim, Wiley-VCH Verlag GmbH & Co.

Reed, A. M. & Gilding, D. K. 1981. Biodegradable Polymers for Use in Surgery - Poly(Glycolic)-Poly(Lactic Acid) Homo and Co-Polymers .2. Invitro Degradation. *Polymer*, 22, 494-498.

Ryu, G. H., Yang, W. S., Roh, H. W., Lee, I. S., Kim, J. K., Lee, G. H., Lee, D. H., Park, B. J., Lee, M. S. & Park, J. C. 2005. Plasma surface modification of poly(D,L-lactic-co-glycolic acid)(65/35) film for tissue engineering. *Surface & Coatings Technology*, 193, 60-64.

Safinia, L., Wilson, K., Mantalaris, A. & Bismarck, A. 2007. Atmospheric plasma treatment of porous polymer constructs for tissue engineering applications. *Macromolecular Bioscience*, 7, 315-327.

Safinia, L., Wilson, K., Mantalaris, A. & Bismarck, A. 2008. Through-thickness plasma modification of biodegradable and nonbiodegradable porous polymer constructs. *Journal of Biomedical Materials Research Part A*, 87A, 632-642.

Shelton, J. R., Lando, J. B. & Agostini, D. E. 1971. Synthesis and Characterization of Poly (Beta-Hydroxybutyrate). *Journal of Polymer Science Part B-Polymer Letters*, 9, 173- & .

Shen, H., Hu, X. X., Bei, J. Z. & Wang, S. G. 2008. The immobilization of basic fibroblast growth factor on plasma-treated poly(lactide-co-glycolide). *Biomaterials*, 29, 2388-2399.

Shen, H., Hu, X. X., Yang, F., Bel, J. Z. & Wang, S. G. 2007. Combining oxygen plasma treatment with anchorage of cationized gelatin for enhancing cell affinity of poly(lactide-co-glycolide). *Biomaterials*, 28, 4219-4230.

Siow, K. S., Britcher, L., Kumar, S. & Griesser, H. J. 2006. Plasma methods for the generation of chemically reactive surfaces for biomolecule immobilization and cell colonization - A review. *Plasma Processes and Polymers*, 3, 392-418.

Storey, R. F. & Taylor, A. E. 1998. Effect of stannous octoate on the composition, molecular weight, and molecular weight distribution of ethylene glycol-initiated poly(epsilon-caprolaccone). *Journal of Macromolecular Science-Pure and Applied Chemistry*, A35, 723-750.

Taylor, M. S., Daniels, A. U., Andriano, K. P. & Heller, J. 1994. 6 Bioabsorbable Polymers - In-Vitro Acute Toxicity of Accumulated Degradation Products. *Journal of Applied Biomaterials*, 5, 151-157.

Tezcaner, A., Bugra, K. & Hasirci, V. 2003. Retinal pigment epithelium cell culture on surface modified poly(hydroxybutyrate-co-hydroxyvalerate) thin films. *Biomaterials*, 24, 4573-4583.

Tserki, V., Matzinos, P., Pavlidou, E., Vachliotis, D. & Panayiotou, C. 2006. Biodegradable aliphatic polyesters. Part I. Properties and biodegradation of poly(butylene succinate-co-butylene adipate). *Polymer Degradation and Stability,* 91, 367-376.

Vaquette, C., Frochot, C., Rahouadj, R. & Wang, X. 2008. An innovative method to obtain porous PLLA scaffolds with highly spherical and interconnected pores. *Journal of Biomedical Materials Research Part B-Applied Biomaterials,* 86B, 9-17.

Vasilets, V. N., Hermel, G., Konig, U., Werner, C., Muller, M., Simon, F., Grundke, K., Ikada, Y. & Jacobasch, H. J. 1997. Microwave CO2 plasma-initiated vapour phase graft polymerization of acrylic acid onto polytetrafluoroethylene for immobilization of human thrombomodulin. *Biomaterials,* 18, 1139-1145.

Vroman, I. & Tighzert, L. 2009. Biodegradable polymers. *Materials,* 2, 307-344.

Wan, Y. Q., Tu, C. F., Yang, J. A., Bei, J. Z. & Wang, S. G. 2006. Influences of ammonia plasma treatment on modifying depth and degradation of poly(L-lactide) scaffolds. *Biomaterials,* 27, 2699-2704.

Wang, E. A., Rosen, V., Dalessandro, J. S., Bauduy, M., Cordes, P., Harada, T., Israel, D. I., Hewick, R. M., Kerns, K. M., Lapan, P., Luxenberg, D. P., McQuaid, D., Moutsatsos, I. K., Nove, J. & Wozney, J. M. 1990. Recombinant Human Bone Morphogenetic Protein Induces Bone-Formation. *Proceedings of the National Academy of Sciences of the United States of America,* 87, 2220-2224.

Wang, S. G., Chen, H. L., Cai, Q. & Bei, J. Z. 2001. Degradation and 5-fluorouracil release behavior in vitro of polycaprolactone/poly(ethylene oxide)/polylactide tri-component copolymer. *Polymers for Advanced Technologies,* 12, 253-258.

Wang, Y. J., Lu, L., Zheng, Y. D. & Chen, X. F. 2006. Improvement in hydrophilicity of PHBV films by plasma treatment. *Journal of Biomedical Materials Research Part A,* 76A, 589-595.

Wang, Y. Q., Qu, X., Lu, J., Zhu, C. F., Wan, L. J., Yang, J. L., Bei, J. Z. & Wang, S. G. 2004. Characterization of surface property of poly(lactide-co-glycolide) after oxygen plasma treatment. *Biomaterials,* 25, 4777-4783.

Xie, Z. W. & Buschle-Diller, G. 2010. Electrospun Poly(D,L-lactide) Fibers for Drug Delivery: The Influence of Cosolvent and the Mechanism of Drug Release. *Journal of Applied Polymer Science,* 115, 1-8.

Yang, J., Bei, J. Z. & Wang, S. G. 2002. Enhanced cell affinity of poly (D,L-lactide) by combining plasma treatment with collagen anchorage. *Biomaterials,* 23, 2607-2614.

Yildirim, E. D., Ayan, H., Vasilets, V. N., Fridman, A., Guceri, S. & Sun, W. 2008. Effect of dielectric barrier discharge plasma on the attachment and proliferation of osteoblasts cultured over poly(epsilon-caprolactone) scaffolds. *Plasma Processes and Polymers,* 5, 58-66.

Zelzer, M., Scurr, D., Abdullah, B., Urquhart, A. J., Gadegaard, N., Bradley, J. W. & Alexander, M. R. 2009. Influence of the Plasma Sheath on Plasma Polymer Deposition in Advance of a Mask and down Pores. *Journal of Physical Chemistry B,* 113, 8487-8494.

Zhu, L. P., Zhu, B. K., Xu, L., Feng, Y. X., Liu, F. & Xu, Y. Y. 2007. Corona-induced graft polymerization for surface modification of porous polyethersulfone membranes. *Applied Surface Science,* 253, 6052-6059.

Zhu, Y. B., Gao, C. Y., Liu, X. Y. & Shen, J. C. 2002. Surface modification of polycaprolactone membrane via aminolysis and biomacromolecule immobilization for promoting cytocompatibility of human endothelial cells. *Biomacromolecules,* 3, 1312-1319.

Zhu, Y. B., Gao, C. Y., Liu, Y. X. & Shen, J. C. 2004. Endothelial cell functions in vitro cultured on poly(L-lactic acid) membranes modified with different methods. *Journal of Biomedical Materials Research Part A,* 69A, 436-443.

Poly(Lactic Acid)-Based Biomaterials: Synthesis, Modification and Applications

Lin Xiao[1,2], Bo Wang[1,2], Guang Yang[*1,2] and Mario Gauthier[3]
[1]College of Life Science and Technology, Huazhong University of Science and Technology,
[2]National Engineering Research Center for Nano-Medicine,
Huazhong University of Science and Technology,
[3]Department of Chemistry, University of Waterloo,
[1,2]China
[3]Canada

1. Introduction

Social and economic development has driven considerable scientific and engineering efforts on the discovery, development, and utilization of polymers. Widespread reliance in everyday life on conventional polymeric materials such as polyolefins has resulted in serious pollution which cannot be resolved in a straightforward fashion. Sustainable development and a green economy both require brand new materials which can avoid the occurrence of these problems.

Poly(lactic acid) (PLA), an aliphatic polyester, has outstanding advantages over other polymers, and may thus be part of the solution. As early as the 1970's, PLA products have been approved by the US Food and Drug Administration (FDA) for direct contact with biological fluids. Four of its most attractive advantages are renewability, biocompatibility, processability, and energy saving (Rasal, 2010). First of all, PLA is derived from renewable and degradable resources such as corn and rice, which can help alleviate the energy crisis as well as reduce the dependence on fossil fuels of our society; PLA and its degradation products, namely H_2O and CO_2, are neither toxic nor carcinogenic to the human body, hence making it an excellent material for biomedical applications including sutures, clips, and drug delivery systems (DDS). Furthermore, PLA can be processed by film casting, extrusion, blow molding, and fiber spinning due to its greater thermal processability in comparison to other biomaterials such as poly(ethylene glycol) (PEG), poly(hydroxyalkanoates) (PHAs), and poly(ε-caprolactone) (PCL) (Rhim et al., 2006). These thermal properties contribute to the application of PLA in industry in fields such as textiles and food packaging. Last but not least, PLA production consumes 25-55% less fossil energy than petroleum-based polymers. Cargill Dow has even targeted a reduction in fossil energy consumption by more than 90% as compared to any of the petroleum-based polymers for the near future, which will surely also lead to significant reductions in air and water pollutant emissions. It is also noteworthy that the total amount of water required for PLA production is competitive with the best performing petroleum-based polymers. This energy-saving feature perfectly caters to the

---

* Corresponding author

new concept of "low-carbon economy" which emerged recently in response to the global warming and energy crisis concerns, and makes investment in PLA a necessary and wise strategy in the future (Vink et al., 2003). Fig. 1 shows the cycle of PLA in nature.

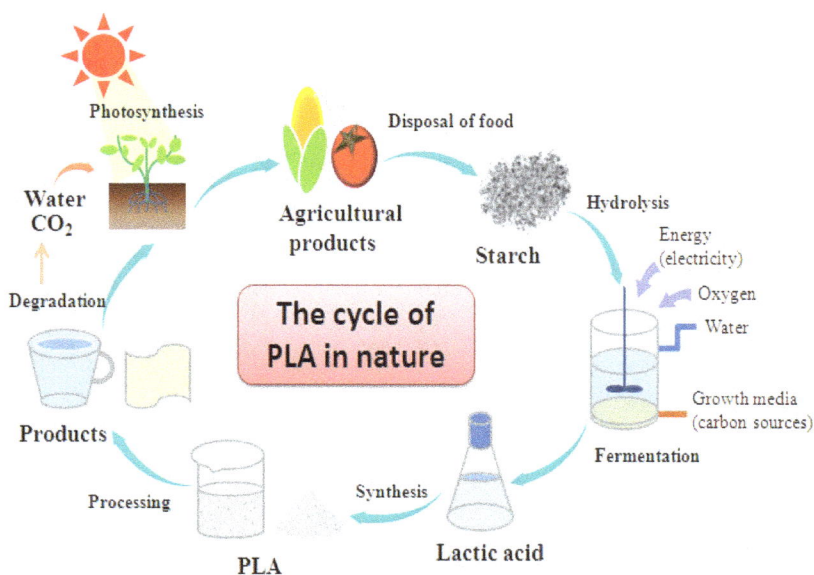

Fig. 1. The cycle of PLA in nature.

While PLA can be considered an eco-friendly biomaterial with excellent properties, it also has many obvious drawbacks when confronted with requirements for certain applications: 1) Its degradation rate through hydrolysis of the backbone ester groups is too slow. This process sometimes takes several years, which can impede its biomedical and food packaging applications (Bergsma et al., 1995). 2) PLA is very brittle, with less than 10% elongation at break, thus it is not suitable for demanding mechanical performance applications unless it is suitably modified (Rasal & Hirt, 2009). 3) PLA is strongly hydrophobic and can elicit an inflammatory response from the tissues of living hosts, because of its low affinity with cells when it is used as a tissue engineering material. 4) Another limitation of PLA towards its wider industrial application is its limited gas barrier properties which prevent its complete access to industrial sectors such as packaging (Singh et al., 2003). Considering the disadvantages of PLA stated above and its high cost (another shortcoming of that material), it is not surprising that PLA has not received the attention it deserves. Nevertheless, researchers have examined different methods for the bulk or surface modification of PLA, the introduction of other components, or the control of its surface energy, surface charge and surface roughness, depending on the requirements of specific applications.

Previous reviews have examined different aspects of PLA chemistry and engineering. Thus Maharana et al. (Maharana et al., 2009) presented a review on the melt-solid polycondensation of lactic acid (LA). Gupta et al. (Gupta et al., 2007) presented an overview of the production of PLA fibers by various methods, along with correlations between the structure and the properties of the fibers. Butterwick et al. (Butterwick et al., 2009) discussed the applications of PLA in Europe and the United States with respect to practitioner experiences and techniques to optimize the outcomes. Rasal et al. (Rasal et al., 2010) examined the chemical modification of PLA, while Graupner et al. (Graupner et al., 2009) assessed the production and the mechanical characteristics of composites prepared from PLA and renewable raw materials including cotton, hemp, kenaf, and man-made cellulose fibres (Lyocell) by compression molding.

In this chapter we will underline novel ideas or technologies introduced over the last 5-10 years, emphasizing some ambitious work which, even though it appears less successful than other mature methods, introduces concepts that may prove extremely positive in the near future. We will also attempt to foretell developmental trends on the basis of social demands and the progress achieved so far. More traditional topics including the synthesis, modification, and applications of PLA in biomedical field will be introduced mainly to provide a more comprehensive picture of PLA as a biomaterial.

2. Physical and chemical properties of PLA

L-lactic acid and D-lactic acid, the two isomers of lactic acid, are shown in Scheme 1. Pure L-lactic acid or D-lactic acid, or mixtures of both components are needed for the synthesis of PLA.

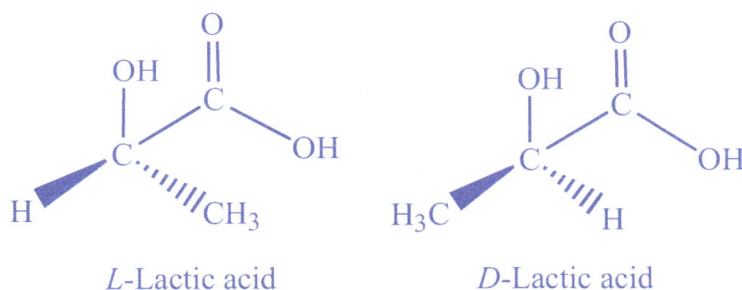

Scheme 1. The stereoisomers of lactic acid.

The homopolymer of LA is a white powder at room temperature with T_g and T_m values of about 55°C and 175°C, respectively. High molecular weight PLA is a colorless, glossy, rigid thermoplastic material with properties similar to polystyrene. The two isomers of LA can produce four distinct materials: Poly(D-lactic acid) (PDLA), a crystalline material with a regular chain structure; poly(L-lactic acid) (PLLA), which is hemicrystalline, and likewise with a regular chain structure; poly(D,L-lactic acid) (PDLLA) which is amorphous; and meso-PLA, obtained by the polymerization of meso-lactide. PDLA, PLLA and PDLLA are soluble in common solvents including benzene, chloroform, dioxane, etc. and degrade by simple hydrolysis of the ester bond even in the absence of a hydrolase. PLA has a degradation half-life in the environment ranging from 6 months to 2 years, depending on the size and shape of the article, its isomer ratio, and the temperature. The tensile properties of PLA can vary widely depending on whether it is annealed or oriented, or its degree of crystallinity (Garlotta et al., 2001). Some of the physical and chemical properties of PLA are summarized in Table 1.

Properties	PDLA	PLLA	PDLLA
Solubility	All are soluble in benzene, chloroform, acetonitrile, tetrahydrofuran (THF), dioxane etc., but insoluble in ethanol, methanol, and aliphatic hydrocarbons		
Crystalline structure	Crystalline	Hemicrystalline	Amorphous
Melting temperature (T_m)/ °C	~180	~180	Variable
Glass transition temperature (T_g)/ °C	50-60	55-60	Variable
Decomposition temperature/°C	~200	~200	185-200
Elongation at break/ (%)	20-30	20-30	Variable
Breaking strength/ (g/d)	4.0-5.0	5.0-6.0	Variable
Half-life in 37°C normal saline	4-6 months	4-6 months	2-3 months

Table 1. Selected physical and chemical properties of PLA.

3. Synthesis of PLA

Two main synthetic methods are used to obtain PLA: Direct polycondensation (including solution polycondensation and melt polycondensation), and ring-opening polymerization (ROP).

3.1 Direct polymerization

Since the LA monomer has both –OH and –COOH groups, necessary for polymerization, the reaction can take place directly by self-condensation (Scheme 2):

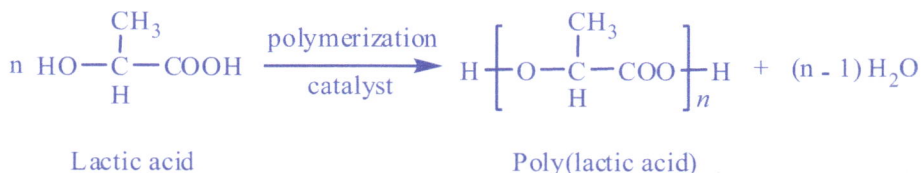

$$n\ HO-\underset{\underset{H}{|}}{\overset{\overset{CH_3}{|}}{C}}-COOH \xrightarrow[\text{catalyst}]{\text{polymerization}} H\left[O-\underset{\underset{H}{|}}{\overset{\overset{CH_3}{|}}{C}}-COO\right]_n H\ +\ (n-1)\,H_2O$$

Lactic acid   Poly(lactic acid)

Scheme 2. Direct polymerization.

Direct polymerization includes solution and melt polycondensation, depending on whether a solvent is used in the reaction to dissolve the PLA or not.

3.1.1 Solution polycondensation

In this case an organic solvent capable of dissolving the PLA without interfering with the reaction is added, and the mixture is refluxed with removal of the water generated in the polycondensation process, which is beneficial to achieve a high molecular weight. Many procedures yield PLA with a weight-average molecular weight (M_w) of over 200,000 by this method (Ohta et al., 1995; Ichikawa et al., 1995). This approach was developed by Carothers and is still used by Mitsui Chemicals. The resultant polymer can be coupled with isocyanates, epoxides or peroxides to produce a range of molecular weights (Lunt et al.,

1998). The reaction proceeds smoothly, however solution polymerization suffers from certain disadvantages such as being susceptible to impurities from the solvent and various side reactions including racemization and trans-esterification. It also consumes large volumes of organic solvents, which are potential pollutants to the environment.

Under optimized conditions, Ajioka et al. obtained PLA with M_w > 300,000 by this method (Ajioka et al., 1995). Characterization data have shown that the glass transition temperatures (T_g) of PLA and polylactide synthesized by the conventional lactide process are essentially identical (T_g = 58°C and 59 °C, respectively), but PLA has a lower melting point (T_m = 163 °C) than polylactide (T_m = 178°C). The mechanical properties of the two polymers are also very similar.

3.1.2 Melt polycondensation

In contrast to solution polycondensation, the melt polycondensation of monomers can proceed without any organic solvent, but only if the temperature of the reaction remains above the T_m of the polymer (Gao et al., 2002). Moon et al. discovered that high M_w PLLA [$M_w \geq$ 100,000] could be produced in this way in a relatively short reaction time ($\leq$ 15 h) (Moon et al., 2000). This method can lower the cost of the synthesis significantly due to the simplified procedure, but major problems still need to be solved before it can be applied industrially because of its sensitivity to reaction conditions (Maharana et al., 2009). Thus Moon et al. worked to develop a melt/solid polycondensation technique using a binary catalyst system (tin dichloride hydrate and p-toluenesulfonic acid) (Moon et al., 2001). Simply put, thermal oligocondensates of LA were first subjected to melt polycondensation to obtain a melt polycondensate, which was then subjected to solid state polycondensation at 105°C. As a consequence, the molecular weight of the PLA was as high as 600,000 after a short reaction time under optimized conditions.

In summary, these one-step polymerization processes are relatively economical and easy to control, but they are equilibrium reactions affected by numerous parameters such as the temperature, the reaction time, catalysts, pressure, and so on. These factors can strongly influence the molecular weight of the products obtained. Besides, the water generated in this process can cause high molecular weight PLA to break down at high reaction temperatures. Thus the polymer resulting from these reactions usually has an unsatisfactorily low molecular weight. Attention must be paid to three aspects of the reaction to obtain a high molecular weight, namely controlling the reaction kinetics, removing the water formed, and preventing the degradation of the PLA chains.

3.2 Ring-opening polymerization

Considering the drawbacks of direct polymerization, PLA is typically synthesized by ring-opening polymerization (ROP) (Scheme 3), an important and effective method to manufacture high molecular weight PLA. This reaction requires strict purity of the lactide monomer, obtained by dimerization of the lactic acid monomer. PLA is obtained by using a catalyst with the monomer under vacuum or an inert atmosphere. By controlling the residence time and the temperatures in combination with the catalyst type and concentration, it is possible to control the ratio and sequence of D- and L-lactic acid (LA) units in the final polymer (Lunt et al., 1998). The polymerization mechanism involved can be ionic, coordination, or free-radical, depending on type of catalyst employed (Penczek et al.,

2000). Most of the researchers are now exploring new and effective catalysts. Köhn et al. (Köhn et al., 2003) first reported that the ROP of *D,L*-lactide could be catalyzed by bis(trimethyl triazacyclohexane) praseodymium triflate [(Me$_3$TAC)$_2$ Pr(OTf)$_3$] (Cat), while Pr(OTf)$_3$ by itself had a poor catalytic activity. Cat was found to catalyze the polymerization of *D,L*-lactide in various solvents (THF, dichloromethane, ethyl acetate, and toluene) without any additional reagents. The optimal polymer yield (95%) and molecular weight (18,000) were obtained after 18 h at 170°C, with a ratio of [LA]:[Cat] = 1000. John et al. (John et al., 2007) produced one of the few reports on lactide polymerization with a Cu-based catalyst. ROP of *L*-lactide catalyzed by {2-[1-(2,6-diethylphenylimino)ethyl]phenoxy}$_2$Cu(II) yielded the highest M_w (26.3 × 10^3) with a monomer conversion of 57%. Two other copper complexes, {2-[1-(2,6-dimethylphenylimino)ethyl]phenoxy}$_2$Cu(II) and {2-[1-(2-methylphenylimino)ethyl]phenoxy}$_2$Cu(II), also catalyzed the reaction under solvent-free melt conditions (160°C) but produced polylactides of moderate molecular weights (M_w = 12.0 × 10^3 and 15.9 × 10^3, respectively).

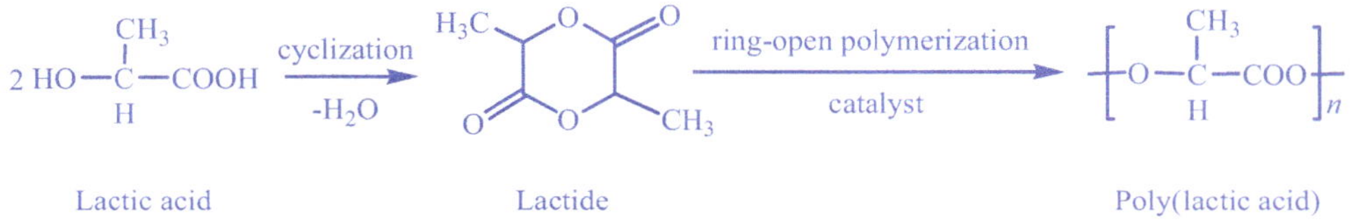

Scheme 3. Ring-opening polymerization.

Numerous studies have examined the influence of different factors such as the concentration and type of catalyst, monomer purity, and temperature on the polymerization of lactide. Special attention has been paid to the catalyst. Currently tin octoate is the most widely used catalyst for the ring-opening polymerization of lactides, but numerous novel efficient metal-free catalytic systems are emerging as valuable alternatives (Jérôme & Lecomte, 2008). The heavy metal-based catalysts are indeed very likely to contaminate the product, which complicates the purification of the PLA obtained and also limits the applications of PLA in the fields of food packaging and biomedicine. Some of the means developed to solve this problem will be addressed later.

3.3 New approaches

The inherent disadvantages of the traditional synthetic methodology have led some researchers to explore solutions such as the development of nontoxic catalysts, unusual polymerization conditions, or other polymerization pathways (Lassalle & Ferreira, 2008).

To solve the potential pollution problems caused by heavy metal catalysts, many nontoxic catalysts derived from magnesium (Wu et al., 2005), calcium (Zhong et al., 2003), zinc (Sarazin et al., 2004), alkali metals (Chisholm et al., 2003), and aluminum (Nomura et al., 2002) have been developed for the ROP of lactides (Deng et al., 2000; Ejfler et al., 2005). For example, Chen et al. (Chen et al., 2007) tested a series of β-diketiminate zinc complexes as initiators for the ROP of lactide and they were all highly active, however the M_w attained was unsatisfactory. It is worth noting that a variety of rare earth derivatives are usually highly reactive, which entitles them to be very promising initiators for the ROP of lactide (Agarwal et al., 2000).

With respect to unusual reaction conditions, supercritical CO_2 ($scCO_2$) technology has attracted much attention because this environmentally friendly, chemically inert, inexpensive, non-toxic, and nonflammable solvent can be substituted for organic solvents (Nalawade et al., 2006). Yoda et al. (Yoda et al., 2004) thus carried out the synthesis of PLLA from an L-lactic acid oligomer in $scCO_2$ with dicyclohexyldimethylcarbodiimide (DCC) as an esterification promoter and 4-dimethylaminopyridine (DMAP) as a catalyst. PLLA with a number-average molecular weight M_n reaching 13,500 was obtained in 95% yield after 24 h at 3500 psi and 80°C. The molecular weight distribution of the products was also narrower than for PLLA prepared by melt–solid phase polymerization under conventional conditions. Not only can $scCO_2$ be used as a medium to synthesize polymers, but it can also serve in the purification and processing of the polymer micro-particles obtained (Kang et al., 2008).

The direct polycondensation of lactic acid has been considered to have a promising future due to its low cost; however it is hard to increase the molecule weight due to the difficulty in removing the water from the system under these conditions. One way to solve this problem is a chain-extension method, although the properties of the PLA obtained in this way can be somewhat affected by the procedure. Simply put, hydroxyl- or carboxyl-terminated low molecular weight PLA obtained by direct polymerization can be linked together through a chain extender, which is a bifunctional compound carrying highly reactive functional groups. Many achievements have been reported in this area, hexamethylene diisocyanate (HDI) being the most widely used chain extender for hydroxyl-terminated prepolymers since the work done by Woo and coworkers (Woo et al., 1995). Finding new and satisfactory chain extenders will remain a major goal in the near future, since HDI is toxic and subject to side reaction in this process.

In addition, LA-polymerizing enzymes functioning in replacement of metal catalysts should enable the biosynthesis of PLA, even though it is enormously challenging both in terms of research and industrial implementation. The best solution could be the development of a PLA-producing microorganism, but this has not been reported so far. Taguchi et al. (Taguchi et al. 2008) have nonetheless obtained encouraging results by developing a recombinant *Escherichia coli* strain allowing the synthesis of LA-based polyesters by introducing the gene encoding polyhydroxyalkanoate (PHA) synthase. This is illustrated in Fig. 2. They thus achieved the one-step biosynthesis of a copolymer with 6 mol% of lactate and 94 mol% of 3-hydroxybutyrate units, having a molecular weight of 1.9 × 10^5. This extremely important result represents a milestone towards the biological synthesis of PLA and confirms that the work is moving in the right direction. At present, the LA fraction in the copolyesters has been enriched up to 96 mol% (Shozui et al., 2011), so the synthesis of homopolymers of LA represents a major goal. To that end, the current microbial cell factory ought to be improved with further evolved LA-polymerizing enzymes (LPE) and metabolic engineering-based optimization (Taguchi, 2010). Matsumura et al. (Matsumura et al., 1997) likewise reported the lipase PC-catalyzed polymerization of cyclic diester-D,L-lactide at a temperature of 80-130°C to yield poly(lactic acid) with molecular masses of up to 12,600. Other novel methods (e.g. metal-free catalysts, non-catalytic systems) are also under development (Zhong et al., 2003; Achmad et al., 2009). The advantages and disadvantages of the PLA synthesis methods mentioned above are summarized in Table 2.

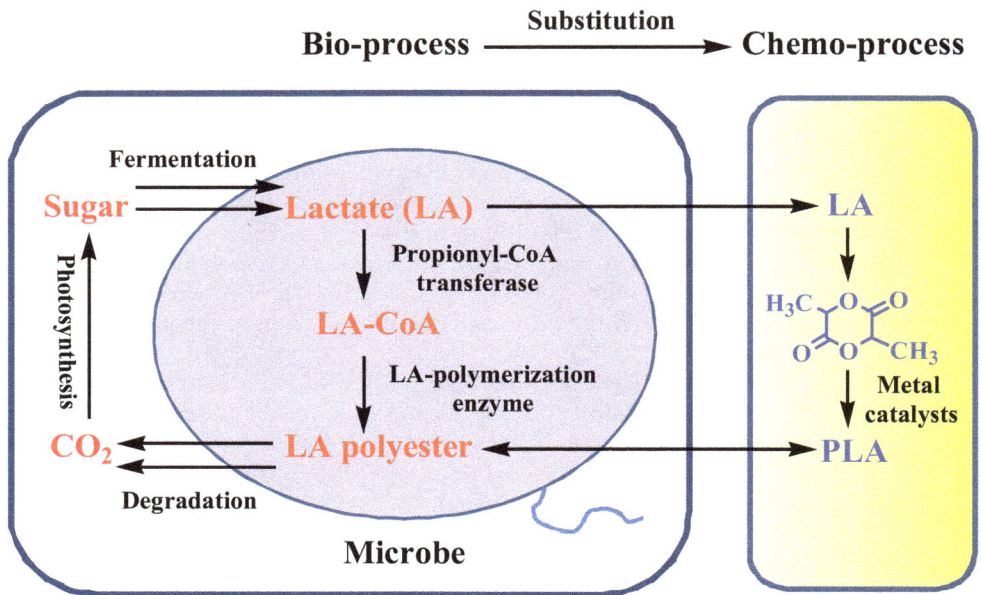

Fig. 2. Mechanism for the bio-synthesis of LA polyester. In the bio-process, the LA monomer is converted into LA-CoA ,which is recognized by the LA-polymerizing enzyme recruited from microbial PHA synthase (Tajima et al., 2009).

Synthesis methods	Advantages	Disadvantages
Solution polycondensation	One-step, economical and easy to control	Impurities, side reactions, pollution, low molecular weight PLA
Melt polycondensation		High reaction temperature, sensitivity to reaction conditions, low molecular weight PLA
Ring-opening polymerization	High molecular weight PLA	Requires strict purity of the lactide monomer, related high cost
New solutions (new catalysts, polymerization conditions, etc)	Efficient, non-toxic, no pollution , high molecular weight PLA, etc.	Under development
Biosynthesis	One-step, efficient, non-toxic, no pollution, low cost, etc.	Under development

Table 2. Comparison of PLA synthesis methods.

4. Modification of PLA

The major drawbacks of PLA limiting its applications are its poor chemical modifiability and mechanical ductility, slow degradation profile, and poor hydrophilicity. In order to be suitable for specific biomedical applications, the PLA has been modified mainly concerning two aspects: Bulk properties and surface chemistry. To achieve this, both chemical modification and physical modification have been tried, involving the incorporation of functional monomers with different molecular architectures and compositions, the tuning of crystallinity and processibility via blending and plasticization, etc., which are described in the following sections.

4.1 Bulk modification

Biomaterials must possess bulk properties, particularly hydrophilic and mechanical properties, meeting special requirements. Critical factors affecting these characteristics include chemical additives, composition, and morphological structure. At present considerable research focuses on a variety of hydrolytic groups, controlling the flexibility and crystallinity of the molecular chains, and the presence of hydrophilic groups.

4.1.1 Physical modification

Blending, plasticization, and composition variations belong to this category.

Blending

Polymer blending is an effective, simple, and versatile method to develop new materials with tailored properties without synthesizing new polymers (Peesan et al., 2005). The properties of different polymers (biodegradable and non-biodegradable) can be combined by blending with PLA, or even new properties can arise in the products due to interactions between the components. Biodegradable components blended with PLA include poly(ethylene glycol) (PEG), poly(β-hydroxybutyrate) (PHB), poly(ε-caprolactone) (PCL), poly(butylene adipate-*co*-terephthalate) (PBAT), chitosan, and starch (Sheth et al., 1997). While blends of PLA and non-biodegradable polymers have not been as extensively studied, low-density polyethylene (LDPE), poly(vinyl acetate) (PVA), and polypropylene (PP) have been examined. Reddy et al. (Reddy et al., 2008) found that PLA in blends obtained from five ratios of PLA/PP had substantially better resistance to biodegradation and hydrolysis, and improved dyeability with dispersed dyes. However most of these blends are immiscible (phase-separated) and display poor mechanical properties due to low interfacial adhesion between the polymer phases.

To improve the processing and mechanical properties of PLA without sacrificing its degradability and biocompatibility, Xu et al. (Xu et al., 2009) blended PLA with a new degradable thermoplastic derived from konjac glucomannan (TKGM), synthesized by graft copolymerization of vinyl acetate and methyl acrylate onto konjac glucomannan (KGM). Dynamic mechanical analysis (DMA) and scanning electron microscopy (SEM) measurements showed that the PLA/TKGM system was miscible due to specific interactions between PLA and TKGM. This led to a maximum elongation at break of 520% for the blend (20/80), as compared to 14% for neat PLA. The impact strength also increased from 11.9 kJ/m² for neat PLA to 26.9 kJ/m² for the 20/80 blend. The synthesis of new polymers, biodegradable or non-biodegradable, to be compatibly blended with PLA, will represent a major task in the future.

Plasticization

PLA is a glassy polymer with poor elongation at break (typically less than 10%). The modification of PLA with different biodegradable and non-biodegradable plasticizers, having a low molecular weight but a high boiling point and a low volatility, has been explored as a mean to lower the T_g and increase the ductility and softness of PLA. This has been achieved by varying the molecular weight, the polarity and functional groups of the plasticizers. Biocompatible molecules such as oligomeric lactic acid, oligomeric citrate ester, oligomeric PEG, and glycerol are all plasticizers of choice for PLA (Martin & Averous, 2001; Ljungberg et al., 2005). Ljungberg et al. (Ljungberg & Wesslén, 2002) have blended PLA with five plasticizers (triacetine, tributyl citrate, triethyl citrate, acetyl tributyl citrate, and acetyl triethyl citrate) and found that triacetine and tributyl citrate were more effective as plasticizers than the others to obtain a significant decrease in T_g for PLA.

Wang et al. (Wang et al., 2009) found that diisononyl cyclohexane-1,2-dicarboxylate (DINCH), a new plasticizer obtained by the hydrogenation of the benzene ring of o-phthalates, had limited compatibility with PLA when compared with tributyl citrate ester (TBC). PLA samples plasticized with 10 and 20 phr DINCH gave a constant T_g of 50°C. They were stiff materials displaying elevated values of elongation at break (129% and 200%, respectively) and impact strength (41.1 MPa and 30.1 MPa, respectively). On the other hand, TBC significantly decreased the T_g and increased the crystallinity of PLA, the PLA/TBC (20 phr TBC) blend being a soft material with a T_g of 24°C. Results from thermogravimetric and thermal analysis also indicated that PLA plasticized with DINCH had good mechanical properties and excellent water resistance (as reflected in time-dependent weight loss data in phosphate buffer) and aging resistance (characterized by the mechanical and thermal properties of specimens exposed to ambient conditions for 4 months).

Composition

Fibers can serve as fillers in the formation of PLA composites processable by compression or injection molding, to enhance the thermal stability, the hydrolysis resistance, or the mechanical properties of PLA. Several investigations on PLA composites prepared from natural and modified cellulose fibers have shown that their mechanical properties scale with the mass fraction of added fibers (Wan et al., 2001; Mathew et al., 2005). Optimization of the natural fiber-reinforced PLA composites, in terms of mechanical and other properties, is critical to minimize their cost, tailor their biodegradability, and broaden their areas of application. Inorganic fillers can also contribute to property modification. Table 3 provides a comparison of some of the organic and inorganic materials tested as PLA fillers.

Graupner et al. (Graupner et al., 2009) prepared composites from different types of natural fibers (cotton, hemp, kenaf) and modified cellulose fibers (Lyocell), with a fiber mass fraction of 40%, by compression molding. The mechanical properties of these composites are summarized in Table 4. Tomé et al. (Tomé et al., 2011) prepared composites from PLA and acetylated bacterial cellulose by mechanical compounding. The composites displayed significant increases in both elastic and Young moduli, as well as in tensile strength (increments of about 100, 40, and 25%, respectively, as compared with neat PLA) at 6% filler loading. Some surface modifiers can enhance adhesion between the fibers and the PLA matrix. For example, 3-aminopropyltriethoxysilane (APS) hydrolyzes in water or solvents to produce silanol groups that are capable of bonding to -OH groups on the kenaf fiber surface (Huda et al., 2008). The -NH$_2$ groups from APS can also bond with -CO$_2^-$ sites formed on the PLA surface by treatment with a sodium hydroxide solution. Thus APS effectively functions as a coupling agent. Yang et al. (Yang et al., 2011) produced a composite from PLA and microcrystalline cellulose modified by L-lactic acid. The tensile strength and the elongation at break of the composite were higher than for neat PLA. The surface modification of the cellulose substrates was considered a key element of the mechanical reinforcement.

Type	Filler	Result	Reference
Organic	Jute	Tensile stress and modulus increase with fiber volume fraction	Khondkeret al., 2006
	Flax fibers	Composite strength about 50% higher than for PP/flax composites	Oksman et al., 2003
	Kenaf fibers	Greatly improved crystallization rate, tensile and storage moduli	Pan et al., 2007
	Bamboo fibers	Increased bending strength and improved thermal properties	Tokoro et al., 2008
	Silkworm silk fibers	Good wettability, increased elasticity modulus and ductility	Cheung et al., 2008
	Microcrystalline cellulose	Poor mechanical properties and adhesion; increased storage modulus	Mathew et al., 2005
	LA-modified microcrystalline cellulose	Higher tensile strength and elongation at break than neat PLA	Yang et al., 2011
	Acetylated bacterial cellulose	Considerable improvement in thermal and mechanical properties	Tomé et al., 2011
Inorganic	Calcium metaphosphate	Narrow pore size distribution and high tensile strength	Jung et al., 2005
	Calcium carbonate	No brittle fracture behavior and comparably high bending strength	Kasuga et al., 2003
	Montmorillonite	Good affinity and improved thermal stability of the nanocomposites	Pluta et al., 2002
	HAP	Improved elastic modulus and unchanged bending strength	Kasuga et al., 2001
	Carbon nanotubes	Dramatic enhancement in thermal and mechanical properties	Wu & Liao, 2007
	Nano/Micro-silica	Increased tensile strength, thermal stability, and hydrolysis resistance	Huang et al., 2009

Table 3. Organic and inorganic fillers for the preparation of PLA composites.

	Tensile strength/ (N/mm^2)	Young's modulus/ (N/mm^2)	Elongation at break/ (%)	Charpy impact strength/ (kJ/mm^2)
Pure PLA sample	30.1	3820.2	0.83	24.4
Cottom-PLA	41.2	4242.3	3.07	28.7
Kenaf-PLA	52.9	7138.6	1.05	9.0
Hemp-PLA	57.5	8064.2	1.24	9.5
Lyocell-PLA	81.8	6783.8	4.09	39.7
Hemp/kenaf-PLA	61.0	7763.8	1.22	11.8
Hemp/Lyocell-PLA	71.5	7034.9	1.65	24.7

Table 4. Mechanical properties of composites and a pure PLA sample (mean values; all the specimens were tested at 0°C; adapted in part from (Graupner et al., 2009).

Kim et al. (Kim et al., 2010) prepared a series of PLA/exfoliated graphite (EG) nanocomposites and confirmed that the graphite nanoplatelets could be dispersed homogeneously within the PLA matrix. Thermogravimetric analysis also showed that the thermal stability of the nanocomposites was improved with incremental amounts of EG up to 3 wt %. For example, the temperature corresponding to a 3% weight loss for a composite with 3.0 wt % EG increased by 14 degrees to ~364 °C vs. pure PLA. Additionally, the Young's modulus of the composites increased with their graphite content and their electrical resistivity was dramatically lowered. Poly(lactic acid)/hydroxyapatite (PLA/HAP) composite scaffolds processed by foaming with supercritical CO_2 were shown to be promising for bone replacement, because their mechanical characteristics closely matched the properties of bone in terms of viscoelasticity and anisotropy (Mathieu et al., 2006).

4.1.2 Chemical modification

The chemical modification of PLA has been achieved mainly through copolymerization and cross-linking.

Copolymerization

The carboxyl and hydroxyl groups of LA make it possible to copolymerize it with other monomers through polycondensation with lactone-type monomers such as ε-caprolactone, which generally leads to low molecular weight copolymers, or alternately through the ring-opening copolymerization of lactide with other cyclic monomers including glycolide, δ-valerolactone, and trimethylene carbonate, as well as with monomers like ethylene oxide (EO) to produce high molecular weight copolymers. The hydrophobicity and crystallinity of the copolymers can be increased for low to moderate comonomer contents. Besides, poly(ethylene oxide) (PEO) and PEG have been most commonly copolymerized with PLA to prepare copolymers on account that it is highly biocompatible, hydrophilic and non-toxic, non-immunogenic and non-antigenic (Metters et al., 2000). Such properties reduce protein adsorption and enhance resistance to bacterial and animal cell adhesion.

Block copolymers are composed of long sequences (blocks) of the same monomer unit, covalently bound to sequences of a different type. The blocks can be connected in a variety of ways. Fig. 3 shows examples of block copolymer structures. Diblock PLA-PEG copolymers and triblock PLA-PEG-PLA copolymers allow modulation of the biodegradation rate, the hydrophilicity, and the mechanical properties of the copolymers, while phase separation can be tailored with PLA-PEG multi-block copolymers of predetermined block lengths (Wang et al., 2005). Star- and dendrimer-like PLA-PEG copolymers have also been synthesized to lower the T_g, T_m, and the crystallinity of the materials (Zhang et al., 2004).

Riley et al. (Riley et al., 2001) prepared a range of PLA-PEG copolymers incorporating a PEG block of constant molecular weight (Mn = 5,000) and varying PLA segment lengths (Mn = 2,000-110,000) by ROP of D,L-lactide catalyzed by stannous octoate; all the dispersions were stable under physiological conditions. In 2003, Li and Vert (Li & Vert, 2003) prepared series of diblock and triblock copolymers by ring-opening polymerization of L(D)-lactide from mono- and dihydroxyl PEO, using zinc metal as a catalyst under vacuum. The copolymers were semicrystalline, their composition and molar mass being determining factors affecting their solubility in water. Fu et al. (Fu et al., 2008) prepared series of LA-based polyurethanes modified by castor oil with controllable mechanical properties. In this work, hydroxyl-terminated prepolymers were synthesized by copolymerization of L-LA and 1, 4-butanediol.

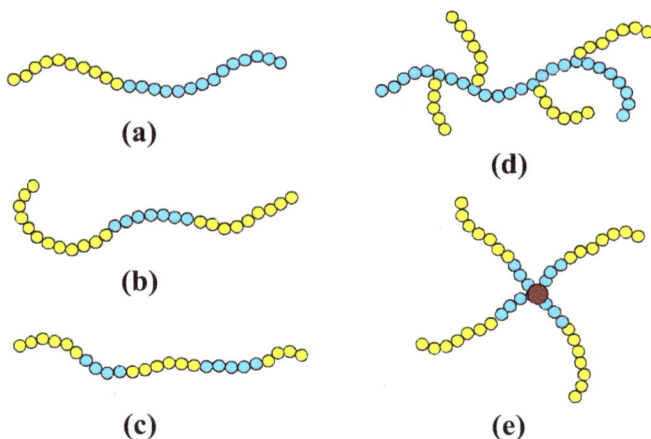

Fig. 3. Schematics of block copolymer structures: (a) diblock; (b) triblock; (c) alternating multiblock; (d); dendrimer-like copolymer; (e) star-like copolymer.

Cross-linking

Cross-linked PLA structures can be formed either by irradiation or through chemical reactions. Electron beam and γ-irradiation have been widely applied to cross-linking PLA in the presence of small amounts of cross-linking agents such as triallyl isocyanurate (TAIC) (Quynh et al., 2008; Phong et al., 2010). The thermal stability of PLA-based materials can be significantly improved in this way (Quynh et al., 2007). Quynh et al. (Quynh et al., 2009) obtained stereocomplexes by cross-linking blends of PLLA and low molecular weight PDLA. Alkaline hydrolysis and enzymatic degradation of the stereocomplex could be controlled by radiation cross-linking, because the alkaline solution as well as proteinase hardly attacked the cross-linked polymer network. Unfortunately, irradiation equipment is expensive and the PLA samples must be processed as thin plates to absorb enough energy from the radiation to initiate cross-linking reactions, which significantly limits its practical application.

Modified PLA with different gel fractions and cross-linking densities can also be obtained through chemical reactions between linking agents and the polymer chains without irradiation (Agrawal et al., 2010). Yang et al. (Yang et al., 2008) thus induced cross-linking via treatment of the PLA melt with small amounts of TAIC and dicumyl peroxide (DCP). The results obtained for samples with different gel fractions and cross-link densities showed that the cross-linking of PLA was initiated at low contents of either TAIC or DCP. The crystallinity of cross-linked PLA samples obtained with 0.5 wt% TAIC and 0.5 wt% DCP decreased from 32% for pure PLA to 24%. Significant increases in tensile modulus from 1.7 GPa to 1.9 GPa, and in tensile strength from 66 GPa to 75 GPa were also observed, and the thermal degradation initiation and completion temperatures were both increased relatively to neat PLA. Additional advantages of this method are that it requires neither extra purification steps nor specialized equipment, since the reaction is carried out in the molten state with only small amounts of cross-linking agent. It is thus economically very advantageous over irradiation, which requires expensive equipment. An increase in brittleness was nevertheless observed following the formation of highly cross-linked structures, which remains a problem to be solved.

4.2 Surface modification

The surface properties of materials play a key role in determining their applications. The presence of specific surface chemical functionalities, hydrophilicity, roughness, surface energy, and topography is crucial for biomedical applications of PLA and its interactions with biomacromolecules. Pure PLA causes a mild inflammatory response if it is implanted into human tissues. It is therefore important to design biomaterials with the required surface properties. The different surface modification strategies examined include physical methods, including surface coating, entrapment and plasma treatment, and chemical methods. Both types of approaches are reviewed.

4.2.1 Physical methods

Surface coating

This is one of the simplest surface modification methods and has been applied to various polymers, but particularly to PLA nanoparticles used for drug delivery. For instance, PEG coating delayed the phagocytosis of PLA nanoparticles and prolonged the circulation time of the nanoparticles in vivo (Gref et al., 1994). Unfortunately the PEG-coated PLA nanoparticles cannot provide specific targeting, which influences their delivery efficiency. One of the most promising alternatives to PEG in this respect is the use of polysaccharides. These materials provide steric protection to the nanoparticles against non-specific interactions with proteins and thereby insure particle stability in the blood circulation system (Ma et al., 2008). Additionally, ligands to achieve active targeting can be conjugated on the surface of these nanoparticles, because many reactive groups are available on the polysaccharides and their derivatives (Gu et al., 2007). Another option is coating of the surface with extracellular matrix (ECM) proteins such as fibronectin, laminin, vitronectin, and collagen, which are conducive to cell adhesion and can greatly improve biocompatibility as well (Lin et al., 2010).

Innovative work was accomplished by Cronin et al. (Cronin et al., 2004), who tested a PLLA fiber scaffold as a substrate for the differentiation of human skeletal muscle cells. Cell attachment (the number of cells attached to the films counted along the center, from one edge to the opposite edge of the film within the field of view) increased significantly on PLLA films coated with ECM gel, fibronectin, or laminin as compared to uncoated or gelatin-coated PLLA films. Myoblasts were able to differentiate into multinucleated myofibers on the ECM gel-coated PLLA fibers and expressed muscle markers such as myosin and α-actinin, as demonstrated by western blot and oligonucleotide microarray analysis.

Entrapment

The entrapment of modifying species (e.g. PEG, alginate, gelatin, etc.) can be achieved through reversible swelling of the PLA surface as illustrated in Fig. 4. This is a simple yet effective method for surface modification requiring no specific functional groups in the polymer chains, as the modifying molecules accumulate merely on the surface of the material without modifying its bulk properties (Lu et al., 2009). Additionally, entrapment can be used to generate different morphologies and thicknesses of 3D scaffolds, which cannot be achieved by other surface modification methods. Finally, entrapment allows the modification of the surface in a controlled fashion because various parameters (e.g. solvent ratio, gelatin concentration, immersion time, and chemical cross-linking) can be varied to tailor the process (Zhu et al., 2003).

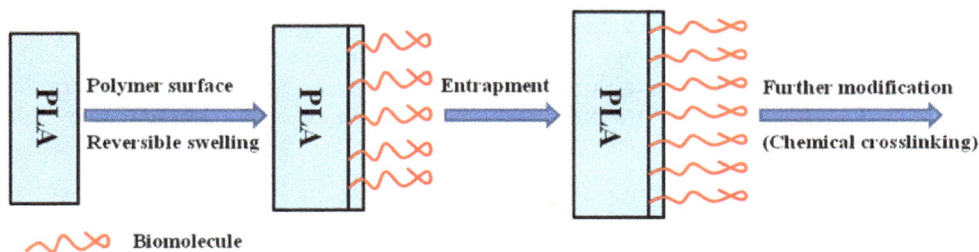

Fig. 4. Schematic illustration of entrapment process.

PEG (M_w = 18,500) and poly(L-lysine) (PLL) (M_w = 29,300) have been trapped on PLA surfaces using 2,2,2-trifluoroethanol (TFE)/water as solvent/nonsolvent mixtures (Quirk et al., 2002). A new entrapment process has also been reported by Liu et al. (Liu et al., 2005), through chemical cross-linking of gelatin with 1-ethyl-3-(3-dimethylaminopropyl) carbodiimide (EDC) HCl and N-hydroxysuccinimide (NHS) (97%) in {2-[N-morpholino] ethanesulfonic acid} (MES) hydrate buffer, after the pretreated PLLA films were immersed in the gelatin solution for a set time. Results in comparison to the control scaffolds have shown that the surface hydrophilicity increased with the amount of entrapped gelatin and that cell attachment and proliferation, the deposition of collagen fibers, and other cell excretion (extracellular matrix, etc.) were also significantly improved.

Plasma treatment

Tests with plasma treatment were initiated in the 1960s and have been since then widely utilized to improve the hydrophilicity and cell affinity of PLA surfaces. The obvious advantages of plasma treatment as compared to other surface modification methods include its ability to control the surface structure, energy and charge, and to uniformly modify the surface without impacting bulk properties (Chu et al., 2002). Functional groups such as –NH$_2$, –COOH, and –OH, which are apt to form covalent bonds with other materials for further modification, are most frequently introduced by plasma treatment (Favia et al., 1998).

Liu et al. thus investigated the influence of the main operation parameters, namely the plasma power, the treatment duration (number of treatment cycles) and the electrode gap on a dielectric barrier discharge (DBD) plasma treatment of PLA films in terms of changes in surface wettability and chemistry (Liu et al., 2004). They further developed equations relating the surface properties (water contact angle and oxygen enrichment, as observed by XPS analysis) to these operational parameters. It was determined that the magnitude of the electrode gap played a dominant role in the treatment of PLA, and the observed wettability improvements were attributed to changes in both surface chemistry and microstructure. Chaiwong and coworkers (Chaiwong et al., 2010) investigated the influence of SF$_6$ plasma on the hydrophobicity and barrier properties of PLA. It was found that the SF$_6$ plasma enhanced the hydrophilicity and increased the water absorption time of PLA two-fold. Plasma treatment did not have any significant influence on the water vapor permeability of PLA, however, since the bulk structure controlling the transport properties are unaffected by the treatment. Other types of plasmas such as oxygen, helium, and nitrogen plasmas have also been investigated (Hirotsu et al., 2002).

While plasma treatment has been successfully applied to improving PLA wettability and cell affinity, its main disadvantage is that surface rearrangements caused by thermally activated macromolecular motions, to minimize its interfacial energy, can also influence the surface modification. Moreover, the potential influence of plasma on the degradation of PLA cannot be ignored.

4.2.2 Chemical modification

PLA does not carry reactive side-chain functional groups. Consequently, the first step of chemical modification is typically a simple surface hydrolysis (with an alkali) or an aminolysis treatment. The hydrophilic –COOH and –OH or reactive –NH$_2$ groups introduced by cleavage of the ester bonds can be used to bind bioactive molecules such as arginine-glycine-aspartic acid (RGD)-containing peptides, chitosan (CS), arginine and lysine, PEG, collagen, and so on to regulate cell adhesion or protein adsorption.

The synthetic RGD-containing peptides could be immobilized on PLA after treatment by hydrolysis or aminolysis (Stupack et al., 2001). Materials prepared by this method provide suitable recognition sites for cell adhesion receptors and biodegradation rates, making them suitable for various applications in fields such as tissue engineering and implant technology. It has also been determined that RGD-conjugated poly(lactic acid-co-lysine)(arginine-glycine-aspartic acid) nanoparticles (PLA-PLL-RGD NP) are non-toxic and bind more efficiently to human umbilical vein endothelial cells (HUVECs) as compared to bare PLA-PLL NP in vitro. Targeted imaging results obtained in vivo showed that PLA-PLL-RGD can selectively bind to BACP-37 breast cancer cells. Lieb et al. also demonstrated largely increased cell densities and cell proliferation on surfaces modified with RGD-anchored monoaminated poly(ethylene glycol)-block-poly(D,L-lactic acid) (H$_2$N-PEG-PLA), mediated through RGD–integrin interactions (Lieb et al., 2005).

Chitosan (CS) is a biopolymer displaying good biocompatibility, non-toxicity and biodegradability, produced by the alkaline N-deacetylation of chitin. The immobilization of this polymer on PLA has been accomplished by coating the surface with chitosan, modified with the photosensitive hetero-bifunctional cross-linking reagent 4-azidobenzoic acid, and irradiation with ultraviolet light to photolyze the azide groups and covalently link the two polymers (Zhu et al., 2002). The -OH and -NH$_2$ groups of chitosan provide further opportunities to introduce a wide range of functional groups on the surface. Thus CS molecules immobilized on PLA were modified with a heparin (Hp) solution to form a polyelectrolyte complex on the surface, which inhibited platelet adhesion and activation, and enhanced cell adhesion.

4.3 Outlook of PLA surface modification

The surface attributes of PLA can be tailored to enhance its hydrophilicity and biocompatiblity through various methods. Unfortunately, all these established methods for surface modification are inherently flawed to some extent. For example, a single plasma treatment can merely improve cell adhesion but cannot accelerate cell growth; non-covalent attachment of a functional material onto a PLA surface is not stable and permanent. An excellent method suggested to solve the second issue is the use of 1,6-hexanediamine for surface aminolysis, followed by conjugation with biocompatible macromolecules such as gelatin, chitosan, or collagen (Zhu et al., 2004). Hong et al. have shown that chondrocyte cells could attach, proliferate, and spread on PLA microspheres coated with collagen in the same way as described above, in particular those having high collagen contents (Hong et al., 2005). It appears that the surface modification of PLA would be best achieved with a combination of distinct approaches, to benefit from the advantages of all the methods. Polysaccharide polyelectrolyte multilayers, including chitosan and dextran sulfate-stabilized silver nano-sized colloids developed by Yu et al. (Yu et al., 2007), were successfully deposited on an aminolyzed PLA membrane in a layer-by-layer self-assembly manner. This seemingly easy process resulted in significant improvements in hydrophilicity, antibacterial activity, hemocompatibility, and cytocompatibility for the PLA membrane, thanks to the different attributes of $-NH_3^+$ (positive charge), chitosan (biocompatibility), and silver nanoparticles (antibacterial activity). The radiation-induced methods are emerging as powerful surface modification techniques, particularly when relying on PLA photoactivation to create reactive groups or moieties useful to graft specific chemical functionalities. The irradiation of PLA with UV (ozone can be generated from molecular oxygen irradiated with UV in this process), for example, is known to increase fiber adhesion to high surface energy components due to the introduction of photo-oxidized polar groups on the surface (Koo & Jang, 2008). Irradiation followed by grafting has also been used extensively to alter PLA surface characteristics, mainly due to the advantages it offers, namely a low operation cost, mild reaction conditions, selectivity to UV light, and the permanent surface chemistry changes induced (Ma et al., 2000).

5. Applications of PLA in the biomedical field

Due to its bioresorbability and biocompatibility in the human body, PLA has been employed to manufacture tissue engineering scaffolds, delivery system materials, or covering membranes, different bioabsorbable medical implants, as well as in dermatology and cosmetics.

5.1 Tissue engineering
Since the introduction of the concept in 1988, tissue engineering, a technique invented to reconstruct living tissues by associating the cells with biomaterials that provide a scaffold on which they can proliferate three-dimensionally and under physiological conditions, has emerged as a potential alternative to tissue or organ transplantation and has thus attracted great attention in science, engineering, and medicine. To meet the diverse needs of tissue engineering, scaffolds made from various materials have been tested in this field. Although certain metals are somewhat good choices for medical implants due to their superior mechanical properties, their lack of degradability in a biological environment makes them disadvantageous for scaffold applications (Liu & Ma, 2004). Inorganic/ceramic materials such as HAP or calcium phosphates, being studied for mineralized tissue engineering with good osteoconductivity, are also limited due to poor processability into porous structures (Ilan & Ladd, 2002). In contrast, polymers have great design flexibility because their composition and structure can be tailored to meet specific needs (Huang et al., 2007). Degradable polymers frequently used for tissue engineering applications are linear aliphatic polyesters such as PGA, PLA, and their copolymers (PLGA), which are fabricated into

scaffolds. These polymers are among the few synthetic polymers approved by the FDA for human clinical applications. The drawbacks of these polyesters include their hydrophobicity and lack of functional groups, which limits cell adhesion, an important factor when constructing polymeric scaffolds. Another drawback is their slow hydrolytic degradation (Iwata & Doi, 1998).

An ideal scaffold used for tissue engineering should possess the following properties: 1) Be biocompatible, so that the scaffold can be well integrated into host tissues without resulting in any immune response; 2) It should be porous with appropriate pore size, size distribution and mechanical function, to allow cell or tissue growth and the removal of metabolic waste; 3) It must be mechanically able to withstand local stress and maintain the pore structure for tissue regeneration; 4) Very importantly, the scaffold should be biodegradable (Ma, 2004). Synthetic scaffolds are considered important components of a successful tissue engineering strategy (Wang, 2007). Hybrid three dimentional porous scaffolds of synthetic and naturally derived biodegradable polymers are particularly promising because they combine the advantages of the two types of materials. They should maintain sufficient mechanical strength while providing specific cell-surface receptors during the tissue remodeling process that stimulate both in vitro and in vivo cell growth (Chen et al., 2002). PLA-based hybrid materials have been successfully tested clinically for that purpose so far, and tests on other tissues including bladder (Engelhardt et al., 2011), cartilage (John et al., 2003), liver (Lv et al., 2007), adipose (Mauney et al., 2007), and bone tissues (Mathieu et al., 2006) have also been reported.

Jiang et al. (Jiang et al., 2010) functionalized chitosan/PLGA by heparin immobilization with controlled loading efficiency. One of the main benefits of introducing chitosan into PLGA microspheres is that chitosan imparts functionality due to its reactive amino groups, so that biomolecules such as heparin could be attached (Jiang et al., 2006). The compressive strength and modulus remained in the range of human trabecular bone after the heparinization process. More importantly, heparinized chitosan/PLGA scaffolds with a low heparin loading (1.71 g/scaffold) showed a stimulatory effect on cell differentiation, as indicated by enhanced osteocalcin expression as compared with a non-heparinized chitosan/PLGA scaffold. Based on these results, Jiang et al. (Jiang et al., 2006) continued to evaluate the novel scaffolds for bone regeneration in vivo. In the rabbit ulnar critical-sized-defect model created, successful bridging of the critical-sized defect on the sides both adjacent to and away from the radius occurred using chitosan/PLGA-based scaffolds. However, the addition of chitosan to PLGA led to somewhat higher inflammation and lower mineralization than for the PLGA counterpart, which is a major problem that remains to be solved.

Three-dimensional (3D) electrospun fibrous scaffolds have been suggested as a potential tissue engineering tool for bone regeneration. Shim et al. (Shim et al., 2010) thus reported a 3D microfibrous PLLA scaffold fabricated using electrospinning techniques with a subsequent mechanical expansion process. The use of these 3D scaffolds for the proliferation of osteoblasts was examined. The 3D scaffolds led to a 1.8-fold higher level of osteoblast proliferation than generally achieved for electrospun 2D nanofibrous membranes. In vivo results further showed that 3D electrospun microfibrous matrices provided a favorable substrate for cell infiltration and bone formation after 2 and 4 weeks when using a rabbit calvarial defect model.

3D printing technology has rapidly expanded in the tissue engineering field since it was first developed at the Massachusetts Institute of Technology. Ge et al. (Ge et al., 2009) developed 3D-printed poly(lactic acid-co-glycolic acid) (PLGA) scaffolds which could support the proliferation and osteogenic differentiation of osteoblasts. Based on their in vitro study, they also evaluated PLGA scaffolds for bone regeneration within a rabbit model (Ge et al., 2009). In both the intra-periosteum and the iliac bone defect models, the implanted scaffolds facilitated new bone tissue formation and maturation over a time period of 24 weeks.

The current clinical use of PLA-based scaffolds nevertheless remains very limited (Iwasa et al., 2009), mainly because of the risk of disease transmission and immune response. This can be illustrated by taking cartilage tissue engineering as an example. Traditional autologous chondrocyte implantation (ACI), first introduced by Brittberg et al. in 1994 (Brittberg et al., 1994), has yielded good clinical results (Bentley et al., 2003). To date, none of the short- or mid-term clinical and histological results using scaffolds were reported to be better than ACI. As for the scaffolds, collagen and hyaluronan-based matrices are among the most popular scaffolds in clinical use nowadays, since they offer substrates which are normally essential elements in native articular cartilage (Iwasa et al., 2009). Among the very few cases of scaffolds in clinical use is the copolymer of PGA/PLA (polyglactin, vicryl) and polydioxanone, which is used for cartilage repair under the trade name of BioSeed®-B and BioSeed®-C (Biotissue Technologies AG, Freiburg, Germany).

In summary, tissue engineering is one of the most exciting interdisciplinary fields today and is growing rapidly with time. The inclusive criteria for studies on scaffolds capable of clinical application were in vivo or clinical studies and thus certain artificially designed scaffold features (such as pore size, interpore connectivity, etc.) are necessary for optimal tissue engineering applications (accelerated tissue regeneration). Suggestions for future directions include the use of designer scaffolds with in vivo experimentation, and coupling scaffold design with cell printing to create designer material/biofactor hybrids to optimize tissue engineering treatments (Hollister, 2005).

5.2 Delivery systems

There has long been a desire to achieve the targeted delivery of bioactive compounds to areas in the body to maximize therapeutic potential and minimize side-effects. Many types of particles have been tested as delivery tools for biomedical applications such as liposomes, solid lipid nanoparticles, and biodegradable polyesters like PLA and PLGA (Torchilin, 2006). With its excellent biocompatibility, biodegradability, mechanical strength, heat processability, and solubility in organic solvents, PLA can be used to produce dosage forms such as pellets, microcapsules, microparticles (MP), nanoparticles (NP), etc. MP and NP of PLA, modified or unmodified, are increasingly investigated for sustained release and targeted drug, peptide/protein, and RNA/DNA delivery applications because of their small size enabling their permeation through biological barriers such as the blood-brain barrier (Roney et al., 2007).

Although PLA-based materials such as PLGA have been FDA-approved and are clinically available, they lack chemical functionalities to facilitate specific cell interactions. Furthermore, their potential for the sustained release of hydrophilic molecules (e.g. proteins) is often limited (Fahmy et al., 2005). Frequent undesired effects include low encapsulation efficiency and high burst release of the encapsulated biomolecule within the first few hours or days, which is mainly due to the desorption of surface-associated hydrophilic molecules having weak interactions with the polymer (Fahmy et al., 2005). To circumvent these limitations and establish therapeutic efficacy, large doses or site-specific administration are often required for devices comprised of polyester biomaterials. In an attempt to address these problems, numerous groups have introduced functional groups (such as amine functionalities) on these materials, either through direct conjugation or device fabrication with additives (Betram et al., 2009).

As for drug release from MPs or NPs, it is generally controlled by both drug diffusion and polymer degradation. To ensure the efficacy of drug delivery, control over the particle size and particle size distribution is critical, since smaller particles and narrower size distributions facilitate the design of targeted drug delivery systems. These involve binding fragments specific to a tumor-associated surface antigen, with a ligand binding to its corresponding receptor on the tumor cell surface, which can be attached on the surface of the PLA-based materials. Furthermore, polymers that display a physicochemical response to changes in their environment are being intensively explored as potential drug and gene delivery systems. The use of stimuli-responsive nanocarriers offers an attractive opportunity for targeted delivery, in which the delivery system becomes an active participant rather than a passive vehicle. The advantage of stimuli-responsive nanocarriers becomes obvious when the stimuli are unique to disease pathology, allowing the nanocarrier to respond specifically to the pathological characteristics. For instance, in solid tumors, the extracellular pH decreases significantly from 7.4 (the pH value under normal physiological conditions) to about 6.5 (Vaupel et al., 1989; Haag, 2004). In addition, the pH ranges from 4.5 in lysosomes to about 8.0 in mitochondria. Given these pH shifts, therapeutic compounds with a pK_a between 5.0 and 8.0 are able to exhibit dramatic changes in physicochemical properties (Ganta et al., 2008). Another option is thermo-sensitive polymeric micelles, containing a hydrophobic core and a thermo-sensitive shell, the later changing from a hydrophilic nature at body temperature to a collapsed hydrophobic polymer at a hyperthermic condition of 42°C (Na et al., 2006). Investigations concerned with this theme include responses induced by chemical substances, changes in temperature (Tyagi et al., 2004) or pH (Sethuraman & Bae, 2007), electric signals (Sawahata et al., 1990) or other environmental conditions (Qiu & Park, 2001).

The use of nucleic acids as therapeutic agents for genetic diseases has been extensively studied (Torchilin, 2008). However, a major limitation of this technique lies in the low delivery efficiency of the therapeutic DNA to the diseased site. To address this issue, various strategies have been explored including vectors engineered from viruses (Brun et al., 2008) and PLGA in NP formulations. PLGA NP have shown particular promise in improving the delivery efficacy (Kocbek et al., 2007). Besides, the physical characteristics of the nanoparticles can be manipulated to escape the degradative endosomal lumen, resulting in cytosolic localization. To develop novel administration paths, hybrid versions of research have been conducted on this subject, yet the results are mostly based on animal models or in vitro results, making it difficult to draw final conclusions. From clinical trials, substantial obstacles to their use, such as immunogenicity and inflammatory potential, have also been demonstrated (Nafee et al., 2007). Therefore, there is still a long way to go before real clinical applications come through.

Some examples of delivery systems incorporating PLA are provided in Table 5 and in Fig. 5 (Chen et al., 2007; Sethuraman & Bae, 2007). Sethuraman et al. (Sethuraman & Bae, 2007) developed a novel drug delivery system for acidic tumors consisting of two components: 1) A polymeric micelle with a hydrophobic core of PLLA and a hydrophilic shell of PEG conjugated to TAT (a cell-penetrating peptide in HIV), and 2) a highly pH-sensitive copolymer of poly(methacryloyl sulfadimethoxine) (PSD) and PEG (PSD-b-PEG). The final carrier, which was able to shield the micelles and expose them at slightly more acidic tumor pH levels, was achieved by complexing PSD with the TAT of the micelles. The results obtained showed significantly higher uptake of TAT micelles at pH 6.6 in comparison with pH 7.4, and that TAT not only translocated into the cells but it was also traced to the surface of the nucleus [see Fig. 5].

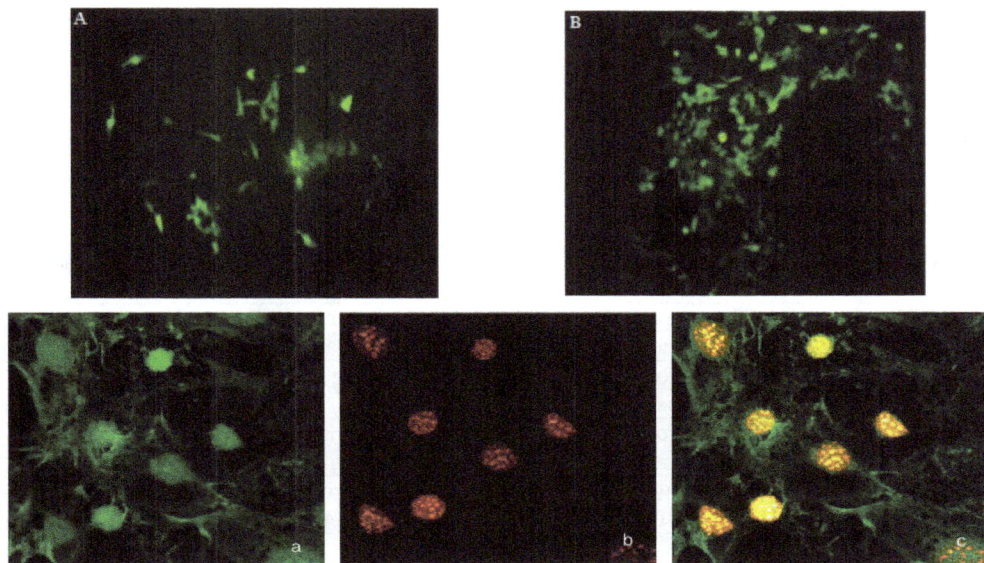

Fig. 5. Test results for PLA-based drug delivery materials. Fluorescent microscopy images are shown on top for COS7 cells transfected by plasmid encoding enhanced green fluorescence protein (EGFP) with different carriers: (A) lipofectamine, (B) methoxypolyethyleneglycol-PLA-chitosan nanoparticles (MePEG-PLA-CS NP); the transfection efficiency, as detected by flow cytometry, is higher in (B) than in (A) (Reproduced with the permission from Chen, J. et al. (2007). Preparation, characterization and transfection efficiency of cationic PEGylated PLA nanoparticles as gene delivery systems, Journal of Biotechnology, Vol.130, No.2, pp.107. Copyright (2007) Elsevier) At the bottom are dual label confocal micrographs for MCF-7 cells incubated with TAT micelles: (a) Cells stained with fluorescein isothiocyanate (FITC) attached to TAT in the micelles; (b) the same nuclei as in (a) were stained with TOPRO-3; (c) superimposed images of (a) and (b); the yellow color shows the localization of TAT within the nuclei (Reproduced with the permission from Sethuraman, V. A. & Bae, Y. H. (2007). TAT peptide-based micelle system for potential active targeting of anti-cancer agents to acidic solid tumors. Journal of Controlled Release, Vol.118, No.2, pp.216. Copyright (2007) Elsevier).

Material	Application	Results	Reference
PLA-PEG particles	Carrier for tetanus toxoid	Enhanced transport across the rat nasal mucosa	Vila et al., 2005
PEG-PLA NP	Conjugated with lactoferrin (Lf)	Increased uptake of the Lf-NP by bEnd.3 cells	Hu et al., 2009
PLA-b-Pluronic-b-PLA	Carrier for oral insulin	Good control over blood glucose concentration	Xiong et al., 2007
PLA NP	Carrier for HIV p24 proteins	Induced seric and mucosal antibody production	Aline et al., 2009
Surfactant-free PLA NP	Carrier for HIV p24 proteins	Elicited strong CTL response and cytokine release	Liggins et al., 2004
PLA microspheres	Carrier for paclitaxel	Reduced inflammation of arthritis rabbit model	Jie et al., 2005
PEO-PLA copolymers	Carrier for 5-FU and paclitaxel	Complete drug release	Zhang & Feng, 2006
PLA-TPGS copolymers	Carrier for paclitaxel	Initial burst followed by sustained release	Freitas et al., 2005
PLA microspheres	Carrier for nimesulide	Initial burst followed by an exponential decrease	Chen et al., 2007
PEGylated PLA NP	Gene delivery systems	Improved transfection activity	Ataman-Önal et al., 2006
PLA-PEG-PLA copolymer	Carrier for 5-FU and paclitaxel	Good control over the release	Venkatraman et al., 2005
AP-PEG-PLA/MPEG-PAE	Drug carrier for cancer therapy	Presented high tumor-specific targeting ability	Wu et al., 2010
PLGA/PEI NP	Carrier for luciferase siRNA	Effective silencing of the gene in cells	Patil & Panyam, 2009
cNGR-PEG-PLA NP	Carrier for DNA	Rapid and efficient nanoparticle internalization	Liu et al., 2011
DMAB coated PLGA NP	Loaded with plasmid DNA	Significantly improved transfection efficiencies	Fay et al., 2010

Table 5. Investigations on PLA-based material as drug delivery systems. AP: peptide, CRKRLDRN; MPEG: methyl ether poly(ethylene glycol); PAE: poly(β-amino ester); PEI: polyethylenimine; cNGR: Cyclic Asn-Gly-Arg; DMAB: dimethyldidodecylammonium bromide.

In summary, some problems still remain to be tackled for this promising novela dministration method. A major issue is the presence of surfactants such as SDS or stabilizers such as PVA in the microparticles, necessary to achieve antigen binding and colloidal stability. Although present only at low concentrations, the acceptability of such components in human vaccines depends on the results of extensive and costly toxicological studies. Biodegradable polymers used for drug delivery to date have mostly been in the form of injectable microspheres or implant systems requiring complicated fabrication processes with organic solvents. In such systems, the organic solvents can denature components such as protein drugs being encapsulated. Besides, these delivery systems have relatively low transfection efficiencies in vitro as compared with reagents commercialized for cell transfection. The last problem concerns the lack of test results for these delivery systems using animal models or in clinical trials, which is of fundamental importance for real applications in biomedical therapy.

5.3 Other fields

Due to its versatility, PLA has been investigated for membrane applications (e.g. wound covers), implants and medical devices (fixation rods, plates, pins, screws, sutures, etc.), and dermatological treatments (e.g. facial lipoatrophy and scar rejuvenation).

With respect to wound treatment, bacterial infections are one of the main factors impacting the healing process. One of the best approaches to treat wound infections is by the immobilization of drugs or antibacterial agents within the nanofibers by electrospinning, or the electrospinning of polymers with intrinsic antibacterial and wound-healing properties. Dozens of patents have been issued on that topic so far (Ghosh et al., 2007; Robinson et al., 2009). Silver nanoparticles (nAg) and the natural polysaccharide chitosan (as well as its quaternized derivatives) are most commonly used as antibacterial agents with a high intrinsic activity against a broad spectrum of bacteria (Rujitanaroj et al., 2008; Ignatova et al., 2009).

Metals are still the most popular materials for fracture fixation, but their disadvantages include stress shielding, accumulation in tissues, hypersensitivity, growth restriction, pain, corrosion, and interference with imaging techniques. Consequently, the focus of research is increasingly shifted to biomaterials like PLA, which offers satisfactory strength during the healing of bone tissue and then degrades over time (Mavrogenis et al., 2009). A commercial product with a proven track record in clinical applications is the VICRYL™ suture material, based on PGA/PLA copolymers (Mehta et al., 2005). The number of applications of PLA as fixation rods, plates, pins, screws, sutures, etc. in orthopaedics and dentistry is also increasing (Raghoebar et al., 2006).

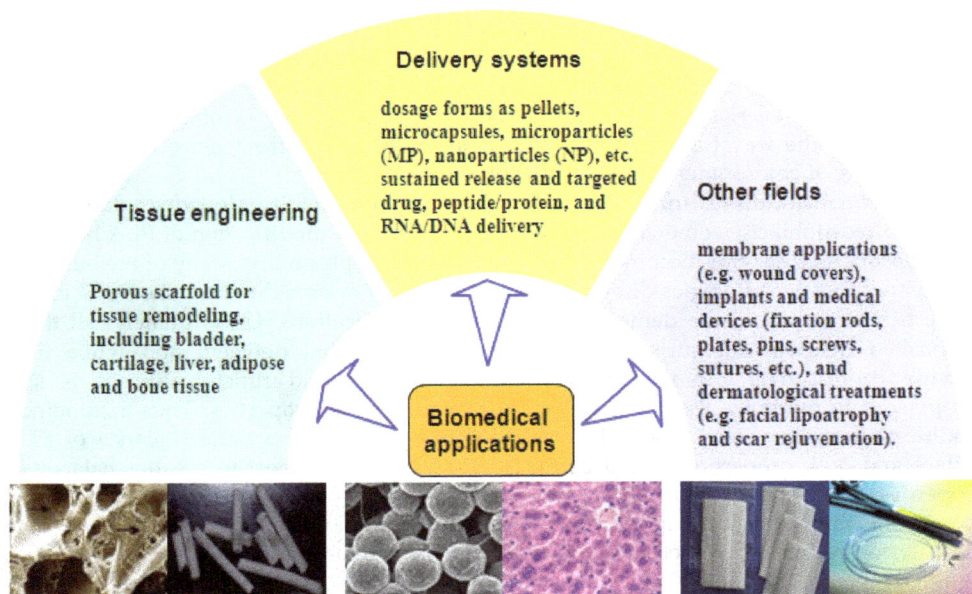

Delivery systems

dosage forms as pellets, microcapsules, microparticles (MP), nanoparticles (NP), etc. sustained release and targeted drug, peptide/protein, and RNA/DNA delivery

Tissue engineering

Porous scaffold for tissue remodeling, including bladder, cartilage, liver, adipose and bone tissue

Biomedical applications

Other fields

membrane applications (e.g. wound covers), implants and medical devices (fixation rods, plates, pins, screws, sutures, etc.), and dermatological treatments (e.g. facial lipoatrophy and scar rejuvenation).

Fig. 6. Applications of PLA in the biomedical field.

In 2004, Sculptra™ [poly(L-lactic acid)] was approved by the FDA as the first injectable facial "volumizer" in the treatment of lipoatrophy due to its significant therapeutic effectiveness (Burgess & Quiroga, 2005). The lipodystrophy syndrome is associated with the usage of highly active antiretroviral therapy (HAART) containing protease inhibitors or nucleoside reverse transcriptase inhibitors for HIV patients. The action mechanism of Sculptra™ is via stimulation of the fibroblastic activity with generation of collagen and other connective tissue fibers. In addition, it acts as dermal matrix adding support by thickening the dermis (Vleggaar & Bauer, 2004). Moreover, PLA can help improve the appearance of scars due to acne, surgery, trauma, or suture (Lowe & Beer, 2005, as cited in Beer & Rendon, 2006).

6. Conclusions

Due to the multiple desirable characteristics of PLA including renewability, biocompatibility, transparency, and thermoplasticity, it is being used or is a potential candidate for many consumer and biomedical applications (Jamshidian et al., 2010). Ever increasing environmental concerns associated with conventional polymers derived from petrochemicals lead to constantly expanding applications for PLA since its discovery in 1932 by Carothers at DuPont.

In previous years, the most negative point of PLA was its higher price as compared with petrochemical-based polymers. Today, by optimizing the LA and PLA production processes, and with increasing PLA demand, a reduction in its price can be achieved. The price of PLA is currently much lower than in previous years. Meanwhile, PLA is mainly synthesized in the industry by ROP employing tin(II) bis(2-ethylhexanoate) ($SnOct_2$) as a catalyst, which has been approved as a food additive by the FDA, but the potential toxicity associated with most tin compounds cannot be ignored for biomedical applications. Scientists all over the world are now exploring novel, well-defined catalysts with good biocompatibility, high catalytic activity, low toxicity, and excellent stereoselectivity. This should remain an everlasting interest area. Finally, the possibility of obtaining 100% biosourced opens the way for PLA to become more independent from petrochemical-based polymers, free of environmental and health concerns.

However, the major disadvantages of PLA such as its poor ductility, slow degradation rate, and poor hydrophilicity somewhat limit its applications. The modification of PLA bulk and surface properties has thus become crucial to increase its applicability. Many of the bulk and surface modification strategies discussed above have been designed to tune the PLA surface properties according to the demands of biomedical applications. Unfortunately, all these established methods for surface modification are somewhat deficient and while they provide control over the wettability, degradation rate, and functionality, it is still compulsory to minimize their negative impact on PLA bulk properties. Thus a combined modification strategy (e.g. irradiation followed by grafting) or a better balance of PLA surface and bulk properties should be sought. Ideally, with respect to a better balance of properties and shorter modification times, one-step approaches need to be developed because it is time-consuming to carry out surface and bulk modifications separately, and the solvents and reagents involved in multiple modification steps tend to affect PLA properties significantly.

All these modification strategies aim at tailoring the properties of PLA-based materials for certain applications. Fortunately, more and more encouraging results have been reported, but the present conclusions from most of these reports cannot be directly generalized to truly biomedical applications since most of the experiments were carried out in vitro. Nevertheless, these findings offer some clues for further improvements. The increasing number of functional PLA-based polyesters provides the opportunity to study the relationships between structure and functionality of these polymers such as cell adhesion and degradability in vitro and in vivo, as well as to develop applications of these materials for delivery system in the form of micro- and nanoparticles or scaffolds for tissue engineering. Finally, cancers and acquired or inherited genetic diseases represent one of the most serious threats to the health of human beings, but no effective therapies are available so far. It is suggested that the development of effective and ideal tools for drug, peptide/protein, and RNA/DNA delivery will represent a good alternative in drug development. It therefore appears that it would be best to focus future research work on the rational design of novel carriers for biomedical uses and targeted delivery systems. Obviously, this requires plenty of relevant experiments on animal model and enough clinical trials before they are widely utilized.

Even though countless studies have focused on the synthesis and the modification of PLA and remarkable progresses has been achieved over the last two decades, vast opportunities as well as challenges remain in terms of exploring the characteristics of PLA-based materials and expanding their domains of applications.

7. Acknowledgement

It was supported by the Natural Science of Hubei Province for Distinguished Young Scholars (No. 2008CDB279) and the Key Technologies R & D Program of Hubei Province (20091933), the Fundamental Research Funds for the Central Universities, Huazhong University of Science and Technology (2010JC016) and the National High Technology Research and Development Program ("863" Program) of China (No. 2008AA10Z339). The authors are also grateful to Dr. Wen-bin Zhang from the University of Akron for his valuable suggestions during the preparation of this manuscript.

8. References

Achmad, F. et al., (2009). Synthesis of polylactic acid by direct polycondensation under vacuum without catalysts, solvents and initiators. *Chemical Engineering Journal*, Vol.151, No.1, (August 2009), pp. 342-350, ISSN 1385-8947

Agrawal, S. K. et al., (2000). Rare earth metal initiated ring-opening polymerization of lactones. *Macromolecular Rapid Communications*, Vol.21, No.5, (March 2000), pp. 195-212, ISSN 1022-1336

Agrawal, S. K. et al., (2010). Energetics of association in poly(lactic acid)-based hydrogels with crystalline and nanoparticle-polymer junctions. *Langmuir*, Vol.26, No.22, (October 2010), pp. 17330-17338, ISSN 0743-7463

Ajioka, M. et al. (1995). The basic properties of poly(lactic acid) produced by the direct condensation polymerization of lactic acid. *Journal of Polymers and the Environment*, Vol.3, No.4, (October 1995), pp. 225-234, ISSN 1566-2543

Aline, F. et al., (2009). Dendritic cells loaded with HIV-1 p24 proteins adsorbed on surfactant-free anionic PLA nanoparticles induce enhanced cellular immune responses against HIV-1 after vaccination. *Vaccine,* Vol.27, No.38, (August 2009), pp. 5284-5291, ISSN 0264-410X

Ataman-Önal, Y. et al., (2006). Surfactant-free anionic PLA nanoparticles coated with HIV-1 p24 protein induced enhanced cellular and humoral immune responses in various animal models. *Journal of Controlled Release,* Vol.112, No.2, (May 2006), pp. 175-185, ISSN 0168-3659

Bentley, G. et al., (2003). A prospective, randomized comparison of autologous chondrocyte implantation versus mosaicplasty for ostechondral defects of the knee. *Journal of Bone and Joint Surgery-British Volume,* Vol.85, No.2, (March 2003), pp. 223-230, ISSN 0301-620X

Bergsma, J. E. et al., (1995). Late degradation tissue response to poly(L-lactide) bone plates and screws. *Biomaterials,* Vol.16, No.1, (1995), pp. 25-31, ISSN 0142-9612

Bertram, J. P. et al., (2009). Functionalized poly(lactic-co-glycolic acid) enhances drug delivery and provides chemical moieties for surface engineering while preserving biocompatibility. *Acta Biomaterialia,* Vol.5, No.8, (October 2009), pp. 2860-2871, ISSN 1742-7061

Brittberg, M. et al., (1994). Treatment of deep cartilage defects in the knee with autologous chondrocyte transplantation. *New England Journal of Medicine,* Vol.331, No.14, (October 1994), pp. 889-895, ISSN 0028-4793

Brun, A. et al., (2008). Antigen delivery systems for veterinary vaccine development. Viral-vector based delivery systems. *Vaccine,* Vol.26, No.51, (December 2008), pp. 6508-528, ISSN 0264-410X

Burgess, C. M. & Quiroga, R. M. (2005). Assessment of the safety and efficacy of poly-L-lactic acid for the treatment of HIV-associated facial lipoatrophy. *Journal of the American Academy of Dermatology,* Vol.52, No.2, (February 2005), pp. 233-239, ISSN 0190-9622

Butterwick, K. & Lowe, N. J. (2009) Injectable poly-L-lactic acid for cosmetic enhancement: Learning from the European experience. *Journal of the American Academy of Dermatology,* Vol.61, No.2, (August 2009), pp. 281-293, ISSN 0190-9622

Chaiwong, C. et al., (2010). Effect of plasma treatment on hydrophobicity and barrier property of polylactic acid. *Surface and Coatings Technology,* Vol.204, No.18-19, (June 2010), pp. 2933-2939, ISSN 0257-8972

Chen, G.; Ushida, T. & Tateishi, T. (2002). Scaffold design for tissue engineering. *Macromolecular Bioscience,* Vol.2, No.2, (February 2002), pp. 67-77, ISSN 1616-5187

Chen, H. Y. et al., (2011). Comparative study of lactide polymerization by zinc alkoxide complexes with a β-diketiminato ligand bearing different substituents. *Journal of Molecular Catalysis A: Chemical,* Vol.339, No.1-2, (April 2011), pp. 61-71, ISSN 1381-1169

Chen, J. et al., (2007). Preparation, characterization and transfection efficiency of cationic PEGylated PLA nanoparticles as gene delivery systems. *Journal of Biotechnology,* Vol.130, No.2, (June 2007), pp. 107-113, ISSN 0168-1656

Cheung, H. Y. et al., (2008) A potential material for tissue engineering: Silkworm silk/PLA biocomposite. *Composites Part B: Engineering,* Vol.39, No.6, (September 2008), pp. 1026-1033, ISSN 1359-8368

Chisholm, M. H. et al., (2003). Binolate complexes of lithium, zinc, aluminium, and titanium; preparations, structures, and studies of lactide polymerization. *Dalton Transactions,* No.3, (July 2003), pp. 406-412, ISSN 1477-9226

Chu, P. K. et al., (2002). Plasma-surface modification of biomaterials. *Materials Science and Engineering: R: Reports,* Vol.36, No.5-6, (March 2002), pp. 143-206, ISSN 0927-796X

Cronin, E. M. et al., (2004).Protein-coated poly(L-lactic acid) fibers provide a substrate for differentiation of human skeletal muscle cells. *Journal of Biomedical Materials Research Part A,* Vol.69A, No.3, (June 2004), pp. 373-381, ISSN 1549-3296

Deng, X. et al., (2000). Polymerization of lactides and lactones VII. Ring-opening polymerization of lactide by rare earth phenyl compounds. *European Polymer Journal,* Vol.36, No.6, (June 2000), pp. 1151-1156, ISSN 0014-3057

Ejfler, J. et al., (2005). Highly efficient magnesium initiators for lactide polymerization. *Dalton Transactions,* No.11, (June 2005), pp. 2047-2050, ISSN 1477-9226

Engelhardt, E. M. et al., (2011). A collagen-poly(lactic acid-co-caprolactone) hybrid scaffold for bladder tissue regeneration. *Biomaterials,* Vol.32, No.16, (June 2011), pp. 3969-3976, ISSN 0142-9612

Fahmy, T. M. et al., (2005). Surface modification of biodegradable polyesters with fatty acid conjugates for improved drug targeting. *Biomaterials,* Vol.26, No.28, (October 2005), pp. 5727–5736, ISSN 0142-9612

Favia, P. & d'Agostino, R. (1998). Plasma treatments and plasma deposition of polymers for biomedical applications. *Surface and coatings Technology,* Vol.98, No.1, (January1998), pp. 1102-1106, ISSN 0257-8972

Fay, F. et al., (2010). Gene delivery using dimethyldidodecylammonium bromide-coated PLGA nanoparticles. *Biomaterials,* Vol.31, No.14, (May 2010), pp. 4214-4222, ISSN 0142-9612

Freitas, M. N. & Marchetti, J. M. (2005). Nimesulide PLA microspheres as a potential sustained release system for the treatment of inflammatory diseases. *International Journal of Pharmaceutics,* Vol.295, No.1-2, (May 2005), pp. 201-211, ISSN 0378-5173

Fu, B. et al., (2008). Preparation of lactic acid based polyurethanes modified by castor oil. *Advanced Materials Research,* Vols. 47-50, (June 2008), pp. 1458-1461, ISSN 1022-6680

Ganta, S. et al., (2008). A review of stimuli-responsive nanocarriers for drug and gene delivery. *Journal of Controlled Release,* Vol.126, No.3, (March 2008), pp. 187-204, ISSN 0168-3659

Gao, Q. W. et al., (2002). Direct synthesis with melt polycondensation and microstructure analysis of poly(L-lactic acid-co-glycolic acid). *Polymer Journal,* Vol.34, No.11, (November 2002), pp. 786-793, ISSN 0032-3896

Garlotta, D. (2001). A literature review of poly(lactic acid). *Journal of Polymers and the Environment,* Vol.9, No.2, (April 2001), pp. 63-84, ISSN 1566-2543

Ge, Z. et al., (2009). Proliferation and differentiation of human osteoblasts within 3D printed poly-lactic-co-glycolic acid scaffolds. *Journal of Biomaterials Applications,* Vol.23, (May 2009), pp.533-547, ISSN 1530-8022

Ge, Z. et al., (2009). Histological evaluation of osteogenesis of 3D-printed poly-lactic-co-glycolic acid (PLGA) scaffolds in a rabbit model. *Biomedical Materials,* Vol. 4, No. 2, (April 2009), 021001, ISSN 1748-6041

Ghosh, D. et al., (2007). Interactive wound cover. US Patent 0,258,958, filed February 12, 2007, issued November 8, 2007

Graupner, N.; Herrmann, A. S. & Müssig, J. (2009). Natural and man-made cellulose fibre-reinforced poly(lactic acid) (PLA) composites: An overview about mechanical characteristics and application areas. *Composites Part A: Applied Science and Manufacturing*, Vol.40, No.6-7, (July 2009), pp. 810-821, ISSN 1359-835X

Gref, R. et al., (1994). Biodegradable long-circulating polymeric nanospheres. *Science*, Vol.263, No.5153, (March1994), pp. 1600-1603, ISSN 0036-8075

Gu, M. et al., (2007). Surface biofunctionalization of PLA nanoparticles through amphiphilic polysaccharide coating and ligand coupling: Evaluation of biofunctionalization and drug releasing behavior. *Carbohydrate Polymers*, Vol.67, No.3, (February 2007), pp. 417-426, ISSN 0144-8617

Gupta, B.; Revagade, N. & Hilborn, J. Poly(lactic acid) fiber: An overview. *Progress in polymer science*, Vol.32, No.4, (April 2007), pp. 455-482, ISSN 0079-6700

Haag, R. (2004). Supramolecular drug-delivery systems based on polymeric core–shell architectures. *Angewandte Chemie International Edition*, Vol.43, No.3, (January 2004), pp. 278-282, ISSN 1433-7851

Hirotsu, T. et al., (2002). Plasma surface treatments of melt-extruded sheets of poly(L-lactic acid). *Polymer Engineering and Science*, Vol.42, No.2, (February 2002), pp. 299-306, ISSN 0032-3888

Hollister, S. J. (2005). Porous scaffold design for tissue engineering. *Nature Materials*, Vol.4, (July 2005), pp. 518-524, ISSN 1476-1122

Hong, Y. et al., (2005). Collagen-coated polylactide microspheres as chondrocyte microcarriers. *Biomaterials*, Vol.26, No.32, (November 2005), pp. 6305-6313, ISSN 0142-9612

Hu, K. et al., (2009). Lactoferrin-conjugated PEG-PLA nanoparticles with improved brain delivery: In vitro and in vivo evaluations. *Journal of Controlled Release*, Vol.134, No.1, (February 2009), pp. 55-61, ISSN 0168-3659

Huang, J. W. et al., (2009), Polylactide/nano and microscale silica composite films. I. Preparation and characterization. *Journal of Applied Polymer Science*, Vol.112, No.3, (May 2009), pp. 1688-1694, ISSN 0021-8995

Huang, L. et al., (2007). Synthesis and characterization of electroactive and biodegradable ABA block copolymer of polylactide and aniline pentamer. *Biomaterials*, Vol.28, No.10, (April 2007), pp. 1741-1751, ISSN 0142-9612

Huda, M. S. et al., (2008). Effect of fiber surface-treatments on the properties of laminated biocomposites from poly(lactic acid)(PLA) and kenaf fibers. *Composites Science and Technology*, Vol.68, No.2, (February 2008), pp. 424-432, ISSN 0266-3538

Ichikawa, F. et al., (1995). Process for preparing polyhydroxycarboxylic acid. US Patent 5,440,008, filed May 19, 1994, issued August 8, 1995

Ignatova, M. et al., (2009). Electrospun non-woven nanofibrous hybrid mats based on chitosan and PLA for wound-dressing applications. *Macromolecular Bioscience*, Vol.9, No.1, (January 2009), pp. 102-111, ISSN 1616-5187

Ilan, D. I. & Ladd, A. L. (2002). Bone graft substitutes. *Operative Techniques in Plastic and Reconstructive Surgery*, Vol.9, No.4, (November 2002), pp. 151-160, ISSN 1071-0949

Iwasa, J. et al., (2009). Clinical application of scaffolds for cartilage tissue engineering. *Knee Surgery, Sports Traumatology, Arthroscopy*, Vol.17, No.6, (June 2009), pp. 561-577, ISSN 0942-2056

Iwata, T. & Doi, Y. (1998). Morphology and enzymatic degradation of poly(L-lactic acid) single crystals. *Macromolecules*, Vol.31, No.8, (April 1998), pp. 2461-2467, ISSN 0024-9297

Jamshidian, M. et al., (2010). Poly-lactic acid: Production, applications, nanocomposites, and release studies. *Comprehensive Reviews in Food Science and Food Safety*, Vol.9, No.5, (September 2010), pp. 552-571, ISSN 1541-4337

Jiang, T. et al., (2006). In vitro evaluation of chitosan/poly(lactic acid-glycolic acid) sintered microsphere scaffolds for bone tissue engineering. *Biomaterials*, Vol.27, No.28, (November 2006), pp. 4894-4903, ISSN 0142-9612

Jiang, T. et al., (2010). Functionalization of chitosan/poly(lactic acid-glycolic acid) sintered microsphere scaffolds via surface heparinization for bone tissue engineering. *Journal of Biomedical Materials Research Part A*, Vol.93A, No.3, (June 2010), pp. 1193-1208, ISSN 1549-3296

Jiang, T. et al., (2010). Chitosan-poly(lactide-co-glycolide) microsphere-based scaffolds for bone tissue engineering: In vitro degradation and in vivo bone regeneration studies. *Acta Biomaterialia*, Vol.6, No.9, (September 2010), pp. 3457-3470, ISSN 1742-7061

Jie, P. et al.,(2005). Micelle-like nanoparticles of star-branched PEO-PLA copolymers as chemotherapeutic carrier. *Journal of Controlled Release*, Vol.110, No.1, (December 2005), pp. 20-33, ISSN 0168-3659

John, A. et al., (2007) Ni (II) and Cu (II) complexes of phenoxy-ketimine ligands: Synthesis, structures and their utility in bulk ring-opening polymerization (ROP) of L-lactide. *Polyhedron*, Vol.26, No.15, (April 2007), pp. 4033-4044, ISSN 0277-5387

Jung, Y. et al., (2005). A poly(lactic acid)/calcium metaphosphate composite for bone tissue engineering. *Biomaterials*, Vol.26, No.32, (November 2005), pp. 6314-6322, ISSN 0142-9612

Jérôme, C. & Lecomte, P. (2008). Recent advances in the synthesis of aliphatic polyesters by ring-opening polymerization. *Advanced drug delivery reviews*, Vol.60, No.9, (June 2008), pp. 1056-1076, ISSN 0169-409X

Kang, Y. et al., (2008). Preparation, characterization and in vitro cytotoxicity of indomethacin-loaded PLLA/PLGA microparticles using supercritical CO_2 technique. *European Journal of Pharmaceutics and Biopharmaceutics*, Vol.70, No.1, (September 2008), pp. 85-97, ISSN 0939-6411

Kasuga, T. et al., (2001). Preparation and mechanical properties of polylactic acid composites containing hydroxyapatite fibers. *Biomaterials*, Vol.22, No.1, (January 2001), pp. 19-23, ISSN 0142-9612

Kasuga, T. et al., (2003). Preparation of poly(lactic acid) composites containing calcium carbonate (vaterite). *Biomaterials*, Vol.24, No.19, (August 2003), pp. 3247-3253, ISSN 0142-9612

Khondker, O. A. et al., (2006). A novel processing technique for thermoplastic manufacturing of unidirectional composites reinforced with jute yarns. *Composites Part A: Applied Science and Manufacturing*, Vol.37, No.12, (December 2006), pp. 2274-2284, ISSN 1359-835X

Kim, I. H. & Jeong, Y. G. (2010). Polylactide/exfoliated graphite nanocomposites with enhanced thermal stability, mechanical modulus, and electrical conductivity.

Journal of Polymer Science Part B: Polymer Physics, Vol.48, No.8, (April 2010), pp. 850-858, ISSN 0887-6266

Kocbek, P. et al., (2007). Targeting cancer cells using PLGA nanoparticles surface modified with monoclonal antibody. *Journal of Controlled Release,* Vol.120, No.1-2, (July 2007), pp. 18-26, ISSN 0168-3659

Koo, G. H. & Jang, J. (2008). Surface modification of poly(lactic acid) by UV/Ozone irradiation. *Fibers and Polymers,* Vol.9, No.6, (December 2008), pp. 674-678, ISSN 1875-0052

Köhn, R. D. et al., (2003). Ring-opening polymerization of,-lactide with bis (trimethyl triazacyclohexane) praseodymium triflate. *Catalysis Communications,* Vol.4, No.1, (January 2003), pp. 33-37, ISSN 1566-7367

Lassalle, V. L. & Ferreira M. L. (2008). Lipase-catalyzed synthesis of polylactic acid: An overview of the experimental aspects. *Journal of Chemical Technology and Biotechnology,* Vol.83, No.11 , (October 2008), pp. 1493-1502, ISSN 0268-2575

Li, S. & Vert, M. (2003). Synthesis, characterization, and stereocomplex-induced gelation of block copolymers prepared by ring-opening polymerization of L (D)-lactide in the presence of poly(ethylene glycol). *Macromolecules,* Vol.36, No.21, (September 2003), pp. 8008-8014, ISSN 0024-9297

Lieb, E. et al., (2005). Mediating specific cell adhesion to low-adhesive diblock copolymers by instant modification with cyclic RGD peptides. *Biomaterials,* Vol.26, No.15, (May 2005), pp. 2333-2341, ISSN 0142-9612

Liggins, R. T. et al., (2004). Intra-articular treatment of arthritis with microsphere formulations of paclitaxel: Biocompatibility and efficacy determinations in rabbits. *Inflammation Research,* Vol.53, No.8, (August 2004), pp. 363-372, ISSN 1023-3830

Lin, Y. M. et al., (2010). Tissue engineering of lung: The effect of extracellular matrix on the differentiation of embryonic stem cells to pneumocytes. *Tissue Engineering Part A,* Vol.16, No.5, (April 2010), pp. 1515-1526, ISSN 1937-3341

Liu, C. et al., (2004). Effects of DBD plasma operating parameters on the polymer surface modification. *Surface and Coatings Technology,* Vol.185, No.2-3, (July 2004), pp. 311-320, ISSN 0257-8972

Liu, C. X. et al., (2011). Enhanced gene transfection efficiency in CD13-positive vascular endothelial cells with targeted poly(lactic acid)-poly(ethylene glycol) nanoparticles through caveolae-mediated endocytosis. *Journal of Controlled Release,* (in press), ISSN 0168-3659

Liu, P. et al., (2010). The research of RGD-conjugated PLA-PLL nanoparticles carriers on targeted delivery to tumor, *Proceedings of Nanoelectronics Conference (INEC), 2010 3rd International,* 978-1-4244-3543-2, Hong Kong, January 2010

Liu, X, & Ma, P. X. (2004). Polymeric scaffolds for bone tissue engineering. *Annals of Biomedical Engineering,* Vol.32, No.3, (March 2004), pp. 477-486, ISSN 0090-6964

Liu, X.; Won, Y. & Ma, P. X. (2005). Surface modification of interconnected porous scaffolds. *Journal of Biomedical Materials Research Part A,* Vol.74A, No.1, (July 2005), pp. 84-91, ISSN 1549-3296

Ljungberg, N. & Wesslén, B. (2002). The effects of plasticizers on the dynamic mechanical and thermal properties of poly(lactic acid). *Journal of Applied Polymer Science,* Vol.86, No.5, (October 2002), pp. 1227-1234, ISSN 0021-8995

Ljungberg, N.; Colombini, D. & Wesslén, B. (2005). Plasticization of poly(lactic acid) with oligomeric malonate esteramides: Dynamic mechanical and thermal film properties. *Journal of Applied Polymer Science*, Vol.96, No.4, (March 2005), pp. 992-1002, ISSN 0021-8995

Lowe, N. & Beer, K. (2005). A new material in subdermal sculpting, In: *Procedures in Cosmetic Dermatology Series; Soft Tissue Augmentation*, Carruthers, J. & Carruthers, A., pp. 143-146, Elsevier Saunders, 1-4160-4214-8, Oxford

Lu, J. W. et al., (2009). Surface engineering of poly(D, L-lactic acid) by entrapment of soluble eggshell membrane protein. *Journal of Biomedical Materials Research Part A*, Vol.91A, No.3, (December 2009), pp. 701-707, ISSN 1549-3296

Lunt, J. (1998). Large-scale production, properties and commercial applications of polylactic acid polymers. *Polymer Degradation and Stability*, Vol.59, No.1, (January 1998), pp. 145-152, ISSN 0141-3910

Lv, Q. et al., (2007). Preparation and characterization of PLA/fibroin composite and culture of HepG2 (human hepatocellular liver carcinoma cell line) cells. *Composites Science and Technology*, Vol.67, No.14, (June 2007), pp. 3023-3030, ISSN 0266-3538

Ma, H.; Davis, R. H. & Bowman, C. N. (1999). A novel sequential photoinduced living graft polymerization. *Macromolecules*, Vol.33, No.2, (January 2000), pp. 331-335, ISSN 0024-9297

Ma, P. X. (2004). Scaffolds for tissue fabrication. *Materials Today*, Vol.7, No.5, (May 2004), pp. 30-40, ISSN 1369-7021

Ma, W. et al., (2008). Evaluation of blood circulation of polysaccharide surface-decorated PLA nanoparticles. *Carbohydrate Polymers*, Vol.72, No.1, (April 2008), pp. 75-81, ISSN 0144-8617

Maharana, T.; Mohanty, B. & Negi, Y. S. (2009). Melt-solid polycondensation of lactic acid and its biodegradability. *Progress in Polymer Science*, (January 2009), Vol.34, No.1, pp. 99-124, ISSN 0079-6700

Martin, O. & Averous, L. (2001). Poly(lactic acid): Plasticization and properties of biodegradable multiphase systems. *Polymer*, Vol.42, No.14, (June 2001), pp. 6209-6219, ISSN 0032-3861

Mathew, A. P.; Oksman, K. & Sain, M. (2005). Mechanical properties of biodegradable composites from poly lactic acid (PLA) and microcrystalline cellulose (MCC). *Journal of applied polymer science*, Vol.97, No.5, (September 2005), pp. 2014-2025, ISSN 0021-8995

Mathieu, L. M. et al., (2006). Architecture and properties of anisotropic polymer composite scaffolds for bone tissue engineering. *Biomaterials*, Vol.27, No.6, (February 2006), pp. 905-916, ISSN 0142-9612

Matsumoto, K. & Taguchi, S. (2010). Enzymatic and whole-cell synthesis of lactate-containing polyesters: Toward the complete biological production of polylactate. *Applied microbiology and biotechnology*, Vol.85, No.4, (December 2009), pp. 921-932, ISSN 0175-7598

Matsumura, S.; Mabuchi, K. & Toshima, K. (1997). Lipase-catalyzed ring-opening polymerization of lactide. *Macromolecular Rapid Communications*, Vol.18, No.6, (April 2003), pp. 477-482, ISSN 1022-1336

Mauney, J. R. et al., (2007). Engineering adipose-like tissue in vitro and in vivo utilizing human bone marrow and adipose-derived mesenchymal stem cells with silk fibroin

3D scaffolds. *Biomaterials*, Vol.28, No.35, (December 2007), pp. 5280-5290, ISSN 0142-9612

Mavrogenis, A. F. et al., (2009). Early experience with biodegradable implants in pediatric patients. *Clinical Orthopaedics and Related Research*, Vol.467, No.6, (June 2009), pp. 1591-1598, ISSN 0009-921X

Mehta, R. et al., (2005). Synthesis of poly(lactic acid): A review. *Journal of Macromolecular Science, Part C: Polymer Reviews*, Vol.45, No.4, (October 2005), pp. 325-349, ISSN 1532-1797

Metters, A. T.; Anseth, K. S. & Bowman, C. N. (2000). Fundamental studies of a novel, biodegradable PEG-b-PLA hydrogel. *Polymer*, Vol.41, No.11, (May 2000), pp. 3993-4004, ISSN 0032-3861

Moon, S. I. et al., (2000). Melt polycondensation of L-lactic acid with Sn(II) catalysts activated by various proton acids: A direct manufacturing route to high molecular weight Poly(L-lactic acid). *Journal of Polymer Science Part A: Polymer Chemistry*, Vol.3, No.89, (May 2000), pp. 1673-1679, ISSN 0887-624X

Moon, S. I. et al., (2001). Synthesis and properties of high-molecular-weight poly(L-lactic acid) by melt/solid polycondensation under different reaction conditions. *High Performance Polymers*, Vol.13, No.2, (June 2001), pp. 189-196, ISSN 0954-0083

Moran, M. J.; Pazzano, D. & Bonassar, L. J. (2003). Characterization of polylactic acid-polyglycolic acid composites for cartilage tissue engineering. *Tissue Engineering*, Vol.9, No.1, (February 2003), pp. 63-70, ISSN 1076-3279

Na, K. et al., (2006). Biodegradable thermo-sensitive nanoparticles from poly(L-lactic acid)/poly(ethylene glycol) alternating multi-block copolymer for potential anti-cancer drug carrier. *European Journal of Pharmaceutical Sciences*, Vol.27, No.2–3, (December 2006), pp. 115–122, ISSN 0928-0987

Nafee, N. et al., (2007). Chitosan-coated PLGA nanoparticles for DNA/RNA delivery: Effect of the formulation parameters on complexation and transfection of antisense oligonucleotides. *Nanomedicine*, Vol.3, No.3, (September 2007), pp. 173-183, ISSN 1743-5889

Nalawade, S. P.; Picchioni F. & Janssen L. P. B. N. (2006). Supercritical carbon dioxide as a green solvent for processing polymer melts: Processing aspects and applications. *Progress in Polymer Science*, Vol.31, No.1, (January 2006), pp. 19-43, ISSN 0079-6700

Nomura, N. et al., (2002). Stereoselective ring-opening polymerization of racemic lactide using aluminum-achiral ligand complexes: Exploration of a chain-end control mechanism. *Journal of the American Chemical Society*, Vol.124, No.21, (May 2002), pp. 5938-5939, ISSN 0002-7863

Ohta, M.; Obuchi, S. &Yoshida, Y. (1995). Preparation process of polyhydroxycarboxylic acid. US Patent, 5,444,143, filed December 21, 1993, issued August 22, 1995

Oksman, K.; Skrifvars, M. & Selin J. F.. (2003). Natural fibres as reinforcement in polylactic acid (PLA) composites. *Composites Science and Technology*, Vol.63, No.9, (July 2003), pp. 1317-1324, ISSN 0266-3538

Pan, P. et al., (2007). Crystallization behavior and mechanical properties of bio-based green composites based on poly(L-lactide) and kenaf fiber. *Journal of Applied Polymer Science*, Vol.105, No.3, (Auguest 2007), pp. 1511-1520, ISSN 0021-8995

Patil, Y. & Panyam, J. (2009). Polymeric nanoparticles for siRNA delivery and gene silencing. *International Journal of Pharmaceutics.* Vol.367, No.1-2, (February 2009), pp. 195-203, ISSN 0378-5173

Peesan, M.; Supaphol, P. & Rujiravanit R. (2005). Preparation and characterization of hexanoyl chitosan/polylactide blend films. *Carbohydrate Polymers,* Vol.60, No.3, (March 2005), pp. 343-350, ISSN 0144-8617

Penczek, S. et al., (2000). What we have learned in general from cyclic esters polymerization. *Macromolecular Symposia,* Vol.153, No.1, (March 2000), pp. 1-15, ISSN 1022-1360

Phong, L. et al. (2010). Properties and hydrolysis of PLGA and PLLA cross-linked with electron beam radiation. *Polymer Degradation and Stability,* Vol.95, No.5, (May 2010), pp. 771-777, ISSN 0141-3910

Pluta, M. et al., (2002) Polylactide/montmorillonite nanocomposites and microcomposites prepared by melt blending: Structure and some physical properties. *Journal of Applied Polymer Science,* Vol.86, No.6, (November 2002), pp. 1497-1506, ISSN 0021-8995

Qiu, Y. & Park, K. (2001). Environment-sensitive hydrogels for drug delivery. *Advanced Drug Delivery Reviews,* Vol.53, No.3, (December 2001), pp. 321-339

Quirk, R. A. et al., (2000). Surface engineering of poly(lactic acid) by entrapment of modifying species. *Macromolecules,* Vol.33, No.2, (January2000), pp. 258-260, ISSN 0024-9297

Quynh, T. M. et al., (2007). Properties of crosslinked polylactides (PLLA & PDLA) by radiation and its biodegradability. *European Polymer Journal,* Vol.43, No.5, (May 2007) pp. 1779-1785, ISSN 0014-3057

Quynh T. M. et al., (2008). The radiation crosslinked films based on PLLA/PDLA stereocomplex after TAIC absorption in supercritical carbon dioxide. *Carbohydrate Polymers,* Vol.72, No.4, (June 2008), pp. 673-681, ISSN 0144-8617

Quynh, T. M. et al., (2009). Properties of radiation-induced crosslinking stereocomplexes derived from poly(L-lactide) and different poly(D-lactide). *Polymer Engineering and Science,* Vol.49, No.5, (May 2009), pp. 970-976, ISSN 0032-3888

Raghoebar, G. M. et al., Resorbable screws for fixation of autologous bone grafts. *Clinical Oral Implants Research,* Vol.17, No.3, (June 2006), pp. 288-293, ISSN 0905-7161

Rasal, R. M. & Hirt, D. E. (2009). Toughness decrease of PLA-PHBHHx blend films upon surface-confined photopolymerization. *Journal of Biomedical Materials Research Part A,* Vol.88A, No.4, (March 2009), pp. 1079-1086, ISSN 1549-3296

Rasal, R. M.; Janorkar, A. V. & Hirt, D. E. (2010). Poly(lactic acid) modifications. *Progress in Polymer Science,* Vol.35, No.3, (March 2010), pp. 338-356, ISSN 0079-6700

Reddy, N.; Nama, D. & Yang, Y. (2008). Polylactic acid/polypropylene polyblend fibers for better resistance to degradation. *Polymer Degradation and Stability,* Vol.93, No.1, (January 2008), pp. 233-241, ISSN 0141-3910

Rhim, J. W. et al., (2006). Effect of the processing methods on the performance of polylactide films: Thermocompression versus solvent casting. *Journal of Applied Polymer Science,* Vol.101, No.6, (September 2006), pp. 3736-3742, ISSN 0021-8995

Riley, T. et al., (2001). Physicochemical evaluation of nanoparticles assembled from poly(lactic acid)-poly(ethylene glycol) (PLA-PEG) block copolymers as drug delivery vehicles. *Langmuir,* Vol.17, No.11, (May 2001), pp. 3168-3174, ISSN 0743-7463

Robinson, T. M.; Kieswetter, K. & McNulty, A. (2009). System and method for healing a wound at a tissue site. US Patent 0,216,170, filed February 27, 2009, issued August 27, 2009

Roney, C. et al., (2005). Targeted nanoparticles for drug delivery through the blood-brain barrier for Alzheimer's disease. *Journal of Controlled Release,* Vol.108, No.1-2, (November 2005), pp. 193-214, ISSN 0168-3659

Rujitanaroj, P.; Pimpha, N. & Supaphol, P. (2008). Wound-dressing materials with antibacterial activity from electrospun gelatin fiber mats containing silver nanoparticles. *Polymer,* Vol.49, No.21, (October 2008), pp. 4723-4732, ISSN 0032-3861

Sarazin, Y.; Schormann, M. & Bochmann, M. (2004). Novel zinc and magnesium alkyl and amido cations for ring-opening polymerization reactions. *Organometallics,* Vol.23, No.13, (May 2004), pp. 3296-3302, ISSN 0276-7333

Sawahata, K. et al., (1990). Electrically controlled drug delivery system using polyelectrolyte gels. *Journal of Controlled Release,* Vol.14, No.3, (December 1990), pp. 253-262, ISSN 0168-3659

Sethuraman, V. A. & Bae, Y. H. (2007). TAT peptide-based micelle system for potential active targeting of anti-cancer agents to acidic solid tumors. *Journal of Controlled Release,* Vol.118, No.2, (April 2007), pp. 216-224, ISSN 0168-3659

Sheth, M. et al., (1997) Biodegradable polymer blends of poly(lactic acid) and poly(ethylene glycol). *Journal of Applied Polymer Science,* Vol.66, No.8, (November 1997), pp. 1495-1505, ISSN 0021-8995

Shin, I. K. et al., (2010). Novel three-dimensional scaffolds of poly(L-lactic acid) microfibers using electrospinning and mechanical expansion: Fabrication and bone regeneration. *Journal of Biomedical Materials Research B: Applied Biomaterials,* Vol.95B, No.1, (October 2010), pp.150-160, ISSN 1552-4981

Singh, R. P. et al., (2003). Biodegradation of poly(ε-caprolactone)/starch blends and composites in composting and culture environments: The effect of compatibilization on the inherent biodegradability of the host polymer. *Carbohydrate Research,* Vol.338, No.17, (August 2003), pp. 1759-1769, ISSN 0008-6215

Stupack, D. G. et al. (October 2001). Apoptosis of adherent cells by recruitment of caspase-8 to unligated integrins. *The Journal of Cell Biology,* Vol.155, No.3, (2001), pp. 459-470, ISSN 0021-9525

Taguchi, S. et al., (2008). A microbial factory for lactate-based polyesters using a lactate-polymerizing enzyme. *Proceedings of the National Academy of Sciences,* Vol.105, No.45, (November 2008), pp. 17323-17327, ISSN 0027-8424

Taguchi, S. (2010). Current advances in microbial cell factories for lactate-based polyesters driven by lactate-polymerizing enzymes: Towards the further creation of new LA-based polyesters. *Polymer Degradation and Stability,* Vol.95, No.8, (January 2010), pp. 1421-1428, ISSN 0141-3910

Tajima, K. et al., (2009). Chemo-enzymatic synthesis of poly(lactate-co-(3- hydroxybutyrate)) by a lactate-polymerizing enzyme. *Macromolecules,* Vol.42, No.6, (February 2009), pp. 1985-1989, ISSN 0024-9297

Tokoro, R. et al., (2008). How to improve mechanical properties of polylactic acid with bamboo fibers. *Journal of Materials Science,* Vol.43, No.2, (January 2008), pp. 775-787, ISSN 0022-2461

Tomé, C. L. et al.,(2011). Transparent bionanocomposites with improved properties prepared from
acetylated bacterial cellulose and poly(lactic acid) through a simple approach. *Green Chemistry*, Vol.13, No.2, (February 2011), pp. 419-427, ISSN 1463-9262

Torchilin, V. P. (2006). Multifunctional nanocarriers. *Advanced Drug Delivery Reviews*, Vol.58, No.14, (December 2006), pp. 1532–1555, ISSN 0169-409X

Torchilin, V. P. (2008). Cell penetrating peptide-modified pharmaceutical nanocarriers for intracellular drug and gene delivery. *Biopolymers*, Vol.90, No.5, (September 2008), pp. 604-610, ISSN 0006-3525

Tyagi, P. et al., (2004). Sustained intravesical drug delivery using thermosensitive hydrogel. *Pharmaceutical Research*, Vol.21, No.5, (May 2004), pp. 832-837, ISSN 0724-8741

Vaupel, P.; Kallinowski, F. & Okunieff, P. (1989). Blood flow, oxygen and nutrient supply, and metabolic microenvironment of human tumors: A review. *Cancer Research*, Vol.49, No.23, (December 1989), pp. 6449–6465, ISSN 0008-5472

Venkatraman, S. S. et al., (2005). Micelle-like nanoparticles of PLA-PEG-PLA triblock copolymer as chemotherapeutic carrier. *International Journal of Pharmaceutics*, Vol.298, No.1, (July 2005), pp. 219-232, ISSN 0378-5173

Vila, A. et al., (2005). PLA-PEG particles as nasal protein carriers: The influence of the particle size. *International Journal of Pharmaceutics*, Vol.292, No.1-2, (March 2005), pp. 43-52, ISSN 0378-5173

Vink, E. T. H. et al., (2003). Applications of life cycle assessment to NatureWorks™ polylactide (PLA) production. *Polymer Degradation and Stability*, Vol.80, No.3, (2003), pp. 403-419, ISSN 0141-3910

Vleggaar, D. & Bauer, U. (2004). Facial enhancement and the European experience with Sculptra. *Journal of Drugs in Dermatology*, Vol.3, No.5, (September-October 2004), pp. 542-547, ISSN 1545-9616

Wan, Y. Z. et al., (2001). Influence of surface treatment of carbon fibers on interfacial adhesion strength and mechanical properties of PLA-based composites. *Journal of Applied Polymer Science*, Vol.80, No., (April 2001), pp. 367-376, ISSN 0021-8995

Wang, M. (2007). Surface modification of biomaterials and tissue engineering scaffolds for enhanced osteoconductivity, *Proceedings of IFMBE*, 978-3-540-68016-1, Kuala Lumpur, Malaysia, December 2006

Wang, R. Y. et al., (2009). Morphology, mechanical properties, and durability of poly(lactic acid) plasticized with di(isononyl) cyclohexane-1, 2-dicarboxylate. *Polymer Engineering and Science*, Vol.49, No.12, (December 2009), pp. 2414-2420, ISSN 0032-3888

Wang, S.; Cui, W. & Bei, J. (2005). Bulk and surface modifications of polylactide. *Analytical and Bioanalytical Chemistry*, Vol.381, No.3, (February 2005), pp. 547-556, ISSN 1618-2642

Woo, S. I. et al., (1995) Polymerization of aqueous lactic acid to prepare high molecular weight poly(lactic acid) by chain-extending with hexamethylene diisocyanate. *Polymer Bulletin*, Vol.35, No.4, (May1995), pp. 415-421, ISSN 0170-0839

Wu, C. S. & Liao, H.T. (2007). Study on the preparation and characterization of biodegradable polylactide/multi-walled carbon nanotubes nanocomposites. *Polymer*, Vol.48, No.15, (July 2007), pp. 4449-4458, ISSN 0032-3861

Wu, J. C. et al., (2005). Ring-opening polymerization of lactide initiated by magnesium and zinc alkoxides. *Polymer*, Vol.46, No.23, (November 2005), pp. 9784-9792, ISSN 0032-3861

Wu, X. L. et al., (2010). Tumor-targeting peptide conjugated pH-responsive micelles as a potential drug carrier for cancer therapy. *Bioconjugate Chemistry*, Vol.21, No.2, (February 2010), pp. 208-213, ISSN 1043-1802

Xiong, X. Y. et al., (2007). Vesicles from Pluronic/poly(lactic acid) block copolymers as new carriers for oral insulin delivery. *Journal of Controlled Release*, Vol.120, No.1-2, (July 2007), pp. 11-17, ISSN 0168-3659

Xu, C. et al., (2009). Preparation and characterization of polylactide/thermoplastic konjac glucomannan blends. *Polymer*, Vol.50, No.15, (July 2009), pp. 3698-3705, ISSN 0032-3861

Yang, G. et al., (2011). Composite from modified biomass fiber and polylactide (PLA) and the method of preparation. CN Patent 201010510498, filed October 15, 2010, and issued January 26, 2011

Yang, S. L. et al., (2008) Thermal and mechanical properties of chemical crosslinked polylactide (PLA). *Polymer Testing*, Vol.27, No.8, (December 2008), pp. 957-963, ISSN 0142-9418

Yoda, S.; Bratton, D. & Howdle, S. M. (2004). Direct synthesis of poly(L-lactic acid) in supercritical carbon dioxide with dicyclohexyldimethylcarbodiimide and 4-dimethylaminopyridine. *Polymer*, Vol.45, No.23, (October 2004), p. 7839-7843, ISSN 0032-3861

Yu, D. G.; Lin, W. C. & Yang, M. C. (2007). Surface modification of poly(L-lactic acid) membrane via layer-by-layer assembly of silver nanoparticle-embedded polyelectrolyte multilayer. *Bioconjugate Chemistry*, Vol.18, No.5, (September-October 2007), pp. 1521-1529, ISSN 1043-1802

Zhang, L. et al., (2004). Camptothecin derivative-loaded poly(caprolactone-co-lactide)-b-PEG-b-poly(caprolactone-co-lactide) nanoparticles and their biodistribution in mice. *Journal of controlled release*, Vol.96, No.1, (April 2004), pp. 135-148, ISSN 0168-3659

Zhang, Z. & Feng, S. S. (2006). Nanoparticles of poly(lactide)/vitamin E TPGS copolymer for cancer chemotherapy: Synthesis, formulation, characterization and in vitro drug release. *Biomaterials*, Vol.27, No.2, (January 2006), pp. 262-270, ISSN 0142-9612

Zhong, Z. et al., (2003). Single-site calcium initiators for the controlled ring-opening polymerization of lactides and lactones. *Polymer Bulletin*, Vol.51, No.3, (November 2003), pp. 175-182, ISSN 0170-0839

Zhu, A. et al., (2002). Covalent immobilization of chitosan/heparin complex with a photosensitive hetero-bifunctional crosslinking reagent on PLA surface. *Biomaterials*, Vol.23, No.23, (December 2002), pp. 4657-4665, ISSN 0142-9612

Zhu, H.; Ji, J. & Shen, J. (2002). Surface engineering of poly(DL-lactic acid) by entrapment of biomacromolecules. *Macromolecular Rapid Communications*, Vol.23, No.14, (October 2002), pp. 819-823, ISSN 1022-1336

Zhu, Y. et al., Immobilization of biomacromolecules onto aminolyzed poly(L-lactic acid) toward acceleration of endothelium regeneration. *Tissue Engineering*, Vol.10, No.1-2, (January-February 2004), pp. 53-61, ISSN 1076-3279

4

Arterial Mass Transport Behaviour of Drugs from Drug Eluting Stents

Barry M. O'Connell and Michael T. Walsh
Centre for Applied Biomedical Engineering Research (CABER),
Department of Mechanical Aeronautical and Biomedical Engineering,
and Materials and Surface Science Institute,
University of Limerick, Limerick,
Ireland

1. Introduction

Coronary artery disease (CAD) is the foremost cause of morbidity in the worlds industrialised nations. Consequently our ability in understanding the treatments available and the mechanisms of their success/failure is of particular importance as we strive to improve procedural success rates through the evolution of existing and the innovation of new interventional technologies. Today, the gold standard for treating CAD is to deploy a drug eluting stent (DES) in the blocked artery to first restore luminal blood flow and second to resist the body's tendency to block the artery once more through its overly aggressive healing response known as restenosis. Intrinsically, the understanding of mass transport is elemental to all aspects of DES design from the type of drug used and the polymer release characteristics to the shape and thickness of the stent struts. Design optimisation of DES is vital to achieving increases in procedural success rates as the technology evolves, as is the case for its newest embodiment, the biodegradable DES.

This chapter addresses issues fundamental to drug mass transport from DES, starting with an overview of CAD and associated interventional procedures. DES mass transport theory is then presented and followed by accounts of mass transport problem classifications employed by researchers under a number of deployment scenarios. The importance of experimental mass transport validation is highlighted and a computational investigation is developed to demonstrate how differences in DES deployment conditions alter drug concentrations in the artery wall.

2. Coronary artery disease

Atherosclerosis is a degenerative disease that affects coronary, carotid and other peripheral arteries in the body. Disease formation can occur as early as childhood with the development of fatty streaks within the artery wall. As the aging process progresses, these fatty streaks accumulate to become larger lipid deposits within the artery. This gradual propagation of plaque can be detrimental to the smooth operation of the vasculature. Occlusions ensuing from aggressive atherosclerotic plaque progression can often culminate in an ischemic attack, such as a stroke or a heart attack. CAD pertains to a blockage or narrowing of the coronary arteries that provide oxygen and nutrients vital to the smooth operation of the heart muscle. Once identified as such, there are a number of interventional procedures available to the cardiologist but the successful emergence of stents, and more recently DES, has seen them become the preferred choice of treatment for CAD, so much so that by the beginning of 2006 more than 8 out of 10 coronary stents were DES (Head et al., 2007) at a cost of between $4 and $5 billion annually (Kaul et al., 2007).

CAD has been intrinsically linked to atherosclerosis since the early 20th century (Chen et al., 2005) and refers to the localisation of disease within the coronary arteries. It is generally asymptomatic and those afflicted often only realise they have the condition when it manifests itself in the form of a heart attack. For this reason CAD is the foremost cause of mortality in the world's industrialised nations (Khakpour &Vafai, 2008). CAD alone is reportedly responsible for approximately 700,000 deaths in the United States of America annually (Kaazempur-Mofrad et al., 2005). Lifestyle choices made by an individual, such as not smoking, regular exercise and a balanced diet, have been shown to influence CAD development but it is the concentration of lipid rich cholesterol in the blood that is considered the most important factor (Sun et al., 2006).

2.1 Traditional interventional procedures

Numerous ways exist to alleviate a stenosis in a coronary artery once detected. The first such interventional procedure commonly practiced was coronary artery bypass graft (CABG) surgery. CABG surgery was successfully performed first by Robert H. Goetz and his team in 1960 at the Albert Einstein College of Medicine (Haller & Olearchyk, 2002). Prior to surgery an angiogram is conducted in order to locate the occlusion within the artery, after which a median sternotomy is performed, which exposes the heart and enables the blocked coronary arteries to be bypassed. This procedure is traumatic for the patient with extensive recovery times and significant scarring to the chest. Furthermore, the long term patency rates of these grafts were moderate and several efforts have been made in vain to optimise downstream graft artery junctions. Despite this, CABG remained the gold standard in the treatment of CAD until 1977 when Andreas Gruntzig first performed percutaneous transluminal coronary angioplasty (PTCA) (Kukreja et al., 2008). CABG is a procedure still used today as not every patient is eligible for minimally invasive surgeries such as PTCA due to highly tortuous or extensively blocked arteries.

Initially PTCA was welcomed by the clinical community due to its minimally invasive approach to arterial stenosis alleviation. A balloon catheter is introduced through an incision in the femoral artery and is manoeuvred through the vasculature until it reaches the stenosis. Once inflated, the balloon pushes the plaque back against the artery wall and enables blood flow to recommence after it has been deflated. The initial success of PTCA was short lived as investigators soon discovered that a substantial percentage of patients, reported to be anywhere between 30% and 60%, experienced recurrent ischemia due to the re-blocking of the artery (restenosis) within the first 6 months. This was attributed to mechanical injury caused by over dilating a device within the vessel (Head et al., 2007).

The next major advance in the field of minimally invasive interventional cardiology came in the early 1990's with the advent of the coronary artery stent (CAS). Prior to surgery a cylindrical metallic scaffold, or stent, is placed on the end of an existing balloon catheter and deployed in the same way as traditional PTCA. Initially, these stents were mounted on the balloon catheter by the physician, however, in more recent times the manufacturers supply the catheters with a stent already in situ. After deployment the stent remains within the artery in an attempt to retain arterial patency. CAS reduced failure rates tobetween 10% and 40% (Duraiswamy et al., 2007; Mongrain et al., 2007) through the elimination of elastic recoil and negative remodelling of the artery associated with PTCA (Costa and Simon, 2005).

2.2 Restenosis and the advent of the drug eluting stent era

Restenosis can best be described as an overly aggressive inflammatory healing response in the artery wall due to the mechanical injury inflicted by balloon/stent expansion. It can be quantified by the reduction of lumen size after an intravascular interventional procedure. The development of restenosis can be described by three processes after PTCA; 1) elastic recoil, 2) arterial negative remodelling and 3) neointimal hyperplasia (Rajagopal&Rockson, 2003). Elastic recoil can occur within an hour of PTCA and is due to passive recoil of the elastic medial layer of the artery. Arterial remodelling on the other hand can be both positive (vessel enlargement) or negative (vessel shrinking) and is characterised as such by a change in vascular dimension. Investigators report contrasting views on the mechanisms behind negative remodelling but whatever the underlying pathology behind vascular remodelling, it is believed to be virtually eliminated when angioplasty is used in conjunction with a stent.

Over inflation of a balloon catheter can result in the fracture of atherosclerotic plaque and in some cases can cause partial fracture of the artery wall (Schwartz et al., 2004). The same crushing/fracturing effect is witnessed when a stent is used in conjunction with an over inflated balloon. However, a stent can also cause excessive injury by penetrating the media which in turn increases neointimal formation. In some extreme cases stents have been known to penetrate as deep as the adventitial layer of the artery (Costa and Simon, 2005). The introduction of the DES to market has gone someway to alleviating the issue of arterial restenosis and excessive vessel injury via stent expansion. Variations of anti-restenotic drugs have been used to coat the stent in order to prevent post-operative in stent restenosis (ISR) and these modern stents can have strut profiles in the order of 80μm which would minimise the possibility of adverse artery wall penetration.

It is generally accepted that one of the main causes of restenosis following BMS implantation is SMC proliferation from the medial artery layer to the injured site. Attempts at systemic drug delivery to inhibit restenosis after stenting failed because effective dosing levels had a toxic effect and could not be tolerated by the patients (Waksman, 2002). Therefore the concept of local drug delivery was developed to redress the issue through the application of a drug eluting coating to the stent platform. This enables site specific local delivery of drugs that can be applied to the injured vessel at the exact location and time that damage occurs. The anti-restenotic coating on DES inhibits the formation of neointimal hyperplasia via suppression of the inflammatory reaction, platelet activation and SMC proliferation, curbing the overly aggressive healing response. Most of the early drugs explored originally were those used as agents for anti-transplant rejection or immunosuppressive drugs (van der Hoeven et al., 2004).

In April 2003 the first DES to gain commercial approval from the Food and Drug Administration (FDA) in the United States was the Cypher stent, which was developed by Cordis Corporation (Miami, FL. USA). The drug used on the Cypher stent is called sirolimus. Boston Scientific's (Natik, MA. USA) TAXUS family of stents were the second DES platform approved by the FDA in March of the following year. The drug employed on the TAXUS stent is called paclitaxel (Venkatraman and Boey, 2007).The first generation DES had a profound effect on reducing restenosis rates compared to bare metal stent (BMS) models. Clinical trials carried out on the Cypher stent (SIRIUS-1) showed restenosis rates of 8.9% after 8 months compared to 36.6% for BMS in the same study. Likewise the TAXUS IV trials heralded a dramatic reduction in restenosis rates when compared to BMS after 9 months, 7.9% versus 26.6% respectively (Venkatraman and Boey, 2007).

A successful DES procedure is defined by its ability to transport the right amount of an appropriate drug within the correct timeframe that will ultimately deem the operation a success through the prevention of ISR. There are aspects of the performance of a DES that can be controlled by the manufacturer which has led to considerable reductions in the instances of ISR, such as the type of drug used and the characteristics of the coating that the drug is stored in. Even the stent shape, thickness and width of the struts can all influence the manner in which the drug is transported. However, the ability for drug uptake within the artery wall is also governed by its interaction with the patient specific arterial environment, making it near impossible to completely eradicate ISR. The degree of initial stenosis, the presence of luminal and abluminal thrombus on the stent and even the advent of re-endothelialisation will all contribute to the DES ability to transport drugs throughout the artery wall.

3. Mass transport theory of drug eluting stents

Mass transport refers to the movement of mass, i.e. the species of interest which is drugs in the case of a DES, within a defined system. This transport of species may be provoked by concentration gradients between two points, but quite often in systems, especially in the vasculature, overpowering complex flow dynamics will ultimately be responsible for the mass transport outcome. In the absence of a free flowing system the presence of these concentration gradients induces diffusion, e.g. between the DES and the artery wall. Mass transport can be broken up into two types within the human vasculature. Firstly blood side mass transport (BSMT) refers to species transport within the vessel lumen and is subject to the haemodynamics therein. Often evanescent due to haemodynamic washout, BSMT can only be effective in transporting anti-proliferative agents to the wall in regions of high recirculation.

The second, and most important, mode of mass transport is in relation to transport within the wall of the artery, referred to as wall side mass transport (WSMT). Along with the properties of the species being transported within the artery wall, WSMT depends on the structural condition of the wall itself, whereby a damaged intimal layer could facilitate accelerated mass transport through to the medial layer. WSMT can be governed by two transport forces, a pressure driven convective force and a diffusive force. The Peclet number (Pe), see equation 11, is a dimensionless parameter that can be used to determine the relative influences of these two forces. A small Pe (i.e. Pe<1) is representative of transport which is dominated by diffusion, while a higher Pe (i.e. Pe>1) indicates convection dominated mass transport (Friedman, 2008).

3.1 Governing equations

Computational Fluid Dynamics has emerged as one of the most powerful numerical tools for engineers, scientists and mathematicians alike. Its foundations are based on theoretical analysis drawn from experimental observations over various branches of physics. The starting point for any computational analysis is the appropriate allocation of the governing equations. These equations are then substituted with equivalent numerical descriptions that are then solved using appropriate mathematical techniques. There are a number of numerical techniques available that will return a solution to a specified problem. Two of the more popular methods are the Finite Volume Method and the Finite Element Method. The assumptions generally applied when modelling fluid flow problems of this nature are as follows:

- The flow is incompressible and isothermal
- The fluid is Newtonian and possesses constant physical properties
- Flow is considered to be laminar

3.1.1 Conservation of mass: Continuity equation

The conservation of mass is a form of continuity equation which states the net mass flow into a control volume is equal to the rate at which mass leaves the control volume. That is providing there are no sinks or sources within the control volume. The differential form of the equation can be obtained by simply considering the flow into and out of elementary control volume. For the Cartesian co-ordinate system, having coordinates x, y, z referenced to a stationary frame with the corresponding velocity components u, v, w (m/s), the continuity equation can be written as:

$$\frac{\partial \rho}{\partial t} + \frac{\partial(\rho u)}{\partial x} + \frac{\partial(\rho v)}{\partial y} + \frac{\partial(\rho w)}{\partial z} = 0 \tag{1}$$

Where the density (ρ, kg/m³) is a constant, as is the case of incompressible flow, this reduces further to a volume continuity equation.

$$\frac{\partial u}{\partial x} + \frac{\partial v}{\partial y} + \frac{\partial w}{\partial z} = 0 \tag{2}$$

3.1.2 Balance of momentum: Navier-Stokes Equations

The balance of momentum is derived from Newton's second law of motion, which states that the rate of change of momentum of a fluid particle is equal to the sum of the forces on the particle. The Navier-Stokes Equations describe the full three dimensional, viscous nature of fluid motion in a control system:

$$\rho \frac{\partial u}{\partial t} = -\frac{\partial P}{\partial x} + \mu \left(\frac{\partial^2 u}{\partial x^2} + \frac{\partial^2 u}{\partial y^2} + \frac{\partial^2 u}{\partial z^2} \right)$$

$$\rho \frac{\partial v}{\partial t} = -\frac{\partial P}{\partial y} + \mu \left(\frac{\partial^2 v}{\partial x^2} + \frac{\partial^2 v}{\partial y^2} + \frac{\partial^2 v}{\partial z^2} \right) \tag{3}$$

$$\rho \frac{\partial w}{\partial t} = -\frac{\partial P}{\partial z} + \mu \left(\frac{\partial^2 w}{\partial x^2} + \frac{\partial^2 w}{\partial y^2} + \frac{\partial^2 w}{\partial z^2} \right)$$

Where μ(Pa s) is dynamic viscosity and P(Pa) is pressure. Due to the porous nature of the artery wall, flow within it must consider the influence of the tissues permeability. Therefore the flow within the wall is assumed to follow Darcy's law and is demonstrated (in the x direction) by the inclusion of the permeability term in equation 4, where K(m²) is the permeability of the arterial tissue.

$$\rho \frac{\partial u}{\partial t} = -\frac{\partial P}{\partial x} + \mu \left(\frac{\partial^2 u}{\partial x^2} + \frac{\partial^2 u}{\partial y^2} + \frac{\partial^2 u}{\partial z^2} \right) - \left(\frac{\mu}{K} \right) u \tag{4}$$

3.1.3 Fick's laws of diffusion

Species transport via diffusion is a process driven by concentration gradients between two locations. Fick's first law can be used to describe the diffusional flux (J_x, mol/m²s) of such species, shown in 1D in equation 5, where D(m²/s) is diffusivity and c is concentration (mol/m³):

$$J_x = -D\frac{\partial c}{\partial x} \tag{5}$$

The negative term in equation 5 indicates that the flux is positive in the presence of a negative concentration gradient. Biological mass transport often requires the application of a time-dependent mass transport process that can predict variations in concentration over time. Fick's second law (equation 6) can provide such a relationship and is defined here in one dimension:

$$\frac{\partial c}{\partial t} = D\frac{\partial^2 c}{\partial x^2} \tag{6}$$

3.1.4 Conservation of species: Convection-diffusion equation

The addition of a convective term, equal to the product of the fluid velocity and the local concentration, to equation 6 demonstrates the 3D transport of species in a flowing solution. This is known as the convection-diffusion equation.

$$\frac{\partial c}{\partial t} + u\frac{\partial c}{\partial x} + v\frac{\partial c}{\partial y} + w\frac{\partial c}{\partial z} = D\left(\frac{\partial^2 c}{\partial x^2} + \frac{\partial^2 c}{\partial y^2} + \frac{\partial^2 c}{\partial w^2}\right) \tag{7}$$

3.1.5 Ratio of convective to diffusive forces

The Peclet number (Pe) is a dimensionless number that determines the relative contribution of convective and diffusive forces to species transport within a defined system. It can be defined as a product of the Reynolds number (Re) and the Schmidt number (Sc).

$$Pe = Re.Sc \tag{8}$$

The Reynolds number is a non-dimensional parameter concerning fluid forces due to viscosity and inertia and is essentially used to determine whether a flow is laminar, transitional or turbulent in nature. For example a Reynolds number of approximately 90 can be obtained for a mean arterial velocity (u) of 0.1m/s in an artery with a diameter (a) of 3mm. When considering transmural flow through the porous artery wall the value a would represent the thickness of the porous wall.

$$Re = \frac{\rho u a}{\mu} \tag{9}$$

The Schmidt number (Sc) is defined as the ratio of kinematic viscosity (v, m²/s) to diffusivity (D).

$$Sc = \frac{v}{D} = \frac{\mu}{\rho D} \tag{10}$$

Substituting equations 9 and 10 into 8 describes how convective and diffusive forces can influence the outcome of the Peclet number.

$$Pe = \frac{ua}{D} \tag{11}$$

3.1.6 Diffusion in porous materials

When considering diffusion in a fluid saturated porous media, as is the case with the artery wall, diffusion takes place over a tortuous path. Because these pores are not straight, the distance over which diffusion takes place becomes effectively longer than for a homogenous material of the same thickness. The effective diffusivity (D_{eff}) can therefore be deduced by considering the impact of the materials structure on the species free diffusivity (D_{free}). The effective diffusivity of a porous material is a function of its porosity (ε) and tortuosity (τ).

$$D_{eff} = \frac{\varepsilon}{\tau} D_{free} \tag{12}$$

One of the more common ways to determine the free diffusivity of a species in a solvent is to use the Stokes-Einstein equation (13), where k(J/K) is the Boltzmann constant, T(K) is temperature and R(m) is the radius of the solute. For the purpose of diffusion in the artery wall, this solvent is considered to be plasma.

$$D_{free} = \frac{kT}{6\pi\mu R} \tag{13}$$

The radius of the solute can be calculated from equation 14 assuming that the particle is spherical in shape, where M(Kg/mol) is the solute molecular weight and Na(mol^{-1}) is Avogadro's number.

$$R = \left(\frac{3M}{4\pi\rho Na} \right)^{\frac{1}{3}} \tag{14}$$

The structure of the porous medium is defined by the tortuosity(τ) of its porous network (15) and by the porosity(ε) (16) of the material itself.

$$\tau = \frac{L}{X} \tag{15}$$

Where L = pore path length and X = distance between beginning and end of the pore path.

$$\varepsilon = \frac{Pore\,Volume}{Total\,Volume} \tag{16}$$

4. Problem classification

In reality the classification of problems of this nature are inherently patient specific and as such no one representation of the problem is correct. However, there are innate similarities between patients. Blood flow within the vasculature is a highly complex 3D process to model given the pulsatile nature of arterial haemodynamics. Coupled with this pulsatile process, the coronary arteries are situated on the surface of the heart and as such are subject to cyclic motion due to the beating of the organ. Therefore the modelling of drug transport from a DES in these arteries is multifaceted in nature, comprising of both luminal and artery wall mass transport, the latter of which may also be subject to a reaction giving that some drug may bind to the arterial tissue. The introduction of a multi-layered artery wall to the model increases the complexity of the domain even further. So to what extent does one go to when modelling DES mass transport?

Comprehensively modelling the behaviour of a DES computationally over a given time period would require the application of the following *in vivo* conditions experience by the device.

- The DES would have to be placed in multi-layered diseased artery.
- Both BSMT and WSMT would have to be considered.
- The real time occurrence of thrombus formation and re-endothelialisation under pulsating flow conditions would need to be modelled.
- The structural deformation of the artery wall due to DES deployment would need to be taken into account.
- Also the movement of the vessel in space due to its location on the surface of the beating heart and how this may alter depending on the extent of the patient's physical activeness would have to be considered.

In light of the computational requirements to undertake such a model it is possible, and almost necessary, to make certain assumptions in order to simplify both BSMT and WSMT models whilst retaining enough detail of the actual model to draw relevant conclusions from the analysis.

The implementation of an arterial pulse and a beating heart are neglected by most researchers. Often the artery is modelled as rigid in space in order to analyse mass transport post DES deployment. This is an effective assumption but one must consider the deformation of the artery wall due to the dynamic expansion of the stent, as this can have an impact on the mass transport outcome due to the porous nature of the wall and the compression it incurs upon stent expansion. As for the application of laminar blood flow, it can be seen that the majority of drug that enters the artery wall from the DES does so via physical contact with the wall and the drugs emanating from the areas of the stent exposed to flow, be it laminar or pulsatile, are predominantly carried downstream.

4.1 Artery wall classification

Arteries transport oxygen rich blood around the body providing essential nutrients to vital organs. The artery wall consists of a complex multilayer porous substructure with an interstitial phase comprising predominantly of plasma. In a healthy artery this substructure (Figure 1) is comprised of three concentric layers; the tunica intima, the tunica media and the tunica adventitia. The tunica intima is the innermost layer, consisting of a single layer of endothelial cells and a subendothelial layer mainly consisting of delicate connective tissues and collagen fibres. The outer boundary of the tunica intima is surrounded by an elastic tissue with fenestral pores known as the internal elastic lamina (IEL). The medial layer consists primarily of concentric sheets of smooth muscle cells (SMC) within a loose connective tissue framework. This configuration of SMC enables the artery wall to contract and relax. The tunica media and the tunica adventitia are separated by another thin band of elastic fibres known as the external elastic lamina (EEL). The outermost layer of the artery, the tunica adventitia, is comprised of connective tissue fibres and some capillaries. These fibres blend into the surrounding connective tissues and aid in stabilising the arteries within the body (Khakpour and Vafai, 2007).

The target layer for the anti-restenotic drugs is the tunica media, where the SMC reside, and quite often computational studies will consider just this arterial layer not only because of this fact but also due to the possible erosion of the tunica intima upon stent deployment. Regardless of the level of complexity modelled, the artery wall is porous in composition and drug transport is facilitated through the surrounding plasma not only via diffusion but there is also the presence of a transmural velocity due to a pressure gradient observed across the artery wall. However, the presence of arterial plaque will reduce the magnitude of this transmural velocity and can even stem it altogether. As DES are deployed in highly occluded arteries it is reasonable to reduce the complexity of the problem by neglecting convection in the wall. Equation 12 gives us an indication of how arterial properties such as porosity, tortuosity and free diffusivity can influence the transport of drugs within the respective artery wall layers. The compression of these layers will alter these properties which in turn may inhibit the transport of species as governed by the mass transport equations. The compression of a porous structure not only reduces the materials porosity but it results in the creation of a more arduous pore path over which mass transport would normally occur. The combination of a reduced porosity with an increased tortuosity, when the artery wall has been compressed, has a net effect of reducing the effective diffusivity thus hindering mass transport within the vessel.

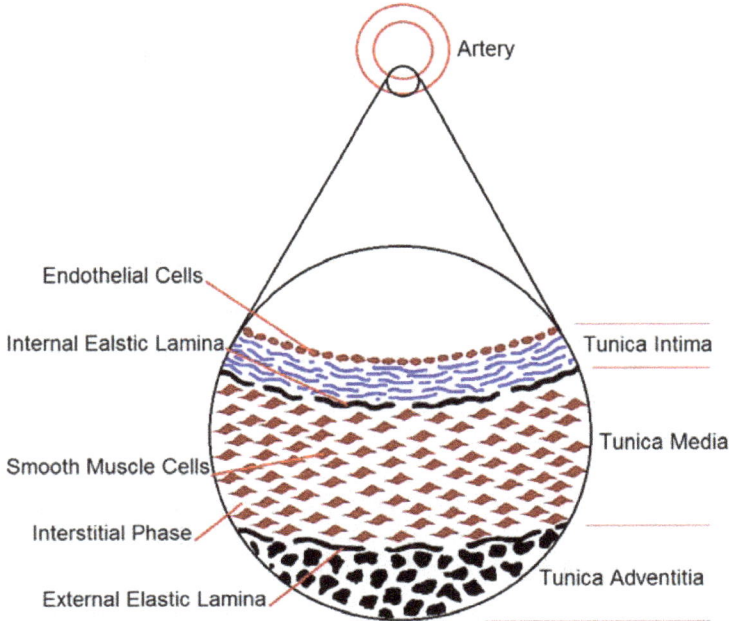

Fig. 1. Illustration of the cross-sectional structure of a healthy artery wall.

4.2 Influence of thrombus

Hwang et al. (2005) were among the first to explore the influence of thrombus height, width and type on the arterial drug uptake. Stents can be deployed at sites of thrombus and after implantation a clot will inevitably develop once the struts become covered with plasma proteins. In most cases this will not be angiographically present or clinically evident but even a fine layer of clotting blood deposited on the surface of the DES can alter drug distribution within the wall. The presence of clot alters the local environment of the stent strut and the physiological transport forces that regulate arterial uptake and retention. Balakrishnan et al.(2008) reported that drug eluting stents clot at a rate of 0.6% each year after implantation for up to 3 years. Strut position within a clot also has a major influence on the arterial uptake. The greater the volume of clot covering the strut, and the closer the strut is to the wall, creates improved conditions for greater drug delivery. Hwang et al. (2005) discovered that in this configuration concentration distribution in the wall can be 30 fold higher than situations where no clot is present. Similarly thrombus or plaque between the strut and the artery wall act as a buffer layer and reduce wall concentrations. Clot diffusivities higher than that of the artery wall will result in drug transfer to the blood at a rate faster than can be absorbed by the wall. Clots with diffusivities equal to or lower than the artery wall can transport drugs to the wall at a rate where the wall can effectively absorb the drugs, thus reducing drug loss to the bloodstream.

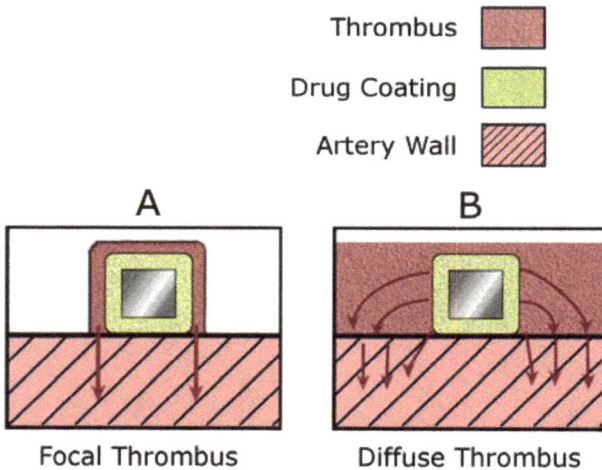

Fig. 2. An illustration of how Focal (A) and Diffuse (B) thrombus, surrounding a DES strut, contribute to drug mass transport within the artery wall.

In 2008, Balakrishnan et al. demonstrated how variations in thrombus size and distribution contribute to fluctuating arterial drug intensities (Figure 2). Their simulations indicate that thrombus cannot influence the slow rate of drug release from the stent because the polymer resistance to drug transport is significantly greater than that of the thrombus, which is consistent with *in vivo* experimental drug release. Focal thrombus, with a thickness of 0.1mm, increased peak average drug concentration by 80%. Greater clot formation between stent struts will also have an effect on arterial concentrations because once the species is transported within the clot it effectively increases the surface area from which the artery wall can absorb drugs. The formation of this interstrut thrombus, described as diffuse thrombus, acts as a shield from drug washout and culminates in an increase in arterial concentration by up to 3.5-fold.

The variability of thrombus can have a major impact on arterial drug concentrations. It can aid in drug uptake and retention within the wall when it covers the stent, but too much thrombus will effectively block the artery, thus creating a problem that the DES aimed to alleviate. Also it can act as a barrier in preventing drugs from reaching the wall when it is located between the strut and the wall. The likely scenario following DES implantation within the vasculature is that at some location along the length of the stent each of these situations will be present. In addition to issues arising from the development of thrombus, the presence of plaque in a freshly stented artery will have a similar effect on mass transport but does so immediately upon implantation, as opposed to thrombus which develops with time. The initial presence of the plaque coupled with the time dependent formation of thrombus, and even the occurrence of re-endothelialisation, would create a realistic stenting scenario if modelled but the volume and formation of these will be different for each person. So to what extent does one model this?

4.3 Polymer and drug characterisation

Stent coatings play a vital role in the regulation of drug release from the stent. Careful consideration must be observed when allocating a polymer as a drug carrier for a DES. If the polymer is not biocompatible an inflammatory response ensues. It has been shown that the development of neointimal hyperplasia can be doubled when certain polymer configurations are used compared with a controlled substance (Granada et al., 2003). Each biologically viable polymer must be able to endure the stresses exerted when stents are deployed, resist cracking, peeling and maintain their physiochemical properties after sterilisation. Drug release rates can be altered with the addition of an extra layer of polymer coating to modulate between slow and fast release formulations. This adds an extra degree of complexity when designing functional DES and indeed analysing them computationally.

Sirolimus, everolimus and paclitaxel were among the first anti-restenotic agents used in DES. Sirolimus, or rapamycin, is a naturally occurring macrocyclic lactone. It was approved by the FDA in 1999 for the prophylaxis of renal transplant rejection because of its potent immunosuppressive properties (Costa & Simon, 2005). Essentially sirolimus inhibits the activation of multiple kinases, associated with cell proliferation, resulting in the cessation of cellular division between the G1/S phases and is said to be cytostatic in nature (Burt & Hunter, 2006). Everolimus is an analogue of the sirolimus immunosuppressant which binds to cytosolic immunophillin FKBP12. It is very similar to sirolimus in that it prevents the cellular division between the G1/S phases of the cell cycle therefore inhibiting SMC proliferation (Stone et al., 2008).

Paclitaxel is a naturally occurring drug that was originally extracted from the pacific yew tree Taxusbrevifolia. It was initially used to treat several types of cancer such as breast and ovarian cancer (Tanabe et al., 2004). Like sirolimus and everolimus, paclitaxel is effective in reducing restenosis but it does so in a manner that results in cell death, suggesting that it works through a cytotoxic mechanism (Parry et al., 2005). Paclitaxel achieves its anti-proliferative effects by binding to aminoterminus β-tubulin thus disrupting microtubule dynamics. This results in the arrest of cells at the M stage and even G2 stages of the cell cycle, leading to cell death (Hara et al., 2006). Due to its lipophilic properties paclitaxel has been known to be loaded directly onto stents without a polymer coating but there could be potential implications with regards to a lack of controlled release (Burt & Hunter, 2006).

4.4 DES problem design and deployment

This problem can be approached in one of two ways, do you a) want to compare the mass transport capabilities between different stent designs or b) want to analyse arterial mass transport for a single stent design under various stenting and artery wall conditions. The latter of which is of more initial interest because a comprehensive understanding of a multifaceted mass transport study with a generic stent design will give a greater understanding of the interactions between drug/polymer characteristics and the arterial condition. Once these interactions are better understood the researcher can revert to comparing stent designs for a predetermined deployment configuration.

The main goal of a DES is to prevent the onslaught of arterial restenosis, which occurs in part due to damage inflicted on the artery during stent deployment. However, researchers to date have generally neglected the artery wall damage induced and its influence on mass transport. An exploratory DES mass transport computational study, even if it is only 2D modelling, should consider both cause and effect. A stent should not just be placed flush with the artery wall, there is going to be some wall indentation and intimal damage, the extent of which is a study onto itself. To this end the resulting artery wall compression will alter the effect that is the transport of the anti-restenotic drugs throughout the artery wall. Stent design and drug/polymer properties, although of significant importance, should be a secondary consideration until these fundamental issues have been addressed.

5. Typical computational boundary conditions for DES models

Once the governing equations have been applied to the model domain the boundary conditions need to be allocated. Often with biological modelling it is necessary to make assumptions when applying boundary conditions. For example treating the artery wall as rigid (Mongrain et al., 2007; Devereux, 2005; Kaazempur-Mofrad and Ethier, 2001) or assuming that mass transport within the wall is modulated solely by diffusion (Balakrishnan et al., 2008, 2007, 2005; Mongrain et al., 2007, 2005) are two examples of ways commonly employed to simplify what is in reality a very complex problem. However, as previously mentioned the fundamental *in vivo* issues should be taken into account as much as possible when applying such simplifications.

5.1 Application of momentum boundary conditions
5.1.1 Inlet: Velocity

The heart is a muscular organ that undergoes repetitive contraction and relaxation of its walls in order to propel blood through the circulatory system. Coupled with the complex geometry of the coronary arteries, the pulsating blood velocity profile is an integral part in the mass transport behaviour of blood borne species. However, when considering the transport of drugs from a DES a common assumption to make is the presence of a steady fully developed flow profile within the lumen (Balakrishnan et al., 2008, 2007, 2005; Mongrain et al., 2007, 2005). When modelling BSMT it may be necessary to incorporate the time-dependent pulsatile nature of blood flow. However, the transient nature of blood side drug transport enables the assumption of a steady flow inlet boundary condition that in most cases will not have a considerable impact on WSMT. Arterial flow can be replicated by applying a pulsatile parabolic velocity profile at the vessels inlet.

5.1.2 Outlet: Pressure

A pressure of zero can be applied to the outlet of DES computational models which reduces the likelihood of encountering backflow through the outlet. The application of such a boundary condition is prevalent in mass transport studies of this nature (Kolachalama et al., 2009; Balakrishnan et al., 2008, 2007, 2005; Rajamohan et al., 2006).

5.1.3 Lumen-wall interface: No-slip

The no-slip boundary condition is standard for a model with a stationary wall and states that the velocity of the fluid is zero relative to the boundary (O'Brien et al., 2005; Walsh et al., 2003).

5.1.4 Axial-symmetry

Due to the idealised nature of some arterial computational models the use of an axial-symmetry boundary condition can greatly reduce the number of degrees of freedom that need to be solved for. This will reduce the computational demand without having to sacrifice the accuracy of the analysis. The Reynolds number for blood flow in DES computational models is normally small (Re<100), indicating laminar flow and as such the application of the axial-symmetry boundary condition should not influence the fluid flow solution.

5.2 Application of mass transport boundary conditions
5.2.1 Luminal inlet: Concentration

Because mass transport within the artery's lumen is convection dominated, it would be virtually impossible for drugs to diffuse in the direction opposing blood flow. Therefore it can be assumed that the luminal inlet concentration has a constant value equal to zero. This boundary condition has been used extensively in mass transport studies on DES (Balakrishnan et al., 2008, 2007, 2005; Mongrain et al., 2007, 2005).

5.2.2 Luminal outlet: Convective flux

A convective flux outflow condition is generally imposed on the outlet of a models lumen, resulting in a zero concentration gradient at the outlet. This boundary condition assumes that all the mass passing through the boundary is convection dominated.

5.2.3 Lumen-stent-wall interface: Continuity

A common assumption for DES mass transport studies is that the intimal layers of the artery are denuded and that the stent is in direct contact with the medial layer of the artery wall (Kolachalama et al., 2009; Balakrishnan et al., 2008, 2007, 2005; Mongrain et al., 2007, 2005). This negates the need to model the endothelial, intima and internal elastic lamina layers. Regardless of the inclusion or exclusion of these layers the continuity equation should be the default setting for all interior boundaries. This condition states that in the absence of sources or sinks, the flux in the normal direction is continuous across the boundary, i.e. the concentration is equal on both sides of the boundary.

5.2.4 External faces: Insulation/symmetry

This condition is specified at the perivascular wall and at the up- and down-stream wall boundaries, which should be a sufficient distance away from the stent. It specifies where the domain is well insulated or it can reduce the size of a model by taking advantage of symmetry. Intuitively this condition states that the gradient across the boundary must be zero, therefore it is impermeable to mass transport.

6. Experimental validation of mass transport behaviour

Historically, experimental validations have been necessary to prove researchers hypotheses across all paradigms of science. In order to validate a theory one must not only be mindful of their goal but also their ability to achieve it. For instance, trying to validate a computational DES model using excised arterial DES would be nice in theory but in practice may prove fruitless because it would be near impossible to obtain the site specific drug concentrations within the artery wall, necessary for the researcher to examine the nuances of stent design that they are interested in analysing. It is often useful to validate a single aspect of the model if possible. With regards to DES, a solitary WSMT model may be acceptable,as a vast body of knowledge pertaining to fluid flow problems already exists, unlike the as of yet mature understanding of WSMT and how the behaviour of the porous artery wall and other pertinent features influence the mass transport therein.

Figure 3 illustrates an example of a validation flowchart for a study of mass transport from DES. The first process in designing a validation experiment is to analyse the problem as a whole and see what you would like to prove. Then certain aspects of the problem that are relatively rudimentary can be neglected from the validation, providing they won't have a fundamental impact on the outcome of the test. The flowchart is divided up into two streams that are developed jointly in order to achieve a desired validation. The experimental mass transport validation is on the left and the development of the corresponding computational mass transport model is on the right. In both streams BSMT is not highlighted for inclusion in this example's validation procedure. This may be due to the readily available examples of validations of this nature in literature or its minimal impact on the outcome of the results. This study was designed to investigate instances of WSMT in relation to DES design and deployment.

How accurate do you want your computational model to be and what assumptions are going to be made? The flowchart (Figure 3) describing this example validation decided to neglect convection-diffusion mass transport in the artery wall and concentrate on validating a diffusion only model. The reason for such a decision could be threefold; 1) an artery requiring a DES would be highly calcified and therefore the plaque can act as a buffer to stop or considerably reduce the flow in the porous wall, 2) an analysis of the Peclet number (equation 8) demonstrates diffusion dominated transport for a give drug or 3) the inability to create and obtain tangible results from an applicable experimental model with both convection and diffusion. The next aspect of the computational model is the characterisation of the artery wall, i.e. should it be modelled as a porous medium. Although flow in the wall has been neglected the application of a porous wall still can have a bearing on the outcome of mass transport due to the characterisation of the effective diffusivity and its propensity to change under varying stenting conditions (equation 12).

In this example a computational model has been identified in which many of the aspects of the *in vivo* stenting conditions remain but more importantly contains the ability to develop and analyse a corresponding experimental validation. Validations of this nature are a powerful tool in a researchers arsenal because once the initial hypotheses has been validated the computational model can be changed to any geometry imaginable to create more realistic stenting scenarios, and when solving the problem using the same physics one can have full confidence in the results.

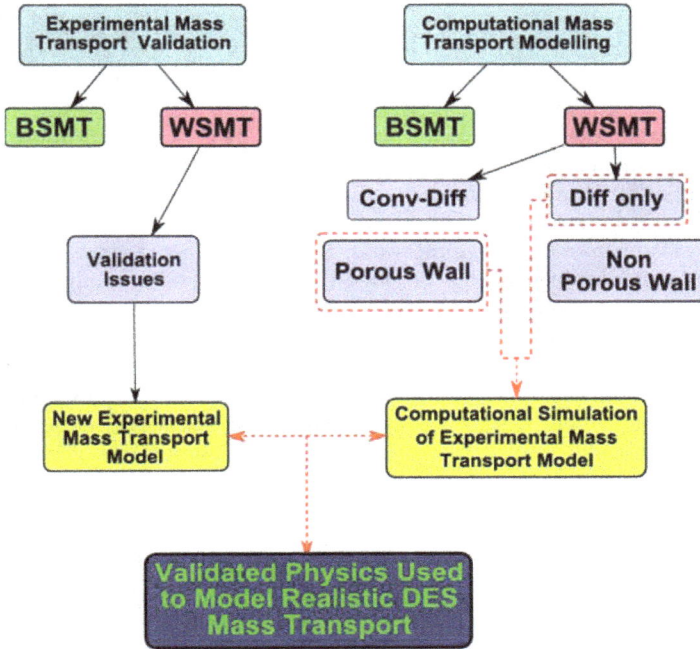

Fig. 3. Flowchart demonstrating the process of identifying an appropriate route for the experimental validation of DES mass transport.

6.1 Historical experimental mass transport validations

Experimental techniques to simulate mass transport in the vasculature have been an integral part in the development of DES. Various techniques can be applied depending on the required outcome from the analysis. For example if lumen transport is to be analysed a common approach would be to introduce a dye to a flowing system. Concentrations are then measured at certain time points and locations by withdrawing fluid samples so that the optical dye intensity can be determined.

Markou et al. (1998) employed this approach when analysing the local transport of anti-restenotic agents from a novel drug delivery device. The device consisted of a porous membrane that lined the artery wall and was the location where drugs would be infused into the artery. Their experimental approach consisted of a simulated artery section with a 1cm slit around its circumference. Dye was then infused from this slit in a radial direction into the lumen at uniform speed. It was found that the radial diffusion of the species was minimal in comparison to the axial convection therefore the majority of species remain in close proximity to the wall. This experiment was validated with the commercially available finite volume solver Fluent (Lebanon, NH). The computational models predicted an increase in dye concentration at the wall with an increase in the infusion rate. The same effect was witnessed experimentally and there was good agreement between the results. This study indicates the dominance of convective transport over diffusive transport in the arteries lumen.

Lutostansky et al. (2003) adopted a similar technique when conducting an experimental analysis of mass transport in the recirculation region downstream of a sudden expansion. Dye concentrations were determined at four equidistant locations after the sudden expansion. Experimental concentrations were taken over the course of one hour and showed very good correlation with predicted results from two numerical codes, the finite volume code Fluent (Lebanon, NH) and a finite element code FTSP (developed by Graz University of Technology). Although these experiments provided a validation of their numerical approach, they are limited in that they cannot be used to analyse mass transport within a porous wall.

In 2001, Hwang et al. found that stent based delivery, from a Palmaz-Schatz Crown stent (Cordis), resulted in large variations in concentration gradients. Drug concentrations were found to vary from zero to several times the mean over a few micrometres. The aforementioned stents were spray coated with a fluorescein sodium/ethylene vinyl acetate copolymer and deployed in excised bovine carotid arteries. The arteries were positioned in an *ex vivo* perfusion apparatus and immersed in a perivascular bath where coronary flow was simulated. After 3 hours the arteries were removed and cut into slices. The fluorescein concentration was then measured with a spectrofluorometer. Although the experiment was not used to validate a computational model, the effects witnessed experimentally were compared to variations in simulated drug physicochemical properties. They concluded that the proximity of the device does not necessarily ensure adequate targeting because transport forces can cause local concentrations to deviate from the mean concentration.

In a later paper Hwang et al. (2005) evaluated the paclitaxel uptake in stented abdominal aortas of adult male Sprague-Dawley rats in the presence and absence of controlled mural thrombus. The in vivo clot dimensions were determined and used as boundary conditions and input parameters for the computational model. The computational analysis predicted an arterial drug ratio of 0.56 which correlates with the 50% decrease in arterial uptake ascertained from the animal experiments. Hwang et al. (2005) discovered that by varying clot size and location, large variations in arterial uptake were witnessed.

In 2007, Balakrishnan et al. deployed a Cypher sirolimus eluting stents in porcine arteries. At the desired time points of 1, 8, 14, 30, 60 and 90 days after implantation the stents were harvested and analysed. In each case the stents were carefully removed from the artery and the remaining drug within the polymer was determined. When subtracted from the amount of drug prior to implantation the release fraction can be calculated. At each time point this fractional drug release was compared to numerical predictions using Fluent (Lebanon, NH). A good correlation validated the Fickian diffusion analysis applied with the numerical solver to approximate the drug transport from the coating. However, validating drug release from the polymer coating does not elucidate subsequent drug uptake within the porous artery wall.

In 2010, O'Connell and Walsh developed an analogous model of artery wall mass transport, examining the hypothesis of how compression of a porous media alters mass transport within. Due to the difficulty in measuring site specific concentrations within the artery wall they developed a scaled up experiment. It consisted of a bed of pH paper that was saturated in a neutral pH solution in order to fill the pore space of the material, similar to that of the artery walls interstitial fluid. The wall is then compressed, in increments up to a maximum of 23.75% of its initial thickness, and then the species of interest, an acid of pH 2.0, is introduced to the system and the resulting colour change was used as a marker for concentration. This enabled the site specific measurement of concentration at different depths throughout the porous wall. These experimental results were then validated computationally using the finite element solver COMSOL Multiphysics. The authors concluded that compression of a porous artery wall contributes significantly to the modulation of arterial WSMT and should be considered in future DES computational studies.

7. Computational modelling of mass transport from drug eluting stents

The following computational models were created to illustrate how changes in stenting deployment conditions can influence drug concentrations within the artery wall, analysed after 30 and 60 minutes for each model. Figure 4 describes the five 2-D axis-symmetric computational models that were analysed. Model 1 in figure 4 depicts the locations of drug concentration measurement through the depth of the artery wall and axially down the artery at a depth of 25% and 50% of the wall thickness (WT) beneath the strut. The models are described as follows:

Model 1. In the absence of a lumen and subsequent BSMT, this model only considered WSMT from a single DES stent strut (150μm) deployed flush against a single layer artery wall (200μm). In this instance WSMT is purely diffusive.

Model 2. Similar to Model 1 but with the inclusion of a steady blood flow profile (mean velocity = 0.1m/s) through the arterial lumen. Here WSMT is purely diffusive and BSMT is modelled using the convection-diffusion equation.

Model 3. Similar to Model 2 but with the inclusion of a 20μm thick layer of plaque along the artery wall.

Model 4. Similar to Model 2 only the stent strut becomes embedded in the artery wall as it compresses it by 25% of the wall thickness.

Model 5. Similar to Model 4 except upon compression of the wall the stent strut doesn't become embedded.

A hypothetical drug was used in the analysis with effective diffusivity values in each respective media defined in table 1. It is the combination of both the drug used and the characteristics of the media within which transport takes place that determines the effective diffusivity value. This fact becomes evident as the effective diffusivity of the drug in the compressed wall is determined. The drug remains the same but, as described in equation 12, changes to the tortuosity (τ) and porosity (ε) of the wall alters the effective diffusivity within.

From equation 15 the pore path (L) remains the same length but the distance (X) has reduced due to the 25% compression of the artery wall. Therefore the tortuosity of the compressed wall (τ_{CW}) can be described as a function of the tortuosity of the wall in its original state (τ_W).

$$\tau_{CW} = \frac{L}{0.75X} = 1.333\frac{L}{X} = 1.333\tau_W$$

Similarly, as the wall is compressed the total volume and pore volume of the wall under compression reduces but the fibre volume remains the same. To this end $\varepsilon_{CW}= 0.787\varepsilon_W$ and the effective diffusivity of the compressed wall (D_{CW}) can be described as follows:

$$D_{CW} = \frac{\varepsilon_{CW}}{\tau_{CW}}D_{free} = \frac{0.787\varepsilon_w}{1.333\tau_W}D_{free} = 0.59D_W = 0.59\times10^{-12}m^2/s$$

Diffusivity in Lumen	$D_L = 1\times10^{-10}$ m^2/s
Diffusivity in Stent Coating	$D_S = 1\times10^{-14}$ m^2/s
Diffusivity in Plaque	$D_P = 1\times10^{-13}$ m^2/s
Diffusivity in Wall	$D_W = 1\times10^{-12}$ m^2/s
Diffusivity in Compressed Wall	$D_{CW} = 0.59\times10^{-12}$ m^2/s

Table 1. Effective diffusivity values of the different layers of arterial DES models.

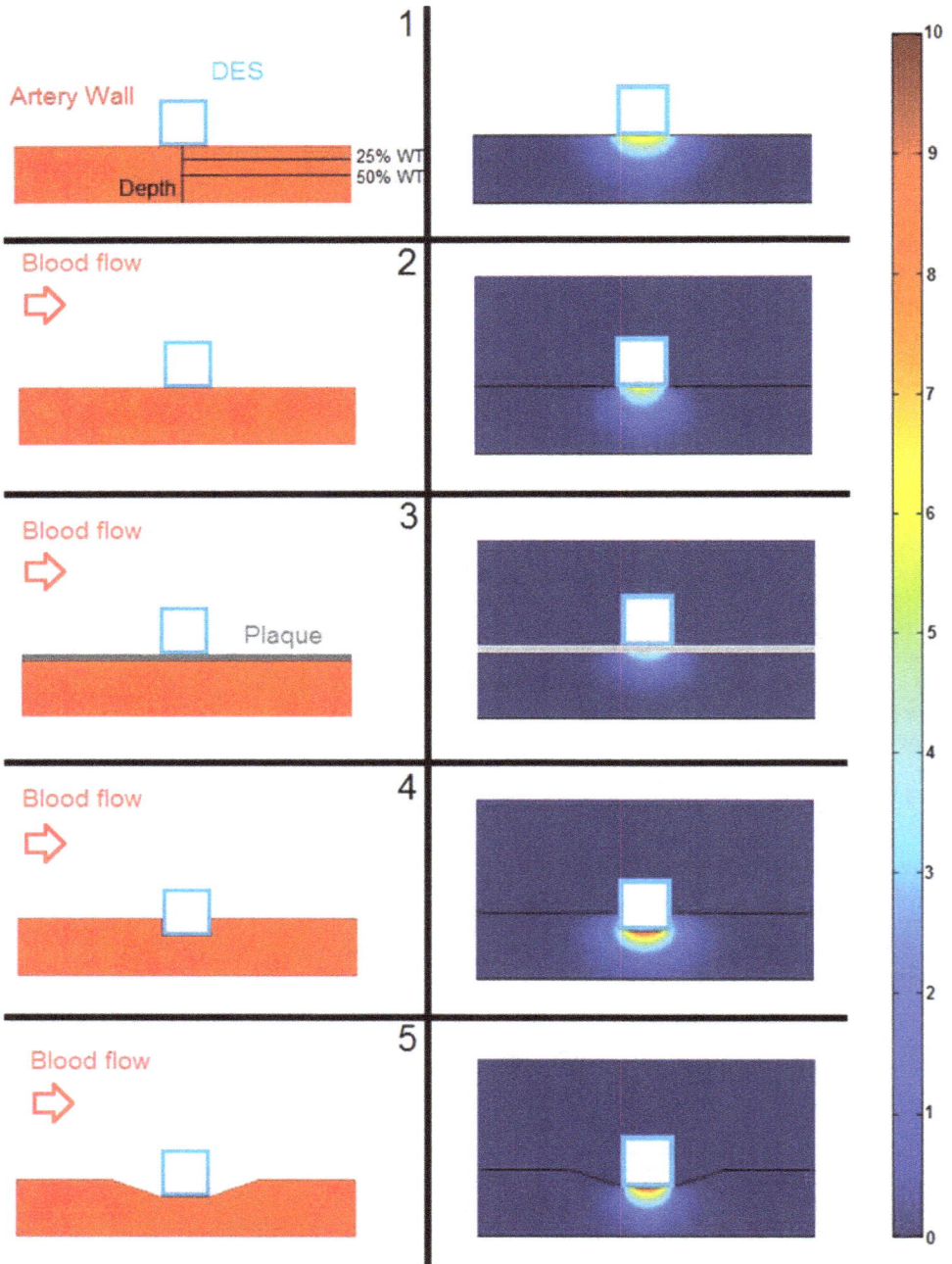

Fig. 4. Illustration of 2-D axis-symmetric computational models used in the DES mass transport analysis and their resulting drug concentration contours after 30 minutes.

7.1 Computational predictions of mass transport in the artery wall

Even though BSMT doesn't significantly contribute to the distal absorption of drugs into the artery wall, it is a necessary element of the modelling process as it provides a more realistic representation of how the drugs in the DES dissipate in the vasculature. Failure to model BSMT (Model 1) culminates in the entire reservoir of antirestenotic drug, from both the luminal and abluminal side of the stent, having no choice but to eventually transport into the wall. This as we know is not the case as a considerable amount of drug is lost to the blood stream. Figures 5-7 demonstrate that for both time points and at each location the drug concentration for Model 1 is greater than that of Model 2 due to the absence of BSMT. Model 3 examined the influence of arterial plaque, which was given a drug diffusivity of $1\times10^{-13}m^2/s$. In reality plaque size and composition will vary from patient to patient and the implication of this is a study in itself where a range of plaque types would need to be modelled in order to quantitatively predict its influence on WSMT. For the purpose of demonstrating the influence that the presence of any plaque may have on mass transport, a drug diffusivity was chosen that is an order of magnitude between the diffusivities in the DES coating and the uncompressed artery wall respectively. What Model 3 demonstrates is that even a $20\mu m$ thick layer of plaque between the stent and the artery wall can significantly reduce uptake within the artery wall, even more so than arterial compression.

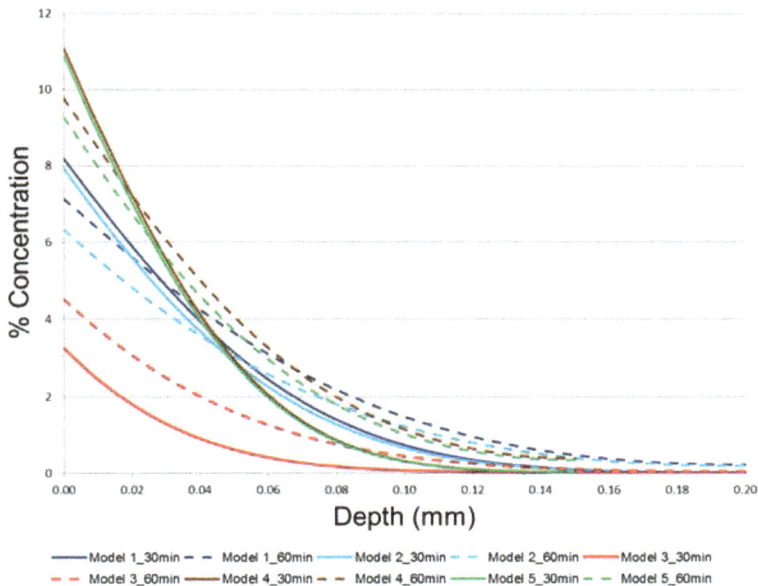

Fig. 5. Drug concentrations through the depth of the artery wall, as illustrated in Fig 4 Model 1. Concentrations are measured after 30 minutes (solid line) and 60 minutes (dashed line) respectively.

Biomedical Science and Engineering

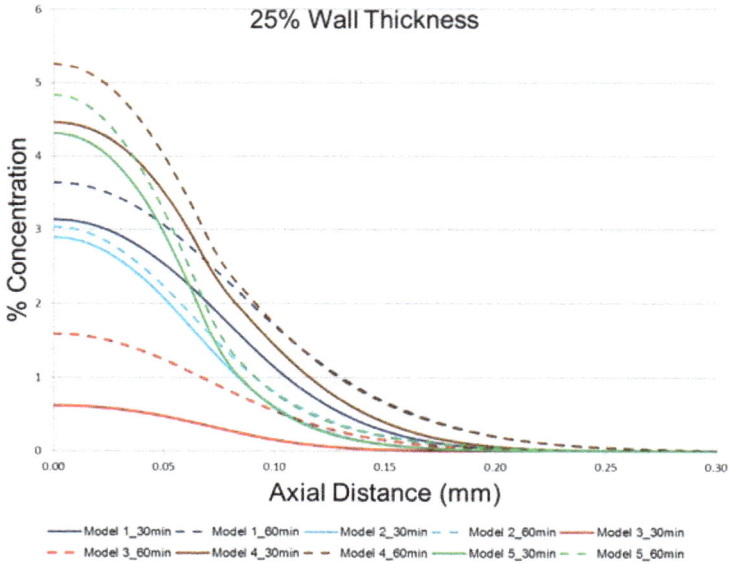

Fig. 6. Axial drug concentrations at 25% of the artery wall depth, illustrated in Fig 4 Model 1 by the 25% WT concentration line. Concentrations are measured after 30 minutes (solid line) and 60 minutes (dashed line) respectively.

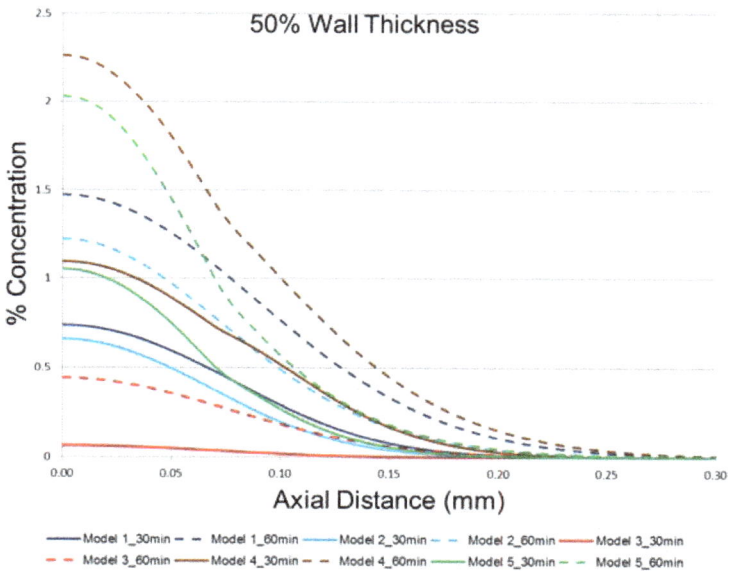

Fig. 7. Axial drug concentrations at 50% of the artery wall depth, illustrated in Fig 4 Model 1 by the 50% WT concentration line. Concentrations are measured after 30 minutes (solid line) and 60 minutes (dashed line) respectively.

The influence of compression on a porous wall in relation to mass transport has been demonstrated experimentally by O'Connell and Walsh (2010) and is evident in the results from Models 4 and 5. Histological evidence demonstrated that upon stent implantation some struts almost cut into the wall (Model 4) while others merely compress it providing gradual recovery either side of the stent strut (Model 5). Contrasting concentration profiles from these models elucidate to the inclusion of compression alone may not necessarily be adequate enough because the surrounding tissue orientation will also play a role in WSMT. Intuitively the greater the DES surface area in contact with arterial tissue the greater its ability to transport drugs into the wall. This holds true when comparing the concentration profiles of Models 4 and 5. The depth concentrations at the early time point for both models are very similar (Figure 5) but there is a notable difference in the axial concentration plots at both 25% and 50% of the wall thickness, whereby the 60% greater wall contact surface area of model 4 results in an increased concentration profile.

8. Conclusion

Understanding the behaviour of DES in the pursuit of improving device functionality is of great importance to clinicians and researchers alike. The modern application of computational techniques have greatly aided in achieving this goal, with researchers continuingly adding to the global understanding of drug mass transport from these devices. However, computational modelling isn't the complete solution because exact recreation of DES deployment and ensuing mass transport is not feasible for a number of reasons. No two patients will have identical stenosis of the coronary artery and therefore a single computational model will not provide the information required to comprehensively assess the viability of a single stent design. Instead a variety of models that will cover the spectrum of DES deployment scenarios is required and to computationally recreate these like for like with *in vivo* stenting conditions would be too computationally demanding for the same rewards that one could yield with a simplified analysis.

However, one must be mindful when simplifying the problem. The computational models developed in this chapter were created as 2-D axis-symmetric models as the study was intent on analysing the influence of relatively simple geometries. However, this may not always be the case and the need may arise where it would be necessary to model the problem in 3-D. The pitfalls of over simplification can be seen in the computational Model 1 where failure to model BSMT results in drug concentrations that are higher than that of the more realistic case, Model 2, that includes the luminal blood flow and mass transport therein.

Computational models where DES struts are flush against a bare artery wall have their merits but a greater degree of complexity needs to be implemented if an improved insight into DES mass transport within the coronary artery environment is to be gained. These arteries are heavily diseased and even a thin layer of plaque between the stent strut and the wall can inhibit WSMT. The compression of the porous artery wall upon stent expansion has an interesting effect on drug concentration within the wall. The reduction in artery wall diffusivity results in higher peak concentrations beneath the stent strut compared to the models where artery wall compression is not present. This pooling of drugs close to the stent strut is undesirable as it retards the early penetration of drugs into the wall, demonstrated in Figure 5 as the uncompressed artery wall of Model 2 recorded concentrations higher than that of Models 4 and 5 from a depth of approximately $50\mu m$

through the rest of the artery wall. There is also the danger of potential toxicity due to prolonged exposure of high drug concentrations close to the strut.

Analysis of mass transport from DES requires a multifaceted approach in order to predict behaviour of these devices and subsequently their response to *in vivo* arterial conditions. Future research that is mindful of preoperative DES design and postoperative environmental conditions will increase our knowledge of the second generation DES and enable us as a community to prepare for the advent of biodegradable DES.

9. Acknowledgements

The authors would like to thank the Irish Research Council for Science, Engineering and Technology (IRCSET), Grant no. RS/2005/159, for funding this body of work. The authors would also like to thank the members of CABER and the MSSI.

10. References

Balakrishnan, B., Dooley, J.F., Kopia, G. & Edelman, E.R. (2007).Intravascular Drug Release Kinetics Dictate Arterial Drug Deposition, Retention and Distribution. *Journal of Controlled Release*, Vol. 123, pp. 100-108.

Balakrishnan, B., Dooley, J.F., Kopia, G. & Edelman, E.R. (2008).Thrombus Causes Fluctuations in Arterial Drug Delivery from Intravascular Stents. *Journal of Controlled Release*. Vol. 131, pp. 173-180.

Balakrishnan, B., Tzafriri, A.R., Seifert, P., Groothuis, A., Rogers, C. & Edelman, E.R. (2005).Strut Position, Blood Flow, and Drug Deposition: Implications for Single and Overlapping Drug-Eluting Stents. *Circulation*, Vol. 111, pp. 2958-2965.

Burt, H.M. & Hunter, W.L. (2006).Drug-Eluting Stents: A Multidisciplinary Success Story. *Advanced Drug Delivery Review*, Vol. 58, pp. 350-357.

Chen, M.C., Liang, H.F., Chiu, Y.L., Chang, Y., Wei, H.J. & Sung, H.W. (2005). A Novel Drug-Eluting Stent Spray Coated with Multi-Layers of Collagen and Sirolimus. *Journal of Controlled Release*, Vol.108, pp. 178-189.

Costa, M.A. & Simon, D.I. (2005).Molecular Basis of Restenosis and Drug-Eluting Stents. *Circulation*, Vol. 111, pp. 2257-2273.

Devereux, P.D. (2005). Mass Transport Disturbances in the Downstream Junction of Peripheral Bypass Grafts. *Ph.D. Thesis, University of Limerick, Limerick, Ireland.*

Duraiswamy, N., Schoephoerster, R.T., Moreno, M.R. & Moore Jr, J.E. (2007). Stented Artery Flow Patterns and their Effects on the Artery Wall. *Annual Review of Fluid Mechanics*, Vol. 39, pp. 357–382.

Friedman, M.H. (2008). *Principles and Models of Biological Transport.* Springer Science+Business Media, ISBN 978-0-387-79239-2, New York.

Granada, J.F., Kaluza, G.L. & Raizner, A. (2003).Drug-Eluting Stents for Cardiovascular Disorders.*Current atherosclerosis reports*, Vol. 5, pp. 308-316.

Haller, J.D. & Olearchyk, A.S. (2002).Cardiology's 10 Greatest Discoveries. *Texas Heart Institution Journal*, Vol. 29, pp. 342–344.

Hara, H., Nakamura, M., Palmaz, J.C. & Schwartz, R.S. (2006).Role of Stent Design and Coatings on Restenosis and Thrombosis. *Advanced Drug Delivery Review.* Vol. 58, pp. 377-386.

Head, D.E., Sebranek, J.J., Zahed, C., Coursin, D.B. & Prielipp, R.C. (2007).A tale of Two Stents: Perioperative Management of Patients with Drug-Eluting Coronary Stents. *Journal of Clinical Anesthesia,* Vol. 19, pp. 386-396.

Hwang, C.W., Levin, A.D., Jonas, M., Li, P.H. & Edelman, E.R. (2005).Thrombosis Modulates Arterial Drug Distribution for Drug Eluting Stents. *Circulation,*Vol. 111, pp. 1619-1626.

Hwang, C.W., Wu, D.& Edelman, E. R. (2001). Physiological Transport Forces Govern Drug Distribution for Stent Based Delivery. *Circulation,* Vol. 104, pp. 600-605.

Kaul, S., Shah, P.& Diamond, G.A. (2007).As Time Goes By:Current Status and Future Directions in the Controversy Over Stenting. *Journal of the American College of Cardiology,* Vol. 50, No. 2, pp. 128–137.

Kaazempur-Mofrad, M.R., Wada S., Myers J.G. &Ethier C.R. (2005). Mass Transport and Fluid Flow in StenoticArteries: Axisymmetric and Asymmetric Models. *International Journal of Heat and Mass Transfer,* Vol. 48, pp. 4510-4517.

Kaazempur-Mofrad, M. R. & Ethier C.R. (2001).Mass Transport in an Anatomically Realistic Human Right Coronary Artery. *Annals of Biomedical Engineering,* Vol. 29, pp. 121-127.

Khakpour, M. & Vafai, K. (2008).A Comprehensive Analytical Solution of Macromolecular Transport within an Artery. *International Journal of Heat and Mass Transfer,* Vol. 51, pp. 2905-2913.

Khakpour, M. & Vafai, K. (2007).Critical Assessment of Arterial Transport Models. *International Journal of Heat and Mass Transfer,* Vol. 51, pp. 807-822.

Kolachalama, V.B., Tzafriri A.R., Arifin, D.Y. & Edelman, E.R. (2009). Luminal Flow Patterns Dictate Arterial Drug Deposition in Stent-Based Delivery. *Journal of Controlled Release,* Vol. 133, pp. 24–30.

Kukreja, N., Onuma, Y., Daemen, J. & Serruys, P.W. (2008).The Future of Drug-Eluting Stents. *Pharmacological Research,* Vol. 57, pp. 171–180.

Lutostansky, E.M., Karner, G., Rappitsch, G., Ku, D.N. & Perktold, K. (2003).Analysis of Hemodynamic Fluid Phase Mass Transport in a Separated Flow Region. *Journal of Biomechanical Engineering,* Vol. 125, pp. 189-196.

Markou, C.P., Lutostansky, E.M., Ku, D.N. & Hanson, S.R. (1998).A Novel Method for Efficient Drug Delivery. *Annals of Biomedical Engineering,* Vol. 26, pp. 502–511.

Mongrain, R., Faik, I., Leask, R., Rodes-Cabau, J., Larose, E. & Bertrand, O.F. (2007).Effects of Diffusion Coefficients and Strut Apposition Using Numerical Simulations for Drug Eluting Coronary Stents. *Journal of Biomechanical Engineering.* Vol. 129, pp. 733-742.

Mongrain, R., Leask, R., Brunette, J., Faik, I., Bulman-Feleming, N. & Nguyen, T. (2005).Numerical Modeling of Coronary Drug Eluting Stents. *Studies in Health Technology and Informatics,* Vol. 113, pp. 443-458.

O'Brien, T., Walsh, M. & McGloughlin, T. (2005).On Reducing Abnormal Haemodynamics in the Femoral End-to-Side Anastomosis: The Influence of Mechanical Factors. *Annals of Biomedical Engineering.* Vol. 33, pp. 309-321.

O'Connell, B.M. & Walsh, M.T. (2010).Demonstrating the Influence of Compression on Artery Wall Mass Transport. *Annals of biomedical Engineering*, Vol. 38, pp. 1354-136.

Parry, T.J., Brosius, R., Thyagarajan, R., Carter, D., Argentieri, D., Falotico, R. &Siekierka, J. (2005). Drug–Eluting Stents: Sirolimus and Paclitaxel Differently Affect Cultured Cells and Injured Arteries. *European Journal of Pharmacology*, Vol. 524, pp. 19-29.

Rajagopal, V. &Rockson, S.G. (2003).Coronary Restenosis: A Review of Mechanisms and Management. *American Journal of Medicine*, Vol. 115, pp. 547-533.

Rajamohan, D., Banerjee, R.K., Back, L.H., Ibrahim, A.A., Jog, M.A. (2006). Developing Pulsatile Flow in a Deployed Coronary Stent. *Journal of Biomechanical Engineering*, Vol. 128, pp. 347-359.

Schwartz, R.S., Chronos, N.A. & Vivmani, R. (2004).Preclinical Restenosis Models and Drug-Eluting Stents: Still Important, Still Much to Learn. *Journal of the American College of Cardiology*. Vol. 44, pp. 1373-1385.

Stone, G.W., Midei, M., Newman, W. et al. (2008).Comparison of an Everolimus-Eluting Stent and a Paclitaxel-Eluting Stent in Patients with Coronary Artery Disease: A Randomized Trial. *The Journal of The American Medicine Association*, Vol. 299, pp. 1903-1913.

Sun, N., Wood, N.B., Hughes. A.D., Thom, S. A. M. & Xu, X.Y. (2006). Fluid-Wall Modelling of Mass Transfer in an Axisymmetric Stenosis: Effects of Shear-Dependent Transport Properties. *Annals of Biomedical Engineering*, Vol. 34, pp. 1119-1128.

Tanabe, K., Regar, E., Lee, C.H., Hoye, A., Van der Giessen, W.J.&Serruys, P.W. (2004). Local Drug Delivery Using Coated Stents: New Developments and Future Perspectives. *Current Pharmaceutical Design*, Vol. 10, pp. 357-368.

Van der Hoeven, B.L., Pires, N.M.M., Warda, H.M., Oemrawsingh, P.V., Van Vlijmen, B.J.M., Quax, P.H.A., Schalij, M.J., Van der Wall, E.E. & Jukema, J.W. (2004). Drug Eluting Stents: Results, Problems and Promises. *International Journal of Cardiology*, Vol. 99, pp. 9-17.

Venkatraman, S. &Boey, F. (2007).Release Profiles in Drug-Eluting Stents: Issues and Uncertainties. *Journal of Controlled Release*, Vol. 120, pp. 149-160.

Waksman, R. (2002). Drug-Eluting Stents: From Bench to Bed. *Cardiovascular Radiation Medicine*, Vol. 3, pp. 226– 241.

Walsh, M.T., Kavanagh, E.G., O'Brien, T., Grace, P.A. & McGloughlin, T.(2003). On the Existence of an Optimum End-to-Side Junctional Geometry in Peripheral Bypass Surgery – A Computer Generated Study. *European Journal of Vascular and Endovascular Surgery*, Vol. 26, pp. 649-656.

5

Biosurfactants and Bioemulsifiers Biomedical and Related Applications – Present Status and Future Potentials

Letizia Fracchia[1], Massimo Cavallo[1],
Maria Giovanna Martinotti[1] and Ibrahim M. Banat[2]
[1]*Department of Chemical, Food, Pharmaceutical and Pharmacological Sciences,*
Drug and Food Biotechnology Center,
Università del Piemonte Orientale "Amedeo Avogadro", Novara,
[2]*School of Biomedical Sciences, Faculty of Life and Health Sciences,*
University of Ulster, Coleraine, N. Ireland,
[1]*Italy*
[2]*UK*

1. Introduction

Many microorganisms are able to produce a wide range of amphipathic compounds, with both hydrophilic and hydrophobic moieties present within the same molecule which allow them to exhibit surface activities at interfaces and are generally called biosurfactants or bioemulsifiers. These surface-active compounds (SAC) are mainly classified according to their mode of action, molecular weight and general physico-chemical properties.

In literature, the terms 'biosurfactants' and 'bioemulsifiers' are often used interchangeably, however in general those that reduce surface and interfacial tension at gas-liquid-solid interfaces are called biosurfactants and those that mainly reduce the interfacial tension between immiscible liquids or at the solid-liquid interfaces leading to the formation of more stable emulsions are called bioemulsifiers or bioemulsans. The former group includes low-molecular-weight compounds, such as lipopeptides, glycolipids, proteins, while the latter includes high-molecular-weight polymers of polysaccharides, lipopolysaccharides proteins or lipoproteins (Smyth et al., 2010a, 2010c).

In heterogeneous systems, biosurfactants tend to aggregate at the phase boundaries or interfaces. They form a molecular interfacial film that affects the properties (surface energy and wettability) of the original surface. This molecular layer, in addition to lowering the surface tension in liquids, also lowers the interfacial tension between different liquid phases on the interfacial boundary existing between immiscible phases and therefore can have an impact on the interfacial rheological behaviour and mass transfer.

When at interfaces (solid- liquid, liquid-liquid or vapour-liquid), the hydrophobic moiety of the surface active molecules aggregates at the surface facing the hydrophobic phase (usually the oil phase) while the hydrophilic moiety is oriented towards the solution or hydrophilic phase (mainly water). Their diverse functional properties namely, emulsification, wetting, foaming, cleansing, phase separation, surface activity and reduction in viscosity of heavy liquids such as crude oil, make them suitable for utilization for many industrial and domestic application purposes (Gautam & Tiagi, 2006; Franzetti et al., 2010a; Perfumo et al., 2010a; Satpute et al., 2010b).

During the past two decades biosurfactants have been under continuous investigation as a potential replacement for synthetic surfactants and are expected to have several industrial and environmental applications mainly related to detergency, emulsification, dispersion and solubilisation of hydrophobic compounds (Banat et al., 2000). In addition, biosurfactants present several advantages over surfactants of a chemical origin, particularly in relation to their biodegradability, environmental compatibility, low toxicity, high selectivity and specific activity at extreme temperatures, pH and salinity (Banat 1995a, 1995b). Due to all these properties, they have steadily gained increased significance in industrial and environmental applications such as bioremediation, soil washing, enhanced oil recovery and other general oil processing and related industries (Perfumo et al., 2010b). Furthermore, potential commercial applications in several other industries including paint, cosmetics, textile, detergent, agrochemical, food and pharmaceutical industries begin to emerge (Banat et al., 2000).

Numerous investigations in the field of biosurfactants/bioemulsifiers are leading to the discovery and description of many interesting chemical and biological properties and potential biomedical therapeutic and prophylactic applications. In this chapter we will focus on the most recent and appealing biomedical and therapeutic applications of biosurfactants and bioemulsifiers with special emphasis on the most recent results in the fields of biotechnology, nanotechnology and bioengineering.

2. Classification, properties and functional mechanisms of microbial surface-active compounds

Microbial surface-active compounds are a range of structurally diverse molecules produced by different microorganisms and are mostly therefore classified by their structural features, the producing organism and their molecular mass. Their hydrophilic moiety is mainly comprised of an acid, peptide cations, or anions, mono-, di- or polysaccharides while their hydrophobic moiety can be an unsaturated or saturated hydrocarbon chains or fatty acids. The structural orientation on the surfaces and inter phases confers the range of properties, such as the ability to lower surface and interfacial tension of liquids and the formation of micelles and microemulsions between these different phases (Chen et al., 2010a, 2010b).

2.1 Low molecular weight compounds
2.1.1 Lipopeptides
Microbial surface-active compounds can be roughly divided into low molecular weight molecules that efficiently reduce surface and interfacial tension (biosurfactants) (Fig. 1.), and high molecular weight polymers that stabilize emulsions but do not lower the surface tension as much (bioemulsans or bioemulsifiers) (Fig. 2.) (Neu, 1996; Rosenberg, 2006; Rosenberg & Ron, 1997; Smyth et al., 2010a, 2010c).

The most studied low-molecular-weight biosurfactant compounds are lipopeptides and glycolipids. Lipopeptides are mainly produced by members of the *Bacillus* species; they are composed of different families and each family is constituted of several variants, which can differ in their fatty acid chain and their peptide moiety (Dastgheib et al., 2008; Jacques, 2010; Thavasi et al., 2008, 2011).

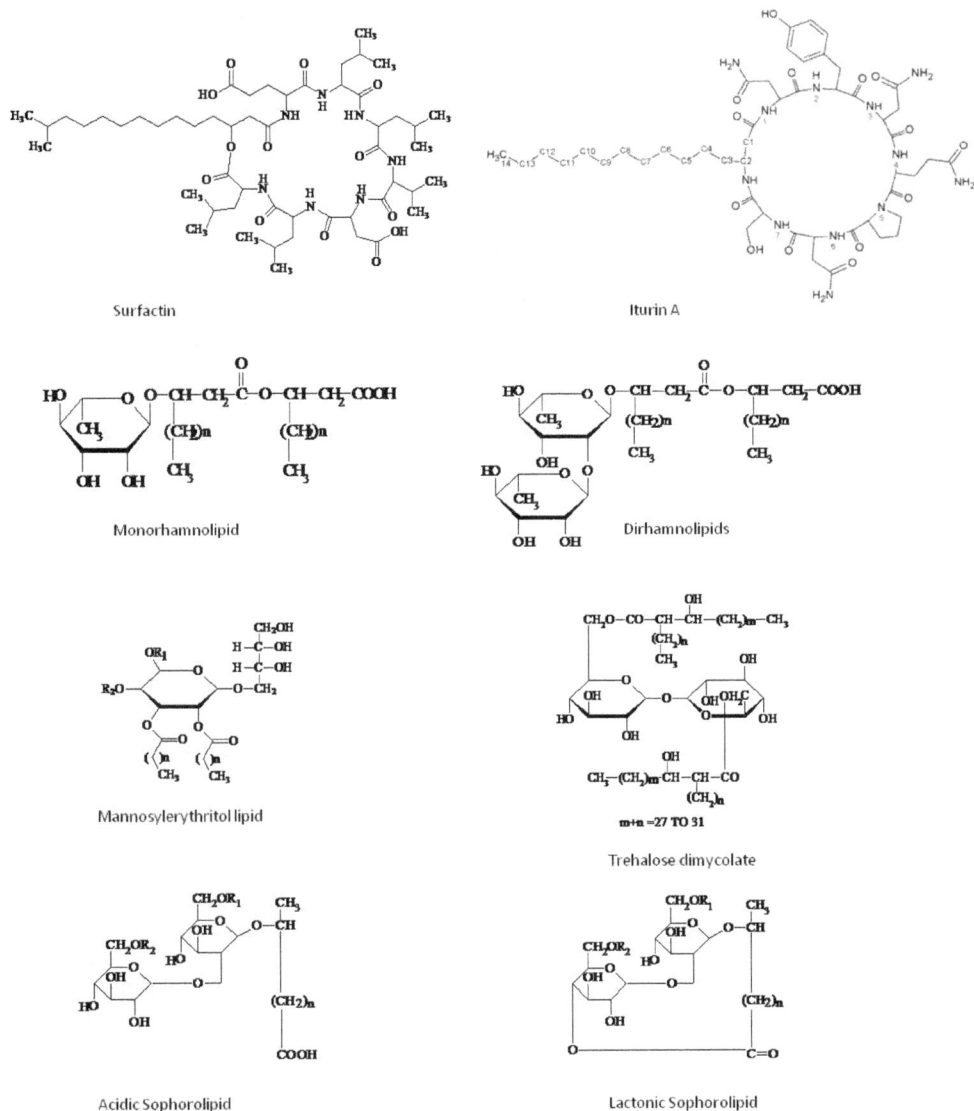

Fig. 1. Chemical structures of the main low molecular weight microbial surface active compounds reported; surfactin, iturin A, mono & di-rhamnolipids, mannosylerythritol lipids, dimycolates trehalose lipids, acidic and lactonic sophorolipids.

Surfactin, a cyclic lipopeptide produced by *Bacillus subtilis* is considered the most active biosurfactant discovered so far (Ron & Rosenberg, 2001). Surfactin was discovered by

Arima et al., (1968) from the culture broth of *Bacillus subtilis* and it was named thus due to its exceptional surfactant activity (Peypoux et al., 1999). Natural surfactins are a mixture of isoforms A, B, C and D which are classified according to the differences in their amino acid sequences and possess various physiological properties (Shaligram & Singhal, 2010). Surfactin is composed of a seven amino-acid ring structure coupled to a fatty-acid chain via a lactone linkage. Surfactin-A has L-leucine, surfactin-B has L-valine and surfactin-C has L-isoleucine at the amino acid position involved in the lactone ring formation with the C14–C15 β-hydroxy fatty acid. The amino-acid residues may vary and the presence of these variants can be related to alterations in the culture conditions such as providing substrate containing some specific amino-acid residues in the culture media (Jacques, 2010).

Another surfactin related compound is lichenysin, a lipopeptide discovered in the supernatant of *Bacillus licheniformis* culture (Horowitz et al., 1990). Its chemical structure and physio-chemical properties are similar to surfactin (McInerney et al., 1990). In particular, lichenysin has Glutamine amino-acid in position 1 while surfactin has Glutamic acid. Other surfactin-like compounds are pumilacidin A, B, C, D, E, F and G, a complex of acylpeptide antibiotics isolated from *Bacillus pumilus* culture supernatants with interesting antiviral properties (Morikawa et al., 1992; Naruse et al., 1990). Among the lipopeptides belonging to the iturin family, iturin A is the most studied compound. It is a heptapeptide interlinked with a β-amino-acid fatty acid with carbon chain length from C14 to C17 (Peypoux, 1978, as cited in Jacques, 2010) produced by *Bacillus subtilis* strains reported to have antifungal activities (Besson et al., 1976).

Other members of the iturin family are iturin C, bacillomycin D, F, and Lc and mycosubtilin (Bonmatin et al., 2003). The family of fengycins includes fengycins A and B, lipodecapeptides which differ by their amino-acid residue in position 6 that can be Alanine or Valine and are known for their interesting fungitoxic and immunomodulating activities (Jacques, 2010). Other interesting lipopeptides are serrawettins, nonionic cyclodepsipeptide biosurfactants produced by *Serratia marcescens* (Matsuyama et al., 2010) and implicated with anti-tumor and anti-nematode activities.

2.1.2 Glycolipids

Are commonly mono or disaccharides compounds acylated with long chain fatty acids or hydroxyl fatty acids. Among them, rhamnolipids, mannosylerythritol lipids (MELs), sophorolipids and trehalolipids are the best-studied structural subclasses.

Rhamnolipids are glycosides, produced mainly by *Pseudomonas aeruginosa* and by the *Burkholderia* genus, that are composed of one (for mono-rhamnolipids) or two (for di-rhamnolipids) rhamnose sugar moieties linked to one or two β-hydroxyfatty acid chains (Perfumo et al., 2006; Raza, 2009). These molecules display high surface activities and many potential applications in the biomedical field due to their antibacterial, antifungal, antiviral, antiadhesive reported properties (Abalos et al., 2001; Cosson et al., 2002; Kim et al., 2000; Remichkova et al., 2008; Sotirova et al., 2008; Yoo et al., 2005). They have also been used in the preparation of nanoparticles (Palanisamy & Raichur, 2009; Xie et al., 2006) and microemulsions (Nguyen & Sabatini, 2009; Xie et al., 2007).

The mannosylerythritol (MELs) glycolipids are produced by the yeasts strains of the genus *Pseudozyma* sp. and *Ustilago* sp. from soybean oil or *n*-alkane (Arutchelvi & Doble, 2010). MELs are a mixture of partially acylated derivative of 4-*O*-β-D-mannopyranosyl-D-erythritol, containing $C_{2:0}$, $C_{12:0}$, $C_{14:0}$, $C_{14:1}$, $C_{16:0}$, $C_{16:1}$, $C_{18:0}$ and $C_{18:1}$ fatty acids as the hydrophobic groups (Bhattacharjee et al., 1970, as cited in Arutchelvi & Doble, 2010). Based on the degree of acetylation at C4 and C6 position, and their order of appearance on the thin layer chromatography, the MELs are classified into MEL-A, -B, -C and –D (Arutchelvi & Doble, 2010). MEL-A representing the diacetylated compound while MEL-B and MEL-C are monoacetylated at C6 and C4, respectively. The completely de-acetylated structure is known as MEL-D (Rau et al., 2005, as cited in Arutchelvi & Doble, 2010).

MELs have recently gained attention due to their environmental compatibility, mild production conditions, structural diversity, self-assembling properties and versatile biochemical functions. In particular, interesting applications have been described in the biomedical field as antimicrobial, antitumor and immunomodulating molecules, in the biotechnological field for gene and drug delivery, and in cosmetic applications as skin moisturizers (Arutchelvi & Doble, 2010).

Sophorolipids are another extracellular glycolipids synthesized by some yeast species including *Candida bombicola*, *Candida apicola*, *Rhodotorula bogoriensis*, *Wickerhaminella domercqiae* and *Candida batistae* (Van Bogaert & Soetaert, 2010). They consist of two glucose units linked β-1,2. The 6- and 6'-hydroxyl groups are generally acetylated. The lipid portion is connected to the reducing end through a glycosidic linkage. The terminal carboxyl group of the fatty acid can be in the lactonic form or hydrolyzed to generate an anionic surfactant (Rosenberg & Ron, 1999). Sophorolipids have been reported suitable for a number of application in the biomedical field including use as antimicrobial, antiviral and anticancer. They also have been used in the synthesis of metal-bound nanoparticles in cosmetic and pharmacodermatological products (Van Bogaert & Soetaert, 2010).

Trehalose lipids are also a glycolipids containing threhalose as the sugar moiety which is a non-reducing disaccharide in which the two glucose units are linked in an α,α-1,1-glycosidic linkage. It is the basic component of the cell wall glycolipids in *Mycobacteria* and *Corynebacteria* (Franzetti et al., 2010b). The most reported trehalose lipid is trehalose 6,6'-dimycolate, which is a α-branched chain mycolic acid esterified to the C6 position of each glucose. Different trehalose containing glycolipids are known to be produced by several other microorganisms belonging to mycolates group, such as *Arthrobacter*, *Nocardia*, *Rhodococcus* and *Gordonia*. *Rhodococcus* genus in particular produced several types of trehalose lipids as reported by Lang & Philp (1998). These glycolipids vary in the number and overall chain length (C20–C90) of the esterified fatty acids. Beside their known industrial applications, trehalose lipids recently attracted attention to their functions in cell membrane interaction and their potential as antitumor therapeutic agents (Aranda et al., 2007, Harland et al., 2009, Imasato et al., 1990, Isoda et al., 1995, as cited in Shao, 2010; Ortiz et al., 2008, 2009; Zaragoza et al., 2009, 2010).

2.2 High molecular weight biosurfactants

These are generally grouped together as polymeric biosurfactants. They are produced by a number of different bacteria and are composed of lipoproteins, proteins, polysaccharides, lipopolysaccharides or complexes containing several of these structural types (Ron & Rosenberg, 2001; Rosenberg & Ron, 1997, 1999). The most commonly studied biopolymer is emulsan (Fig. 2.), a lipopolysaccharide isolated from *Acinetobacter calcoacetius* RAG-1 ATCC 31012 with a molecular weight of around 1,000 kDa (Rosenberg et al., 1979).

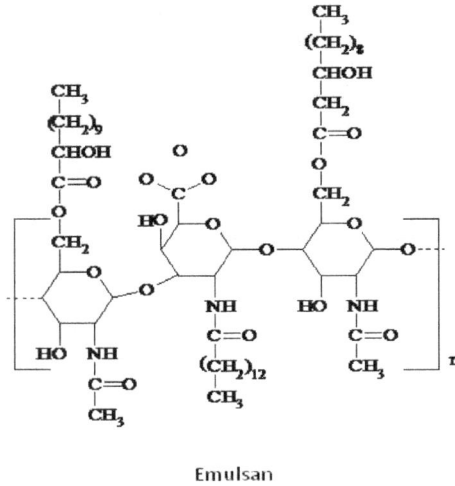

Emulsan

Fig. 2. Chemical structure of most known high molecular weight microbial surface active compound; emulsan.

RAG-1 emulsan is a complex of an anionic heteropolysaccharide and protein (Rosenberg & Kaplan, 1987, as cited in Rosenberg & Ron, 1999). Its surface activity is due to the presence of fatty acids, comprising 15% of the emulsan dry weight, which are attached to the polysaccharide backbone via O-ester and N-acyl linkages (Belsky et al., 1979, as cited in Rosenberg & Ron, 1999).

Another high molecular weight biosurfactant is alasan, a complex of an anionic polysaccharide and a protein with a molecular weight of around 1,000 kDa isolated from *Acinetobacter radioresistens* (Navon-Venezia et al., 1995, as cited in Smyth et al., 2010c). These high molecular weight biosurfactants generally possess effective emulsifying activity and are called bioemulsifiers. A large number of other polymeric compounds have been discovered but remain partially or totally uncharacterized (Smyth et al., 2010c). Little is known in general about these bioemulsifiers other than the producing organism and the overall chemical composition of the crude mixture. *Halomonas eurihalina* produces an extracellular sulfated heteropolysaccharide (Calvo et al., 1998, as cited in Rosenberg & Ron, 1999). *Pseudomonas tralucida* produced an extracellular acetylated polysaccharide that was effective in emulsifying several insecticides (Appaiah & Karanth 1991, as cited in Rosenberg & Ron, 1999).

Several bioemulsifiers are effective at high temperature, including the protein complex from *Methanobacterium thermoautotrophium* (De Acevedo et al., 1996, as cited in Rosenberg & Ron, 1999) and the protein-polysaccharide-lipid complex of *Bacillus stearothermophilus* ATCC 12980 (Gunjar et al., 1995, as cited in Rosenberg & Ron, 1999). Yeasts produce a number of emulsifiers, which are particularly interesting because of the food-grade status of several yeasts which allows use in food related industries. Liposan is an extracellular emulsifier produced by *Candida lipolytica* (Cirigliano & Carman, 1985, as cited in Rosenberg & Ron, 1999). It is composed of 83% carbohydrate and 17% protein. Mannanprotein emulsifiers are produced by *Saccharomyces cerevisiae* (Cameron et al., 1988, as cited in Rosenberg & Ron, 1999). Many of these bioemulsifiers have been used in the food, cosmetic, and petroleum industries (Rosenberg & Ron, 1999).

2.3 Properties and functions of biosurfactants

There is a growing interest in the study of the physicochemical and biological properties of biosurfactants because of their potential industrial applications (Cameotra & Makkar, 2004; Desai & Banat, 1997; Lang, 2002; Rodrigues et al., 2006a; Singh & Cameotra, 2004). The interesting biological activities displayed by these compounds constitute an added value to their potential uses (Lang et al., 1989; Lang & Wagner, 1993; Stanghellini & Miller, 1997, as cited in Sánchez et al., 2010). Due to these reasons, an intense research activity is currently directed toward identification of new biosurfactants and characterization of their chemical and biological properties (Biria et al., 2010; Morita et al., 2009a; Satpute et al., 2010a; Singh & Cameotra, 2004; Singh et al., 2007).

The most obvious property of biosurfactants compounds is their ability to effectively lower water surface tension, and a number of approaches that measure directly the surface activity of biosurfactants can be used as screening methods for their detection. Among them, the most frequently used as quick and simple techniques are the drop collapse (Bodour and Miller-Maier, 1998) and the oil spreading tests (Morikawa et al., 2000) (Fig. 3).

Fig. 3. Oil spreading test. This technique measures the diameter of clear zones caused when a drop of a biosurfactant-containing solution is placed on an oil–water surface (Morikawa et al., 2000). Photo courtesy by Fabrizio Rivardo.

An efficient biosurfactant can reduce the surface tension between pure water and air from 72mN/m to less than 30mN/m. Surfactin, as one of the most powerful biosurfactants, can reduce the surface tension of water from 72mN/m to 27mN/m (which is close to the minimum detectable value) at a concentration as low as 10 µM (Seydlová & Svobodová, 2008). Rhamnolipids can similarly achieve such level of reduction (Hisatsuka et al., 1971, as cited in Muthusamy et al., 2008; Mohammad Abdel-Mawgoud et al., 2010). The sophorolipids from *T. bombicola* have been reported to reduce the surface tension to 33mN/m (Muthusamy et al., 2008) while MELs and trehalose lipids to less than 30mN/m (Arutchelvi & Doble, 2010; Shao et al., 2010).

As surfactant monomers are added into solution, the surface or interfacial tension will decrease until the biosurfactant reaches the critical micelle concentration (CMC). The CMC is defined as the minimum concentration necessary to initiate micelle formation (Becher, 1965). Above the CMC no further reduction in surface or interfacial tension is observed. At the CMC, surfactant monomers begin to spontaneously associate into structured aggregates such as micelles, vesicles or continuous bilayers. These aggregates are produced as a result of numerous weak chemicals interactions such as hydrophobic, van der Waals and hydrogen bonding (Maier, 2003; Raza et al., 2010). Since no chemical bonds are formed, these structures are fluid-like and are easily transformed from one state to another as conditions such as electrolyte concentration and temperature are changed (Lin, 1996). The aggregate structure depends on the polarity of the solvent in which the surfactant is dissolved. In an aqueous solution, the polar head groups of a micelle will be oriented outward toward the aqueous phase, and the hydrophobic tails will associate in the core of the micelle within an oil-in-water micelle. In contrast, in oil, the polar head groups will associate in the center of the micelle while the hydrophobic tails will be oriented toward the outside within the water–in-oil micelle (Soberón-Chávez & Maier, 2010). Efficient surfactants have a low CMC, i.e. less surfactant is necessary to decrease the surface tension (Seydlová & Svobodová, 2008).

Biosurfactants are most effective and efficient at their CMC which can be 10–40 times lower than that of chemical surfactants, i.e. less surfactant is necessary to get a maximum decrease in surface tension (Desai & Banat, 1997). Another important property for industrial and biotechnological applications is that most biosurfactants surface activities are not affected by environmental conditions such as temperature and pH (Muthusamy et al., 2008) particularly those of glycolipids composition. It has been reported that lichenysin from B. licheniformis JF-2 was not affected by temperature (up to 50 °C), pH (4.5–9.0) and NaCl and Ca concentrations up to 50 and 25 g/l respectively. A lipopeptide from B. subtilis LB5a was also stable after autoclaving (121°C/20 min) and after 6 months at –18°C; the surface activity did not change from pH 5 to 11 and NaCl concentrations up to 20% (Muthusamy et al., 2008).

Moreover, unlike synthetic surfactants, microbial-produced compounds are easily degraded and are generally considered as low or non-toxic products and therefore, appropriate for pharmaceutical, cosmetic and food uses. Although little is known about the toxicity of microbial surfactants, some data in the literature suggest that they are less toxic than synthetic surfactants (Muthusamy et al., 2008). The synthetic anionic surfactant (Corexit) for example had an LC50 (lethal concentration to 50% of test species) against Photobacterium phosphoreum at approximately ten times lower concentrations than that for rhamnolipids, demonstrating the higher toxicity of the chemical-derived surfactant. It was also reported that biosurfactants needed higher effective concentration to decrease 50% of test population values (EC50) and were degraded faster than commercial dispersants. In another report, biosurfactants from P. aeruginosa were noted to have much less toxic and mutagenic activities when compared to synthetic surfactant Marlon A-350, which is widely used in the industry.

Understanding the functional mechanisms of biosurfactants and bioemulsifiers is of great help to discover interesting applications. Surfactin, one of the most powerful biosurfactants, is known to destabilize membranes disturbing their integrity and permeability (Bernheimer et al., 1970). This is due to changes in physical membrane structure or through disrupting protein conformations which alter important membrane functions such as transport and energy generation (Ortiz et al., 2008, 2009; Sánchez et al., 2009, 2010; Sotirova et al., 2008; Van Hamme et al., 2006; Zaragoza et al., 2009).

The molecular mechanisms of surfactin interactions with membrane structures have been described by Shaligram & Singhal (2010) and by Seydlová & Svobodová (2008). A key step for membrane destabilization and leakage is the dimerization of surfactin into the bilayer (Carrillo et al., 2003). The hypothetical mechanisms of surfactin interactions with membranes exhibit a complex pattern of effects such as insertion into the lipid bilayers, chelating mono- and divalent cations, modification of membrane permeability by channel formation or membrane solubilization by a detergent-like mechanism. *In vitro*, the incorporation of surfactin into the membrane gives rise to dehydration of the phospholipid polar head groups and the perturbation of lipid packing which strongly compromise the bilayer stability, leading to the disturbance of the membrane barrier properties.

These structural fluctuations may well explain the primary mode of the antibiotic action and the other important biological effects of this lipopeptide (Carrillo et al., 2003). The extent of perturbation of the phospholipid bilayer correlates with the concentration of surfactin. At low concentrations surfactin penetrates readily into the cell membrane, where it is completely miscible with the phospholipids and forms mixed micelles. At moderate concentrations, the lipopeptide forms domains segregated within the phospholipid bilayer that may contribute to the formation of ion-conducting pores in the membrane leading to membraine disruption and permeabilization at high concentrations, showing a stronger activity than that of Triton (Heerklotz et al., 2007, as cited in Seydlová & Svobodová, 2008).

As biological amphiphilic molecules, biosurfactants naturally tend to self-assemble into hierarchically ordered structures using hydrogen bonding, hydrophobic and van der Waals interactions (Kitamoto et al., 2009). Glycolipids, and in particular MELs, are well known for their self-assembling properties, that are influenced by the stereochemistry of the saccharide head groups (Kitamoto et al., 2005). Some of glycolipid type surfactants, which possess relatively large hydrophilic head groups as compared to the hydrophobic parts, generally form micelles in a dilute aqueous solution. Other than spherical micelles, they also form oblate (disk-like) and prolate (rod-like) structures (Söderman, 2000, as cited in Kitamoto et al., 2009). As the surfactant concentration further increases, glycolipid/water systems start to form a range of liquid crystalline phases. In particular, glycolipid biosurfactants spontaneously self-assemble into a variety of molecular assemblies with well-defined and/or unique structures, such as sponge (L3), cubic (V2), hexagonal (H2), or lamellar (L_α) configurations (Imura et al., 2007, as cited in Kitamoto et al., 2009).

Among these molecular assemblies, vesicles are one of the most intensively studied ones. MELs in particular, due to their efficient molecular orientation property and effective balance between hydrophilic and hydrophobic groups, are able to form giant vesicles of diameter larger than 10 μm (Kitamoto et al., 2002). In comparison, rhamnolipids show a pH-sensitive conversion of molecular assemblies due to the presence of a carboxyl group on the side chain (Kitamoto et al., 2002). This leads rhamnolipids to form micelles at pH more than 6.8, lipid particles at pH 6.6-6.2, lamella structures at pH 6.5-6.0, and finally vesicles in the size of 50-100 nm at pH 5.8-4.3. Glycolipid biosurfactant-based vesicles or bilayer membranes appear, thus, to be very promising for exploiting useful nanostructured materials and/or systems.

Another function of microbial surface-active molecules with interesting biotechnological potential is the ability to form stable emulsions (Fig. 4.). High molecular-mass biosurfactants are in general better emulsifiers than low-molecular-mass biosurfactants and are thus called bioemulsifiers. Bioemulsifiers, can form and stabilize oil in water or water in oil emulsions, but are not necessarily efficient detergents that are able to demonstrate remarkable surface tension reduction (Dastgheib et al., 2008). Emulsions can be produced with prolonged lifespan of months and years. Liposan, for example, does not reduce surface tension, but has been used successfully to emulsify edible oils (Cirigliano & Carman, 1985). Emulsan is an effective emulsifier at low concentrations (0.01-0.001%) representing emulsan-to-hydrocarbon ratios of 1:100 to 1:1000, while exhibiting considerable substrate specificity (Ron & Rosemberg, 2001). Polymeric surfactants offer additional advantages because they coat droplets of oil, thereby forming stable emulsions. This property is especially useful for making oil/water emulsions for cosmetics and food industries (Muthusamy et al., 2008).

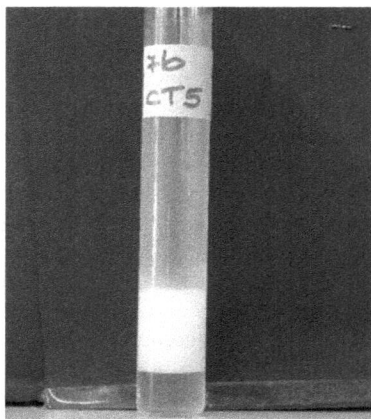

Fig. 4. Example of emulsion produced by the bioemulsifier extracted from the bacterial strain 7bCT5, isolated from a Cambodian soil. This emulsion is stable since 2009.

2.3.1 Natural roles

Although biosurfactants are produced by a large number of microorganisms and are clearly significant in many aspects of growth, it is difficult to generalize on their roles in microbial physiology. Due to their very diverse chemical structures and surface properties, different groups of biosurfactants may have different natural roles in the growth of the producing microorganisms and probably provide advantages in a particular ecological niche. Ron & Rosenberg, (2001) and Van Hamme et al., (2006) recently reviewed the physiological roles of microbial surface-active compounds. Some biosurfactants are essential for the motility of the microorganisms (gliding and swarming). For example, serrawettin plays a fundamental role for surface locomotion and access to water repelling surfaces for *Serratia marcescens* whereas surfactin, together with flagellar biosynthesis, are crucial for swarming motility in *B. subtilis* (Arutchelvi et al., 2008; Van Hamme et al., 2006). Bioemulsifiers also play an important role in regulating the attachment-detachment of microorganisms to and from surfaces (Van Hamme et al., 2006).

In addition, bioemulsifiers are involved in cell-to-cell interactions such as bacterial pathogenesis, quorum sensing and biofilm formation, maintenance and maturation. Rhamnolipids, for example, are essential to maintain the architecture of the biofilms and are considered as one of the virulence factors in *Pseudomonas* sp. (Arutchelvi et al., 2008; Ron & Rosenberg, 2001; Van Hamme et al., 2006). Rhamnolipids, mannosylerythritol lipid and surfactin show antimicrobial and antibiotic properties thus conferring a competitive advantage to the organism during colonization and cell–cell competition. In addition cellular differentiation, substrate accession and resistance to toxic compounds are all roles attributed to microbial surface-active compounds. Their most widespread role however is believed to be the interaction between microbes and insoluble substrates such as hydrocarbons. Some biosurfactants/bioemulsifiers enhance the growth of bacteria on hydrophobic water-insoluble substrates by increasing their bioavailability, presumably by increasing their surface area, desorbing them from surfaces and increasing their apparent solubility (Neu, 1996; Ron & Rosenberg, 2001; Van Hamme et al., 2006).

3. Biomedical applications of microbial surface-active compounds

The use and potential commercial applications of biosurfactants in the medical field have increased during the past decade. Their antibacterial, antifungal and antiviral activities make them relevant molecules for applications in combating many diseases and as therapeutic agents. Furthermore, biosurfactants are generally considered safer than synthetic pharmaceuticals, due to their biological origin. Their pertinence in these fields is related to their biological properties such as the ability to disrupt membranes leading to cell lysis and metabolite leakage through increased membrane permeability and hence antimicrobial activity. Moreover, similarly to organic-conditioning films, their ability to partition at the interfaces can affect the adhesion properties of cells/microorganisms. Biomedical applications of biosurfactants have been thoroughly described (Banat et al., 2010; Cameotra & Makkar, 2004; Rodrigues et al., 2006a; Rodrigues & Teixeira, 2010; Seydlová & Svobodová, 2008; Singh & Cameotra, 2004).

3.1 Antimicrobial activity of biosurfactants

The search for new antimicrobial drugs remains a major concern nowadays because of the newly emerged pathogenic microorganisms and traditional others which have become virtually unresponsive to existing antibiotics. In fact, no novel or effective chemical antibiotics have been discovered during the last few decades (Hancock & Chapelle, 1999). Microbial metabolites have been recognized as a major source of compounds endowed with ingenious structures and potent biological activities (Donadio et al., 2002). Among these, some biosurfactants have been reported to be suitable alternatives to synthetic medicines and antimicrobial agents and may therefore be used as effective and safe therapeutic agents (Banat et al., 2000; Cameotra & Makkar, 2004; Singh & Cameotra, 2004).

3.1.1 Antibacterial activity

Lipopeptides have the most potent antimicrobial activity and have been a subject of several studies on the discovery of new antibiotics. The antibiotic activity is due to the ability of molecules of lipopeptide biosurfactants to self-associate and form a pore-bearing channel or micellular aggregate inside a lipid membrane (Carrillo et al., 2003; Deleu et al., 2008). Surfactin, in particular, has been associated with several physical and biological actions, such as antimicrobial, antiviral, anti-mycoplasma and haemolytic activities. It can penetrate into the membrane through hydrophobic interactions, thus influencing the ordering of the hydrocarbon chains and thus varying the membrane thickness (Bonmatin et al., 2003). Such membrane disruptions are a nonspecific mode of action and are advantageous for action on different cell membranes of both Gram-positive and Gram-negative bacteria (Lu et al., 2007). It has been suggested that such action by surfactin type peptides on membrane integrity rather than other vital cellular processes may perhaps constitute the next generation of antibiotics (Rodrigues & Teixeira, 2010). Similar bioactive fractions from the marine *Bacillus circulans* biosurfactant had antimicrobial action against various Gram-positive and Gram-negative pathogenic and semi-pathogenic bacteria including *Micrococcus flavus, Bacillus pumilis, Mycobacterium smegmatis, Escherichia coli, Serratia marcescens, Proteus vulgaris, Citrobacter freundii, Proteus mirabilis, Alcaligenes faecalis, Acetobacter calcoaceticus, Bordetella bronchiseptica, Klebsiella aerogenes* and *Enterobacter cloacae* (Das et al., 2008). The chemical identity of this bioactive biosurfactant fraction showed overlapping patterns with that of surfactin lipopeptides and lichenysin. Mild antimicrobial action was also observed against methicillin-resistant *Staphylococcus aureus* (MRSA) and other MDR pathogenic strains. The biosurfactant was also found to be nonhaemolytic in nature thus indicating possible use as a drug in antimicrobial chemotherapy.

Very recently Huang et al., (2011) evaluated antimicrobial activity of surfactin and polylysine against *Salmonella enteritidis* in milk using a response surface methodology and showed *S. enteritidis* to be very sensitive to both molecules with minimum inhibitory concentrations of 6.25 and 31.25 µg/mL, respectively. The optimization of antimicrobial activity indicated that *S. enteritidis* could be reduced by 6 orders of magnitude at a temperature of 4.45°C, action time of 6.9 h, and concentration of 10.03 µg/mL (surfactin/polylysine weight ratio, 1:1).

In addition to surfactin, *Bacillus subtilis* strains produce a broad spectrum of bioactive peptides with great potential for biomedical applications, such as fengycin (Vanittanakom et al., 1986) and the iturin compounds: iturins (Besson et al., 1978; Peypoux et al., 1978), mycosubtilins (Peypoux et al., 1986), and bacillomycins (Peypoux et al., 1984), all of which are amphiphilic surface and membrane-active compounds with potent antimicrobial activities. Huang et al., (2007) reported that a lipopeptide antimicrobial substance produced by *B. subtilis* fmbj strain, which is mainly composed of surfactin and fengycin, was able to inactivate endospores of *B. cereus* through damaging the surface structure of the spores as seen by Transmission Electron Microscopy.

Lichenysin, pumilacidin and polymyxin B (Grangemard et al., 2001; Landman et al., 2008; Naruse et al., 1990; Yakimov et al., 1995) are other antimicrobial lipopeptides produced by *Bacillus licheniformis, Bacillus pumilus* and *Bacillus polymyxa*, respectively. Polymyxin B, in particular, due to its high affinity for the lipid moieties of lipopolysaccharide, has shown antibacterial activities against a wide variety of Gram-negative pathogens. Being a cationic agent, it binds to the anionic bacterial outer membrane, leading to a detergent effect that disrupts membrane integrity. Important nosocomial pathogens such as *Escherichia coli, Klebsiella* spp., *Enterobacter* spp., *Pseudomonas aeruginosa*, and *Acinetobacter* spp. are usually susceptible to polymyxins and considerable activity has been reported against *Salmonella* spp., *Shigella* spp., *Pasteurella* spp., and *Haemophilus* spp. (Landman et al., 2008).

Another promising example of an antimicrobial lipopeptide that is under commercial development is daptomycin (Cubicin®). It has been approved for the treatment of skin infections by the FDA in 2003 (Giuliani et al., 2007, as cited in Seydlová & Svobodová, 2008). Daptomycin produced by *Streptomyces roseosporus* has been shown to be highly active against multiresistant bacteria such as MRSA (Tally & De Bruin, 2000, as cited in Seydlová & Svobodová, 2008). Another lipopeptide with antimicrobial activity and other interesting biological properties is viscosin, a cyclic lipopeptide from *Pseudomonas* (Saini et al., 2008).

Glycolipids, both rhamnolipids (Abalos et al., 2001; Benincasa et al., 2004) and sophorolipids (Kim et al., 2002; Van Bogaert et al., 2007) also have shown interesting antimicrobial activities (Fig. 5). Benincasa et al., (2004) reported that a mixture of six rhamnolipides homologues performed very well against *Bacillus subtilis* with a MIC of 8 µg/mL. Mannosylerythritol lipids (MEL-A and MEL-B) produced by *Candida antarctica* strains have also been reported to exhibit antimicrobial action against Gram-positive bacteria (Kitamoto et al., 1993).

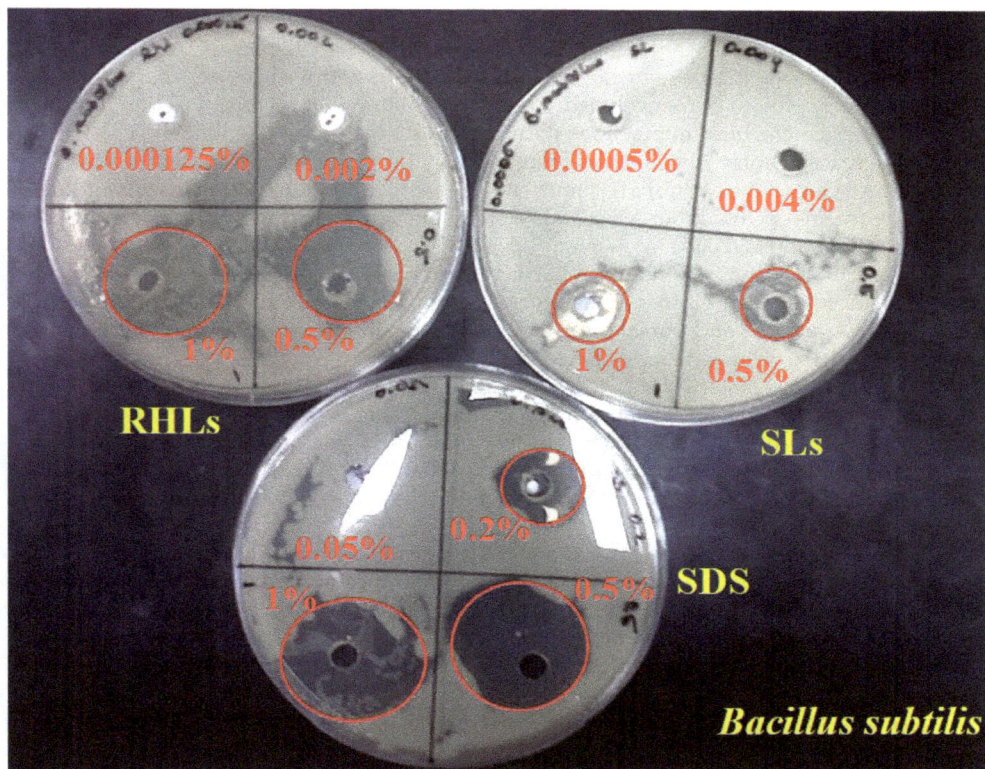

Fig. 5. Measuring antimicrobial activity for rhamnolipids, sophorolipids and SDS at various concentrations above and below the CMC for these surface active molecules against *Bacillus subtilis*, red circles showing clearing/inhibition zones.

Very recently, Nitschke et al., (2010) reported rhamnolipids produced by *P. aeruginosa* LBI with antimicrobial activity against several bacteria and fungi, including *Bacillus cereus*, *Staphylococcus aureus*, *Micrococcus luteus*, *Mucor miehei* and *Neurospora crassa*. Another glycolipid, flocculosin, a cellobiose lipid produced by the yeast-like fungus *Pseudozyma flocculosa*, was particularly effective against *Staphylococcus* species, including MRSA. Its antibacterial activity was not influenced by the presence of common resistance mechanisms against methicillin and vancomycin and it was able to eliminate *C. albicans* cells in a very short period of time (Mimee et al., 2009).

Trehalose lipids produced by *Tsukamurella* sp. strain DSM 44370 together with trisaccharide and tetrasaccharide lipids also showed some activity against Gram-positive bacteria, with the exception of the pathogenic strain *Staphylococcus aureus*, whereas Gram-negatives were either slightly or not inhibited at all (Vollbrecht et al., 1999, as cited in Franzetti et al., 2010b). Studies carried out to elucidate the molecular interactions between this biosurfactant and the lipidic component of the membrane showed that trehalose lipid increased the fluidity of phosphatidylethanolamine and phosphatidylserine membranes and formed domains in the fluid state and did not modify the macroscopic bilayer organization (Ortiz et al., 2008, 2009).

3.1.2 Antiviral activity

Antiviral effects have also been reported for surfactin and its analogues (Naruse et al., 1990). More effective inactivation of enveloped viruses, such as retroviruses and herpes viruses, were noted compared to non-enveloped viruses, suggesting that inhibitory action links may be due to physico-chemical interactions with the virus envelope (Vollenbroich et al., 1997a). Antiviral activity of some lipopeptides therefore may take place as a result of the viral lipid envelope and capsid disintegration due to ion channels formation, with consequent loss of the viral proteins involved in virus adsorption and/or penetration (Jung et al., 2000; Seydlová & Svobodová, 2008).

In vitro experiments showed that both surfactin and fengycin produced by *B. subtilis* fmbj were able to inactivate cell-free virus stocks of porcine parvovirus, pseudorabies virus, newcastle disease virus and bursal disease virus and could effectively inhibit infections and replication of these viruses (Huang et al., 2006).

Sophorolipids are also claimed to have activity against human immunodeficiency virus (Shah et al., 2005) and trehalose lipids (namely trehalose dimycolate, TDM) conferred higher resistance to intranasal infection by influenza virus in mice though inducing proliferation of T-lymphocytes bearing gamma/delta T-cell receptors, associated with the maintenance of acquired resistance to the infection (Hoq et al., 1997, as cited in Franzetti et al., 2010b).

Rhamnolipid alginate complex also showed significant antiviral activity against herpes simplex virus types 1 and 2. In particular, they significantly inhibited the herpesvirus cytopathic effect in the Madin-Darby bovine kidney cell line (Remichkova et al., 2008). The suppressive effect of the compounds on herpes simplex virus replication was dose-dependent and occurred at concentrations lower than the critical micelle concentration.

3.1.3 Anti mycoplasma activity

Some investigations have shown interesting anti-mycoplasma effects for surfactins. Mycoplasma contamination in cell culture is a frequently occurring serious limitation to biomedical research, particularly when it affects the irreplaceable cell lines which ultimately ends up destroyed. Earlier studies showed that surfactin treatment of mammalian cells that had been contaminated with mycoplasmas permitted specific inactivation of mycoplasmas without significantly damaging effects on cell metabolism in the culture (Vollenbroich et al., 1997b). In a more recent study, surfactin was used to eliminate mycoplasma from an extensively infected irreplaceable hybridoma cell line (Kumar et al., 2007). There were apparent indications of limited elimination, suggesting the possible use of surfactin in achieving total decontamination. However, it was observed that surfactin was toxic to the infected hybridoma cells plated at various cell densities and exposure times, therefore it was suggested that preliminary tests should be carried out to determine the cytotoxicity of surfactin before use in decontamination.

Another study confirmed surfactin potential to eliminate mycoplasma cells independently of the target cell, which is a significant advantage over the mode of action of conventional antibiotics (Fassi et al., 2007). This study also reported that surfactin exhibited a synergistic effect in combination with enrofloxacin, and resulted in mycoplasma-killing activity of about two orders of magnitude greater than when the molecules were used separately.

3.1.4 Antifungal activity

The antifungal activities of biosurfactants have long been known, although their action against human pathogenic fungi has been rarely described (Abalos et al., 2001; Chung et al.,

2000; Tanaka et al., 1997). The previously mentioned cellobiose lipid flocculosin isolated from *Pseudozyma flocculosa*, was shown to display *in vitro* antifungal activity against several pathogenic yeasts, associated with human mycoses, including *Candida lusitaniae*, *Cryptococcus neoformans*, *Trichosporon asahii* and *Candida albicans* (Mimee et al., 2005). This product positively inhibited all pathogenic strains tested under acidic conditions and showed synergistic activity with amphotericin B. Moreover, no significant cytotoxicity was detected when tested against human cell lines. In nature, flocculosin is part of *P. flocculosa* biocontrol arsenal against other fungi. Recent reports however have suggested that flocculosin is also used by *P. flocculosa* as a nutrient source when experiencing food limitation and the molecule is rapidly deacylated under alkaline conditions losing its antimicrobial activity which may explain conflicting results concerning the antimicrobial activity of this class of glycolipids (Mimee et al., 2009).

Other antifungal activity of biosurfactants against phytopathogenic fungi has also been described. It has been recently demonstrated that glycolipids, such as cellobiose lipids (Kulakovskaya et al., 2009, 2010) and rhamnolipids (Debode et al., 2007, Banat et al, 2010) and cyclic lipopeptides (Tran et al., 2007, 2008, as cited in Banat et al, 2010), including surfactin, iturin and fengycin (Kim et al., 2010; Arguelles-Arias et al., 2009, Chen et al., 2009, Grover et al., 2010, Mohammadipour et al., 2009, Snook et al., 2009) can all have varying degrees of antimicrobial activities.

3.2 Antiadhesion activity of biosurfactants

Microbial biofilms formation on medical and technical equipment is an important and mostly hazardous occurrence, especially as the bacteria within such biofilms usually become highly resistant to antibiotics and adverse environmental challenges. Several approaches have been adopted in order to limit pathogen colonization. Strict hygienic practices by healthcare personnel such as hand washing and regular disinfection of equipment and environment become of grave importance. However, it should be noted that routine disinfection is becoming controversial as frequent application becomes less effective (Dettenkofer et al., 2004, 2007, Kramer et al., 2006, as cited in Falagas & Makris, 2009).

3.2.1 Biofilms on medical devices

Device-related infections are often identified as having a biofilm aetiology and biofilm formation can occasionally be facilitated by the host inflammatory response molecules which can make adhesion to the surface of the device easier (Hall-Stoodley et al., 2004). Almost all kind of surfaces are suitable to be colonized by biofilms (Donlan & Costerton, 2002). Biomedical devices are not the exception, biofilms are often found on the surface of urinary catheters (Stickler, 2008), central venous catheters (Petrelli et al., 2006), heart valves (Litzler et al., 2007), voice prostheses (Buijssen et al., 2007), contact lenses (Imamura et al., 2008), hip prostheses (Dempsey et al., 2007) and intrauterine devices (Chassot et al., 2008). Current biofilm preventive strategies are essentially aimed at coating medical surfaces with antimicrobial agents, a process not always successful (Basak et al., 2009; von Eiff et al., 2005). Bacteria in biofilms become highly resistant to antibiotics, and so they evade host defenses withstanding antimicrobial chemotherapy (Morikawa, 2006). Since nosocomial infections remain an important problem even for hospitals with strict infection control programmes, infection control measures remain highly sought after (Falagas & Makris, 2009). Development of successful technologies based on the biofilm formation and growth control is expected to be a major breakthrough in the clinical practice and preventive medicine.

To eliminate biofilm formation novel compounds capable of specifically targeting biofilm growth while causing no adverse toxicity to the environment of application are needed. Several reports have suggested that, in addition to their direct action against pathogens, biosurfactants are able to interfere with biofilm formation, modulating microbial interaction with interfaces (Federle & Bassler 2003; Merk et al., 2005; Neu, 1996; Rasmussen & Givskov, 2006; Rodrigues et al., 2006b, 2006c; Rodrigues et al., 2007).

Surfactin for example has shown to be an important part of a list of biofilm controlling agents. Surfactin is able to inhibit biofilm formation of *Salmonella typhimurium, S. enterica, E. coli* and *Proteus mirabilis* in polyvinyl chloride wells, as well as vinyl urethral catheters (Mireles et al., 2001). Many *Salmonella* species are important opportunistic pathogens of the urinary tract system. Recently, two lipopeptide biosurfactants, produced by *B. subtilis* V9T14 and *B. licheniformis* V19T21, showed the ability to selectively inhibit biofilm formation by pathogenic strains on polystyrene (Rivardo et al., 2009). In particular, *S. aureus* ATCC 29213 and *E. coli* CFT073 biofilm formation were decreased by 97% and 90%, respectively. V9T14 biosurfactant was active on the Gram-negative strain yet ineffective against the Gram-positive while the opposite was observed for V19T21 biosurfactant. These effect were observed either by coating the polystyrene surface with these compounds or by adding the biosurfactant to the inoculum. The chemical characterization of V9T14 lipopeptide biosurfactant carried out by LC/ESI-MS/MS revealed that it was composed of 77% of surfactin and of 23% of fengycin (Pecci et al., 2010).

The activity of $AgNO_3$ combined with the lipopeptide biosurfactant V9T14 has also been studied against a preformed *E. coli* biofilm on the Calgary Biofilm device (Rivardo et al., 2010). Results indicated that the activity of silver can be synergistically enhanced by the presence of V9T14, allowing a reduction in the quantity of silver used to achieve greater antimicrobial impact. The concentrations of silver in the silver–biosurfactant solutions were 129 to 258 fold less than the concentrations needed when silver was used alone. In another study, the V9T14 biosurfactant in association with antibiotics led to a synergistic increase in the efficacy of antibiotics in *E. coli* CFT073 biofilm innhibition and, in some combinations, to total eradication of the uropathogenic strain biofilm (Rivardo et al., 2011).

In another recent work, marine bacterial culture supernatants of *Bacillus pumilus* and *B. indicus* significantly inhibited the initial attachment process and biofilm formation and dispersal of mature biofilms of *Vibrio* spp. strains (Nithya & Pandian, 2010). The bacterial supernatants also reduced the surface hydrophobicity of *Vibrio* spp. which is one of the important requirements for biofilm development. Valle et al., (2006) observed that distinct serotypes of group II capsular polysaccharides, produced by the uropathogenic *E. coli* (UPEC strain CFT073) behaved like surface-active polymers that displayed anti-adhesion properties. The treatment of abiotic surfaces with group II capsular polysaccharides drastically reduced both initial adhesion and biofilm development of important nosocomial pathogens.

More recently, the effect of different temperatures on the anti-adhesive activity of surfactin and rhamnolipid biosurfactants was tested on polystyrene surfaces, regarding the attachment of *Staphylococcus aureus, Listeria monocytogenes*, and *Micrococcus luteus* (Zeraik & Nitschke, 2010). Surfactin inhibited bacterial adhesion at all tested conditions, and its activity increased with the decrease in temperature, giving a 63–66% adhesion reduction in the bacterial strains at 4°C. Rhamnolipids promoted a slight decrease in the attachment of *S. aureus* but were not as effective. The ability of rhamnolipid biosurfactant to inhibit adhesion of microorganisms to silicone rubber was also investigated in a parallel-plate flow chamber (Rodrigues et al., 2006c). The results showed an effective reduction in the initial deposition rates and in the number of bacterial cells adhering after 4h, for several microorganisms. Moreover, perfusing the flow chamber with biosurfactant containing solution followed by the passage of a liquid–air interface produced high detachment (96%) of adhered cells for several microorganisms. These capabilites have a lot of implications regarding biofilm formation and microbial contamination and establishments on such biomedical devices made of such compounds.

Antibiofilm activity was also reported for a glycolipid biosurfactant isolated from another marine bacterium *Brevibacterium casei* MSA19 against pathogenic biofilms *in vitro* (Kiran et al., 2010a). The purified glycolipid disrupted the biofilm formation under dynamic conditions and the biofilm-forming capacity of both mixed culture and individual human and fish pathogenic strains was significantly inhibited at 30 mg/mL glycolipid. Raya et al., (2010) analyzed the effects of rhamnolipids and shear on initial attachment of *Pseudomonas aeruginosa* PAO1 in glass flow chambers. The presence of rhamnolipids significantly reduced the initial attachment of PAO1, even at the low concentration of 13 mg/L. Prewashing the cells with a 100 mg/L rhamnolipid solution, however, did not affect the attachment significantly. The initial cell attachment increased with increasing shear at the very low shear range (up to 3.5–5.0 mN/m²), however the attachment could be minimized with further increase of the shear.

The biosurfactant Lunasan produced by the yeast *Candida sphaerica* UCP0995 completely inhibited the adhesion of *Streptococcus agalactiae, Streptococcus sanguis* 12, *Streptococcus mutans, Streptococcus mutans* NS, *Staphylococus epidermidis, Staphylococus aureus* and *Candida albicans* on plastic tissue culture plates at a concentration of 10 mg/ml and ≈92% inhibition of adhesion occurred for *Pseudomonas aeruginosa* (Luna et al., 2011). Lunasan, tested at the same concentration, also showed antimicrobial activity against the strains *Streptococcus oralis* (68%), *Candida albicans* (57%), and *Staphylococcus epidermidis* (57.6%). The same research group also described antiadhesive and antimicrobial activities of Rufisan, a biosurfactant produced by the yeast *Candida lipolytica* UCP 0988 (Rufino et al., 2011). Crude biosurfactant showed anti-adhesive activity at ≥0.75 mg/L against most of the microorganisms tested (*Staphylococus aureus, Streptococcus agalactiae, Streptococcus mutans* NS) and the anti-adhesive property was proportional to the concentration of the biosurfactant while antimicrobial activities were also observed at higher biosurfactant concentrations.

In conclusion, the anti-adhesive activity of biosurfactants against several pathogens indicates their potential utility as coating agents for medical insertional materials that may lead to a reduction in a large number of hospital infections without the need for use of synthetic drugs and chemicals.

3.2.2 Biofilms on food processing surfaces

In addition to the treatment of biomaterials used for medical devices, biosurfactants have also been used in the pre-treatment of material surfaces found in food-processing environments. Pathogenic bacteria implicated in food-borne illness outbreaks are able to form biofilms on food contact surfaces that are more resistant to sanitation than free-living cells (Kalmokoff et al., 2001; Kim et al., 2006; Stepanovic et al., 2004). The pre-conditioning of surfaces using microbial surface-active compounds may be an interesting strategy for preventing the adhesion of food-borne pathogens to solid surfaces. Meylheuc et al., (2006b) demonstrated that the preconditioning of stainless steel surfaces with an anionic biosurfactant produced by *Pseudomonas fluorescens* reduced the number of *L. monocytogenes* LO28-adhering cells and thus favoured the bactericidal activities of the disinfectants sodium hypochlorite (NaOCl) and peracetic acid/hydrogen peroxide (PAH).

Similarly, biosurfactants obtained from *Lactobacillus helveticus* and *P. fluorescens* were able to inhibit the adhesion of four *Listeria* strains to stainless steel (Meylheuc et al., 2006a). Whichever strain of *L. monocytogenes* used in combination with biosurfactants, the anti-adhesive biological coating developed both reduced the total adhering flora and the viable and culturable adherent bacteria on stainless steel surfaces. More recently, another group investigated the effect of rhamnolipid and surfactin biosurfactants on the adhesion of the food pathogens *E. sakazakii*, *L. monocytogenes* and *S. enteritidis* to polypropylene and stainless steel surfaces (Nitschke et al., 2009). Preconditioning with surfactin, rather than rhamnolipid, caused a reduction in the number of adhering cells particularly of *L. monocytogenes* and to some extent *E. sakazakii* on stainless steel. Surfactin showed a significant decrease in the adhesion on polypropylene of all strains. The adsorption of surfactin on polystyrene also reduced the adhesion of *S. enteritidis*- and *L. monocytogenes*-growing cells. In addition, surfactin was able to delay bacterial adhesion within short contact periods using non-growing cells or longer contact periods using growing cells.

Other antimicrobial and antiadhesive properties of a biosurfactant produced by *Lactobacillus paracasei* ssp. *paracasei* A20 isolated from a Portuguese dairy plant were also described (Gudiña et al., 2010). The biosurfactant had antimicrobial activity against a broad range of microorganisms including the pathogenic *C. albicans*, *E. coli*, *S. aureus*, *S. epidermidis* and *Streptococcus agalactiae* while exhibiting a considerable antiadhesive activity against a wide range of microorganisms.

The activity demonstrated by biosurfactants suggests that they could be considered as new tools in developing strategies to prevent or delay microbiological colonization of industrial plant surfaces used in foodstuffs preparation.

3.3 Probiotics biosurfactants activity

Probiotics are: "Live microorganisms which when administered in adequate amounts confer a health benefit on the host". They have been reported to have positive effects on the maintenance of human health (Gupta & Garg 2009). Interest in probiotics has gained great significance due to the increasing antimicrobial resistance of bacteria worldwide. Evidence suggests that probiotic organisms may have a role in lowering the incidence or the duration of antibiotic-related diarrhea, contributing to the prevention or treatment of vaginal candidiasis, bacterial vaginosis and recurrent lower urinary tract infections. Furthermore, they encourage improved immunological defense responses and can decrease the activity of numerous toxic antimetabolites (Falagas et al., 2006a, 2006b, 2007).

Probiotics mechanisms of action vary, however, some are known to produce various antimicrobial agents such as organic acids, hydrogen peroxide, carbon peroxide, diacetyl, low molecular weight antimicrobial substances and bacteriocins (Merk et al., 2005). In addition, probiotics have long been known also for the capacity to interfere with the adhesion and formation of biofilms of pathogens to epithelial cells of urogenital and intestinal tracts (Reid et al., 1998, 2001). The mechanisms of this interference include the release of surface active molecules (Gudiña et al., 2010; Rodrigues et al., 2006d). Hong et al., (2005) reported the production of antimicrobial lipopeptides by *Bacillus* probiotics products as the main mechanisms by which they inhibit the growth of pathogenic microorganisms in the gastrointestinal tract. Similarly, competition with other microorganisms for adherence to epithelial cells as well as biosurfactants production are well known mechanisms used by *Lactobacillus* probiotics to interfere with vaginal pathogens (Barrons & Tassone, 2008; Cribby et al., 2008; Falagas et al., 2007).

Several investigators have pointed to evidence that probiotic type microorganisms and their biosurfactants may antagonize the growth of nosocomial pathogens on inanimate surfaces (Rodrigues et al., 2004a, 2004b, 2006b, 2006c; Walencka et al., 2008). Falagas & Makris, (2009) reviewed studies involving *in vitro* experiments on the potential role of probiotics microorganisms and their products in the inhibition of bacterial or fungal colonisation of artificial surfaces, such as vinyl urethral catheters and silicon rubber voice prostheses (Busscher et al., 1997, 1998; Velraeds et al., 1996, 1997, 2000; Rodrigues et al., 2004a, 2004b, 2006b, Van der Mei et al., 2000). The majority of the investigators examined the preconditioning of the materials surfaces with probiotic biosurfactants, while others added probiotic biosurfactant producing strains to examin adhesion or biofilm development. It was generally demonstrated that both probiotics microorganisms alone (mainly *Streptococcus thermophilus* and *Lactobacillus* spp. strains) or their biosurfactants were able to antagonize growth and development of potentially pathogenic microorganisms including *Staphylococcus aureus, Staphylococcus epidermidis, Streptococcus* spp., *Enterococcus faecalis, Candida albicans, Candida tropicalis* (Busscher et al., 1997; Van Hoogmoed et al., 2000).

Rodrigues et al., (2004a) demonstrated that the biosurfactant obtained from the probiotic bacterium *Lactococcus lactis* 53 was able to inhibit the adhesion of bacterial pathogens to silicone rubber with an adsorbed biosurfactant layer. Adhesion of yeasts was also decreased in the presence of biosurfactant, but to a lesser extent. In another work, using an artificial throat model, the same authors showed that biosurfactants obtained from probiotic strains greatly reduced microbial numbers on voice prostheses and induced a decrease in the airflow resistance of voice prostheses after biofilm formation, which may prolong the lifetime of indwelling silicone rubber voice prostheses (Rodrigues et al., 2004b).

In a more recent work, it was demonstrated that the preconditioning of silicon rubber with a biosurfactant produced by the strain *Streptococcus thermophilus* A reduced adhering bacterial pathogens by up to 97% and adhering *Candida* spp. by up to 70% (Rodrigues et al., 2006b). Velraeds et al., (1996) also reported on the inhibition of adhesion of pathogenic enteric bacteria by a biosurfactant produced by a *Lactobacillus* strain and later showed that the biosurfactant caused an important dose-related inhibition of the initial deposition rate of *E. coli* and other bacteria adherent on both hydrophobic and hydrophilic substrata (Velraeds et al., 1997).

Another interesting application area that is gaining increased interest relates to probiotics use in preventing oral infections (Çaglar et al., 2005; Hatakka et al., 2007; Kõll et al., 2008; Meurman, 2005; Meurman & Stamatova, 2007). Van Hoogmoed et al., (2004) demonstrated that *Streptococcus mitis* biosurfactant inhibited adhesion of *Streptococcus sobrinus* HG 1025 and *Streptococcus mutans* ATCC 25175 to bare enamel, while *S. mitis* biosurfactant was able to inhibit the adhesion of *S. sobrinus* HG 1025 to salivary pellicles. The authors later reported that these reductions may be attributed to increased electrostatic repulsion between the bacteria and the biosurfactant-coated pellicles (Van Hoogmoed et al., 2006).

New biosurfactant molecules produced by probiotic bacteria are reported from dairy products and environment. Recent work by Walencka et al., (2008) demonstrated that surfactants obtained from three *Lactobacillus acidophilus* strains inhibited *S. epidermidis* and *S. aureus* biofilm integrity and formation. Moreover, surfactant addition to preformed mature biofilms accelerated their dispersal and altered the characteristics of the biofilm morphology. A novel xylolipid biosurfactant from *Lactococcus lactis*, a probiotic strain isolated from a traditional Indian fermented dairy product, showed a good antibacterial activity against clinical pathogens of *E. coli* and MRSA strains (Saravanakumari & Mani, 2010). Xylolipid was non-pathogenic and safe for oral consumption and dermal applications, suggesting that it could be safely used as a therapeutic agent or as a preservative in food or cosmetic products.

In another recent work, a biosurfactant producing strain, *Lactobacillus* sp. CV8LAC, isolated from fresh cabbage, showed interesting antiadhesive activity against two *C. albicans* pathogenic biofilm-producing strains (CA-2894 and DSMZ 11225) (Fracchia et al., 2010). The CV8LAC biosurfactant significantly inhibited the adhesion of fungal pathogens to polystyrene microtiter plates in pre-coating and co-incubation experiments. In pre-coating assays, biofilm formation of strain CA-2894 was reduced by 82% at concentration of 312.5 µg/mL while that of strain DSMZ 11225 was reduced by 81% at 625 µg/mL. In co-incubation assays, biofilm formation of the two strains was inhibited by 70% at 160.5 µg/well and by 81% at 19.95 µg/well, respectively. It was interesting to note that no inhibition of both *C. albicans* planktonic cells was observed, thus indicating that the biosurfactant displayed specific anti-biofilm formation but not antimicrobial activity.

Considering their importance for human health and their recognized safety, environmental probiotic organisms may, thus, represent a safe and effective intervention for infection control purposes. Probiotics themselves or their products (biosurfactants), could be applied to patient care equipment, such as tubes or catheters, with the aim of decreasing the colonisation of these sites by nosocomial pathogens and potentially impede a central step in the pathogenesis of nosocomial infections (Falagas & Makris, 2009).

3.4 Other promising biological activities

Biosurfactants have been shown to have many other roles in biomedical application. Some of the most powerful molecules (eg. surfactin, mannosylerythritol lipids (MELs), trehalose lipids) are known to have anti-inflammatory, anti-tumour, immunosuppressive and immunomodulating functions, in addition to other properties such as self-assembling, human cells stimulation and differentiation, interaction with stratum corneum lipids, cell-to-cell signaling, hemolytic activity.

3.4.1 Anti-tumor activity

Recently, it has been demonstrated that these interesting microbial products can control a variety of mammalian cell functions. They are considered to participate in various intercellular molecular recognitions such as signal transduction, cell differentiation, cell immune response, etc. (Osada, 1998). Cao et al., (2010) demonstrated that surfactin induces apoptosis in human breast cancer MCF-7 cells through a ROS/JNK-mediated mitochondrial/caspase pathway. In a more recent work, they investigated the reactive oxygen species (ROS) and Ca^{2+} impact on mitochondria permeability transition pore (MPTP) activity, and MCF-7 cell apoptosis induced by surfactin (Cao et al., 2011). The results showed that surfactin initially induced the ROS formation, leading to the MPTP opening accompanied with the collapse of mitochondrial membrane potential which lead to an increase in the cytoplasmic Ca^{2+} concentration. In addition, cytochrome c was released from mitochondria to cytoplasm through the MPTP which activated caspase-9, eventually inducing apoptosis.

In another study, viscosin, an effective surface-active cyclic lipopeptide recovered from *Pseudomonas libanensis* M9-3, inhibited the migration of the metastatic prostate cancer cell line, PC-3M, without visible toxicity effects (Saini et al., 2008). More recently, lipopeptides (namely isoforms of surfactins and fengycins) derived from a marine *Bacillus circulans* DMS-

2 showed interesting cytotoxic activity against cancer cell lines (Sivapathasekaran et al., 2010). The purified lipopeptides at a concentration of 300 µg/mL showed more than 90% inhibition of proliferation on both colon cancer cell lines HCT 15 and HT 29 after 24 h treatment and the antiproliferative activity of lipopeptides was observed in a dose dependent manner.

Significant effects against both tumor cell lines as compared to non-tumor cell line were also observed, thus indicating the selective inhibitory activity of these molecules. Serratamolide AT514, cyclic depsipeptide from *Serratia marcescens*, belonging to the group of serrawettins, has also been reported to be a potent inducer of apoptosis of several cell lines derived from various human tumors and B-chronic lymphocytic leukemia cells, primarily involving the mitochondria-mediated apoptotic pathway and interference with Akt/NF-kB survival signals (Escobar-Díaz et al., 2005, as cited in Matsuyama et al., 2010). Biological studies of AT514 using human B-lymphocytes are now in progress for clinical applications of AT514 in the field of medical oncology.

Interesting anti-tumor activities has also been reported for glycolipids. Mannosylerythritol lipids (MELs) are among the most promising biosurfactants known due to their versatile interfacial and biochemical actions. Interesting studies, thoroughly reviewed by Kitamoto et al. (2002) and by Arutchelvi & Doble, (2010), have shown that MEL-A and MEL-B display excellent growth inhibition and differentiation-inducing activities against human leukemia cells including myelogenous leukemia cell K562, promyelocytic leukemia cell HL60, and the human basophilic leukemia cell line KU812, as well as growth inhibition activity of mouse melanoma B 16 cells. Recently Chen et al., (2006) also demonstrated that a sophorolipid produced from the yeast *Wickerhamiella domercqiae* induced apoptosis in H7402 human liver cancer cells by blocking cell cycle at G1 phase and partly at S phase, activating caspase-3, and increasing Ca^{2+} concentration in cytoplasm.

3.4.2 Anti-inflamatory activity

Byeon et al., (2008) observed that surfactin was able to downregulate LPS-induced nitric oxide production in RAW264.7 cells and primary macrophages by inhibiting NF-κB activation, suggesting a good potential as a bacterium-derived anti-inflammatory agent. Selvam et al., (2009) studied the effect of *B. subtilis* PB6, a natural probiotic, on plasma cytokine levels in inflammatory bowel disease and colon mucosal inflammation. The strain was found to secrete surfactins which are known to inhibit phospholipase A2, involved in the pathophysiology of inflammatory bowel disease. In animal experiments carried out in rat models for trinitrobenzene sulfonic acid-induced colitis, oral administration of PB6 as a probiotic suppressed colitis as measured by mortality rate and changes in colon morphology and weight gain. Plasma levels of pro-inflammatory cytokines were also significantly lowered and the anti-inflammatory cytokine significantly increased after the oral administration of PB6, supporting the concept that PB6 inhibits PLA2 by secreting surfactins.

In another work, surfactin isomers derived from the mangrove bacterium *Bacillus* sp. (No. 061341) showed interesting anti-inflammatory activities (Tang et al., 2010). In particular, this class of cyclic lipopeptide showed strong inhibitory properties on the overproduction of nitric oxide and the release of IL-6 in LPS-induced murine macrophage cell RAW264.7. Moreover, structure-activity relationship studies revealed that the existence of the free carboxyl group in the structure of surfactin isomer was crucial as to the anti-inflammatory activities. An interesting recent study explored the mechanisms responsible for surfactin-induced anti-inflammatory actions in the context of periodontitis caused by *Porphyromonas gingivalis,* the major pathogen of periodontal disease (Park et al., 2010). The Authors observed that surfactin significantly reduces pro-inflammatory cytokines, including tumor necrosis factor-α, interleukin (IL)-1β, IL-6, and IL-12, through suppression of nuclear factor-κB activity in *P. gingivalis* LPS-stimulated THP-1 human macrophage cells, in a Heme oxygenase-1 (HO-1)-dependent fashion. Furthermore, surfactin treatment effectively induces HO-1 expression, a major defense in response to oxidative stress.

These observations support the potential of surfactin as a candidate in strategies to prevent caries, periodontitis, or other inflammatory diseases.

3.4.3 Immuno-modulatory action

Park & Kim, (2009) studied the role of surfactin in the inhibition of the immunostimulatory function of macrophages through blocking the NK-κB, MAPK and Akt pathway. This provided a new insight into the immunopharmacological role of surfactin in autoimmune disease and transplantation. Their work indicated that surfactin has potent immunosuppressive capabilities which suggested important therapeutic implications for transplantation and autoimmune diseases, including allergy, arthritis and diabetes.

A biosurfactant glycolipid complex from *Rhodococcus ruber* was also shown to activate the production of IL-1beta and TNF-alpha cytokines without modifying the production of IL-6, thus suggesting good prospects for further studies of immunomodulating and antitumor activities (Kuyukina et al., 2007).

3.4.4 Other biomedical related properties

Han et al., (2008) observed that high surfactin micelle concentration affected the aggregation of amyloid β-peptide (Aβ (1-40)) into fibrils, a key pathological process associated with Alzheimer's disease. Fengycin, another lipopeptide biosurfactant is also able to cause membrane perturbations (Deleu et al., 2008). Recent results by Eeman et al., (2009) emphasized the ability of fengycin to interact with the lipid constituents of the stratum corneum extracellular matrix and with cholesterol. Another interesting property of surfactin and its synthetic analogues is the ability to alter the nanoscale organisation of supported bilayers and to induce nanoripple structures with intriguing perspectives for biomedical and biotechnological applications (Bouffioux et al., 2007; Brasseur et al., 2007; Francius et al., 2008).

Morita et al., (2010) investigated the cell activating property of MELs using cultured fibroblast and papilla cells, and a three dimensional cultured human skin model. The di-acetylated MEL (MEL-A) produced from soybean oil significantly increased the viability of the fibroblast and of the papilla cells over 150% compared with that of control cells, suggesting potential use as new hair growth agent stimulating the papilla cells. Using a three-dimensional cultured human skin model, Morita et al., (2009b) observed that the viability of the SDS damaged cells was markedly improved by the addition of MEL-A in a dose-dependent manner. This demonstrated that MEL-A also had a ceramide-like moisturising activity toward the skin cells. Similarly, (Kitagawa et al., 2007, as cited in Worakitkanchanakul et al., 2008) reported that MEL-B shows excellent moisturizing properties, equivalent to those of natural ceramides, toward human skin.

Trehalose lipids also display various interesting biological activities mainly due to their great tendency to partition into phospholipid membranes (Ortiz et al., 2008, 2009). In particular, the trehalose lipid was suggested to incorporate into the membrane bilayers and produce structural perturbations, which might affect the function of both phosphatidylethanolamine and phosphatidylserine membranes. Zaragoza et al., (2010) observed that a succinoyl trehalose lipid produced by *Rhodococcus* sp. caused the swelling of human erythrocytes followed by hemolysis at concentrations well below its critical micellar concentration. They concluded that trehalose lipid caused the hemolysis of human erythrocytes by a colloid-osmotic mechanism, most likely by formation of enhanced permeability domains, or "pores" enriched with biosurfactant, within the erythrocyte membrane.

Permealization of biological and artificial membranes was also reported to be induced by *Pseudomonas aeruginosa* dirhamnolipid (Sánchez et al., 2010). In particular, it caused the hemolysis of human erythrocytes through a lytic mechanism, as shown by the similar rates of K[+] and hemoglobin leakage, and by the absence of effect of osmotic protectants. Scanning electron microscopy showed that the addition of the biosurfactant changed the usual disc shape of erythrocytes into that of spheroechinocytes.

4. Biotechnological and nanotechnological applications of surface-active compounds

Biosurfactants, have been increasingly attracting attention in the field of nanotechnology (Kitamoto et al., 2005, 2009). During the last decade, unique properties of biosurfactants, like versatile self-assembling and biochemical properties, which are not usually observed in conventional chemical surfactants, have been reported (Kitamoto et al., 2005, 2009). In recent years, the development of new functional structures and/or systems using self-assembly of amphiphilic molecules has evolved into a dynamic and rapidly growing area of nanotechnology (Ariga et al., 2007, Shimizu et al., 2005, as cited in Kitamoto et al., 2009) due to their ability to self-assemble into hierarchically ordered structures using hydrogen bonding, hydrophobic and van der Waals interactions as mentiond earlier.

Mannosylerythritol lipids (MELs) show the most interesting self-assembling properties and numerous related potential applications (Kitamoto et al., 2009). Konishi et al., (2007), Imura et al., (2007, 2008), and Ito et al., (2007), for example, developed and studied the kinetics of interactions in carbohydrate ligand systems composed of self-assembled monolayers of mannosylerythritol lipid-A (MEL-A) serving as a high-affinity, easy to handle and low-cost ligand system for immunoglobulin G and M and lectins..

Table 1 below lists the latest discoveries in the biotechnological and nanotechnological fields applicable to biosurfactants, and in particular the latest successful results of mannosylerythritol lipids (MELs) application in the enhancement of the gene transfection efficiency of cationic liposomes as well as some interesting applications of glycolipids and other biosurfactants in drug delivery and gene therapy. Biosurfactants use as a "green" alternative for high-performance nanomaterials production and, in particular, for the synthesis and stabilization of metal-bound nanoparticles will also be described.

Biosurfactant type	Activity/application	Study
Mannosylerythritol lipids-A	Ligand system for immunoglobulin G and M and lectins	Konishi et al., (2007); Imura et al., (2007, 2008), Ito et al., (2007)
	DNA capsulation and membrane fusion with anionic liposomes	Ueno et al., (2007a)
	In vitro promotion of gene transfection mediated by cationic liposomes	Inoh et al., (2001, 2004, 2010); Igarashi et al., (2006); Ueno et al., (2007b)

Biosurfactant type	Activity/application	Study
	In vivo promotion of liposome-mediated gene transfection	Inoh et al., (2009)
	Herpes simplex virus thymidine kinase gene therapy	Maitani et al., (2006)
	Water-in-oil microemulsions	Worakitkanchanakul et al., (2008)
	Increase membrane fluidity of monolayers composed of L-α-dipalmitoylphosphatidylcholine (DPPC)	Kitamoto et al., (2009)
Mannosylerythritol lipids-B	Self-assembling and vesicle-forming activity	Worakitkanchanakul et al., (2008)
Rhamnolipids and sophorolipids	Deuterated rhamnolipids and sophorolipids	Smyth et al., (2010b)
	Cadmium sulfide nanoparticles	Singh et al., (2011)
	Biocompatible microemulsions of lecithin/rhamnolipid/sophorolipid biosurfactants	Nguyen et al., (2010)
Rhamnolipids	Silver nanoparticles with antibioticmicrobial activity	Kumar et al., (2010)
	Nickel oxide nanoparticles by microemulsion technique	Palanisamy & Raichur, (2009)
	Silver nanoparticles	Xie et al., (2006)
	ZnS nanoparticles	Narayanan et al., (2010)
	Microemulsions	Xie et al., (2005, 2007)
	Alcohol-free microemulsions	Nguyen & Sabatini, (2009)
Sophorolipids	Cobalt nanoparticles	Kasture et al., (2007)
	Silver nanoparticles	Kasture et al., (2008)
	Sophorolipid-coated silver and gold nanoparticles with antibacterial activity	Singh et al., (2009, 2010)
	Biocompatible microemulsions of lecithin/rhamnolipid/sophorolipid biosurfactants	Nguyen et al., (2010)
Glycolipid biosurfactant	Silver nanoparticles	Kiran et al., (2010b)
Fengycin and surfactin	Enhancers for the skin accumulation of aciclovir	Nicoli et al., (2010)
Surfactin	Surfactin-mediated synthesis of gold nanoparticles	Reddy et al., (2009)
	Cadmium sulfide nanoparticles	Singh et al., (2011)

Table 1. Examples of recent biosurfactant applications in the biotechnological and nanotechnological fields.

4.1 Liposomes and gene transfection

Gene transfection into the cells is a fundamental technology not only for molecular and cellular biology processes but also a clinical gene therapy (Ueno et al., 2007b). Although several methods for gene transfection have been investigated (Felgner et al., 1989, Fujiwara, 2000, Gao & Huang, 1991, Hatakeyama et al., 2007, Nishiyama et al., 2005, Ueno et al., 2007b), more efficient and safe systems are still needed (Ueno et al., 2007b). Among the various methods, lipofection using cationic liposomes is considered to be a promising method for introducing foreign gene to the targeted cells due to their high transfection efficiency, low toxicity and immunogenicity, ease of preparation and targeted application (Farhood et al., 1992, Felgner et al., 1989, Kogure et al., 2007, Lasic, 1998, Nakanishi, 2003, Inoh et al., 2010). The physicochemical properties of cationic liposomes, such as lipid packing density, shape, and zeta-potential, have a significant effect on gene transfection efficiency (Lin et al., 2003, Takeuchi et al., 1996, Wittenberg et al., 2008, Xu et al., 1999, as cited by Inoh et al., 2010).

Inoh et al., (2001) reported that MEL-A promoted DNA transfection efficiency mediated by cationic liposomes. Confocal laser scanning microscopic analysis showed the distribution of lipids and oligonucleotide DNA in MEL-A-containing liposome–DNA complex in the plasma membrane and the nucleus of target cells at 1 h after the addition of complex (Inoh et al., 2004). This suggests that MEL-A induces the membrane fusion between the target cells and the cationic liposomes, accelerating the efficiency of gene transfection significantly. Similarly, Igarashi et al., (2006) reported that MEL-A significantly increased the cellular association and the efficiency of gene transfection mediated by cationic liposomes in human cervix carcinoma Hela cells. Analysis of flow cytometric profiles clearly indicated that the amount of DNA associated with the cells was rapidly increased and sustained by addition of MEL-A to the liposome. Confocal microscopic observation also indicated that the MEL-lipoplex distributed widely in the cytoplasm and DNA presence was intensely detected in cytoplasm around the nucleus.

The above results suggested that MEL-A increased gene expression by enhancing the association of the lipoplexes with the cells in serum and, thus, MEL-liposome may prove a significant nonviral vector for gene transfection and gene therapy.

In an attempt to explain how MEL-A-containing liposomes could accelerate gene transfection, Ueno et al., (2007a) examined MEL-containing liposomes properties such as their activity for DNA capsulation and membrane fusion abilities of cationic liposomes with artificial anionic liposomes. They observed that MEL-A-containing liposomes exhibited high activity in DNA incapsulation and membrane fusion with anionic liposomes, which are important properties for gene transfection. On the other hand, MEL-B- and MEL-C-containing liposomes only increased either the incapsulation or the membrane fusion. Ueno et al., (2007b) further examined the mechanism of the transfection mediated by cationic liposomes with NBD-conjugated MEL-A and reported that MEL-A distributed on the intracellular membranes through the plasma membranes of target cells, while the cationic liposomes with MEL-A fused to the plasma membranes within 20–35 min. Thereafter, they noted that the oligonucleotide released from the vesicles was immediately transferred to the nucleus. They therefore suggested that MEL-A was capable of promoting the transfection efficiency of target cells by inducing membrane fusion between liposomes and the plasma membrane of these cells.

Recently Kitamoto et al., (2009) demonstrated that monolayers composed of L-α-dipalmitoylphosphatidylcholine (DPPC) containing MEL-A had greater membrane fluidity than those containing only DPPC. It was also reported that unsaturated fatty acids in MEL-A significantly influenced surface pressure and packing density in the monolayer and thus the physicochemical properties of MEL-A and MEL-A/lipids (Imura et al., 2008). Transfection efficiency of nano vectors with MEL-A was investigated *in vivo* on tumor cells in the mouse abdominal cavity (Inoh et al., 2009). When a complex of the nano vectors with MEL-A and plasmid DNA was injected intraperitoneally into C57BL/6J mice bearing B16/BL6 tumors, the biosurfactant significantly increased liposome-mediated gene transfection to the mouse tumor cells. The transfection efficiency of the plasmids into the solid tumors by the cationic liposomes of cholesteryl-3beta-carboxyamidoethylene-N-hydroxyethylamine (OH-Chol) with MEL-A increased by approximatley 100-fold compared to that by the commercially available DC-Chol cationic liposomes without MEL-A. This suggests that nonviral vectors with MEL-A are very useful for gene transfection *in vivo*. The mechanisms of gene delivery by nano vectors with MEL-A and the numerous biological activities of these biosurfactants have been described by Nakanishi et al., (2009) and Kitamoto et al., (2009).

Inoh et al., (2010) further investigated the effects of unsaturated fatty acid ratio within the MEL-A compound on the physicochemical properties and gene delivery into cells of cationic liposomes using MEL-A with three different unsaturated fatty acid (USF) component ratios. Gene transfer efficiency of cationic liposomes containing MEL-A (containing 21.5% USF) was much higher than that of those containing MEL-A (containing 9.1%USF) and MEL-A (containing 46.3%USF). In particular, MEL-A (21.5% USF)-containing cationic liposomes induced highly efficient membrane fusion after addition of anionic liposomes and led to subsequent DNA release.

Imaging analysis revealed that MEL-A (21.5% USF)-containing liposomes fused with the plasma membrane and delivered DNA into the nucleus of NIH-3T3 cells, MEL-A (46.3% USF)-containing liposomes fused with the plasma membrane did not deliver DNA into the nucleus, and MEL-A (9.1% USF)-containing liposomes neither fused with the plasma membrane nor delivered DNA into the nucleus. These results suggest that the MEL-A unsaturated fatty acid ratio significantly affects transfection efficiency due to changes in membrane fusion activity and the efficiency of DNA release from the liposomes.

Mannosylerythritol lipid-B (MEL-B) with a different configuration of the erythritol moiety was found to self-assemble into a lamellar phase over remarkably wide concentration and temperature ranges; furthermore it showed great potential as a vesicle-forming lipid, suggesting its potential application in drug and gene delivery as well as in transdermal delivery systems (Worakitkanchanakul et al., 2008). In another work, a liposome vector containing betasitosterol beta-D-glucoside biosurfactant-complexed DNA was successfully used for herpes simplex virus thymidine kinase gene therapy (Maitani et al., 2006).

4.2 Biosurfactants potential in drug delivery

Properties such as detergency, emulsification, foaming and dispersion make biosurfactants interesting molecules with potential application in the field of drug delivery (Faivre & Rosilio, 2010). MEL-A for example has much higher emulsifying activity with soybean oil and tetradecane than polysorbate 80 (Kitamoto et al., 2009) and is able to form stable water-in-oil microemulsions without addition of co-surfactant or salt (Worakitkanchanakul et al., 2008).

Rhamnolipids and sophorolipids have also been mixed with lecithins to prepare biocompatible microemulsions in which the phase behavior was unaffected by changes in temperature and electrolyte concentration, making them desirable for cosmetic and drug delivery applications (Nguyen et al., 2010). In 1988, rhamnolipid liposomes were patented as drug delivery systems, useful as microcapsules for drugs, proteins, nucleic acids, dyes and other compounds, as biomimetic models for biological membranes and as sensors for detecting pH variations. These novel liposomes were described as safe and biologically decomposable, with suitable affinity for biological organisms, stable and with long service and shelf life.

The potential of lipopeptides, fengycin and surfactin to act as enhancers for the transdermal penetration and skin accumulation of aciclovir was also recently investigated (Nicoli et al., 2010) to elucidate any possible synergistic effect between surfactin and fengycin associated with anodal iontophoresis. It was demonstrated that these lipopeptides did not enhance aciclovir transport across the skin (not even when associated with iontophoresis) although they increased aciclovir concentration in the epidermis by a factor of 2 (Nicoli et al., 2010).

Microemulsion produced using biosurfactant are thermodynamically stable and their isotropic systems that form spontaneously-consisting of microdomains of oil or water stabilized by an interfacial film - in addition to their long-term stability, easy preparation and high solubilization capacity are considered to be very promising liquid vehicles for future drug delivery systems (Date et al., 2008, as cited in Faivre & Rosilio, 2010).

4.3 Nanoparticles

Another interesting application for natural surfactant is the the synthesis of metal-bound nanoparticles as an alternative environmentally friendly technology (Sharma et al., 2009). Nanomaterials synthesis and use has been an active research area due to interesting properties of the nanomaterials as compared to bulk material use (Palanisamy & Raichur, 2009). Metal nanoparticles have beening explored in various fields such as catalysis, mechano- and electrical applications and biomedical uses (Van Bogaert & Soetaert, 2010). The reduction in size gives rise to size dependent effects such as high surface to volume ratio, lower melting point, changes in electronic structure and changes in lattice structure and interatomic distances which in turn affect the processing parameters (Liveri, 2006, as cited in Palanisamy & Raichur, 2009).

The use of gold nanoparticles, in particular, is currently undergoing a dramatic expansion in the field of drug and gene delivery, targeted therapy and imaging technologies (Boisselier & Astruc, 2009; Pissuwan et al., 2009, 2011). Potential therapeutic applications of gold compound and gold nanoparticles also include anti-HIV activity, anti-angiogenesis, anti-malarial activity, anti-arthritic activity and biohydrogen production (Kalishwaralal et al., 2010). Silver nanoparticles are also been reported to possess anti-fungal activity, anti-inflammatory effect, anti-viral, anti-angiogenesis and anti-platelet activity (Kalishwaralal et al., 2010).

Reddy et al., (2009) successfully synthesized surfactin-mediated gold nanoparticles and investigated the effects of proton concentrations and temperature on the morphology of the obtained nanoparticles. It was demonstrated that the nanoparticles synthesized at pH 7 and 9 remained stable for 2 months, while aggregates were observed at pH 5 within 24 h. Moreover, the nanoparticles formed at pH 7 were uniform in shape and size and were polydispersed and anisotropic at pH 5 and 9. The nanoparticles synthesized produced at room temperature were monodispersed and were more uniform when compared to those formed at 4°C. More recently they also carried out a biological synthesis of gold and silver nanoparticles using the bacteria *Bacillus subtilis*. Gold nanoparticles were synthesized both intra- and extracellularly, while silver nanoparticles were exclusively formed extracellularly (Reddy et al., 2010). According to the Authors the nanoparticles were stabilized by the surface-active molecules i.e., surfactin or other biomolecules released into the solution by *B. subtilis*.

Surfactin produced by *Bacillus amyloliquefaciens* KSU-109 was also used for the synthesis of cadmium sulfide nanoparticles which remained stable up to six months without compromising their functionality (Singh et al., 2011). This kind of nanoparticles works as semiconductors with unique optical properties and tunable photo-luminescence allowing potential applications in solar energy conversion, nonlinear optical, photoelectrochemical cells and heterogeneous photocatalysis (Singh et al., 2011). In addition, surfactin produced by strain KSU-109 was easily extracted and used without further purification for nanoparticles stabilization under ambient conditions (Singh et al., 2011). Such simple, inexpensive and environmental friendly procedure of obtaining surfactin offers a further advantage of use in nanobiotechnology for the large-scale production of highly stable metal nanoparticles.

Both rhamnolipids and sophorolipids have also been successfully used for the synthesis and stabilization of metal-bound nanoparticles. Purified rhamnolipids from *P. aeruginosa* strain BS-161R were used to synthesize silver nanoparticles which exhibited good antibiotic activity against both Gram-positive and Gram-negative pathogens and *Candida albicans*, suggesting their broad spectrum antimicrobial activity (Kumar et al., 2010). In another work, a glycolipid biosurfactant produced from sponge-associated marine bacteria *Brevibacterium casei* MSA19, using agro-industrial and industrial waste as substrate, were used as a "green" stabilizer for the synthesis of stable and uniform silver nanoparticles (Kiran et al., 2010b). The biosurfactant acted as stabilization agent and prevented the formation of aggregates.

Palanisamy & Raichur, (2009) also described a simple and eco-friendly method for synthesizing spherical nickel oxide nanoparticles by microemulsion technique using rhamnolipids as alternative surfactant. The synthesized nanoparticles were found to be fully crystalline and spherical in shape with uniform distribution and increasing the pH of the solution decreased the size of the nanoparticles. Xie et al., (2006) were also able to synthesize silver nanoparticles in rhamnolipid reverse micelles while in another study rhamnolipids were used as capping agents for the synthesis of ZnS nanoparticles in aqueous medium (Narayanan et al., 2010).

Sophorolipids were also tested for use in nanoparticles synthesis and reported to be good reducing and capping agents for cobalt and silver particles (Kasture et al., 2007, 2008, as cited in Van Bogaert & Soetaert, 2010). Singh et al., (2009) demonstrated the antibacterial activity of sophorolipid-coated silver and gold nanoparticles against both Gram-positive and –negative bacteria. They also verified that sophorolipid-coated gold nanoparticles were more cyto and geno-compatible with respect to silver nanoparticles (Singh et al., 2010). They also plan to investigate these nanoparticles suitability for medical and diagnostic applications.

Recently, methodologies for the biological synthesis of metal nanoparticles using microbes have also been described (Narayanan & Sakthivel, 2010; Kalishwaralal et al., 2010; Reddy et al., 2010). In addition Smyth et al., (2010b) reported on the production of selectively deuterated rhamnolipids and sophorolipids using deuterated substrates. The production of such deuterated biosurfactants, in particular, or other bioactive microbial products in general, in which distinct pattern of labeling could be achieved resulting in varing molecular weight products and or stereochemistry unrecognised by existing degradative enzymes is very improtant. Such molecules would have great future implications with regards to efficacy and/or persistence or the development of resistance for some bioactives particularly in biomedical related applications.

4.4 Microemulsions

Microemulsions are thermodynamically stable, isotropic dispersions of oil, water and surfactant (Rosen, 1989, as cited in Nguyen et al., 2010). Microemulsion systems produce high solubilization capacity and ultralow interfacial tensions of oil and water, making them desirable in practical applications such as enhanced oil recovery, drug delivery, food and cosmetic applications (Bourrel & Schechter, 1988, Kogan & Garti, 2006, Komesvarakul et al., 2006, Lawrence & Rees. 2000, Vandamme, 2002, Yuan et al., 2008, as cited in Nguyen et al., 2010). Xie et al., 2005 demonstrated that rhamnolipids could be successfully used to form microemulsions using medium chain alcohols as cosurfactant. Subsequently, the same Authors observed that the phase behavior and microstructure of these microemulsions were rational to the conformational changes of rhamnolipid molecules at the interface of oil/water (Xie et al., 2007). Microemulsion technique using oil–water–surfactant mixture has also emerged as a promising method for nanoparticle synthesis and can be used to synthesize different types of particles (Eastoe et al., 2006, as cited in Palanisamy & Raichur, 2009). Palanisamy & Raichur, (2009), for example, successfully used rhamnolipids as the surfactant to synthesize spherical nickel oxide nanoparticles by microemulsion technique. In another work, Nguyen & Sabatini, (2009) were able to formulate alcohol-free microemulsions using rhamnolipid biosurfactant and rhamnolipid mixtures.

Lecithin-based microemulsions have proven to be desirable in biocompatible formulations due to their tendency to mimic the phospholipid nature of cell membranes (Nguyen et al., 2010). In a recent report Nguyen et al., (2010) formulated and evaluated microemulsions of lecithin/rhamnolipid/sophorolipid biosurfactants with a range of oils. Sophorolipid played an important role as the hydrophobic component in these formulations and the phase behavior of these biocompatible microemulsions did not change significantly with changing temperature and electrolyte concentration, making them desirable for cosmetic and drug delivery applications.

4.5 A survey over biotechnological commercial applications and patents of biosurfactants and bioemulsifiers

Due to their broad-range of functional properties and the diverse synthetic capabilities of microbes, biological surfactants and emulsifiers have been recently used in various industries like detergents and soaps, petroleum, textile, agriculture, cosmetic, medicine and food (Banat et al., 2000, 2010). Due to their environmental acceptability, biodegradability and lower toxicity, they are generally accepted as good candidates to substitute synthetic surfactants. Commercial applications of biosurfactants and bioemulsifiers in the biotechnological field are mainly related to the oil industry, enhanced oil recovery and bioremediation technologies (Desai & Banat, 1997). However, interesting marketable products and patents have been issued in the last few years in the health care and cosmetic industries, reviewed by Shete et al., (2006) and Banat et al., (2010).

Sugar-based biosurfactants, sophorolipids in particular, are very attractive in these fields, because of their good detergency, emulsifying, foaming and dispersing properties (Faivre & Rosilio, 2010). Sophorolipids are better solubilizers than emulsifiers, but their derivatives containing propylene glycol have excellent hygroscopic properties and are applied as moisturizer or softener in cosmetic products (Faivre & Rosilio, 2010). For example, a product containing 1 mol of sophorolipid and 12 mol of propylene glycol has excellent skin compatibility and is used commercially as a skin moisturizer (Yamane, 1987, as cited in Desai & Banat 1997). Sophorolipid is commercially used by Kao Co. Ltd. as a humectant for cosmetic makeup brands such as Sofina. This company has developed a fermentation process for sophorolipid production, and after a two-step esterification process, the product finds application in lipstick and as moisturizer for skin and hair products (Inoue et al., 1979 a, 1979b, as cited in Desai & Banat, 1997). Moreover, sophorolipids are also believed to stimulate the leptin synthesis through adipocytes, in this way reducing the subcutaneous fat overload (Pellecier & André, 2004, as cited in Van Bogaert & Soetaert, 2010).

The French company Soliance (http://www.groupesoliance.com) produces sophorolipid-based cosmetics for the body and skin and the Korean MG Intobio Co. commercializes Sopholine cosmetics (Van Bogaert & Soetaert, 2010). They are also found in cleaning soap mixtures (Ecover™ products). Despite the high number of scientific publications and patents, industrial surfactin applications still remain quite limited (Jacques, 2010). Sold by SIGMA and SHOWA DENKO for analytical or laboratory purposes, the compound is also available in several Japanese cosmetic products.

During the last decades, many patents have been issued worldwide in relation with applications of biosurfactants and bioemulsifiers in the health care field (Shete et al., 2006). Bioemulsifiers produced by *Acinetobacter calcoaceticus*, for examples, have been used in shampoos and soaps against acne and eczema and in personal care products. The skin cleansing cream and lotion containing these bioemulsifiers have, among other properties, the ability to interfere with microbial adhesion on skin or hair (Hayes et al., 1989, 1990, 1991, 1992, as cited in Shete et al., 2006). Viscosin and analogues have been patented as antibacterial, antiviral, antitrypanosomal therapeutic compounds that inhibit the growth of *Mycobacterium tuberculosis*, Herpes simplex virus 2 and/or *Trypanosoma cruzi* (Burke et al., 1999, as cited in Shete et al., 2006). *Lactobacillus* biosurfactants have also been patented as inhibitors of adherence and colonization of bacterial pathogens on medical devices (Reid et al., 2000).

Another interesting patented area is related to antimicrobial biosurfactant peptides produced by probiotic strains able to selectively bind to collagen and inhibit infections around wounds at the site of implants and biofilms associated with infections in mammals (Howard et al., 2002, as cited in Shete et al., 2006). Sophorolipids, in particular have been the object of many patents as moisturizing agents and for the amelioration of skin physiology, skin restructuration and repair (Shete et al., 2006). Sophorolipids are also used for the treatment of skin, as an activator of macrophages, and as agent in fibrinolytic healing, desquamating and depigmenting process (Maingault, 1999 as cited in Shete et al., 2006). A germicidal composition containing fruit acids, a surfactant and a sophorolipid biosurfactant, able to kill in 30 seconds 100% of *E. coli*, *Salmonella* and *Shigella*, has been patented for cleaning fruits, vegetable, skin and hair (Pierce & Heilman, 2001).

Rhamnolipids in comparison have been patented in a process to make some liposomes and emulsions (Ishigami & Suzuki 1997; Ramisse et al., 2000) both important in the cosmetic industry. More recently an activator and anti-aging agent containing MEL as active ingredient has been patented (Suzuki et al., 2010). Another recent invention is directed to polymeric acylated biosurfactants that can self-assemble or auto-aggregate into polymeric micellar structures useful in topically-applied dermatologic products (Owen & Fan, 2010).

Another patent has been deposited about a biosurfactant composition produced by a new *B. licheniformis* strain, with anti-adhesion activity against biofilm producer microbial pathogens (Martinotti et al., 2009).

5. Conclusions and perspectives

As evidenced by the growing number of publications on the topic of biosurfactants, there is an increasing interest in the study of these molecules and their potential applications. The demand for new specialty surfactants in the agriculture, cosmetic, food, pharmaceutical, and environmental industries is steadily increasing and biosurfactants, as effective and environmentally compatible compounds, perfectly meet this demand (Banat et al., 2000, 2010; Mukherjee et al., 2006).

The most important limitation for the commercial use of biosurfactants is the complexity and high cost of production, which has limited the development of their use on a large scale (Soberón-Chávez & Maier, 2010). However, the proven antimicrobial, anti-adhesive, immune-modulating properties of biosurfactants and the recent successful applications in gene therapy, immunotherapy and medical insertion safety suggest that it is worth persisting in this field. Moreover, in pharmaceutical and biomedical sectors, the high cost of production could be compensated for by the small amounts of product required. In fact, it has been elucidated that biosurfactants used as pharmaceutical agents are needed only in very low concentrations (Cameotra & Makkar, 2004). Prerequisites for making biosurfactant production more profitable and economically feasible include optimized growth/production conditions and novel and efficient multi-step downstream processing methods as well as the use of recombinant varieties of microorganisms or selected hyperproducing mutants, which can grow on a wide range of cheap renewable substrates (Muthusamy et al., 2008).

Recent advances in the area of biomedical application are probably going to take the lead due to higher potential economic returns. Moreover, due to their self-assembly properties, new and fascinating applications in nanotechnology are predicted for biosurfactants (Kitamoto et al., 2009; Palanisamy, 2008; Reddy et al., 2009). In-depth studies of their natural roles in microbial competitive interactions, cell-to-cell communication, pathogenesis, motility and biofilm formation and maintenance could suggest improved and interesting future applications.

6. Acknowledgements

This work was partially supported by the Local Research funding of the Italian *Ministero dell'Istruzione, dell'Università e della Ricerca*.

7. References

Abalos A, Pinazo A, Infante MR, Casals M, García F & Manresa A (2001) Physicochemical and antimicrobial properties of new rhamnolipids produced by *Pseudomonas aeruginosa* AT10 from soybean oil refinery wastes. *Langmuir*. 17:1367–1371

Arima K, Kakinuma A & Tamura G (1968) Surfactin, a crystalline peptide-lipid surfactant produced by *Bacillus subtilis*: isolation, characterization and its inhibition of fibrin clot formation. *Biochem Biophys Res Commun*. 31:488–494

Arutchelvi J & Doble M (2010) Mannosylerythritol lipids: microbial production and their applications, In: *Biosurfactants: From Genes to Applications*, Soberón-Chávez G Ed., pp. 145-177, Springer, Münster, Germany

Arutchelvi JI, Bhaduri S, Uppara PV & Doble M (2008) Mannosylerythritol lipids: a review. *J Ind Microbiol Biotechnol.* 35:1559-1570

Banat IM (1995a) Biosurfactants production and use in microbial enhanced oil recovery and pollution remediation: A review. *Bioresource Technol.* 51:1-12

Banat IM (1995b) Biosurfactants characterization and use in pollution removal; state of the art. A review. *ACTA Biotechnologica.* 15:251-26

Banat IM, Franzetti A, Gandolfi I, Bestetti G, Martinotti MG, Fracchia L, Smyth TJ & Marchant R (2010) Microbial biosurfactants production, applications and future potential. *Appl Microbiol Biotechnol.* 87:427-44.

Banat IM, Makkar RS & Cameotra SS (2000) Potential commercial applications of microbial surfactants. *Appl Microbiol Biotechnol.* 53:495-508

Barrons R & Tassone D (2008) Use of *Lactobacillus* probiotics for bacterial genitourinary infections in women: a review. *Clin Ther.* 30:453-468

Basak P, Adhikari B, Banerjee I & Maiti TK (2009) Sustained release of antibiotic from polyurethane coated implant materials. *J Mater Sci Mater Med.* 20:S213-S221

Becher P (Ed.) (1965) *Emulsions, Theory and practice*, Reinhold Publishing, New York, USA.

Benincasa M, Abalos A, Oliveira I & Manresa A (2004) Chemical structure, surface properties and biological activities of the biosurfactant produced by *Pseudomonas aeruginosa* LBI from soapstock. *Anton Leeuw Int J G.* 85:1–8

Bernheimer AW & Avigad LS (1970) Nature and properties of a cytolytic agent produced by *Bacillus subtilis. J Gen Microbiol.* 6:361-366

Besson, F, Peypoux F, Michel G, & Delcambe L (1978). Identification of antibiotics of iturin group in various strains of *Bacillus subtilis. J. Antibiot.* (Tokyo) 31:284–288

Besson F et al. (1976) Characterisation of iturin A in antibiotics from various strains of *Bacillus subtilis. J. Antibiot* 29: 1043–1049

Biria D, Maghsoudi E, Roostaazad R, Dadafarin H, Lotfi AS & Amoozegar MA (2010) Purification and characterization of a novel biosurfactant produced by *Bacillus licheniformis* MS3. *World J Microbiol Biotechnol.* 26:871–878

Bodour AA, Miller-Maier RM (1998) Application of a modified drop-collapse technique for surfactant quantitation and screening of biosurfactant-producing microorganisms. *J. Microbiol Methods.* 32:273-280

Boisselier E & Astruc D (2009) Gold nanoparticles in nanomedicine: preparations, imaging, diagnostics, therapies and toxicity. *Chem Soc Rev.* 38:1759–1782

Bonmatin JM, Laprevote O & Peypoux F (2003) Diversity among microbial cyclic lipopeptides: iturins and surfactins. Activity-structure relationships to design new bioactive agents. *Comb Chem High Throughput Screen.* 6:541-556

Bouffioux O, Berquand A, Eeman M, Paquot M, Dufrêne YF, Brasseur R & Deleu M (2007) Molecular organization of surfactin-phospholipid monolayers: effect of phospholipid chain length and polar head. *Biochim Biophys Acta Biomembr.* 1768:1758-1768

Brasseur R, Braun N, El Kirat K, Deleu M, Mingeot-Leclercq MP & Dufrêne YF (2007) The biologically important surfactin lipopeptide induces nanoripples in supported lipid bilayers. *Langmuir* 23:9769–9772

Buijssen KJ, Harmsen HJ, van der Mei HC, Busscher HJ & van der Laan BF (2007) Lactobacilli: important in biofilm formation on voice prostheses. *Otolaryngol Head Neck Surg.* 137:505-507

Busscher HJ, Bruinsma G, van Weissenbruch R, et al. (1998) The effect of buttermilk consumption on biofilm formation on silicone rubber voice prostheses in an artificial throat. *Eur Arch Otorhinolaryngol.* 255:410-413

Busscher HJ, van Hoogmoed CG, Geertsema-Doornbusch GI, van der Kuijl-Booij M & van der Mei HC. (1997) *Streptococcus thermophilus* and its biosurfactants inhibit adhesion by *Candida* spp. on silicone rubber. *Appl Environ Microbiol.* 63:3810-3817.

Byeon SE, Lee YG, Kim BH, Shen T, Lee SY, Park HJ, Park SC, Rhee MH & Cho JY (2008) Surfactin blocks NO production in lipopolysaccharide-activated macrophages by inhibiting NF-κB activation. *J Microbiol Biotechnol.* 18:1984–1989

Çaglar E, Kargul B & Tanboga I (2005) Bacteriotherapy and probiotics role on oral health. *Oral Dis.* 11:131–137

Cameotra SS & Makkar RS (2004) Recent applications of biosurfactants as biological and immunological molecules. *Curr Opin Microbiol.* 7:262–266

Cao XH, Wang AH, Wang CL, Mao DZ, Lu MF, Cui YQ & Jiao RZ (2010) Surfactin induces apoptosis in human breast cancer MCF- 7 cells through a ROS/JNK-mediated mitochondrial/caspase pathway. *Chem Biol Interact* 183:357–362

Cao XH, Zhao SS, Liu DY, Wang Z, Niu LL, Hou LH & Wang CL (2011) ROS-Ca(2+) is associated with mitochondria permeability transition pore involved in surfactin-induced MCF-7 cells apoptosis. *Chem Biol Interact.* 190(1):16-27

Carrillo C, Teruel JA, Aranda FA & Ortiz A (2003) Molecular mechanism of membrane permeabilization by the peptide antibiotic surfactin. *Biochem Biophys Acta.* 1611: 91-97

Chassot F, Negri MF, Svidzinski AE, Donatti L, Peralta RM, Svidzinski TI & Consolaro ME (2008) Can intrauterine contraceptive devices be a *Candida albicans* reservoir? *Contraception.* 77:355-359

Chen J, Song X, Zhang H, Qu Y & Miao J (2006) Sophorolipid produced from the new yeast strain *Wickerhamiella domercqiae* induces apoptosis in H7402 human liver cancer cells. *Appl Microbiol Biotechnol.* 72:52–59

Chen ML, Penfold J, Thomas RK, Smyth TJP, Perfumo A, Marchant R, Banat IM, Stevenson P, Parry A, Tucker I & Grillo I (2010a) Solution self-assembly and adsorption at the air-water interface of the mono and di-rhamnose rhamnolipids and their mixtures. *Langmuir.* 26:18281-18292

Chen ML, Penfold J, Thomas RK, Smyth TJP, Perfumo A, Marchant R, Banat IM, Stevenson P, Parry A, Tucker I, & Grillo I (2010b) Mixing behaviour of the biosurfactant, rhamnolipid, with a conventional anionic surfactant, sodium dodecyl benzene sulfonate. *Langmuir.* 26:17958-17968

Chung YR, Kim CH, Hwang I & Chun J (2000) *Paenibacillus koreensis* sp. nov. A new species that produces an iturin-like antifungal compound. *Int J Syst Evol Microbiol.* 50:1495–1500

Cirigliano, MC & Carman, GM, (1985) Purification and characterization of liposan, a bioemulsifier from *Candida lipolytica*. *Appl Environ Microbiol.* 50:846–850

Cosson P, Zulianello L, Join-Lambert O, Faurisson F, Gebbie L, Benghezal M, van Delden C, Curty LK & Köhler T (2002) Pseudomonas aeruginosa *virulence analyzed in a* Dictyostelium discoideum *host system.* J Bacteriol. *184(11):3027-3033*

Cribby S, Taylor M & Reid G (2008) Vaginal microbiota and the use of probiotics. *Interdisciplinary perspectives on infectious diseases.* Article ID 256490, 9 pages.

Das P, Mukherjee S & Sen R (2008) Antimicrobial potential of a lipopeptide biosurfactant derived from a marine *Bacillus circulans.* J Appl Microbiol. 104:1675–1684

Dastgheib SMM, Amoozegar MA, Elahi E, Asad S & Banat IM (2008) Bioemulsifier production by a halothermophilic *Bacillus* strain with potential applications in microbially enhanced oil recovery. *Biotechnol Lett.* 30(2):263-270

Deleu M, Paquot M & Nylander T (2008) Effect of fengycin, a lipopeptide produced by *Bacillus subtilis,* on model biomembranes. *Biophys J.* 94:2667–2679

Dempsey KE, Riggio MP, Lennon A, Hannah VE, Ramage G, Allan D & Bagg J (2007) Identification of bacteria on the surface of clinically infected and noninfected prosthetic hip joints removed during revision arthroplasties by 16S rRNA gene sequencing and by microbiological culture. *Arthritis Res Ther.* 9:R46

Desai JD & Banat IM (1997) Microbial production of surfactants and their commercial potential. *Microbiol Mol Biol Rev.* 61:47–64

Donadio S, Monciardini P, Alduina R, Mazza P, Chiocchini C, Cavaletti L, Sosio M & Puglia AM (2002) Microbial technologies for the discovery of novel bioactive metabolite. J Biotechnol. 99:187–198

Donlan RM & Costerton JW (2002) Biofilms: survival mechanisms of clinically relevant microorganisms. *Clin Microbiol Rev.* 15: 167-193

Eeman M, Francius G, Dufrêne YF, Nott K, Paquot M & Deleu M (2009) Effect of cholesterol and fatty acids on the molecular interactions of fengycin with stratum corneum mimicking lipid monolayers. *Langmuir.* 25:3029–3039

Faivre V & Rosilio V (2010) Interest of glycolipids in drug delivery: from physicochemical properties to drug targeting *Expert Opin Drug Deliv.* 7(9):1031-1048

Falagas ME, Betsi GE, Tokas T & Athanassiou S (2006a) Probiotics for prevention of recurrent urinary tract infections in women: a review of the evidence from microbiological and clinical studies. *Drugs.* 66:1253-1261.

Falagas ME, Betsi GI & Athanasiou S (2007) Probiotics for the treatment of women with bacterial vaginosis. *Clin Microbiol Infect.* 13: 657–664

Falagas ME, Betsi GI, Athanasiou S (2006b) Probiotics for prevention of recurrent vulvovaginal candidiasis: a review. *Antimicrob Chemother.* 58:266-272.

Falagas ME & Makris GC (2009) Probiotic bacteria and biosurfactants for nosocomial infection control: a hypothesis. *J Hosp Infect.* 71(4):301-306

Fassi FL, Wroblewski H & Blanchard A (2007) Activities of antimicrobial peptides and synergy with enrofloxacin against *Mycoplasma pulmonis. Antimicrob Agents Chemother.* 51:468-74

Federle MJ & Bassler BL (2003) Interspecies communication in bacteria. *J Clin Invest.* 112:1291–1299

Fracchia L, Cavallo M, Allegrone G & Martinotti MG (2010) A *Lactobacillus*-derived biosurfactant inhibits biofilm formation of human pathogenic *Candida albicans* biofilm producers, In: *Current Research, Technology and Education Topics in Applied*

Microbiology and Microbial Biotechnology (vol. 2), Mendez Vilas A Ed., pp. 827-837, FORMATEX, Spain

Francius G, Dufour S, Deleu M, Paquot M, Mingeot-Leclercq MP & Dufrêne YF (2008) Nanoscale membrane activity of surfactins: influence of geometry, charge and hydrophobicity. *Biochim Biophys Acta.* 1778:2058–2068

Franzetti A, Tamburini E & Banat IM (2010a) Applications of biological surface active compounds in remediation technologies, In: *Biosurfactants, BIOSURFACTANTS Book Series: Advances in Experimental Medicine and Biology*, Sen R Ed., Volume: 672, pp. 121-134, Landes Bioscience, Austin, TX

Franzetti A, Gandolfi I, Bestetti G, Smyth TJP & Banat IM (2010b) Production and applications of trehalose lipid biosurfactants. *Eur J Lipid Sci Technol.* 112:617–627

Gautam KK & Tiagi VK (2006) Microbial surfactants: A review. *J Oleo Sci.* 55:155-166

Grangemard I, Wallach J, Maget-Dana R & Peypoux F (2001) Lichenysin: a more efficient cation chelator than surfactin. *Appl Biochem Biotechnol.* 90:199–210

Gudiña EJ, Teixeira JA & Rodrigues LR (2010) Isolation and functional characterization of a biosurfactant produced by *Lactobacillus paracasei*. *Colloids Surf B Biointerfaces.* 76:298–304

Gupta V & Garg R (2009) Probiotics. *Indian J Med Microbiol.* 27:202–209

Hall-Stoodley L, Costerton JW & Stoodley P (2004) Bacterial biofilms: from the natural environment to infectious diseases. *Nat Rev Microbiol.* 2:95-108

Han Y, Huang X, Cao M & Wang Y (2008) Micellization of surfactin and its effect on the aggregate conformation of amyloid β(1-40). *J Phys Chem B.* 112:15195–15201

Hancock, REW & Chapelle DS (1999) Pepedide antibiotics. *Antimicrob Agents Chemother.* 43:1317–1323

Hatakka K, Ahola AJ, Yli-Knuuttila H, Richardson M, Poussa T & Meurman JK (2007) Probiotics reduce the prevalence of oral *Candida* in the elderly — a randomized controlled trial. *J Dent Res.* 86:125–130

Hong HA, Duc LH & Cutting SM (2005) The use of bacterial spore formers as probiotics. *FEMS Microbiol Rev.* 29:813–835

Horowitz S, Gilbert JN & Griffin WM (1990) Isolation and characterization of a surfactant produced by *Bacillus licheniformis* 86. *J Ind Microbiol Biot.* 6(4):243-248

Huang X, Lu Z, Bie X, Lü F, Zhao H & Yang S (2007) Optimization of inactivation of endospores of *Bacillus cereus* by antimicrobial lipopeptides from *Bacillus subtilis* fmbj strains using a response surface method. *Appl Microbiol Biotechnol.* 74:454–461

Huang X, Lu Z, Zhao H, Bie X, Lü FX & Yang S (2006) Antiviral activity of antimicrobial lipopeptide from *Bacillus subtilis* fmbj against pseudorabies virus, porcine parvovirus, newcastle disease virus and infectious bursal disease virus in vitro. *Int J Pept Res Ther.* 12:373–377

Huang X, Suo J & Cui Y (2011) Optimization of antimicrobial activity of surfactin and polylysine against *Salmonella enteritidis* in milk evaluated by a response surface methodology. *Foodborne Pathog Dis.* 8(3):439-43

Igarashi S, Hattori Y & Maitani Y (2006) Biosurfactant MEL-A enhances cellular association and gene transfection by cationic liposome. *J Control Release.* 112:362–368

Imamura Y, Chandra J, Mukherjee PK, Lattif AA, Szczotka-Flynn LB, Pearlman E, Lass JH, O'Donnell K & Ghannoum MA (2008) *Fusarium* and *Candida albicans* biofilms on soft contact lenses: model development, influence of lens type, and susceptibility to lens care solutions. *Antimicrob Agents Chemother.* 52: 171-182

Imura T, Ito S, Azumi R, Yanagishita H, Sakai H, Abe M & Kitamoto D (2007) Monolayers assembled from a glycolipid biosurfactant from *Pseudozyma* (*Candida*) *antarctica* serve as a high-affinity ligand system for immunoglobulin G and M. *Biotechnol Lett.* 29:865-870

Imura T, Masuda Y, Ito S, Worakitkanchanakul W, Morita T, Fukuoka T, Sakai H, Abe M & Kitamoto D (2008) Packing density of glycolipid biosurfactant monolayers give a significant effect on their binding affinity toward immunoglobulin G. *J Oleo Sci.* 57:415-22

Inoh Y, Furuno T, Hirashima N & Nakanishi M (2009) Nonviral vectors with a biosurfactant MEL-A promote gene transfection into solid tumors in the mouse abdominal cavity. *Biol Pharm Bull.* 32:126−128

Inoh Y, Furuno T, Hirashima N, Kitamoto D & Nakanishi M (2010) The ratio of unsaturated fatty acids in biosurfactants affects the efficiency of gene transfection. *Int J Pharmaceut.* 398:225-230

Inoh Y, Kitamoto D, Hirashima N & Nakanishi M (2001) Biosurfactants of MEL-A increase gene transfection mediated by cationic liposomes. *Biochem Biophys Res Commun.* 289:57-61

Inoh Y, Kitamoto D, Hirashima N & Nakanishi M (2004) Biosurfactant MEL-A dramatically increases gene transfection via membrane fusion. *J Control Release* 94:423-431.

Ishigami Y & Suzuki S (1997) Development of biochemicals−functionalization of biosurfactants and natural dyes. *Prog Org Coatings.* 31:51-61.

Ito S, Imura T, Fukuoka T, Morita T, Sakai H, Abea M & Kitamoto D (2007) Kinetic studies on the interactions between glycolipid biosurfactant assembled monolayers and various classes of immunoglobulins using surface plasmon resonance. *Colloids Surf B Biointerfaces.* 58:165-171

Jaques P (2010) Surfactin and other lipopeptides from *Bacillus* spp. In: *Biosurfactants: From Genes to Applications,* Soberón-Chávez G Ed., pp. 57-91, Springer, Münster, Germany

Jung M, Lee S & Kim H (2000) Recent studies on natural products as anti-HIV agents. *Curr Med Chem.* 7:649-661

Kalishwaralal K, Deepak V, Pandiana SBRK, Kottaisamy M, BarathManiKanth S, Kartikeyan B & Gurunathan S (2010) Biosynthesis of silver and gold nanoparticles using *Brevibacterium casei. Colloids Surf B Biointerfaces.* 77:257–262

Kalmokoff ML, Austin JW, Wan XD, Sanders G, Banerjee S & Farber JM (2001) Adsoption, attachment and biofilm formation among isolates of *Listeria monocytogenes* using model conditions. *J Appl Microbiol.* 91:725–734

Kim BS, Lee JY & Hwang BK (2000) *In vivo* control and *in vitro* antifungal activity of rhamnolipid B, a glycolipid antibiotic, against *Phytophthora capsici* and *Colletotrichum orbiculare. Pest Manag Sci.* 56:1029-1035

Kim H, Ryu JH & Beuchat LR (2006) Attachment of and biofilm formation by *Enterobacter sakazakii* on stainless steel and enteral feeding tubes. *Appl Environ Microbiol.* 72:5846–5856

Kim K, Yoo D, Kim Y, Lee B, Shin D & Kim E-K (2002) Characteristics sophorolipid as an antimicrobial agent. *J Microbiol Biotechnol.* 12:235-241

Kim PI, Ryu J, Kim YH & Chi YT (2010) Production of biosurfactant lipopeptides iturin A, fengycin, and surfactin A from *Bacillus subtilis* CMB32 for control of *Colletotrichum gloeosporioides*. *J Microbiol Biotechnol*. 20(1):138–145

Kiran GS, Sabarathnam B & Selvin J (2010a) Biofilm disruption potential of a glycolipid biosurfactant from marine *Brevibacterium casei*. *FEMS Immunol Med Microbiol*. 59:432–438

Kiran GS, Sabu A & Selvin J (2010b) Synthesis of silver nanoparticles by glycolipid biosurfactant produced from marine *Brevibacterium casei* MSA19. *J Biotechnol*. 148:221–225

Kitamoto D, Isoda H, Nakahara T. (2002) Functions and potential applications of glycolipid biosurfactants-from energy-saving materials to gene delivery carriers. *J Biosci Bioeng*. 94(3):187-201.

Kitamoto D, Morita T, Fukuoka T, Konishi M & Imura T (2009) Self-assembling properties of glycolipid biosurfactants and their potential applications. *Curr Opin Colloid Interface Sci*. 14:315–328

Kitamoto D, Toma K & Hato M (2005) Glycolipid-based nanomaterials, In: Handbook of Nanostructured Biomaterials and Their Applications in Nanobiotechnology, vol. 1, Nalwa HS Ed., p. 239–271, American Science Publishers, California, USA

Kitamoto D, Yanagishita H, Shinbo T, Nakane T, Kamisawa C & Nakahara T (1993) Surface active properties and antimicrobial activities of mannosylerythritol lipids as biosurfactants produced by *Candida antarctica*. *J Biotechnol*. 29:91–96

Kõll P, Mändar R, Marcotte H, Leibur E, Mikelsaar M & Hammarström L (2008) Characterization of oral lactobacilli as potential probiotics for oral health. *Oral Microbiol Immunol*. 23:139–147

Konishi M, Imura T, Fukuoka T, Morita T & Kitamoto D (2007) A yeast glycolipid biosurfactant, mannosylerythritol lipid, shows high binding affinity towards lectins on a self-assembled monolayer system. *Biotechnol Lett*. 29:473–480

Kumar A, Ali A & Yerneni LK (2007) Effectiveness of a mycoplasma elimination reagent on a mycoplasma-contaminated hybridoma cell line. *Hybridoma (Larchmt)*. 26(2):104-106

Kumar CG, Mamidyala SK, Das B, Sridhar B, Devi GS & Karuna MS (2010) Synthesis of biosurfactant-based silver nanoparticles with purified rhamnolipids isolated from *Pseudomonas aeruginosa* BS-161R. *J Microbiol Biotechnol*. 20:1061-1068

Kuyukina MS, Ivshina IB, Gein SV, Baeva TA & Chereshnev VA (2007) *In vitro* immunomodulating activity of biosurfactant glycolipid complex from *Rhodococcus ruber*. *Bull Exp Biol Med*.144:326-30.

Landman D, Georgescu C, Martin DA & Quale J (2008) Polymyxins revisited. *Clin Microbiol Rev*. 21:449–465

Lang S (2002) Biological amphiphiles (microbial biosurfactants). *Curr Opin Coll Int Sci*. 7:12–20

Lang S & Philp JC (1998) Surface-active lipids in *Rhodococci*. *Anton Leeuw Int. J G*. 74:59–70.

Lin SC (1996) Biosurfactants: recent advances. *J Chem Tech Biotechnol*. 66:109-120

Litzler PY, Benard L, Barbier-Frebourg N, Vilain S, Jouenne T, Beucher E, Bunel C, Lemeland JF & Bessou JP (2007) Biofilm formation on pyrolytic carbon heart valves: influence of surface free energy, roughness, and bacterial species. *J Thorac Cardiovasc Surg*. 134:1025-1032

Lu JR, Zhao XB & Yaseen M (2007) Biomimetic amphiphiles: biosurfactants. *Curr Opin Colloid Interface Sci.* 12:60-67

Luna JM, Rufino RD, Sarubbo LA, Rodrigues LR, Teixeira JA & de Campos-Takaki GM (2011) Evaluation antimicrobial and antiadhesive properties of the biosurfactant lunasan produced by *Candida sphaerica* UCP 0995. *Curr Microbiol.* 62:1527-1534

Maier RM (2003) Biosurfactants: evolution and diversity in bacteria. *Adv Appl Microbiol.* 52: 101-121

Maitani Y, Yano S, Hattori Y, Furuhata M & Hayashi K (2006) Liposome vector containing biosurfactant-complexed DNA as herpes simplex virus thymidine kinase gene delivery system. *J Liposome Res.* 16:359-72.

Martinotti MG, Rivardo F Allegrone G, Ceri H, Turner R (2009) Biosurfactant composition produced by a new *Bacillus licheniformis* strain, uses and products thereof. International patent PCT/IB2009/055334, 25 November

Matsuyama T, Tanikawa T, & Nakagawa Y (2010) Serrawettins and other surfactants produced by *Serratia*. In: *Biosurfactants: From Genes to Applications*, Soberón-Chávez G Ed., pp. 93-120, Springer, Münster, Germany

McInerney MJ, Javaheri M & Nagle DP (1990) Properties of the biosurfactant produced by *Bacillus liqueniformis* strain JF-2. I. *J. Microbiol Biotechnol.* 5:95–102

Merk K, Borelli C & Korting HC (2005) Lactobacilli—bacteria–host interactions with special regard to the urogenital tract. *Int J Med Microbiol.* 295:9–18

Meurman JH (2005) Probiotics: do they have a role in oral medicine and dentistry? *Eur J Oral Sci.* 113:188–196

Meurman JH & Stamatova I (2007) Probiotics: contributions to oral health. *Oral Dis.* 13:443–445

Meylheuc T, Methivier C, Renault M, Herry JM, Pradier CM & Bellon-Fontaine MN (2006a) Adsorption on stainless steel surfaces of biosurfactants produced by gram-negative and gram-positive bacteria: consequence on the bioadhesive behavior of *Listeria monocytogenes. Colloids Surf B Biointerfaces.*52:128–137

Meylheuc T, Renault M & Bellon-Fontaine MN (2006b) Adsorption of a biosurfactant on surfaces to enhance the disinfection of surfaces contaminated with *Listeria monocytogenes. Int J Food Microbiol.* 109:71–78

Mimee B, Labbé C, Pelletier R & Bélanger RR (2005) Antifungal activity of flocculosin, a novel glycolipid isolated from *Pseudozyma flocculosa. Antimicrob Agents Chemother.* 49:1597–1599

Mimee B, Pelletier R & Bélanger RR (2009) In vitro antibacterial activity and antifungal mode of action of flocculosin, a membrane-active cellobiose lipid. *J Appl Microbiol.* 107:989–996

Mireles JR II, Toguchi A & Harshey RM (2001) *Salmonella enterica* serovar *typhimurium* swarming mutants with altered biofilm forming abilities: surfactin inhibits biofilm formation. *J Bacteriol.* 183:5848–5854

Mohammad Abdel-Mawgoud A, Hausmann R, Lépine F, Müller MM & Déziel E (2010) Rhamnolipids: detection, analysis, biosynthesis, genetic regulation, and bioengineering of production, In: *Biosurfactants: From Genes to Applications*, Soberón-Chávez G Ed., pp. 13-55, Springer, Münster, Germany

Morikawa M (2006) Beneficial biofilm formation by industrial bacteria *Bacillus subtilis* and related species. *J Biosci Bioeng.* 101:1–8

Morikawa M, Hirata Y, Imanaka T (2000). A study on the structure–function relationship of the lipopeptide biosurfactants. *Biochim Biophys Acta*. 1488:211-218

Morikawa M, Ito M & Imanaka T (1992) Isolation of a new surfactin producer *Bacillus pumilus* A-1, and cloning and nucleotide sequence of the regulator gene, psf-1. *J Ferment Bioeng*. 74:255-261

Morita T, Fukuoka T, Konishi M, Imura T, Yamamoto S, Kitagawa M, Sogabe A & Kitamoto D (2009a) Production of a novel glycolipid biosurfactant, mannosylmannitol lipid, by *Pseudozyma parantarctica* and its interfacial properties. *Appl Microbiol Biotechnol*. 83:1017–1025

Morita T, Kitagawa M, Suzuki M, Yamamoto S, Sogabe A, Yanagidani S, Imura T, Fukuoka T & Kitamoto D (2009b) A yeast glycolipid biosurfactant, mannosylerythritol lipid, shows potential moisturizing activity toward cultured human skin cells: the recovery effect of MEL-A on the SDS-damaged human skin cells. *J Oleo Sci*. 58:639–642

Morita T, Kitagawa M, Yamamoto S, Suzuki M, Sogabe A, Imura T, Fukuoka T & Kitamoto D (2010) Activation of fibroblast and papilla cells by glycolipid biosurfactants, mannosylerythritol lipids. *J Oleo Sci*. 59:451-5

Mukherjee S, Das P & Sen R (2006) Towards commercial production of microbial surfactants. *Trends Biotechnol*. 24:509-515

Muthusamy K, Gopalakrishnan S, Ravi TK & Sivachidambaram P (2008) Biosurfactants: Properties, commercial production and application. *Curr Sci*. 94:736-747

Nakanishi M, Inoh Y, Kitamoto D & Furuno T (2009) Nano vectors with a biosurfactant for gene transfection and drug delivery. *J Drug Delivery Sci Technol*. 19:165–169.

Narayanan J, Ramji R, Sahu H & Gautam P (2010) Synthesis, stabilisation and characterisation of rhamnolipid-capped ZnS nanoparticles in aqueous medium. *IET Nanobiotechnol*. 4:29-34.

Narayanan KB & Sakthivel N (2010) Biological synthesis of metal nanoparticles by microbes. *Adv Colloid Interfac*. 156:1–13

Naruse N, Tenmyo O, Kobaru S, Kamei H, Miyaki T, Konishi M & Oki T (1990) Pumilacidin, a complex of new antiviral antibiotics: production, isolation, chemical properties, structure and biological activity. *J Antibiot*. (Tokyo) 43:267–280

Neu TR (1996) Significance of bacterial surface-active compounds in interaction of bacteria with interfaces. *Microbiol Rev*. 60:151–166

Nguyen TT & Sabatini DA (2009) Formulating alcohol-free microemulsions using rhamnolipid biosurfactant and rhamnolipid mixtures. *J Surfact Deterg*. 12:109–115

Nguyen TTL, Edelen A, Neighbors B & Sabatini DA (2010) Biocompatible lecithin-based microemulsions with rhamnolipid and sophorolipid biosurfactants: formulation and potential applications. *J Colloid Interface Sci*. 348:498-504

Nicoli S, Eeman M, Deleu M, Bresciani E, Padula C & Santi, P (2010) Effect of lipopeptides and iontophoresis on aciclovir skin delivery. *J Pharm Pharmacol*. 62:702–708

Nithya C & Pandian SK (2010) The in vitro antibiofilm activity of selected marine bacterial culture supernatants against *Vibrio* spp. *Arch Microbiol*. 192:843–854

Nitschke M, Araújo LV, Costa SG, Pires RC, Zeraik AE, Fernandes AC, Freire DM & Contiero J (2009) Surfactin reduces the adhesion of food-borne pathogenic bacteria to solid surfaces. *Lett Appl Microbiol*. 49:241–247

Nitschke M, Costa SG & Contiero J (2010) Structure and applications of a rhamnolipid surfactant produced in soybean oil waste. *Appl Biochem Biotechnol.* 160(7):2066-74

Ortiz A, Teruel JA, Espuny MJ, Marqués A, Manresa Á & Aranda FJ (2008) Interactions of a *Rhodococcus* sp. biosurfactant trehalose lipid with phosphatidylethanolamine membranes. *Biochim Biophys Acta.* 1778:2806–2813

Ortiz A, Teruel JA, Espuny MJ, Marqués A, Manresa Á & Aranda FJ (2009) Interactions of a bacterial biosurfactant trehalose lipid with phosphatidylserine membranes. *Chem Phys Lipids.* 158:46–53

Osada, H (1998) Bioprobe for investigating mammlian cell cycles control. *J. Antibiotics* 51:973-982

Owen D & Fan L (2010) Polymeric Biosufactants. US Patent 20100144643, 6 October

Palanisamy P (2008) Biosurfactant mediated synthesis of NiO nanorods. *Mat Lett.* 62:743–746

Palanisamy P & Raichur AM (2009) Synthesis of spherical NiO nanoparticles through a novel biosurfactant mediated emulsion technique. *Mater Scie Eng C.* 29:199–204

Park SY & Kim Y (2009) Surfactin inhibits immunostimulatory function of macrophages through blocking NK-κB, MAPK and Akt pathway. *Int Immunopharmacol.* 9:886–893

Park SY, Kim YH, Kim EK, Ryu EY & Lee SJ (2010) Heme oxygenase-1 signals are involved in preferential inhibition of pro-inflammatory cytokine release by surfactin in cells activated with *Porphyromonas gingivalis* lipopolysaccharide. *Chem Biol Interact.* 188:437-45

Pecci Y, Rivardo F, Martinotti MG & Allegrone G (2010) LC/ESI-MS/MS characterisation of lipopeptide biosurfactants produced by the *Bacillus licheniformis* V9T14 strain *J Mass Spectrom.* 45:772–778

Perfumo A, Banat IM, Canganella F & Marchant R (2006) Rhamnolipid production by a novel thermotolerant hydrocarbon-degrading *Pseudomonas aeruginosa* AP02-1. *J Appl Microbiol.* 75:132-138

Perfumo A, Smyth TJP, Marchant R & Banat IM (2010a) Production and roles of biosurfactants and bioemulsifiers in accessing hydrophobic substrates, In: *Handbook of Hydrocarbon and Lipid Microbiology*, Timmis KN Ed., pp. 1501-1512, Springer-Verlag, Berlin Heidelberg, Germany

Perfumo A, Rancich I & Banat IM (2010b). Possibilities and challenges for biosurfactants uses in petroleum industry, In: *Biosurfactants, BIOSURFACTANTS Book Series: Advances in Experimental Medicine and Biology*, Sen R Ed., Volume: 672, pp. 135-145, Landes Bioscience, Austin, TX

Petrelli D, Zampaloni C, D'Ercole S, Prenna M, Ballarini P, Ripa S & Vitali LA (2006) Analysis of different genetic traits and their association with biofilm formation in *Staphylococcus epidermidis* isolates from central venous catheter infections. *Eur J Clin Microbiol Infect Dis.* 25:773-781

Peypoux F, Bonmatin JM & Wallach J (1999) Recent trends in the biochemistry of surfactin. *Appl Microbiol Biotechnol.* 51:553–563

Peypoux F, Guinand M, Michel G, Delcambe L, Das BC & Lederer E (1978) Structure of iturin A, a peptidolipid antibiotic from *Bacillus subtilis*. *Biochemistry.* 17:3992–3996

Peypoux F, Pommier MT, Marion D, Ptak M, Das BC & Michel G 1986. Revised structure of mycosubtilin, a peptidolipid antibiotic from *Bacillus subtilis*. *J. Antibiot.* (Tokyo) 39:636–641

Peypoux F, Pommier MT, Das BC, Besson F, Delcambe L & Michel G (1984) Structures of bacillomycin D and bacillomycin L peptidolipid antibiotics from *Bacillus subtilis*. *J. Antibiot.* (Tokyo) 77:1600–1604

Pierce D & Heilman TJ (2001) Germicidal composition. US Patent 6262038, 17 July

Pissuwan D, Niidome T & Cortie MB (2011) The forthcoming applications of gold nanoparticles in drug and gene delivery systems. *J Control Release.* 149:65–71

Pissuwan D, Valenzuela SM, Miller CM, Killingsworth MC & Cortie MB (2009) Destruction and control of *Toxoplasma gondii* tachyzoites using gold nanosphere/antibody conjugates. *Small.* 5:1030–1034.

Ramisse F, Delden C, Gidenne S et al. (2000) Decreased virulence of a strain of *Pseudomonas aeruginosa* O12 overexpressing a chromosomal type 1 β-lactamase could be due to reduced expression of cell-to-cell signalling dependent virulence factors. *FEMS Immunol Med Microbiol.* 28:241–245

Rasmussen TB & Givskov M (2006) Quorum-sensing inhibitors as antipathogenic drugs. *Int J Med Microbiol.* 296:149–161

Raya A, Sodagari M, Pinzon NM, He X, Newby BZ & Ju LK (2010) Effects of rhamnolipids and shear on initial attachment of *Pseudomonas aeruginosa* PAO1 in glass flow chambers. *Environ Sci Pollut Res.* 17:1529–1538

Raza ZA, Khalid ZM & Banat IM (2009) Characterization of rhamnolipids produced by a *Pseudomonas aeruginosa* mutant strain grown on waste oils. *J Environ Sci Health A Tox Hazard Subst Environ Eng.* 44:1367-1373

Raza ZA, Khalid ZM, Khan MS, Banat IM, Rehman A, Naeem A & Saddique MT (2010) Surface properties and sub-surface aggregate assimilation of Rhamnolipid surfactants in different aqueous system. *Biotechnol Lett.* 32:811-816

Reddy AS, Chen CY, Chen CC, Jean JS, Chen HR, Tseng MJ, Fan CW & Wang JC. (2010) Biological synthesis of gold and silver nanoparticles mediated by the bacteria *Bacillus subtilis*. *J Nanosci Nanotechnol.* 10:6567-74

Reddy AS, Chen CY, Chen CC, Jean JS, Fan CW, Chen HR, Wang JC & Nimje VR (2009) Synthesis of gold nanoparticles via an environmentally benign route using a biosurfactant. *J Nanosci Nanotechnol.* 9:6693–6699

Reid G, Bruce A & Smeianov V (1998) The role of Lactobacilli in preventing urogenital and intestinal infections. *Int Dairy J.* 8:555–562

Reid G, Bruce AW, Busscher HJ & Van der Mei HC (2000) *Lactobacillus* therapies. US Patent 6051552 18 April

Reid G, Bruce AW, Fraser N, Heinemann C, Owen J & Henning B (2001) Oral probiotics can resolve urogenital infections. *FEMS Immunol Med Microbiol.* 30:49–52

Remichkova M, Galabova D, Roeva I, Karpenko E, Shulga A & Galabov AS (2008) Anti-herpesvirus activities of *Pseudomonas* sp. S-17 rhamnolipid and its complex with alginate. *Z Naturforsch C.* 63:75–81

Rivardo F, Martinotti MG, Turner RJ & Ceri H (2010) The activity of silver against *Escherichia coli* biofilm is increased by a lipopeptide biosurfactant. *Can. J. Microbiol.* 56:272-278

Rivardo F, Martinotti MG, Turner RJ & Ceri H (2011) Synergistic effect of lipopeptide biosurfactant with antibiotics against *Escherichia coli* CFT073 biofilm. *Int J Antimicrob Agents.* 37:324-331

Rivardo F, Turner RJ, Allegrone G & Ceri H, Martinotti MG (2009) Anti-adhesion activity of two biosurfactants produced by *Bacillus* spp. prevents biofilm formation of human bacterial pathogens. *Appl Microbiol Biotechnol.* 83:541–553

Rodrigues L, Banat IM, Teixeira J & Oliveira R (2006a) Biosurfactants: potential applications in medicine. *J Antimicrob Chemother.* 57:609–618

Rodrigues L, van der Mei H, Banat IM, Teixeira J & Oliveira R (2006b) Inhibition of microbial adhesion to silicone rubber treated with biosurfactant from Streptococcus thermophilus A. *FEMS Immunol Med Microbiol.* 46:107–112

Rodrigues L, van der Mei HC, Teixeira J & Oliveira R (2004a) Biosurfactant from *Lactococcus lactis* 53 inhibits microbial adhesion on silicone rubber. *Appl Microbiol Biotechnol.* 66:306-311.

Rodrigues L, van der Mei HC, Teixeira J & Oliveira R (2004b) Influence of biosurfactants from probiotic bacteria on formation of biofilms on voice prosthesis. *Appl Environ Microbiol* 70:4408-4410.

Rodrigues L, Banat IM, van der Mei HC, Teixeira JA, Oliveira R & Oliveira R (2006c) Interference in adhesion of bacteria and yeasts isolated from explanted voice prostheses to silicone rubber by rhamnolipid biosurfactants. *J Appl Microbiol.* 100:470-480.

Rodrigues L, Banat IM, Teixeira J & Oliveira R (2007) Strategies for the prevention of microbial biofilm formation on silicone rubber voice prostheses. *J Biomed Mater Res B Appl Biomater.* 81B:358-370.

Rodrigues LR, Teixeira JA, van der Mei HC & Oliveira R (2006d) Physicochemical and functional characterization of a biosurfactant produced by *Lactococcus lactis* 53. *Colloids Surf B Biointerfaces.* 49:79–86

Rodrigues LR & Teixeira JA (2010) Biomedical and therapeutic applications of biosurfactants. *Adv Exp Med Biol.* 672:75-87

Ron EZ & Rosenberg E (2001) Natural roles of biosurfactants. *Environ Microbiol.* 3:229–236

Rosenberg E & Ron EZ (1997) Bioemulsans: microbial polymeric emulsifiers. *Curr Opin Biotechnol.* 8:313–316

Rosenberg E & Ron EZ (1999) High- and low-molecular-mass microbial surfactants. *Appl Microbiol Biotechnol* 52:154–162

Rosenberg E, Zuckerberg A, Rubinovitz C & Gutnick DL (1979) Emulsifier of *Arthrobacter* RAG-1:isolation and emulsifying properties. *Appl Environ Microbiol.* 37:402–408

Rosenberg M (2006) Microbial adhesion to hydrocarbons: twenty-five years of doing MATH. *FEMS Microbiol Lett.* 262:129–134

Rufino RD, Luna JM, Sarubbo LA, Rodrigues LR, Teixeira JA & Campos-Takaki GM (2011) Antimicrobial and anti-adhesive potential of a biosurfactant Rufisan produced by *Candida lipolytica* UCP 0988. *Colloids Surf B Biointerfaces.* 84:1-5

Saini HS, Barragán-Huerta BE, Lebrón-Paler A, Pemberton JE, Vázquez RR, Burns AM, Marron MT, Seliga CJ, Gunatilaka AA & Maier RM (2008) Efficient purification of the biosurfactant viscosin from *Pseudomonas libanensis* strain M9-3 and its physicochemical and biological properties. *J Nat Prod.* 71:1011-1015

Sánchez M, Aranda FJ, Teruel JA, Espuny MJ, Marqués A, Manresa Á & Ortiz A (2010) Permeabilization of biological and artificial membranes by a bacterial dirhamnolipid produced by *Pseudomonas aeruginosa. J Colloid Interface Sci.* 341:240-247

Sánchez M, Aranda FJ, Teruel JA & Ortiz A (2009) Interaction of a bacterial dirhamnolipid with phosphatidylcholine membranes: a biophysical study. *Chem Phys Lipids.* 161:51–55

Saravanakumari P & Mani K (2010) Structural characterization of a novel xylolipid biosurfactant from *Lactococcus lactis* and analysis of antibacterial activity against multi-drug resistant pathogens. *Bioresour Technol.* 101:8851–8854

Satpute SK, Banat IM, Dhakephalkar PK, Banpurkar AG, Chopade BA (2010a) Biosurfactants, bioemulsifiers and exopolysaccharides from marine microorganisms. *Biotechnol Adv.* 28:436-450

Satpute SK, Banpurkar AG, Dhakephalkar PK, Banat IM & Chopade BA (2010b) Methods for investigating biosurfactants and bioemulsifiers: a review. *Crit Rev Biotechnol.* 30:127-144

Selvam R, Maheswari P, Kavitha P, Ravichandran M, Sas B & Ramchand CN (2009) Effect of *Bacillus subtilis* PB6, a natural probiotic on colon mucosal inflammation and plasma cytokines levels in inflammatory bowel disease. *Indian J Biochem Biophys.* 46:79–85

Seydlová G & Svobodová J (2008) Review of surfactin chemical properties and the potential biomedical applications. *Cent Eur J Med.* 3:123–133

Shah V, Doncel GF, Seyoum T, Eaton KM, Zalenskaya I, Hagver R, Azim A & Gross R (2005) Sophorolipids, microbial glycolipids with anti-human immunodeficiency virus and sperm-immobilizing activities. *Antimicrob Agents Chemother.* 49:4093–4100

Shaligram NS & Singhal RS (2010) Surfactin – A review, on biosynthesis, fermentation, purification and applications. *Food Technol Biotechnol.* 48:119–134

Shao Z (2010) Trehalolipids, In: *Biosurfactants: From Genes to Applications,* Soberón-Chávez G Ed., pp. 121-143, Springer, Münster, Germany

Sharma VK., Yngard RA & Lin Y (2009) Silver nanoparticles: Green synthesis and their antimicrobial activities. *Adv Colloid Interfac.* 145:83–96

Shete AM, Wadhava G, Banat IB & Chopade BA (2006) Mapping of patents on bioemulsifiers and biosurfactants : A review. *J Sci Ind Res* (India). 65:91-115.

Singh A, van Hamme JD & Ward OP (2007) Surfactants in microbiology and biotechnology: part 2. Application aspects. *Biotechnol Adv.* 25:99–121

Singh BR, Dwivedi S, Al-Khedhairy AA & Musarrat J (2011) Synthesis of stable cadmium sulfide nanoparticles using surfactin produced by *Bacillus amyloliquifaciens* strain KSU-109. *Colloids Surf B Biointerfaces.* In press

Singh P & Cameotra SS (2004) Potential applications of microbial surfactants in biomedical sciences. *Trends Biotechnol.* 22:142–146

Singh S, D'Britto V, Prabhune AA, Ramana CV, Dhawan A & Prasad BLV (2010) A Cytotoxic and genotoxic assessment of glycolipid-reduced and -capped gold and silver nanoparticles. *New J Chem.* 34:294-301

Singh S, Patel P, Jaiswal S, Prabhune AA, Ramana CV & Prasad BLV (2009) A direct method for the preparation of glycolipid–metal nanoparticle conjugates: sophorolipids as reducing and capping agents for the synthesis of water re-dispersible silver nanoparticles and their antibacterial activity. *New J Chem.* 33:646-652

Sivapathasekaran, C, Das P, Mukherjee S, Saravanakumar J, Mandal M & Sen R (2010) Marine bacterium derived lipopeptides: characterization and cytotoxic activity against cancer cell lines. *Int J Pept Res Ther.* 16:215–222

Smyth TJP, Perfumo A, Marchant R & Banat IM (2010a) Isolation and analysis of low molecular weight microbial glycolipids, In: *Handbook of Hydrocarbon and Lipid Microbiology*, Timmis KN Ed., pp. 3705-3723, Springer, Berlin,

Smyth TJ, Perfumo A, Marchant R, Banat IM, Chen M, Thomas RK, Penfold J, Stevenson PS & Parry NJ (2010b) Directed microbial biosynthesis of deuterated biosurfactants and potential future application to other bioactive molecules. *Appl Microbiol Biotechnol.* 87:1347-1354

Smyth TJP, Perfumo A, McClean S, Marchant R & Banat IM (2010c) Isolation and analysis of lipopeptides and high molecular weight biosurfactants, In: *Handbook of Hydrocarbon and Lipid Microbiology*, Timmis KN Ed., pp. 3689–3704, Springer, Berlin, Germany

Soberón-Chávez G & Maier RM (2010) Biosurfactants: a general overview, In: *Biosurfactants: From Genes to Applications*, Soberón-Chávez G Ed., pp. 1-11, Springer, Münster, Germany

Sotirova AV, Spasova DI, Galabova DN, Karpenko E & Shulga A (2008) Rhamnolipid-biosurfactant permeabilizing effects on gram-positive and gram-negative bacterial strains. *Curr Microbiol.* 56:639–644

Stepanovic S, Cirkovic I, Ranin L & Svabic-Vlahovic M (2004) Biofilm formation by *Salmonella* spp. and *Listeria monocytogenes* on plastic surface. *Lett Appl Microbiol.* 38:428–432

Stickler DJ (2008) Bacterial biofilms in patients with indwelling urinary catheters. *Nat Clin Pract Urol.* 5:598-608

Suzuki M, Kitagawa M, Yamamoto S, Sogabe A, Kitamoto D, Morita T, Fukuoka T & Imura T (2010) Activator including biosurfactant as active ingredient, mannosyl erythritol lipid, and production method publication. Patent application number: 20100168405, 7 January

Tanaka Y, Tojo T, Uchida K, Uno J, Uchida Y & Shida O (1997) Method of producing iturin A and antifungal agent for profound mycosis. *Biotechnol Adv.* 15:234–235

Tang JS, Zhao F, Gao H, Dai Y, Yao ZH, Hong K, Li J, Ye WC & Yao XS (2010) Characterization and online detection of surfactin isomers based on HPLC-MS^n analyses and their inhibitory effects on the overproduction of nitric oxide and the release of TNF-α and IL-6 in LPS-induced macrophages. *Mar Drugs.* 8:2605-2618.

Thavasi R, Jayalakshmi S, Balasubramanian T & Banat IM (2008) Production and characterization of a glycolipid biosurfactant from *Bacillus megaterium* using economically cheaper sources. *World J Microbiol Biotechnol.* 24:917-925

Thavasi R, Jayalakshmi S & Banat IM (2011) Effect of biosurfactant and fertilizer on biodegradation of crude oil by maring isolates of *Bacillus megaterium* and *Corynebacterium kutscheri* and *Pseudomonas aeruginosa*. *Bioresouce Technol.* 102:772-778

Ueno Y, Hirashima N, Inoh Y, Furuno T & Nakanishi M (2007a) Characterization of biosurfactant-containing liposomes and their efficiency for gene transfection. *Biol Pharm Bull.* 30:169–172

Ueno Y, Inoh Y, Furuno T, Hirashima N, Kitamoto D & Nakanishi M (2007b) NBD-conjugated biosurfactant (MEL-A) shows a new pathway for transfection. *J Control Release.* 123:247–253

Valle J, Da Re S, Henry N, Fontaine T, Balestrino D, Latour-Lambert P & Ghigo JM (2006) Broad-spectrum biofilm inhibition by a secreted bacterial polysaccharide. *Proc Natl Acad Sci USA*. 103:12558–12563

Van Bogaert INA & Soetaert W (2010) Sophorolipids. In: *Biosurfactants: From Genes to Applications,* Soberón-Chávez G Ed., pp. 179-210, Springer, Münster, Germany

Van Bogaert INA, Saerens K, De Muynck C, Develter D, Wim S & Vandamme EJ (2007) Microbial production and application of sophorolipids. *Appl Microbiol Biotechnol.* 76:23–34

Van der Mei HC, Free RH, Elving GJ, van Weissenbruch R, Albers FWJ & Busscher HJ (2000) Effect of probiotic bacteria on prevalence of yeasts in oropharyngeal biofilms on silicone rubber voice prostheses *in vitro*. *J Med Microbiol*. 49:713-718.

Van Hamme JD, Singh A & Ward OP (2006) Physiological aspects Part 1 in a series of papers devoted to surfactants in microbiology and biotechnology. *Biotechnol Adv*. 24:604–620

Van Hoogmoed CG, Dijkstra RJB, van der Mei HC & Busscher HJ (2006) Influence of biosurfactant on interactive forces between mutans streptococci and enamel measured by atomic force microscopy. *J Dent Res*. 85:54–58

Van Hoogmoed CG, van Der Kuijl-Booij M, van der Mei HC & Busscher HJ (2000) Inhibition of *Streptococcus mutans* NS adhesion to glass with and without a salivary conditioning film by biosurfactant-releasing *Streptococcus mitis* strains. *Appl Environ Microbiol*. 66:659-663.

Van Hoogmoed CG, Van der Mei HC & Busscher HJ (2004) The influence of biosurfactants released by *S. mitis* BMS on the adhesion of pioneer strains and cariogenic bacteria. *Biofouling,* 20:261–267

Vanittanakom, N, Loeffler W, Koch U & Jung G (1986) Fengycin–a novel antifungal lipopeptide antibiotic produced by *Bacillus subtilis* F-29-3. *J. Antibiot.* (Tokyo). 39:888–901

Velraeds MMC, van de Belt-Gritter B, Busscher HJ, Reid G & van der Mei HC (2000) Inhibition of uropathogenic biofilm growth on silicone rubber in human urine by lactobacilli - a teleologic approach. *World J Urol*. 18:422-426.

Velraeds MMC, Van der Mei HC, Reid G & Busscher HJ (1996) Inhibition of initial adhesion of uropathogenic *Enterococcus faecalis* by biosurfactants from *Lactobacillus* isolates. *Appl Environ Microbiol*. 62:1958-1963.

Velraeds MMC, Van der Mei HC, Reid G & Busscher HJ (1997) Inhibition of initial adhesion of uropathogenic *Enterococcus faecalis* to solid substrata by an adsorbed biosurfactant layer from *Lactobacillus* acidophilus. *Urology*. 49:790-794.

Vollenbroich D, Ozel M, Vater J, Kamp RM & Pauli G (1997a) Mechanism of inactivation of enveloped viruses by the biosurfactant surfactin from *Bacillus subtilis*. *Biologicals*. 25:289-297

Vollenbroich D, Pauli G, Ozel M & Vater J (1997b) Antimycoplasma properties and application in cell culture of surfactin, a lipopeptide antibiotic from *Bacillus subtilis*. *Appl. Environ. Microbiol*. 63:44-49

von Eiff C, Kohnen W, Becker K & Jansen B (2005) Modern strategies in the prevention of implant-associated infections. *Int J Artif Organs*. 28:1146–1156

Walencka E, Różalska S, Sadowska B & Różalska B (2008) The influence of *Lactobacillus acidophilus* derived surfactants on staphylococcal adhesion and biofilm formation. *Folia Microbiol*. 53:61–66

Worakitkanchanakul W, Imura T, Fukuoka T, Morita T, Sakai H, Abe M, Rujiravanit R, Chavadej S, Minamikawa H & Kitamoto D (2008) Aqueous-phase behavior and vesicle formation of natural glycolipid biosurfactant, mannosylerythritol lipid-B. *Colloids Surf B Biointerfaces*. 65:106–112

Xie Y, Li Y & Ye R (2005) Effect of alcohols on the phase behavior of microemulsions formed by a biosurfactant—rhamnolipid. *J Dispers Sci Technol*. 26:455–461

Xie Y, Ye R & Liu H (2006) Synthesis of silver nanoparticles in reverse micelles stabilized by natural biosurfactant. *Colloid Surf A-Physicochem Eng Asp*. 279:175–178

Xie YW, Ye RQ & Liu HL (2007) Microstructure studies on biosurfactant-rhamnolipid/*n*-butanol/water/*n*-heptane microemulsion system. *Colloid Surf A-Physicochem Eng Asp*. 292:189–195

Yakimov MM, Timmis KN, Wray V & Fredrickson HL (1995) Characterization of a new lipopeptide surfactant produced by thermotolerant and halotolerant subsurface *Bacillus licheniformis* BAS 50. *Appl Environ Microbiol*. 61:1706–1713

Yoo DS, Lee BS & Kim EK (2005) Characteristics of microbial biosurfactant as an antifungal agent against plant pathogenic fungus *J Microbiol Biotechnol*. 15:1164-1169

Zaragoza A, Aranda FJ, Espuny MJ, Teruel JA, Marqués A, Manresa Á & Ortiz A (2009) Mechanism of membrane permeabilization by a bacterial trehalose lipid biosurfactant produced by *Rhodococcus* sp. *Langmuir*. 25:7892–7898

Zaragoza A, Aranda FJ, Espuny MJ, Teruel JA, Marqués A, Manresa Á & Ortiz A (2010) Hemolytic activity of a bacterial trehalose lipid biosurfactant produced by *Rhodococcus* sp.: evidence for a colloid-osmotic mechanism. *Langmuir*. 26(11):8567–8572

Zeraik AE & Nitschke M (2010) Biosurfactants as agents to reduce adhesion of pathogenic bacteria to polystyrene surfaces: effect of temperature and hydrophobicity. *Curr Microbiol*. 61:554–559

6

Contact Lenses Characterization by AFM MFM, and OMF

Dušan Kojić, Božica Bojović, Dragomir Stamenković,
Nikola Jagodić and Đuro Koruga
University of Belgrade, Faculty of Mechanical Engineering,
Department of BioMedical Engineering
Serbia

1. Contact lense industry

The contact lens (CL) industry and market have displayed a high level of dynamism in the past few decades, and have evolved into a rapidly changing field in which science and everyday practice constantly interact, not only through broadening of material and product portfolio, but through innovative therapeutic and diagnostic solutions as well.

Stable market growth with numerous rearrangements in different product segments is constantly taking place, mainly stirred by innovative material and optical design. The standardly used hydrogel materials are being rapidly replaced by silicone doped hydrogel materials. The analyses of customer CL usage and satisfaction indicate continued market growth in future, however with many changes in product profile and significant increase in multifocal and daily disposable lenses market share.

The main impulse behind the dynamism of CL industry stems from results of scientific and technological improvements, which are enhancing medical field and reminding us that the focal point of sustainable development lies in scientific investigations.

2. Contact lenses in present, past and future

The technology of materials used in CL production has improved vastly in the past decades starting from glass and moving to polymer based materials (PMMA) with, eventually, major steps being taken in including hydrogel and doped-hydrogel based materials, shifting the functionality of CLs from rigid gas-impermeable (RGP) to soft gas-permeable materials represented by silicone hydrogel materials that are now in use.

This chapter will focus on multimodal applied research of rigid gas-permeable contact lenses (CL) that are manufactured from fluorosilicone acrylate based material. Our multimodal research comprises measurement of intermolecular interactions on the basis of optical, mechanical, morphological and magnetic properties of CL material.

The role of our research in such a complex system of CL industry was introduction of new diagnostic modalities through improved material characterization.

In the course of last decade, scientists have developed different possibilities of "on eye"CL application that are not related to its optical capabilities for which they were invented in the first place – correcting eye's refractive error. Furthermore, improvement in CL material manufacturing, both soft and rigid gas permeable, are mostly directed towards increasing oxygen permeability and wearing comfort. Rarely today, CL producers are dedicated to improving CL material properties for the purpose of enhancing the quality of vision, on the contrary, by doping them with silicone, for example, the optical properties become even worse.

Apart from it's properties to correct eye's refractive anomaly (dioptric power), the most frequent factors influencing quality of vision while wearing RGP CL are those related to the fact that visible light, on it's way to the eye's "perception area"– the macula, must pass through CL material itself, and all it's characteristics can seriously modify it.

Geometrical optics and related functions of the eye (vision acuity etc.)   should not be considered as the only one mechanism of light interaction with human organism. We also consider physical inputs that influence the functioning of the central nervous system (CNS) on the basis of optical-neuronal interactions, and point to perspectives of investigating the role of CLs in modifying the influence of light for therapeutic purposes, or using the CLs as a potential diagnostic tool in monitoring the state of other systemic parameters (such as serum glucose level etc). As a rule, when comparing "naked eye"vision with the vision aided by CL it is inevitable to conclude that vision with CLs is of lower quality, due to reduced contrast sensitivity, sub-normal color's perception, spherical and chromatic light aberrations. All these are considered mostly consequence of CL material imperfectness. The aim of our research is to organize a setup for development of a novel material for RGP CL production which should, after adequately lathe cut and polished, improve it's optical properties in transmitting visible and "near visible"light, while increasing contrast sensitivity for "on eye"usage and improving color perception with simulaneous reduction in both higher and lower order light aberrations.

New, advanced CL materials, are still needed because, regarding biocompatibility and oxygen permeability, the advantages offered by high oxygen delivery have solved many hypoxia-related clinical problems, but the complications related to inflammation, infection and mechanical insult to the cornea still occur. The good news is that the industry continues to work on the next innovative materials and designs while our patients, and us, enjoy the silicone hydrogel lenses and the big step they represent toward safer and more effective extended (toward continuous) contact lens wear.

3. Physiological considerations related to contact lenses wearing

3.1 Tear layer considerations

Tear layer is constantly renewed tripartite film that covers conjunctival and corneal surface. All layers are separated:

• **Mucin layer** is the innermost and it is anchored to the corneal and conjunctival epithelial glycocalix of the microplicae.

• **Aqueous layer** is the middle one and represents 90% of the tear film thickness. It consists of water and salts dissolved in it, together with dissolved glucose, lysosime, for tear's specific pre-albumin, lactoferin and other.

• **Lipid layer** is the outermost, secreted by the Meibomian glands and it retards evaporation of the tears.

The tear layers have many different functions:

• Keeping the surface of cornea smooth and that way making it optically clear.

• During blinking tear film lubricates the friction area between lids and ocular surface.

• Many of its constituents prevents ocular infection.

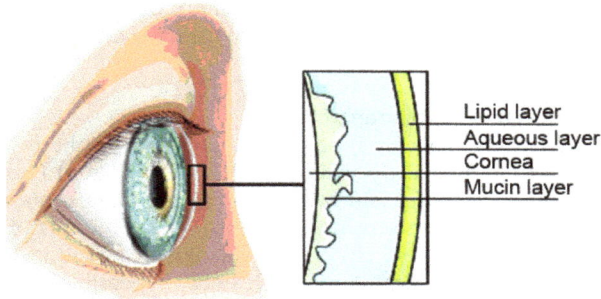

Fig. 1. Structure of sublayers that constitute tear layer.

Presence of any contact lens on the front surface of the eye influences physical and chemical properties of the tears and disturbs tear film stability. These changes in the healthy tear film are most likely caused by the following effects on the eye:

- Contact lens surface
- Contact lens design (mostly edges)
- Contact lens material
- Instillation of contact lens solutions with the lens

In comparison to the normal pre-ocular tear film, conventional (hydrogel) soft contact lens wearers have pre-lens tear film with all three layers of the tear film reduced. Tear film quickly deteriorates and renders poor surface wettability capability (weak attachment of ocular surface epithelium to the mucus layer). Pre-rigid gas permeable lens tear film is even thinner and more unstable then that of soft lens. Its lipid layer is often absent so the aqueous layer quickly dehydrates. In silicone-hydrogels, pre-lens tear film is something like combination of the previous two tear films although lipid layer is always present. As in the case of pre-ocular tear layer, good quality pre-lens tear film layer is *conditio sine qua non* for adequate optical properties of the eye.

Each eye consists of light receptor layer in the retina, optically active tissues that focuses the light on the receptor layer and nerve fiber system (retinal ganglion cell's fibers and optic nerve) which conducts electrical impulses created during electro-chemical reaction in the receptor layer (provoked by light absorption) all the way to the visual cortex and other parts of the brain.

Optically, eye functions very similar to photo-camera, inside the eye it is totally dark and this is provided by its outermost protection layer – the *sclera*, and pigmented parts of the middle layer – the *choroid*, which also provides most of the eye's blood supply. Light transmitting into the eye is absolutely controlled by the very agile diaphragm – the *pupil*. In order for light to be focused at the center of the eyes' reception layer it has to be refracted by *cornea*: absolutely transparent convex-concave lens and further refracted by another agile tissue in the eye - *the crystalline lens*. In order to prevent light reflection from behind the receptor layer to interfere with and disturb the primary light stimulation of the receptor cells, a pigmented cell layer is situated just behind the receptor layer and absorbs all the residual light.

Emmetropisation of the eye is a very sensitive process of dosing the eye's optical activity in relation to its axial growth. Any disturbances in this results in refractive errors of the eye (*myopia, hyperopia* and *astigmatism*). Eye, just as any other optical device, is also prone to higher order light aberrations that influence the stimulation of the receptor cells therefore influencing the quality of vision.

3.2 Light perception and visual signal transmission

Eye's light receptor cells (rods and cones) have large amounts of photo-sensitive pigments (*opsin* and *retinen*). Once stimulated by visible light (397 – 723 nm) these photo-sensitive pigments are changing their structure which leads to the chain of events that are ending in nerve activity. Rods are much more sensitive to the smallest amounts of light stimulation and they are most active in the dark environment (*scotopic* conditions) when pupil (the diaphragm) is dilated in order to receive as much light as possible. On the other hand, cones are much less light-sensitive and are active only in *photopic* conditions when the pupil shrinks in order to prevent too much light entering the eye which can disturb highly sophisticated visual functioning like color vision and small detail discrimination. Light stimulation blocks Na^+/K^+ channels in the receptor cell and disturbs the balance of ions in the sense of hyper-polarization of the cell. This results in reduction of the amount of the synaptic neuro-transmitter (normally released in certain amounts without light stimulation) that triggers electrical activity in the retinal ganglion cell which is conducted to the brain. Before it reaches the optical nerve, electrical activity created in the receptor cells is changed by the activity of the modulation cells present in different retinal layers.

4. Structure-function relationship of conventional and novel CL materials

Newer, soft lenses (hydrogel) are made from polymers that are inherently flexible or may become flexible through absorption of fluid into the polymer matrix. In general, when speaking of rigid CLs, most commonly used polymers are poly(methyl methacrylate) or PMMA, polyacrylamide (PAA), cellulose acetate butyrate (CAB), and in order to make them gas-permeable various mixtures based on PMMA doped with silicone or fluorine are used. Also, many modification of PMMA based materials are also present in the market (such as poly (2-hydroxyethyl methacrylate (PHEMA), poly (2-hydroxypropyl methacrylate) (PHPMA) etc.).

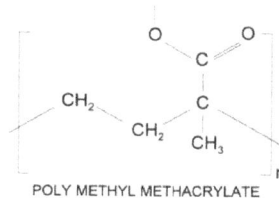
POLY METHYL METHACRYLATE

Fig. 2. Structure of poly methyl methacrylate.

The functions that contemporary contact lense materials are most often required to fulfill, encompass:

- **Dimensional stability** rendered through sufficient strenght and stiffnes – these properties are obtained by matrix of methylmethacrylate, which provides hardness and strength, and ethylene glycol dimethacrylate (EGDMA) which acts as a cross-linking agent adding to dimensional stability and stiffness but reducing water content.
- **Normoxia** rendered through material flexibility and gas permeability – these propeties are obtained by addition of several components, such as silicone (increases flexibility and gas

permeability through the material's silicon-oxygen bonds ($Si - O - Si$); however, it also brings in the disadvantage of poor wettability), fluorine (improves gas-permeability (less than Si) and improves wettability and deposit resistance in Si-containing lenses).

- **Proper adhesion** through controlled wettability – adequate level of adhesion is controlled by inclusion of hydroxyethyl-methacrylate – the basic water-absorbing monomer of most hydrogel-based soft lenses; methacrylic acid and n-vinyl pyrolidone (NVP) monomers are also added, both of which absorb high amounts of water and are usually adjuncts to hydroxyethyl methacrylate to increase lens water content.

Rigid contact lenses made of PMMA are recognized by excellent mechanical properties of dimensional stability (flexure resistance, stiffness and resistance to breakage) but practically have no oxygen permeability. Inclusion of Si and F has introduced gas-permeability properties but have compromised surface characteristics and stiffness. By chemically balancing silicone acrylate (SA) with stiff crosslinking monomers (esters) manufacturers have been able to achieve sufficieng gas-permeability without significantly compromising stiffness. Most contemporary gas-permeable lenses are composed of fluorosilicone acrylate that, due to oxygen's preference to dissolve into fluorinated materials, practically draws in the oxygen from the atmosphere and transports it towards the corneal surface, utilizing its $Si - O - Si$ component.

Hydrogel based lenses draw their advantage from the fact that hydrogels are materials that absorb and hold water inside their polymer matrices causing the spaces between the polymer chains to expand. Anything dissolved in the water can potentially enter the hydrogel matrix, depending on the molecular size and the matrix pore size. The pore size ranges from $0.5\mu m$ to $3.5\mu m$ for low and high-water content lenses, respectively. The oxygen permeability of hydrogel based contact lenses originates from their water content. Because various hydrogel polymers can greatly alter its chemical and physical properties, they may react differently to changes in pH, osmolarity, temperature and the components of the various lens care products that are used.

Silicone hydrogel based lenses use Si-doping which simultaneously decreases water content and increases permeability. Having in mind that the solubility of oxygen is very high in Si when compared with water, silicone doped hydrogel lenses permeability dramatically increases with lowering water content – which is a logical step towards improving functionality of CL materials. However optimal transport conditions for water and ion through a silicone lens, require an adequate amount of water. Synthesizing Si-hydrogel lenses is challenging since involves mixing non-polar (oxygen-rich silicone) and polar component (water) to produce lenses that renders high oxygen permeability, good wettability, flexibility, good optics, reasonable lens movement on the eye and general biocompatibility.

4.1 Conventional materials used in our investigation

We used two types of CL materials: gas-permeable CLs that are manufactured from fluorosilicone acrylate based material (Soleko $SP40^{TM}$) and standard non-doped PMMA CLs. The aim was to test the response of this material's surface roughness quality on the nano-level using standard nanotechnological methods and new nano-photonic method.

5. Characterization of contact lenses

Introduction of effective improvements in production quality as well as therapy efficiency requires the application of non-destructive surface analysis methods on the nanometer scale with minimal sample preparation. Many interesting studies have been performed on CL (Lim et al., 2001), (Kim et al., 2002), (Bruinsma et al., 2003), so that two research approaches can be perceived from the literature: one aimed at surface modification and the other that was concerned with varying or simulating exploitation parameters. In both cases morphology of CL surface served as the main parameter used to describe behavior and quality of CL material under deterioration influences.

Production engineering aspects. Conformation states of polymers constituting CL surface are changed during final stages of manufacturing process (polishing) which presents a complex problem because surface molecules and their orientation influence the final state of surface quality. In general, over-polished internal surface renders a dysfunctional product since it prevents adequate adherence to corneal tissue. CL with over-polished surface slides over the cornea and cannot maintain its initial position therefore disrupting the geometry and function of the optical system (human eye + CL). However, too rough a surface will eventually lead to irritation and possibly damage of corneal tissue.

Exploitation aspects. During CL exploitation a lacrymal film is formed between inner CL surface and cornea. A good fit of CL and cornea depends on the adequacy of CL geometry and surface roughness. Moreover, every single RGP CL wearer provides a unique ambient conditions in which these CL biosurfaces have to function properly. Since CL surfaces become significantly rougher after prolonged wear, they become more prone to bacterial adhesion and protein and lipid deposits. CL can lose functionality due to accumulated proteins, lipids, and other tear components on CL surface, despite routine cleaning activities. The loss of RGP CL functionality needs to be investigated and related to measurable parameters in order to recommend replacement based on significant changes in surface properties.

5.1 Multimodal, complementary and non-invasive approach

Answering the questions of optimal processing parameters directly influences surface quality of CLs which in turn reaches all aspects of its functionality: optical, medical and patient-comfort. Since our measurement confirmed that surface roughness of end-product ranges in the order of several tens of nanometers, the adequate tools for such samples must be derived from applied nanotechnology.

Experimental design, aimed at quantifying surface quality on the level of nano scale, utilized three methods of characterization that are meant to complement each other in interpreting and quantifying the measurement results. Moreover, we have applied a new method that is concerned with obtaining magnetic properties of samples from the interaction of material with visible light, named Opto-Magnetic Fingerprint of matter (OMF), with the aim to improve the speed and accuracy of analysis. Testing the product quality, reliably and accurately, in production environment is an essential characteristic of maintaining high level of product quality control.

We have chosen to investigate near-surface magnetic and optical properties of chosen CL materials since they are the physical quantities that most closely correlate with subtle modifications in material structure and composition. Therefore, we investigated five types of CL materials (two conventional and three nanophotonic materials) by selected three approaches (techniques):

1. **Classical methods**: spectroscopic examination in the region of ultraviolet and visible light (UV-Vis spectroscoscopy).

2. **Nanotechnology based methods**: Phase-Contrasted Atomic Force Microscopy (PC-AFM) with extended mode of Magnetic Force Microscopy (MFM).

3. **Nanomedicine based methods**: Novel method named Opto-Magnetic Fingerprint (OMF) that obtains magnetic properties of materials on the basis of interaction with visible light.

The experiment was designed to provide proofs for three interrelated phenomena:

- Change in topographical and conformation states of surface and polymer via PC-AFM,

- Related change in nanomagnetic behavior of near surface layers via MFM,

- The change of paramagnetic/diamagnetic and optomagnetic state that is recorded by OMF.

All three methods possess very high sensitivity (nanonewton for forces, nanotesla for magnetization), which is a necessity for this type of measurements.

5.2 Ultraviolet-visible spectroscopy

In order to investigate optical properties of RGP CLs, we have performed spectroscopic measurements in the range of ultraviolet-visible (UV-Vis) light in the range of wavelengths between 280–800 nm, using the UV-Vis scanner produced by Horiba JobinYvon, USA. The measurements were conducted in diffuse reflectance regime and the result is displayed as a graph of reflected energy vs. wavelength of emitted radiation. The spectroscopic measurements are used as a reference guide and were performed on two locations: the central point on the outer surface and the same point measured from inner surface in order to determine the difference and hence the degree of eye protection to different ranges of wavelengths and radiation intensities.

5.3 Phase-contrasted atomic force microscopy – PC-AFM

Morphology of CL surfaces was obtained by atomic force microscope (PC-AFM) that can measure sample surface roughness with high precision (less than $10^{-12}m$ or $\approx 1pm$ and confirm sample surface state as belonging to either group adequate or inadequate roughness. Topography of lens surface is important for determining the connection between surface morphology (conformational states of surface layer polymers) and corresponding optical properties that are influenced by the processing parameters.

Basically, AFM is a scanning probe microscopy technique based on point-to-point examination of the specimen made by a sharpened tip probe (Binnig et al., 1986). AFM probe is a micro-cantilever with sharpened conical or pyramidal tip whose radius can range from 2 – 90 nm, depending on the application. All samples were imaged using phase-contrast technique in tapping mode and in ambient air. The AFM system used in this study was JSPM-5200, JEOL, Japan. The cantilever were type PPP-MFMR, produced by Nanosensors, Switzerland.

The principle of operation of AFM is shown in figure 3. In AFM, the probe (cantilever tip) is vibrated at near-resonant frequency and brought ino interaction with the sample by the mechanism of intermolecular attractive/repulsive forces that are distance-dependent. The cantilever is maintained in the close proximity from the sample so that probe tip is within reach of attractive/repulsive forces. A typical AFM system is able to detect intermolecular forces in the order of $10^{-11} - 10^{-13}N$ which makes it an extremely sensitive device. The intermolecular interactions belong to the class of van der Waals type forces, usually modeled by Lennard-Jones potentials. These forces cause the cantilever to deflect from the initial equilibrium position making it possible to derive the distance from the sample on the basis of force field gradient change that modulates the vibration of the cantilever.

A diode laser is directed to the back sided surface of the tip and is reflected to photo-diode detector. During the scanning movement, the angle of reflected beam is changed due to deflection of the cantilever (that is, in turn, due to interactions with the sample) and this movement is precisely recorded via photo-diode detector. A feedback loop is used to adjust the z-position of the sample so that a constant interaction is maintained as the tip is scanned across the sample in x and y directions. The tip-sample interaction equilibrium is constantly disrupted by changes in sample profile height which generates the control signal applied

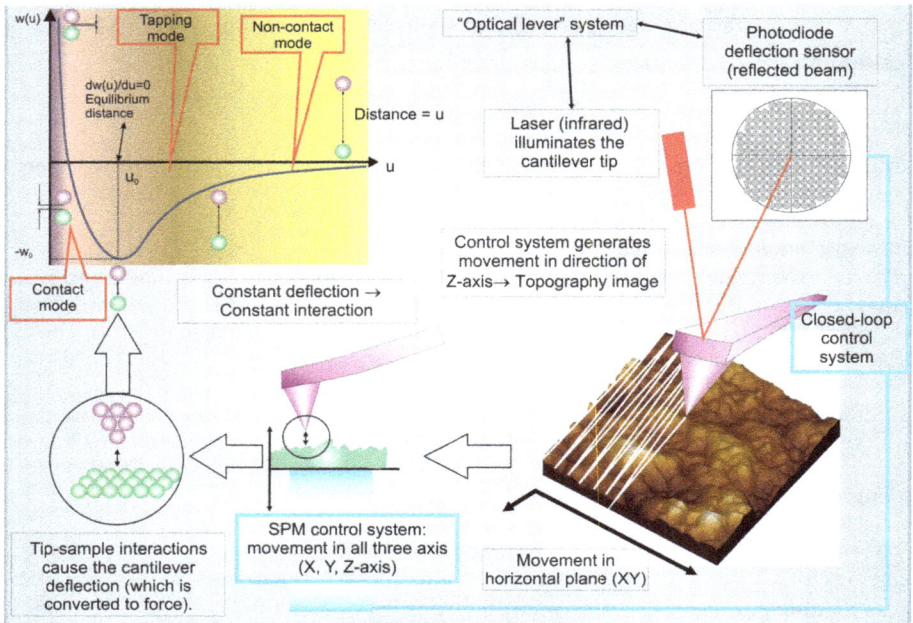

Fig. 3. Principle of operation of atomic force microscope.

to the z-scanner stage. This disturbance represents the changes in surface topography and the action of the control system is recorded and represents the actual data represented on topography image of the sample.

The PC-AFM imaging technique is based on the fact that intermolecular force gradients have certain physico-chemical specificity. The variations in the sample generate variations in the slopes of Lennard-Jones potential curves because of the different intermolecular forces acting on the AFM sensor tip. These differences modulate the vibration amplitude of the AFM cantilever, creating a higher-harmonics in the feedback signal of AFM system. By recording higher harmonics in the oscillation signal (see 3) we can detect different force-distance relationships, hence, different components of a material can be visualized. However, the PC-AFM signal correlation to separate chemical species is still unknown, which disables the AFM from making exact chemical characterization of the sample. The data obtained by PC-AFM are rather of a qualitative nature and are suitable for two or three component systems.

5.4 Magnetic Force Microscopy – MFM

Magnetic Force Microscopy is an extended operation mode of AFM that was used to obtain the distribution of magnetic properties of the surface that are previously imaged with topography mode. This tecnique was used to measure the magnetic properties in para- and diamagnetic range because we are also probing the magnetic properties in the same range with a novel technique that is introduced in the next section (Opto-Magnetic Fingerprint). MFM will is used as a comparative method for OMF. The MFM technique utilizes special cantilever sensors that are coated with a thin film of cobalt ($\approx 50nm$) that renders its ferromagnetic properties and ensures magnetic interaction with the sample.

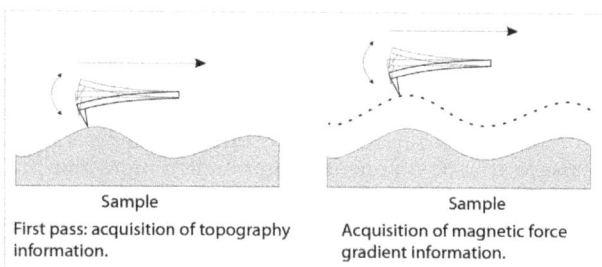

| Sample | Sample |
| First pass: acquisition of topography information. | Acquisition of magnetic force gradient information. |

Fig. 4. Principle of operation of atomic force microscope.

This thin film creates magnetic interaction with the sample that is recorded via two-pass technique or "lift scanning"(see 4) during which the sample is scanned twice, with second scan being performed with a gap distance in order to physically filter the slowly decaying magnetic from all other more rapidly decaying intermolecular forces. The sensitivity of magnetic force measurements goes around $10^{-11}N$ which is able to detect very weak magnetic interaction that encompasses para- and diamagnetic range. The cantilevers which we used in this study are produced by Nanosensors (Switzerland), with a lateral resolution under 50 nm and coercivity of 300 Oe.

The cantilever has been magnetized previously to the measurement in order to ensure its magnetic interaction with the sample. Although MFM, in its current stage of development, still represents a qualitative measurement of magnetic properties, we attempted to improve the accuracy of this analysis by measuring the remnant magnetization of the cantilever, using JR-5 spinner magnetometer (AGICO, Brno, Czech Republic), and identified that our series of measurements fitted within the range of $72 \pm 2nT$, with a standard deviation of 0.3%. The measured axis of cantilever remnant magnetization was positioned in the vertical direction, so it was perpendicular to sample surface. The data on components of vectors of remnant magnetization enabled us to confirm that we have measured the magnetic field component that is perpendicular to the sample surface.

5.5 Opto-Magnetic Fingerprint – OMF

OptoMagnetic Fingerprint (OMF) is a novel method in investigating optical and magnetic properties of materials that is based on electron properties of matter (covalent bonds, hydrogen bonds, ion-electron interaction, van der Walls interaction) and its interaction with light (Koruga et al., 2008). The method was originally developed for early skin cancer and melanoma detection by MySkin, Inc., USA (Bandić et al., 2002). Bearing in mind that the orbital velocity of valence electron in atoms is about $10^6 m/s$, this gives the ratio between magnetic force (F_M) and electrical force (F_E) of matter, of around $\frac{F_M}{F_E} \approx 10^{-4}$. Since force (F) is directly related to quantum action – Planck's action – defined as:

$$h = F \times d \times t = 6.626 \times 10^{-34} Js \qquad (1)$$

where F is force, d is displacement and t is time of action) this means that the action of magnetic forces is four orders of magnitude closer to quantum action than the electrical ones. Since quantum state of matter is primarily responsible for conformational changes on the molecular level, this means that detecting differences between matter states is by far more likely to give greater sensitivity on the level of magnetic forces than it would be on the level of measurement of electrical forces (Koruga et al., 2002).

Picture of surface that is taken by classical optical microscope is based on electromagnetic property of light, while OMF is based on difference between diffuse white light (like that of daily light) and reflected polarized light. Reflected polarized light is produced when source of diffuse light irradiates the surface of matter under certain angle (Brewster's angle, see figure 5). Each type of matter has different angle value of light polarization. Since reflected polarized light contains electrical component of light-matter interaction, taking the difference between white light (electromagnetic) and reflected polarized light (electrical) yields magnetic properties of matter based on light-matter interaction. Because such measurement can identify the conformational state and change in tissue on molecular level we named this method the opto-magnetic fingerprint of matter (OMF).

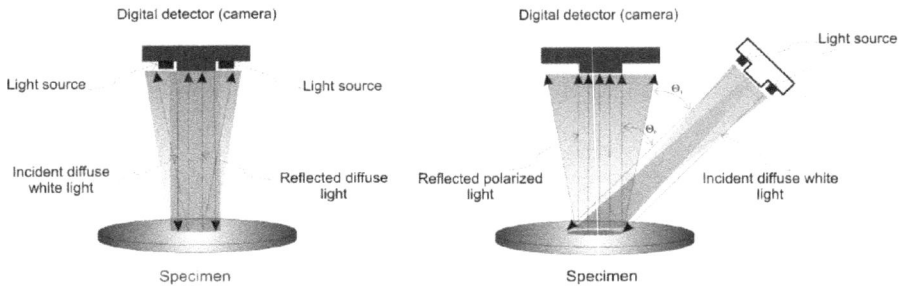

Fig. 5. Incident white light can give different information about thin layer of matter (surface) properties of sample depending on the angle of light incidence. When the incident white light is diffuse, the reflected white light is then composed of electrical and magnetic components, whereas diffuse incident light that is inclined under certain angle will produce reflected light which contains only electrical component of light. For each type of matter there is a characteristic angle of incidence (Bandić et al., 2002) for obtaining the appropriate reflected polarized light.

We used digital images in RGB (R-red, G-green, B-blue) system in our analysis, therefore we chose basic pixel data in red and blue channels for white diffuse light (W) and reflected polarized white light (P). Algorithm for data analysis is based on chromaticity diagram called "Maxwell's triangle"and spectral convolution operation according to ratio of (R-B)& (W-P) (Koruga et al., 2008). The abbreviated designation means that Red minus Blue wavelength of White light and reflected Polarized light are used in spectral convolution algorithm to calculate data for opto-magnetic fingerprint of matter. Therefore, method and algorithm for creating a unique spectral fingerprint are based on the convolution of RGB color channel spectral plots generated from digital images that have captured single and multi-wavelength light-matter interaction (Koruga et al., 2008). Preparation of digital pictures for OMF was made by usage of dermoscopic imaging device (MySkin, USA) that has previously been successfully used in biophysical skin characterization (skin photo type, moisture, conductivity, etc) (Bandić et al., 2002).

The final purpose of our research in applying OMF is the construction of quality control method which would be purely optical and able to, on the basis of digital image analysis and processing, detect both the morphology and functionality parameters in a quicker and more accurate manner. In order to do so we need to construct quantification parameters but, in this stage of research, primarily integrate results from morphological research and opto-magnetic properties.

6. Results

6.1 Topography measurement using Atomic Force Microscope – AFM

Topography measurements were routinely conducted in tapping mode in ambient air using uniform scanning surface of $5 \times 5\mu m$. The CL inner surface has been examined as shown in figure 6. The curvature of the surface prevents the AFM probe to reach its center, unless the sample is destroyed. Nevertheless, a good approach was able to the area that is approximately on the half distance of CLs radii. A total of four points were selected for scanning on each CL in two perpendicular directions. In each point two scans were conducted: one with $60 \times 60\mu m$ surface size and the other with the $5 \times 5\mu m$ surface size. The purpose of larger area scans were to confirm the uniform character of surface morphology while smaller scans were further analysed by fractal analysis.

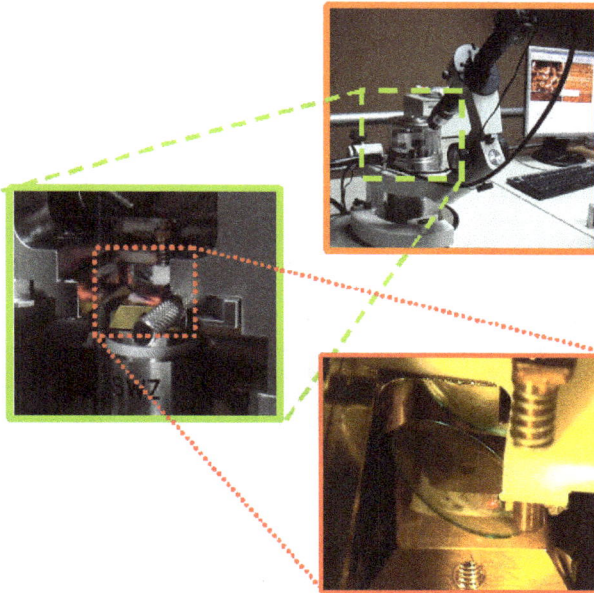

Fig. 6. Experimental setup used in AFM imaging of CL surfaces.

6.2 Phase-contrasted AFM

Phase contrast images are combined with topography images since these two measurements are performed simultaneously. Topography image yields information about surface shape and relative positions and dimensions. Phase contrast image enrichens that information by differentiating between different force gradients which in turn point to different conformational states of polymers, under the assumption that material constituents are homogenously distributed throughout the sample. Phase-Contrasted image of the CL is shown on figure 7.

6.3 Magnetic force microscopy

The AFM/MFM measurements display the features of surface morphology and magnetism on the nanometer scale and are shown on two images in figure 8. Since the sensitivity of

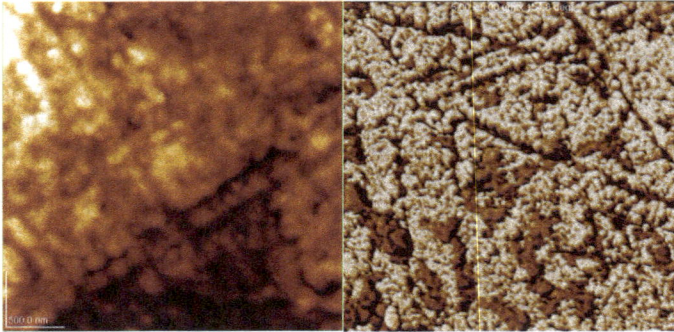

Fig. 7. Phase-Contrasted Atomic Force Microscopy: scan size is $5 \times 5\mu m$. Left: topography image with maximum profile height of 150.9 nm. Right: phase contrasted image for the same portion of the sample that is shown on the left image. Phase contrast image shows granular inhomogeneity in the sample that is synthesized as homogenous (on the nano-scale). The inhomogeneities may originate only as a consequence of processing, showing that certain parts of surface have polymers with altered conformation states, thereby expressing different slope of intermolecular interactions.

forces measured by cantilever go well below nanonewton range, this means that we are able to record paramagnetic and diamagnetic behavior of material. The brighter image areas mark more highly responsive magnetic behavior or paramagnetic areas while than darker areas correspond to diamagnetic areas of the sample.

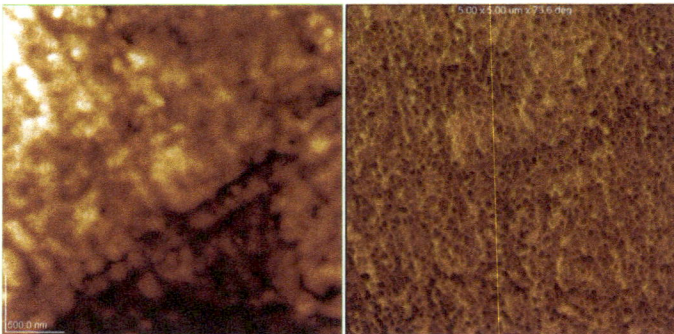

Fig. 8. Magnetic Force Microscopy: scan size is $5 \times 5\mu m$. Left: topography image with maximum profile height of 150.9 nm. Right: the magnetic force gradient image of the same area. This image shows changes in magnetic force gradient, exhibiting that magnetic behaviour exist on the para- and diamagnetic levels and that they are inhomogenous in our samples (that are chemically homogenous).

6.4 Opto-magnetic fingerprint and UV-vis spectroscopy

Digital images of contact lenses were analyzed in terms of their separate color channels (red, blue and green color components) and subsequently processed by spectral convolution algorithm to give the final result – OMF diagram – which shows the intensity values of paramagnetic (positive) and diamagnetic (negative), properties in comparison to wavelength difference. The diagrams on figure 9 show comparison of OMF diagrams with classical UV-Vis-NIR for two samples with different surface qualities.

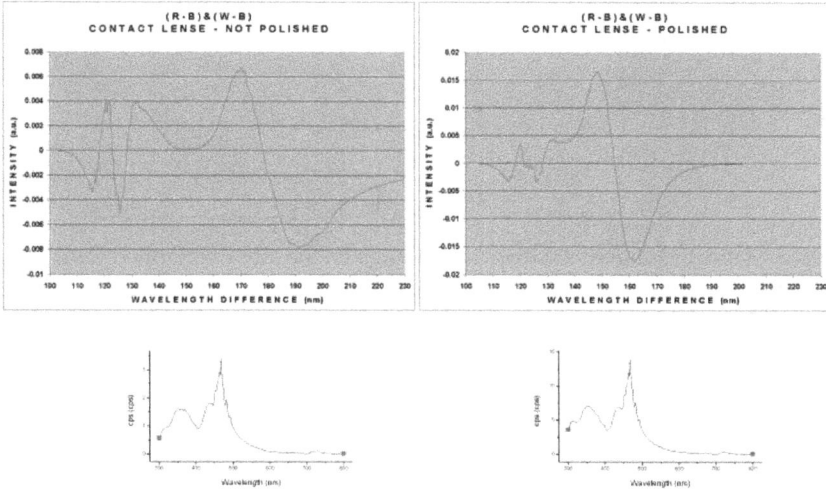

Fig. 9. Opto-Magnetic Fingerprint: comparisons of OMF (upper two diagrams) and UV-Vis spectras (lower two diagrams) for differently surface-treated contact lenses that were produced from the same, standard, material. The characteristic diagrams for lathe processed (left) and polished (right) CL surfaces. UV-Vis spectras show almost identical patterns of absorption while OMF diagrams show visible differences in the wavelength difference range between 100 and 150 nm. These differences are due to different processing and can be quantized by ratio of paramagnetic to diamagnetic properties.

The OMF diagrams show the distribution of energies of reflected light that are distributed over a wavelength difference range. We can observe a qualitatively same pattern, except in the subregion between 100–150 nm, which shows variations in wavelengths and intensities. These diagrams show very high sensitivity of OMF imaging and software analysis, which is the exact purpose this method was intended to achieve.

Moreover, we have tested the OMF response for two inner surfaces of RGP CLs of type $Boston^{TM}$ with respect to changes in the surface qualities. One pair of lenses was processed differently (one was while the other was not polished) while the other pair was processed identically (same polishing times).

6.5 Fractal analysis of contact lenses' roughness profiles

Currently we have no clear answer to the question what surface standard parameter should be used for critical limit determination. The result of the study, reported in (Bruinsma et al., 2003), also confirms the need for quantitatively establishing the replacement schedule of RGP CL. The water contact angle, percentage of elemental surface composition and deposit rate of bacteria was related to standard average roughness parameter R_a. It was stated in (Kim et al., 2002) that surface roughness was the most influential lens surface property after 10 days of wear.

Our research has utilized fractal analysis of roughness profiles to quantify the texture properties of CL inner surface. Standard surface parameters (roughness descriptors) fails

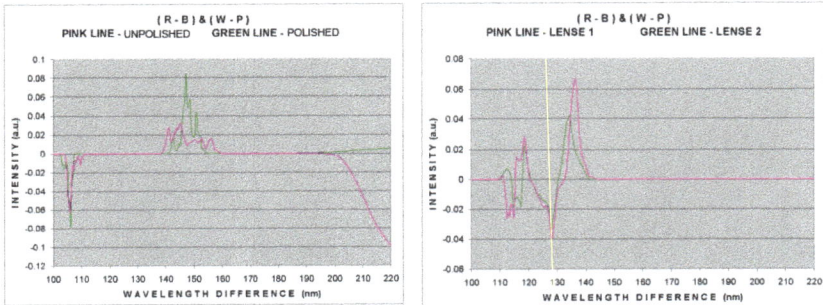

Fig. 10. Opto-Magnetic Fingerprint diagrams. Image on the left: overlapped OMF diagrams of two differently processed CLs that show different behaviour in the wavelength difference range of 130–150nm and sensitivity to changes in surface qualities. Image on the right: overlapped OMF diagrams of two identically processed CLs that show minor change in diagrams and surface qualities.

to describe functional nature of the surface. Moreover, use of more than one roughness parameter exhibits more shortcomings. This is mainly due to the partial information contained in each descriptor.

Fractal analysis of biomedical surface topography is influenced by growing interest in biomaterials surface technology. Fractal geometry provides a useful tool for the analysis of complex and irregular structures such as biomedical surface topography based on image analysis methods that consider an image as a 3D surface.

Fractal dimension calculation is based on "slit-island"and "skyscrapers"methods that were proposed in (Bojović, 2008). This method stipulates that surface recording data is obtained as an image, by using scanning probe microscope. Fractal analysis consists of the following steps:

1. Conversion of AFM image to numerical data in the form of matrix with subsequent conversion of matrix with 256 levels to matrixs with 2^{16} levels of intensity $[0, 65535]$ needed for further calculations.

2. Calculation of image surface area by well known method called "skyscrapers"method. This method approximates surface area of image A with sum of top squares that represent skyscrapers' roofs and the sum of exposed lateral sides of skyscrapers, according to (Chappard et al., 1998). The roof of skyscrapers are increased subsequently by grouping of adjacent pixel grouping. Thus, the intensities of grey scale are averaged. The square size ε is 2^n and the formula is as follows:

$$A(\varepsilon) = \sum \varepsilon^2 + \sum \varepsilon[|z(x,y) - z(x+1,y)| + |z(x,y) - z(x,y+1)|] \tag{2}$$

3. According to (Bojović, 2008), the fractal dimension D can be generated from relation 3 for Hausdorff-Besicovitch dimension where $N(\varepsilon)$ is the number of self-similar structures of linear size ε needed to cover the entire structure. Number $N(\varepsilon)$ can be represented as shown in 4 and used for the area vs. square size relationship 5 resulting in equation 6. Using of logarithmic rules on relation 6 result in a linear equation, expressed as 7. Fractal dimension D is obtained as the slope of fitted line, determined by using relation 7 in the

custom-made procedure for calculation 11.

$$D = \lim_{\varepsilon \to \infty} \frac{\log N(\varepsilon)}{\log \frac{1}{\varepsilon}} \tag{3}$$

$$N(\varepsilon) = c_1 \varepsilon^{-D} \tag{4}$$

$$A(\varepsilon) = N(\varepsilon)\varepsilon^2 \tag{5}$$

$$A(\varepsilon) = c_1 \varepsilon^{2-D_s} \tag{6}$$

$$\log A = (2 - D_s) \log \varepsilon + c \tag{7}$$

4. The log-log graph of surface vs. square size ε is then produced and line is fitted for each image (see figure 11).
5. Finally the fractal dimension is generated by skyscrapers method for AFM topography image as slope of the fitted line. This slope is steeper for rougher contact lens surface.

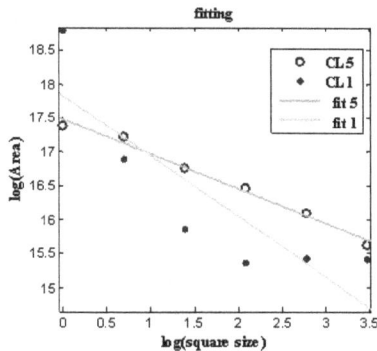

Fig. 11. The log-log graph of image area vs. square size for two CLs. Samples are two RGP CLs made of ML 92 Siflufocon A. The first lens (CL5) was worn for about 3 years while the second lens was worn over a period of more than 5 years (CL1). We can see that fractal dimensions can easily distinguish between two levels of surface roughness created by wear of CL surface.

The procedure is schematically presented in the figure 12 .
Mandelbrot claimed that nature has a fractal face and scholars proved that engineering surfaces have fractal geometry. Compilations of a man-made surface with a tear component on it also show fractal behaviour, proven by power law of area vs. scale relationship. According to (Russ, 1998) a surface with fractal dimension 2.5 would be the optimum as an engineering surface for certain applications.
The fractal dimension generated by skyscrapers method for topography image offers additional and appropriate information about surface roughness. Fractal dimension, as roughness parameter, adequately explains surface functional behaviour. Fractal dimension for new contact lens surface could be an adequate behaviour prediction parameter.

7. Discussion

By performing UV-Vis spectroscopy we have shown that UV-Vis spectra do not change with respect to changes in surface quality (see figure 9) because they measure bulk response of

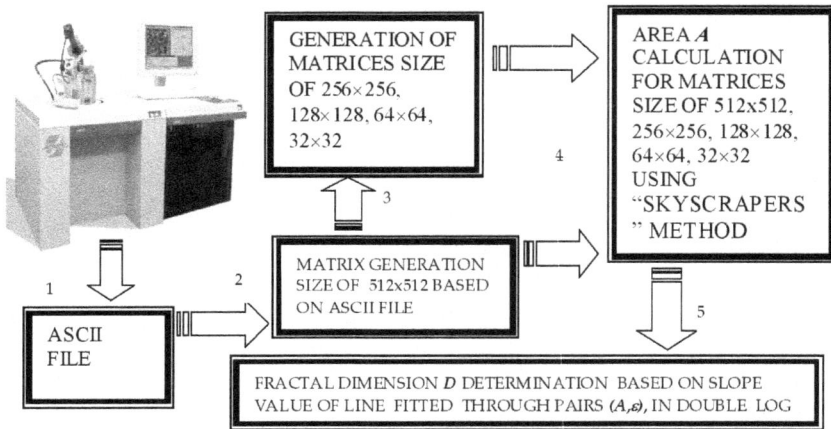

Fig. 12. Diagram of steps involved in fractal analysis of AFM scans.

contact lense materials while opto-magnetic diagrams (OMF) displayed informations that are more sensitive to changes in near-surface properties of material (which are primarily altered during contact lense production). OMF has shown as method that can detect very sensitive to differences in topographical features of contact lenses.

Since, conformation changes in near-surface polymer molecules generate quantum effects they might influence magnetic properties as well. Because of that, we have investigated near-surface regions of contact lenses samples from magnetic and optomagnetic point of view to see whether is there exist a measurable difference in surface magnetic and optical properties. Our aim was to explore the relationship between surface morphology on one side and optical and magnetic properties on the other side.

Measurements on the nano-scale have shown that phase-contrasted atomic force microscopy (PC AFM) and magnetic force microscopy imaging carry additional information that is not contained in morphology scans. However, PC AFM and MFM data need to be integrated with results of OMF in order to obtain quick quantitative assessment of changes in nano- and pico-magnetism that can be related to change in surface structure and its optical properties. Elucidating the origin of these kinds of surface behavior requires further investigations and inclusion of other polishing process parameters on one side and quantitative MFM measurements on the other side. It is our opinion that this kind of analysis will be able to precisely determine parameters of final shape and performance of CL surface.

Since conformation states of RGP CL surface determine paramagnetic diamagnetic properties that can be detected by novel OMF technique, we consider the OMF method and molecular level approach to investigation of optical properties of CL quality as a very promising field for both basic research and technology, with direct influence on application in biomedicine.

New methods of investigation require novel data processing techniques. Fractal analysis has offered more sophisticated tool that, on the basis of nano-scale precision information, generates finer estimate of surface roughness quality.

8. Conclusion

Light has influence on brain activity with very complex pathway (see figure 13). Since light is composed of electrical and magnetic spectra it is very important to know how light interact with contact lenses. These aspects of brain functioning have been thoroughly investigated;

vast amount of informations is available but the most subtle biophysical aspects are still not completely understood.

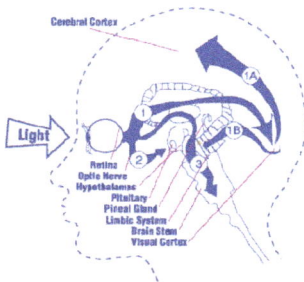

Fig. 13. Nerve pathways from the eyes to the brain goes not only to the visual cortex, but also to deeper brain areas, concerned with neurotransmiters, neurohormones, emotions, etc.

Visual perception is the ability to interpret information from information contained in visible light that reaches the eye. The act of seeing starts when the lens of the eye focuses an image of its surroundings onto a light-sensitive membrane in the back of the eye, called the retina. Since visible light is composed of electrical and magnetic spectra, which have different influence on brain activity (EEG and MEG signals, see figure 14), we investigated magnetic property of contact lenses, as optical material, which have influence on electrical and magnetic light signals properties.

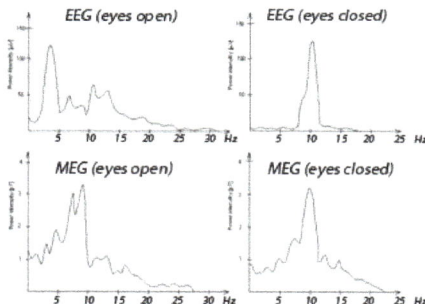

Fig. 14. Brain activity (EEG and MEG) under light influence when eyes are open and closed.

The findings of our measurements enable us to couple optical and magnetic behavior analysis in determination of mechanical properties of surfaces. The dynamical structure of CL materials and its behavior under dynamical mechanical and thermal load is expressed in changes in paramagnetic/diamagnetic behavior. Our results show that these changes are measurable and can be quantified by a simple, quick and accurate method of OMF.

We believe that this combination of intertwining methods could yield an optimal approach towards investigating phenomena in material synthesis and its behavior under mechanical and thermal stress that is still not well understood. The grounds for our propositions are proven relationships between optical, magnetic and mechanical properties of matter. It is our intention to further improve the application of all three used methods and customize their parameters in order to combine them into a new device for opto-magneto-spectroscopic characterization of matter.

Furthermore, our future research will involve nanomaterials as a new doping material influencing CL physical properties such as light transmission, these changes will be investigated by by UV-Vis-NIR spectroscopy as well as optomagnetism. The potential application of nanomaterials might bring significant biophysically based implications for contact lenses industry, biomedical application industry and applied optical science.

9. References

Lim, H., Lee, Y., Han, S., Cho J., Kim, K.J., Surface treatment and characterization of PMMA, PHEMA, and PHPMA, *Journal of Vacuum Science and Technology A*, Vol. 19, No. 4, 2001, pp. 1490–1496, ISSN 0734–2101.

Kim S.H., Opdahl. A., Marmo, C., Somorjai, G.A., AFM and SFG studies of pHEMA-based hydrogel contact lens surfaces in saline solution: adhesion, friction, and the presence of non-crosslinked polymer chains at the surface, *Biomaterials*, Vol. 23, No. 7, 7 September 2001, pp 1657–1666, ISSN 0142–9612.

Bruinsma, G.M., Rustema-Abbinga M., de Vriesa, J., Busschera, H.J., van der Lindenb, M.L., Hooymansc, J.M.M., van der Meia, H.C., Multiple surface properties of worn RGP lenses and adhesion of Pseudomonas aeruginosa, *Biomaterials*, Vol. 24, No. 7, 2003, pp 1663–1670, ISSN 0142–9612.

Binnig, G., Quate, C.F., Gerber, C. (1986), Atomic Force Microscope, *Physical Review Letters*, Vol. 56, No. 9, 3 March 1986, pp. 930–933, ISSN 0031–9007.

Koruga, Đ., Tomić, A., Ratkaj, Z., Matija, L., Classical and Quantum Information Channels in Protein Chain,*Material Science Forum*, Vol. 518, 2006, pp. 491–496, ISSN 1662–9752.

Koruga Đ., Tomić A, Method and algorithm for analysis of light-matter interaction based on spectral convolution, US Pat. App. No.61/061,852, 2008, PCT/US2009/030347, Publication No: WO/2009/089292, Publication Date: 2009-07-16.

Bandić, J., Koruga, Đ., Mehendale, R., Marinkovich, S., System, device and method for dermal imaging, US Pat. App. No. PCT/US2008/050438, Publication No: WO/2008/086311, Publication Date: 2008-07-17.

Bojović, B., Miljković, Z., Babić, B., Fractal Analysis of AFM Images of Worn-Out Contact Lens Surface, *Faculty of Mechanical Engineering Transaction*, 36, 4, 175–180, 2008, ISSN 1451–2092.

Bojović, B., Investigation of Interaction of Engineering Surfaces Condition and Fractal Geometry, PhD Thesis in Serbian, University of Belgrade, 2009.

Chappard, D., Degasne, I., Hure, G., Legrand, E., Audran, M., Basle, M.F., Image analysis measurements of roughness by texture and fractal analysis correlate with contact profilometry, *Biomaterials*, Vol. 24, 2003, pp. 1399–1407, ISSN 0142–9612.

Russ, J.C., Fractal Dimension Measurement of Engineering Surface, *International Journal of Machine Tools and Manufacturing*, Vol. 38, No. 5-6, 1998, pp. 567–571, ISSN 0890-6955.

7

Synthesis and Characterization of Amorphous and Hybrid Materials Obtained by Sol-Gel Processing for Biomedical Applications

Catauro Michelina and Bollino Flavia
Department of Aerospace and Mechanical Engineering,
Second University of Naples, Aversa,
Italy

1. Introduction

An interesting research field with medical applications is represented today by ceramics, as they can be used to obtain useful biomaterials for the production of implants (Vallet-Regí, 2001, 2006a, 2006b, 2006c); many parts of the human body, in fact, can be replaced or repaired with biomaterials and more specifically with bioceramics (Black & Hastings, 1998). Regardless of the ceramic type and the application procedure, the introduction of an implant in a living body always causes inflammation phenomena and frequently infection processes as well. Those problems can be overcome by using local drug delivery methods to confine pharmaceuticals such as antibiotics, anti-inflammatory, anti-carcinogens, etc. (Arcos et. al., 2001; Ragel & Vallet-Regì, 2000; Vallet-Regì et al., 2000). The possibility of introducing certain drugs into the ceramic matrices employed for bone and teeth repair is undoubtedly an added value to be taken into account.

The traditional use of high temperature procedures to model glasses and ceramics to the desired shape is very well known; on the other hand, the degradation temperature of a pharmaceutical compound is usually around 100°C, which is very low if compared with the high ones needed to compact the components (around 1000°C). Consequently, the main problem is how to include pharmaceuticals in conventional glass and ceramic implants. The scientific community is currently investigating new procedures to incorporate drugs into implantable biomaterials.

The sol-gel process, among others, has proved to be a versatile one and has been widely used in the preparation of amorphous and or hybrid materials (Hench & West, 1990; Judeinstein & Sanchez, 1996; Novak, 1993), with applications, for example, in non-linear optical materials (Hsiue et al., 1994) and mesomoporous materials (Wei et al., 1999). The family of organic-inorganic hybrid materials has attracted considerable attention because of its interesting properties such as molecular homogeneity, transparency, flexibility and durability. A key issue that remains unresolved in these organic-modified materials is the degree of mixing of the organic-inorganic components, i.e., phase homogeneity. The high optical transparency to visible light indicates that the organic-inorganic phase separation, if any, is on a scale of $\leq$ 400nm. Such hybrids are promising materials for various applications, e.g.: solid state lasers (optical components), replacements for silicon dioxide as insulating materials in the microelectronic industry, anti-corrosion and scratch resistant coatings, contact lenses or host materials for chemical sensors. In the recent years interest in those materials is connected to their possible applications as biomaterials (Gigant et al., 2002; Joshua et al., 2001; Klukowska et al., 2002; Mackenzie & Bescher, 1998; Matsuura et al., 2001; Spanhel et. al., 1995). One indirect advantage of including polymers is that it is possible to obtain synergistic effects that combine the best properties of polymers with the best properties of inorganic materials. These materials are considered as biphasic materials, where the organic and inorganic phase is mixed at the nm to sub-μm scales. Nevertheless, it is obvious that the properties of these materials are not just the sum of the individual contributions from both phases; the role of the inner interfaces could be predominant. The nature of the interface has recently been used to divide these materials into two distinct classes (Sanchez & Ribot, 1994). In class I, organic and inorganic compounds are embedded and only the weak bonds (hydrogen, van der Waals bonds) give the cohesion to the whole structure. In class II materials, the phases are linked together through strong chemical bonds (covalent or ionic-covalent bonds). Both class I and class II hybrids were prepared by sol-gel technique (Young, 2002).

The aim of the present chapter is to summarize the synthesis via sol-gel and the characterisation methods of amorphous and hybrid materials for biomedical applications. Therefore, the emphasis of our discussion will be focussed on the science, rather than on the technology, of sol-gel processing. The controlled release of pharmaceuticals such as anti-inflammatory agents and antibiotics from strong and biocompatible hosts has relevant applications: they include implantable therapeutic systems, filling materials for bone or teeth repair, which curtail inflammatory or infectious side effects of implant materials when coatings of biocompatible materials containing anti-inflammatory or antibiotic drugs are applied.

2. General processing methods

Different types of colloids can be used to produce polymers or particles from which we can obtain a ceramic material: for example, sols (suspensions of solid particles in a liquid), aerosols (suspensions of particles in a gas) or emulsions (suspensions of liquid droplets in another liquid). The sol-gel chemistry is based on the hydrolysis and polycondensation of molecular precursors such as metal alkoxides $M(OR)x$, where M = Si, Sn, Ti, Zr, Al, Mo, V, W, Ce and so forth. The following sequence of reactivity is usually found as $Si(OR)_4 << Sn(OR)_4 = Ti(OR)_4 < Zr(OR)_4 = Ce(OR)_4$ (Novak, B.M.,1993). Fig. 1 presents a schema of the procedures which one could follow within the scope of sol-gel processing. In the sol-gel process, the precursors for the preparation of a colloid consist of a metal or metalloid element surrounded by various ligands. A list of the most commonly used alkoxy ligands is presented in Tab. 1.

Alkyl		Alkoxy	
methyl	•CH_3	Methoxy	•OCH_3
ethyl	•CH_2CH_3	Ethoxy	•OCH_2CH_3
n-propryl	•$CH_2CH_2CH_3$	n-propoxy	•$OCH_2CH_2CH_3$
Iso-propyl	H_3C(•C)HCH_3	Iso-propyl	H_3C(•O)$CHCH_3$
n-butyl	•CH_2(•CH_2)$_2CH_3$	n-butoxy	•$O(CH_2)_3CH_3$
Sec-butyl	H_3C(•C)HCH_2CH_3	Sec- propoxy	H_3C(•O)$CHCH_2CH_3$
Iso- butyl	•$CH_2CH(CH_3)_2$	Iso- propoxy	•$OCH_2CH(CH_3)_2$
Tert-butyl	•$C(CH_3)_3$	Tert- propoxy	•$OC(CH_3)_3$

Table 1. Commonly used ligands in sol-gel process.

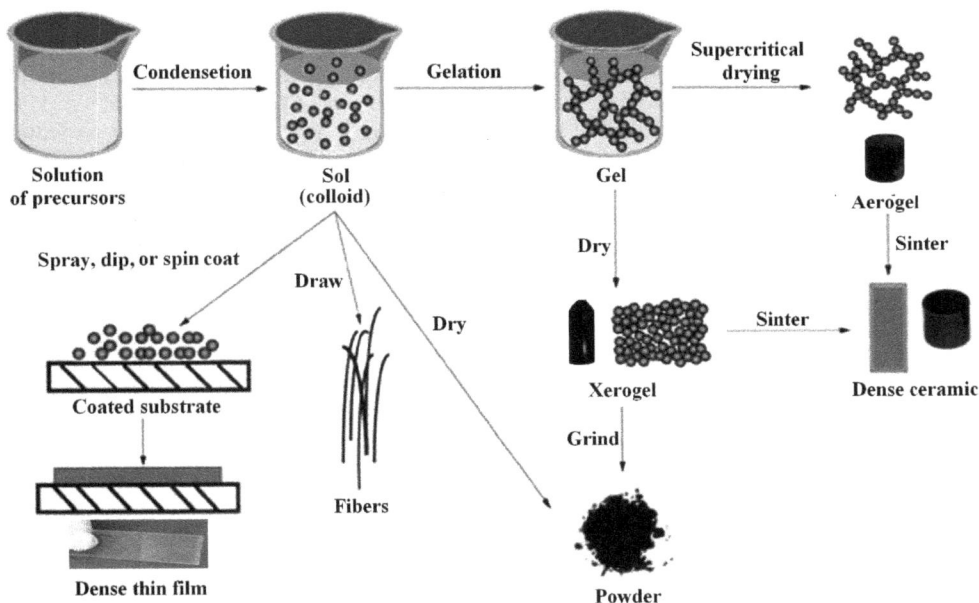

Fig. 1. Schematic of sol-gel processing.

2.1 Hydrolysis and condensation

The alkoxydes used as ligands can be organometallic compounds, where direct metal-carbon bonds are present, or also members of the family or metalloid atoms, the so called metal alkoxides, among which the most widely known, as it has been extensively studied, is the silicon tetraethoxide (or tetrathoxy-silane, or tetraethyl orthosilicate, TEOS), $Si(OC_2H_5)_4$. Silicate gels are most often synthesized by hydrolyzing monomeric, tetrafunctional alkoxide precursors employing a mineral acid (e.g., HCl) or base (e.g. NH_3) as a catalyst. At the functional group level, the sol-gel process starts with the following reaction:

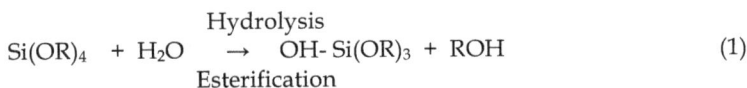

$$Si(OR)_4 + H_2O \xrightarrow{\text{Hydrolysis}}_{\text{Esterification}} OH\text{-}Si(OR)_3 + ROH \qquad (1)$$

which can even be stopped while the metal is only partially hydrolyzed, $Si(OR)_{4-n}(OH)_n$. Then, two partially hydrolyzed molecules can link together in a condensation reaction, such as one of the following:

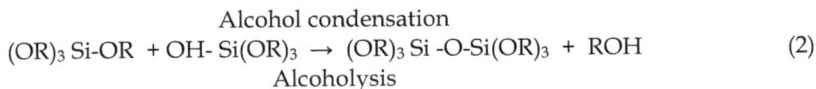

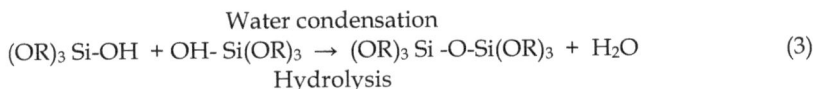

$$(OR)_3 Si\text{-}OR + OH\text{-}Si(OR)_3 \xrightarrow{\text{Alcohol condensation}}_{\text{Alcoholysis}} (OR)_3 Si\text{-}O\text{-}Si(OR)_3 + ROH \qquad (2)$$

$$(OR)_3 Si\text{-}OH + OH\text{-}Si(OR)_3 \xrightarrow{\text{Water condensation}}_{\text{Hydrolysis}} (OR)_3 Si\text{-}O\text{-}Si(OR)_3 + H_2O \qquad (3)$$

where R is an alkyl group, C_xH_{2x+1}. The hydrolysis reaction (eq. 1) replaces alkoxide group (OR) with hydroxyl group (OH). Subsequent condensation reactions involving the silanol group produce siloxane bonds (Si-O-Si) and the by-products alcohol (ROH) (eq. 2) or water (eq. 3). Under most conditions, condensation starts (eqs. 2 and 3) before hydrolysis (eq. 1) is complete. A solvent such as an alcohol is normally used as a homogenizing agent, as water and alkoxysilanes are immiscible (Fig. 2). However, a gel can be prepared from silicon alkoxide-water mixtures without adding a solvent (Avnir & Kaufman, 1987), since the alcohol produced as the by-product of the hydrolysis reaction is sufficient to homogenize the initially phase separated system. It should be noted that the alcohol is not simply a solvent. As indicated by the reverse of eqs. 1 and 2, it can participate in esterification or alcoholysis reactions.

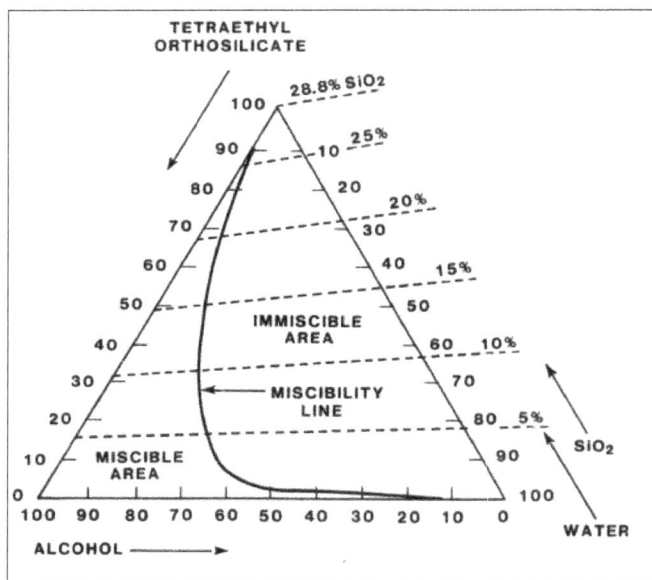

Fig. 2. TEOS , H_2O, Synasol (95% EtOH, 5% water) ternary-phase diagram at 25°C . For pure ethanol the miscibility line is slightly shifted to the right (Cologan & Setterstrom, 1946).

The H_2O:Si molar ratio (r) in eq. 1 has been made to vary from less than one to over 50, and the concentration of acid or bases from less than 0.01 (Brinker et al., 1982) to 7M (Stober et al., 1968) depending on the desired end product. Typical gel-synthesis procedures used to produce bulk gels, films, fibres, and powders are listed in Tab 2. Hydrolysis occurs by the nucleophilic attack of the oxygen contained in water on the silicon atom as shown by the reaction of isotopically labelled water with TEOS that produces only unlabelled alcohol in both acid-base-catalyzed systems (Voronkov & et al., 1978). Hydrolysis is facilitated in the presence of homogenizing agents (alcohols, dioxane, THF, acetone, etc.) that are especially beneficial in promoting the hydrolysis of silanes containing bulk organic or alkoxy ligands. It should be emphasized, however, that the addition of solvents may promote esterification or depolymerization reactions according to the reverse of eqs. 1 and 2.

SiO$_2$ Gel Types	% Mole					
	TEOS	EtOH	H$_2$O	HCl	NH$_3$	H$_2$O/Si(r)
Bulk	6.7	25.8	67.3	0.2	-	10
Fibers	11.31	77.26	11.31	0.11	-	1.0
Films	5.32	36.23	58.09	0.35	-	10.9
Monodisperse Spheres	0.83	33.9	44.5	-	20.75	53.61

Table 2. Sol-gel Silicate compositions for bulk gels, fibres, film and powder.

The Hydrolysis is more rapid and complete when catalysts are employed (Voronkov et al., 1978). Although mineral acids or ammonia are most generally used in sol-gel processing, other known catalysts are acetic acid, KOH, amines , KF, HF, titanium alkoxides, and vanadium alkoxides and oxides (Voronkov et al , 1978). In the literature mineral acids are reported to be more effective catalysts than the equivalent base concentrations. However, neither the increasing acid of silanol groups with the extent of hydrolysis and condensation (Keefer, 1984) nor the generation of unhydrolyzed monomers via base-catalyzed alcoholic or hydrolytic depolymerization processes have generally been taken into account. Aelion et al., (1950a, 1950b) investigated the hydrolysis of TEOS under acid and basic conditions using several cosolvents: ethanol, methanol, and dioxane. The extent of hydrolysis (eq. 1) was determined by distillation of the ethanol by-product. Karl Fischer titration was used to follow the consumption of water by hydrolysis (eq.1) and its production by condensation (eq.3). Aelion et al. observed that the rate and extent of the hydrolysis reaction was mostly influenced by the strength and concentration of the acid or base catalyst. As under acid conditions, the hydrolysis of TEOS in base media was a function of the catalyst concentration (Aelion et al., 1950a, 1950b).

Steric (spatial) factors exert the greatest effect on the hydrolytic stability of organoxysilanes (Voronkov et al., 1978). Any complication of the alkoxy group delays the hydrolysis of alkoxysilanes, but the hydrolysis rate is lowered at most by the branched alkoxy group (Voronkov et al., 1978). The effects of alkyl length and the degree of branching observed by (Aelion et al., 1950a, 1950b) are illustrated in Tab. 3 for the hydrolysis of tetralkoxysilanes.

R	k 10^2 (1 mol^{-1} s^{-1} [H$^+$]$^{-1}$
C$_2$H$_5$	5.1
C$_4$H$_9$	1.9
C$_6$H$_{13}$	0.83
(CH$_3$)$_2$CH(CH$_2$)$_3$CH(CH$_3$)CH$_2$	0.30

Table 3. Rate constant k for acid hydrolysis of tetralkoxysilanes (RO)$_4$Si at 20°C

Fig. 3 compares the hydrolysis of TEOS and TMOS under acid and basic conditions. The delaying effect of the bulkier ethoxide group is clearly evident. According to (Voronkov et al., 1978) in the case of mixed alkoxides, (RO)$_x$(R'O)$_{4-x}$Si where R'O is a higher (larger) alkoxy group than RO, if the R'O has a normal (i.e. linear) structure, its retarding effect on the hydrolysis rate is manifest only when x= 0 or 1. If R'O is branched, its delaying effect is evident even when x= 2. The hydrolysis of the n-propoxide group was observed to be slower than the ethoxide group during the second hydrolysis step under both acid and basic conditions. This result suggests that a delaying effect of a higher, normal alkoxide group is realized regardless of the extent of substitution.

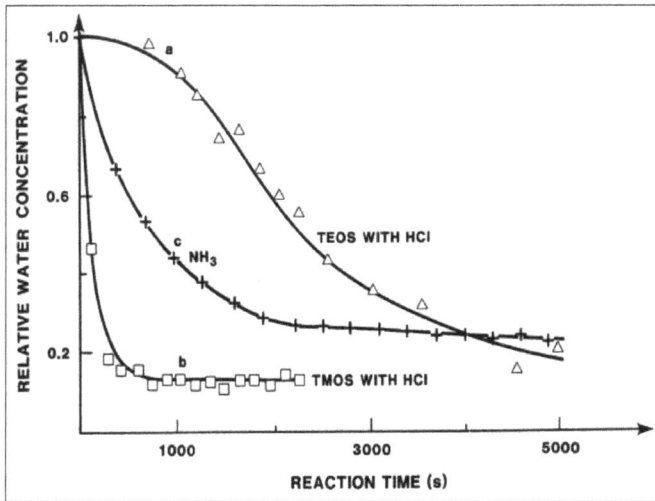

Fig. 3. Relative water concentration versus time during acid- or base-catalzed hydrolysys of □: TMOS with HCl ; X: TEOS and TMOS with NH₃. Δ:TEOS with HCl (Shih et al., 1987)

The substitution of one alkyl group with alkoxy groups increases the electron density on the silicon. Conversely, hydrolysis (substitution of OH for OR) or condensation (substitution of OSi for OR or OH) decreases the electron density on the silicon Fig. 4. Inductive effects are evident from investigations on the hydrolysis of methylethoxysilanes (Schimdt et al., 1984), $(CH_3)_x(C_2H_5O)_{4-x}Si$ where x varies from 0 to 3. Fig. 5 shows that under acidic (HCl) conditions, the hydrolysis rate increases with the degree of substitution x, of electron-providing alkyl group, whereas under basic (NH₃) conditions the reverse trend is clearly observed.

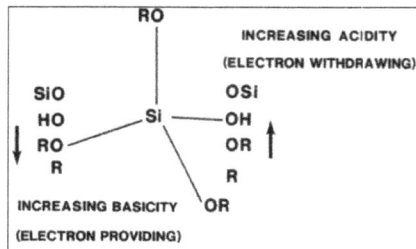

Fig. 4. Inductive effects of substituents attached to silicon,R, OR, OH or OSi (Brinker, 1988)

Fig. 5 also shows the accelerating effect of methoxide substitution on the hydrolysis rate (TMOS versus TEOS). The acceleration and retardation of hydrolysis with increasing x under acid and basic conditions respectively, suggest that the hydrolysis mechanism is sensitive to inductive effects and is apparently unaffected by the extent of alkyl substitution. Because increased stability of the transition state will increase the reaction rate, the inductive effects are evident for positively and negatively charged transition states or intermediates under acid and basic conditions respectively. This reasoning leads to the hypothesis that

Synthesis and Characterization of Amorphous and Hybrid Materials Obtained by Sol-Gel Processing for
Biomedical Applications

173

under acid conditions, the hydrolysis rate decreases with each subsequent hydrolysis step
(electron withdrawing), whereas under basic conditions the increased electron-withdrawing
capabilities of OH (and OSi) compared to OR may establish a condition in which each
subsequent hydrolysis step occurs more quickly as hydrolysis and condensation proceed.

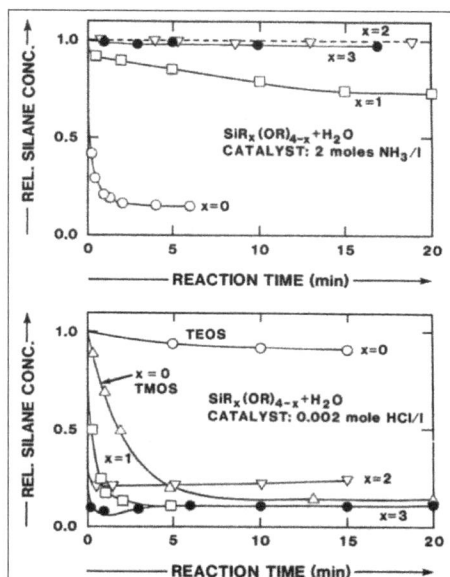

Fig. 5. Relative silane concentration versus time during acid- and base-catalyzed hydrolysis
of different silanes in ethanol (volume ratio to EtOH=1:1). •: $(CH_3)_3SiOC_2H_5$.
∇:$(CH_3)_2Si(OC_2H_5)_2$. □: $(CH_3)_2Si(OC_2H_5)_3$. ○:$Si(OC_2H_5)_4$. Δ:$Si(OCH_3)_4$. (Shih et al., 1987)

From the standpoint of organically modified alkoxysilanes, $R_xSi(OR)_{4-x}$, the inductive effects
indicate that acid-catalyzed conditions are preferable (Schimdt et al., 1984), since acids are
effective in promoting hydrolysis both when x=0 and x>0. As indicated in Tab. 2, the
hydrolysis reaction has been performed with r values ranging from <1 to over 25
depending on the desired polysilicate product, for example, fibers, bulk gel or colloidal
particles. From eq. 1, an increased value of r is expected to promote the hydrolysis reaction.
(Aelion et al., 1950a, 1950b) found the acid-caltayzed hydrolysis of TEOS to be first-order in
[H_2O]; however, they observed an apparent zero-order dependence of the water
concentration under base-catalyzed conditions. As explained, this is probably due to the
production of monomers by siloxane bond hydrolysis and redistribution reactions.
Solvents are usually added to prevent liquid-liquid phase separation during the initial
stages of the hydrolysis reaction and to control the concentrations of silicate and water that
influence the gelation kinetics. More recently, the effects of solvents have been studied
primarily in the context of drying control chemical additives (DCCA) used as cosolvents
with alcohol in order to facilitate rapid drying of monolithic gels without cracking (Hench et
al., 1986). Solvents can be classified as polar or nonpolar and as protic or aprotic . The dipole
moment of a solvent determines the length over which the charge of one species can be "felt"
by surrounding species. The lower the dipole moment, the larger this length becomes. This

is important in electrostatically stabilized systems and when considering the distance over which a charged catalytic species, for example an OH- nucleophile or H_3O^+ electrophile, is attracted to or repelled from potential reaction sites, depending on their charge. The availability of labile protons determines whether anions or cations are solvated more strongly through hydrogen bonding. Because hydrolysis is catalyzed either by hydroxyl (pH>7) or hydronium ions (pH<7), solvent molecules that hydrogen bonds to hydroxyl or hydronium ions reduce the catalytic activity under basic or acid conditions respectively. Therefore, aprotic solvents that do not form a hydrogen bond to hydroxyl ions have the effect of making hydroxyl ions more nucleophilic, whereas protic solvents make hydronium ions more electrophilic (Morrison & Boyd, 1966).

Hydrogen bonding may also influence the hydrolysis mechanism, hydrogen bonding with the solvent can sufficiently activate weak leaving group to realize a bimolecolar, nucleophilic (S_N2-Si) reaction mechanism (Voronkov et al., 1978). The availability of labile protons also influences the extent of the reverse reactions, reesterification (reverse eq. 1) or siloxane bond alcoholysis or hydrolysis (reverse of eqs. 2 and 3). Aprotic solvents do not participate in reverse reactions such as reesterification or hydrolysis, because they lack a sufficiently electrophilic proton and are unable to be deprotonated to form sufficiently strong nucleophiles (OH- or OR-) necessary for reaction 4. Therefore compared to alcohol or water, aprotic solvents such as THF or dioxane are considerably more "inert" (they do not formally take part in sol-gel processing reactions), they may influence reaction kinetics by increasing the strength of nucleophiles or decreasing the strength of electrophiles.

$$R - \overset{..}{\underset{..}{O}} - H$$
$$-\underset{|}{\overset{|}{Si}} - \overset{..}{\underset{..}{O}} - \underset{|}{\overset{|}{Si}} - \quad \leftrightharpoons \quad -\underset{|}{\overset{|}{Si}} - \overset{..}{O}R \quad + \quad H\overset{..}{\underset{..}{O}} - \underset{|}{\overset{|}{Si}} - \tag{4}$$

2.2 Gelation

Clusters resulting from the hydrolysis and condensation reactions eventually collide and link together into a gel, which is often defined as "strong" or "weak" according to whether the bonds connecting the solid phase are permanent or reversible; however, as noted by (Flory, 1974), the difference between weak and strong ones is a matter of time scale. Even covalent siloxane bonds in silica gel can be cleaved, allowing the gel to exhibit a slow and irreversible (viscous) deformation. Thus the chemical reactions that bring about gelation continue long beyond the gelation point, permitting flow and producing gradual changes in the structure and properties of the gel. The outline of the aging of a gel is as follows:

- Phenomenology
- Classical theory
- Percolation theory
- Kinetic model

The simplest picture of gelation is that clusters grow by condensation of polymers or aggregation of particles until the clusters collide; then, links form between the clusters to produce a single giant cluster that is called gel. When the gel forms, many clusters will be present in the sol phase, entangled in but not attached to the spanning cluster; with time,

Synthesis and Characterization of Amorphous and Hybrid Materials Obtained by Sol-Gel Processing for
Biomedical Applications

175

they progressively become connected to the network and the stiffness of the gel will increase. The gel appears when the last link is formed between two large clusters to create the spanning cluster. This bond is no different from innumerable others that form before and after the gel point, except that it is responsible for the onset of elasticity by creating a continuous solid network. The sudden change in rheological behaviour is generally used to identify the gel point in a crude way.

The classic theory explains the theory developed by Flory (1953) and Stockmayer (1945) to account for the gel point and the molecular-weight distribution in the sol. The most important deficency of this model is that it neglects the formation of closed loops within the growing clusters, and this leads to unrealistic predictions about the geometry of the polymers. The percolation theory offers a description that does not exclude the formation of closed loops and so does not predict a divergent density for large clusters. The disadvantage of the theory is that it generally does not lead to analytical solutions for such properties as the percolation threshold or the size distribution of polymers. However, these features can be determined with great accuracy from computer simulations, and the results are often quite different from the predictions of the classical theory. Excellent reviews of percolation theory and its relation to gelation have been written by Zallen (1983) and Stauffer et al. (1982); and the kinetic models are based on Smoluchowski' s analysis of the growth and aggregation of clusters.

The Smoluchowski equation describes the rate at which the number, n_s, of clusters of size s changes with time t, during an aggregation process :

$$\frac{dn_s}{dt} = \frac{1}{2}\Sigma_{i+j=s} K(i,j)n_i n_j - n_s \Sigma_{j=1}^{\infty} K(s,j)n_j \tag{5}$$

The coagulation kernel, K(i,j) is the rate coefficient for aggregation of a cluster of size i with another and of size j. The first term in eq. 5 gives the rate of creation of size s by aggregation of two smaller clusters, and the second term gives the rate at which clusters of size s are eliminated by further aggregation. For this equation to apply, the sol must be so diluted that collisions between more than two clusters can be neglected, and the clusters must be free to diffuse so that the collisions occur at random. Further, since K depends only on i and j, ignoring the range of structures that could be present in a cluster of a given size, this is a mean-field analysis that replaces structural details with averages.

2.3 Drying

The drying of a porous material is a process which can be divided into several stages. At first the body shrinks by an amount equal to that volume of the evaporated liquid and the liquid-vapor interface remains at the exterior surface of the body. The second stage begins when the body becomes too stiff to shrink and the liquid recedes into the interior, leaving air-filled pores near the surface. Even as air invades the pores, a continuous liquid film supports flow to the exterior, so evaporation continues to occur from the surface of the body. Eventually, the liquid becomes isolated into pockets and drying can proceed only by evaporation of the liquid within the body and diffusion of the vapor to the outside. In the specialized literature the factors affecting stress development are discussed and various strategies to avoid warping and cracking are described. The outline is as follows:

- Phenomenology
- Drying stress
- Avoiding fracture

The first stage of drying is called the constant rate period (CRP), because the rate of evaporation per unit area of the drying surface is uniform (Fortes & Okos, 1980; Macey, 1942; Moore, 1961). The evaporation rate is close to that of an open dish of liquid, as indicated by the data for the drying of alumina gel (Dwivedi, 1986), shown in Fig. 6. The rate may differ slightly, depending on the texture of the surface. For example, as sand beds dry, the water conforms to the shapes of the particles, so the wet area is larger than the planar one pertaining to the surface of the body, and the rate of evaporation is correspondingly higher (Ceaglske & Hougen, 1937). The distribution of a spreading liquid is illustrated schematically in Fig.6. The chemical potential, μ, of the liquid in the adsorbed film is equal to the one under the concave meniscus, otherwise liquid would flow from one to the other to balance the potential. The chemical potential μ is lower than bulk liquid because of disjoining and capillary forces, therefore the vapour pressure (p_v) decreases according to:

$$\frac{p_v}{p_0} = Exp\ (\Delta\mu/R_gT) \tag{6}$$

where p_0 is the vapour pressure of bulk liquid, R_g is the ideal gas constant, T is the temperature and $\Delta\mu$ is the increment of the chemical potential. The rate of evaporation, V_E, is proportional to the difference between p_v and the ambient vapour pressure, p_A:

$$V_E = k\ (p_v - p_A) \tag{7}$$

where k is a coefficient that depends on the design of the drying chamber, draft rate, etc. It appears reasonable to conclude that the surface of the body must be covered with a film of liquid (as in Fig. 6a), because the rate would decrease as the body shrinks if evaporation occurrs only from the menisci, Fig. 6b.

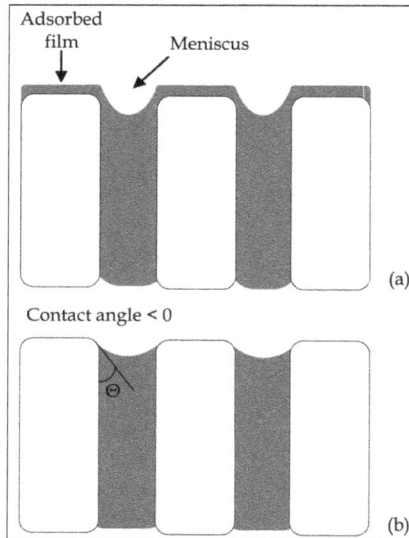

Fig. 6. Distribution of liquid at the surface of a drying porous body, when liquid is (a) spreading (contact angle θ=0°) or (b) wetting, but not spreading (90°> θ >0°). The chemical potential of the liquid in the adsorbed film is equal to that under the meniscus.

3. Experimental procedures

3.1 Sol-gel synthesis of organic – inorganic hybrid materials

Hybrid organic-inorganic biomaterials were prepared by means of a sol-gel process from an analytical reagent grade of metal alkoxides $M(OR)_x$, in an ethanol, organic polymer like poli-ε-caprolactone (PCL Mw = 65,000), water and solvent ($CHCl_3$) mixture. Water, diluted with ethanol was added to the solution under a vigorous stirring. A flow-chart of hybrids (MO_2 + PCL x wt%) can show the synthesis by the sol-gel method. MO_2/PCL, all mixed with drugs (y wt%), were also prepared by using an analytical reagent grade as a precursor material.

In this study SiO_2/PCL (PCL 0, 6, 12, 50 wt%) materials were used as support matrices for controlled drug release. Silica gel, originally developed for engineering applications, is also currently being studied as a polymer for the entrapment and sustained release of drugs (Teoli et al., 2006). In the present study the sol-gel method was applied to encapsulate Ketoprofen (5, 10, 15 wt%) as a model drug. The drug loaded amorphous bioactive materials were studied in terms of their drug release kinetics

The hybrid inorganic-organic materials (PCL 0, 6, 12, 50 wt%) were prepared by means of sol-gel process from an analytical reagent grade of tetraethyl orthosilicate (TEOS) in an ethanol, poly-ε-caprolactone (PCL), water, and chloroform ($CHCl_3$) mixture. Water, diluted with ethanol was added to the solution under vigorous stirring. Fig. 7 shows the flow chart of hybrid (SiO_2 + %PCL + %Ketoprofen) synthesis by the sol-gel method. As it is shown in the same Fig. 7, SiO_2/PCL (PCL 0, 6, 12, 50 wt%) all mixed with ketoprofen (5, 10, 15 %) were prepared by using an analytical reagent grade as precursor material.

After the addition of each reactant the solution was stirred and the resulting sols were uniform and homogeneous. The gelification time was controlled by varying the concentration of PCL, as shown in Tab. 4. After gelification the gels were air dried at 50°C for 24h to remove the residual solvent; as this treatment does not modify the stability of ketoprofen, glassy pieces were obtained (Fig. 8). Discs with a diameter of 13 mm and a thickness of 2 mm were obtained by pressing a fine (<125 μm) gel powder into a cylindrical holder.

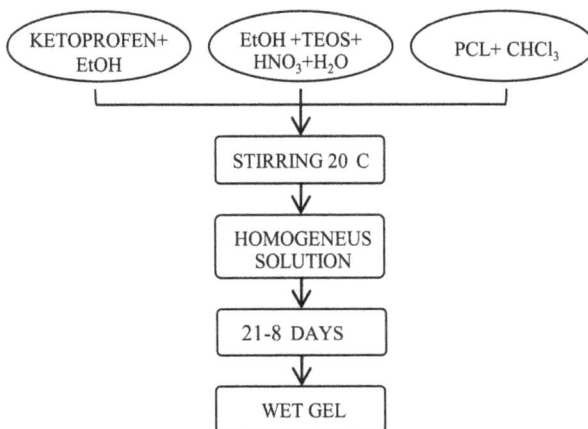

Fig. 7. Flow chart of SiO_2/PCL gel synthesis.

Materials prepared	Gelation time (20-25°C)
SiO_2 + 0%PCL + 5%Ketoprofen	21
SiO_2 + 6%PCL + 5%Ketoprofen	16
SiO_2 + 12%PCL + 5%Ketoprofen	14
SiO_2 + 50%PCL + 5%Ketoprofen	8
SiO_2 + 0%PCL + 10%Ketoprofen	19
SiO_2 + 6%PCL + 10%Ketoprofen	15
SiO_2 + 12%PCL + 10%Ketoprofen	12
SiO_2 + 50%PCL + 10%Ketoprofen	9
SiO_2 + 0%PCL + 15%Ketoprofen	18
SiO_2 + 6%PCL + 15%Ketoprofen	14
SiO_2 + 12%PCL + 15%Ketoprofen	12
SiO_2 + 50%PCL + 15%Ketoprofen	10

Table 4. Variation in the gelification time, controlled by changing the concentration of PCL.

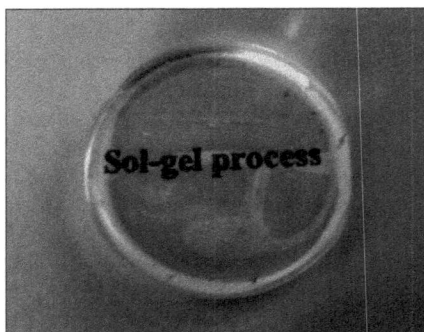

Fig. 8. SiO_2/PCL gel after drying.

Chromatographic experiments were carried out on a Shimadzu HPLC system, equipped with a Class-VP 5.0 software, an UV spectrophotometric detector SPD-10AVvp and two pumps LC-10ADvp, with low-pressure gradient systems. Samples of solutions were injected by a syringe via a Rheodyne loop injector; the loop volume was 20 µl, the analytical column was a Phenomenex C18 (150 × 4.60 mm; 5 µ); the flow rate of the mobile phase A (water) was set at 0.8 ml/min and that of the mobile phase B (methanol) was set at 0.2 ml/min. The total runtime was 10 minutes. HPLC grade methanol was obtained by Sigma-Aldrich. HPLC grade water was prepared using a Millipore (0.22 µm) system. A standard solution of ketoprofen 3 mM in a simulated body fluid (SBF) was prepared and the samples were taken at the end of the release from the materials.

The nature of SiO_2 gel, poly-ε-caprolactone (PCL) and PCL/SiO_2 hybrid materials were ascertained by X-ray diffraction (XRD) analysis using a Philips diffractometer. The presence of hydrogen bonds between organic-inorganic components of the hybrid materials was ascertained by FTIR analysis. Fourier transform infrared (FTIR) transmittance spectra were recorded in the 400-4000 cm^{-1} region using a Prestige 21 Shimatdzu system, equipped with a DTGS KBr (Deuterated Tryglicine Sulphate with potassium bromide windows) detector, with resolution of 2 cm^{-1} (45 scans). KBr pelletized disks containing 2 mg of sample and 200 mg KBr were made. The FTIR spectra were elaborated by IR solution software. The microstructure of the synthesized gels was studied by a scanning electron microscopy (SEM) Cambridge model S-240 on samples previously coated with a thin Au film and by a Digital Instruments Multimode atomic force microscopy (AFM) in contact mode in air.

3.2 Study of in vitro bioactivity

In order to study their bioactivity, samples of the studied hybrid materials were soaked in a simulated body fluid (SBF) with ion concentrations nearly equal to those of the human blood plasma, as reported elsewhere and shown in Tab. 5 (Hench & Clark, 1978; Ohtsuki et al., 1992; Paul, 1992). During soaking, the temperature was kept fixed at 37°C. Taking into account that the ratio of the exposed surface to the volume solution influences the reaction, a constant ratio of 50 mm^2 ml^{-1} of solution was respected (Hutmacher et al., 2001).

Ion	Ions concentration (mM)	
	Human blood plasma	SBF
Na^+	142.0	142.0
K^+	5.0	5.0
Mg^{2+}	1.5	1.5
Ca^{2+}	2.5	2.5
Cl^-	103.0	148.0
HCO_3^-	27.0	4.2
HPO_4^{2-}	1.0	1.0
SO_4^{2-}	0.5	0.5
pH	7.2 - 7.4	7.4

Table 5. Simulated body fluid (SBF) ionic concentration (mM).

It was shown that SBF reproduces *in vivo* bonelike apatite formation on bioactive glass and ceramic (Hench, 1991). *In vitro* studies using SBF have suggested that bioactive glass and glass-ceramics form bonelike apatite on their surface (see Fig. 9) by providing surface functional groups of silanol (Si$-$OH), which are effective for apatite nucleation. These groups combine with Ca^{2+} ions present in the fluid imposing an increase of positive charge on the surface. In addition, Ca^{2+} ions combine with the negative charge of the phosphate ions to form amorphous phosphate, which spontaneously transforms into hydroxyl-apatite [$Ca_{10}(PO_4)_6(OH)_2$] where the atomic ratio Ca/P is 1.60 (Ohtsuki et al., 1992). The SBF is already supersaturated with respect to the apatite under normal conditions. Once the apatite nuclei are formed, they can grow spontaneously by consuming the calcium and phosphate ions from the surrounding body fluid. It is known from literature (Hench, 1991; Ohtsuki et al., 1992) that CaO, SiO$_2$-based glasses, CaO, P$_2$O$_5$- based glasses, sodium silicate glasses are more bioactive then ion free glass and ceramic. That is due to the dissolution of appreciable amounts of calcium and phosphate ions, which increase the degree of supersaturation of the surrounding body fluid with respect to the apatite. Moreover, the released ions are exchanged with H$_3$O$^+$ ions in SBF forming silanol groups on their surface; this reaction causes a pH increase of SBF solution and, consequently, Si–OH groups are dissociated into negatively charged units Si–O$^-$ that interacts with the positively charged calcium ions to form the calcium silicate.

Fig. 9. Apatite layer on SiO_2/PCL surface.

3.3 Study of in-vitro release

For the study of drug release, the discs of the investigated material were soaked in 15 ml of SBF, continuously stirred, at 37°C. The SBF was previously filtered with a Millipore (0.22 μm) system, to avoid any bacterial contamination. Drug release measurements were carried out by means of UV-VIS spectroscopy. Absorbance values were taken at a wavelength λ corresponding to an absorbance maximum value. The calibration curve was determined by taking absorbance versus drug concentration between 0 and 3 mM as parameters. For that interval the calibration curve fits the Lambert and Beers' law (Wang & Pantano, 1992):

$$A = 1,26 \cdot C \tag{8}$$

where A is the absorbance and C is the concentration (mM).

4. Characterization

4.1 Sol-gel characterization

Gelification is the result of hydrolysis and condensation reactions according to the following reactions:

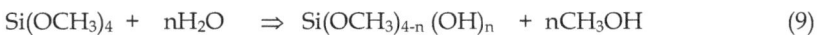

$$Si(OCH_3)_4 + nH_2O \Rightarrow Si(OCH_3)_{4-n}(OH)_n + nCH_3OH \tag{9}$$

$$\text{-SiOH} + CH_3O\text{-Si-} \Rightarrow \text{-Si-O-Si-} + CH_3OH \tag{10}$$

$$\text{-Si-OH} + \text{OH-Si} \Rightarrow \text{-Si-O-Si-} + H_2O \tag{11}$$

Reaction 12 shows the formation of hydrogen bonds between the carbossilic group of organic polymer and the hydroxyl group of inorganic matrix.

$$- \underset{\underset{O-H}{|}}{\overset{\overset{OCH_3}{|}}{Si}} - O - \underset{\underset{OH}{|}}{\overset{\overset{OCH_3}{|}}{Si}} -$$

$$- O - \underset{\overset{\|}{:O:}}{C} - (CH_2)_5 - + - \underset{\underset{OCH_3}{|}}{\overset{\overset{OH}{|}}{Si}} - O - \underset{\underset{OCH_3}{|}}{\overset{\overset{OH}{|}}{Si}} - \rightarrow - O - \underset{\overset{\|}{:O:}}{C} - (CH_2)_5 - \tag{12}$$

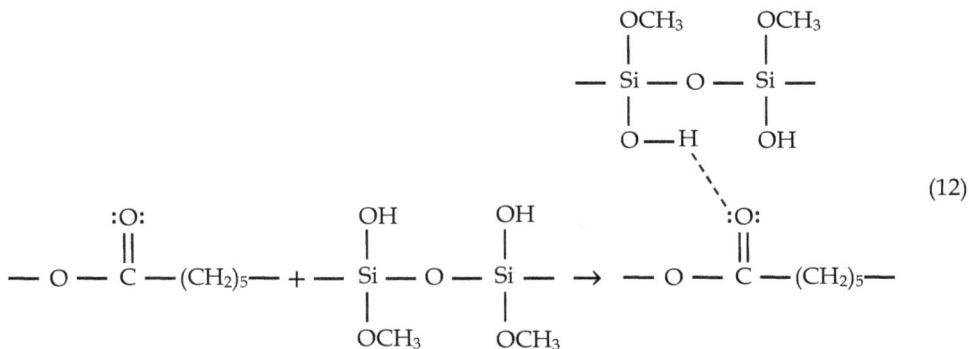

The existence of hydrogen bonds was proved by FTIR measurements. Fig. 10 shows the infrared spectrum of the SiO_2 gel (Fig. 10a), the SiO_2/PCL (6, 12, 50%wt) gels (Fig. 10b, 10c, 10d) and PCL (Fig. 10e). In Fig. 10a the bands between 3400 and 1600 cm^{-1} are attributed to water (Sanchez & Ribot, 1994). The bands at 1080 and 470 cm^{-1} are due to the stretching and bending modes of SiO_4 tetrahedra (Sanchez & Ribot, 1994). In the Fig. 10b, 10c, 10d and 10e, the bands at 2928 and 2840 cm^{-1} are attributed to a symmetric stretching of -CH$_2$- of policaprolattone. The band at 1730 cm^{-1} is due to the characteristic carboxylic group shifted to low wave numbers. The broad band at 3200 cm^{-1} is the characteristic O-H group of hydrogen bonds.

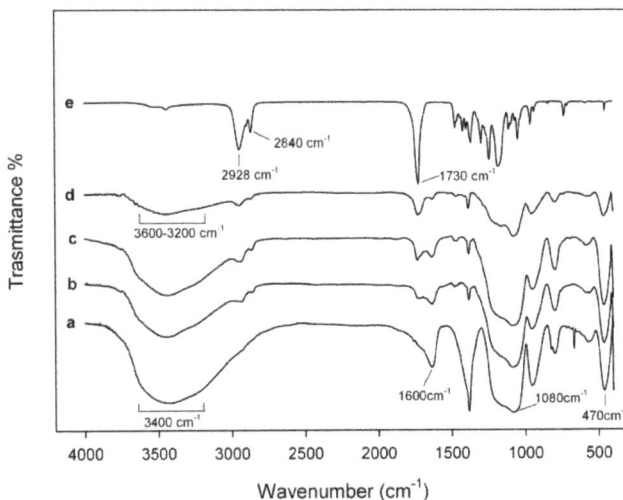

Fig. 10. FTIR of (a) SiO_2 gel, (b) SiO_2+ PCL 6wt% (c) SiO_2+ PCL 12wt% (d) SiO_2+ PCL 50wt% gels and (e) PCL.

The nature and the microstructure of the hybrid materials have been studied by X-ray diffraction (XRD), scanning electron microscopy (SEM) and atomic force microscopy (AFM). The diffractograms in Fig. 11a show that SiO_2 gel exhibits the broad humps which are characteristic for amorphous materials, while the sharp peaks that can be detected on the diffractogram of poly-ε-caprolactone and ketoprofen are typical of crystalline materials (Fig. 11b and 11c). On the other hand the XRD spectrum of hybrid SiO_2/PCL exhibits the broad humps characteristic of amorphous materials (Fig. 11d), as well that of SiO_2 gel.

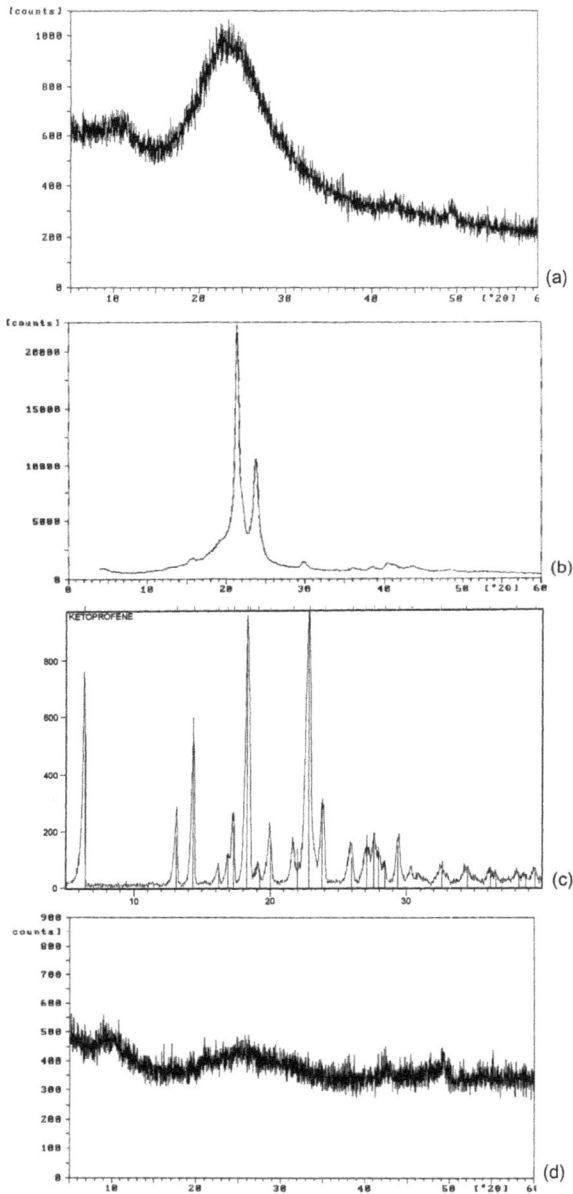

Fig. 11. XRD diffractogram of (a) SiO_2 gel, (b) PCL, (c) Ketoprofen, (d) SiO_2/PCL gel.

SEM micrographs show that no appreciable difference can be observed between the morphology of the four amorphous materials. The samples appear as shown in Fig. 12. The degree of mixing of the organic-inorganic components, i. e. the phase homogeneity, has been ascertained by applying the atomic force microscopy (AFM) in the analysis of the sol-gel hybrid material.

Fig. 12. SEM micrograph of SiO$_2$/PCL gel.

The AFM contact mode image can be measured in the height mode or in the force mode. Force images (z range in nN) have the advantage that they appear sharper and richer in contrast and that the contours of the nanostructure elements are clearer. In contrast, height images (z range in nm) show a more exact reproduction of the height itself. In this work the height mode has been adopted to evaluate the homogeneity degree of the hybrid materials.

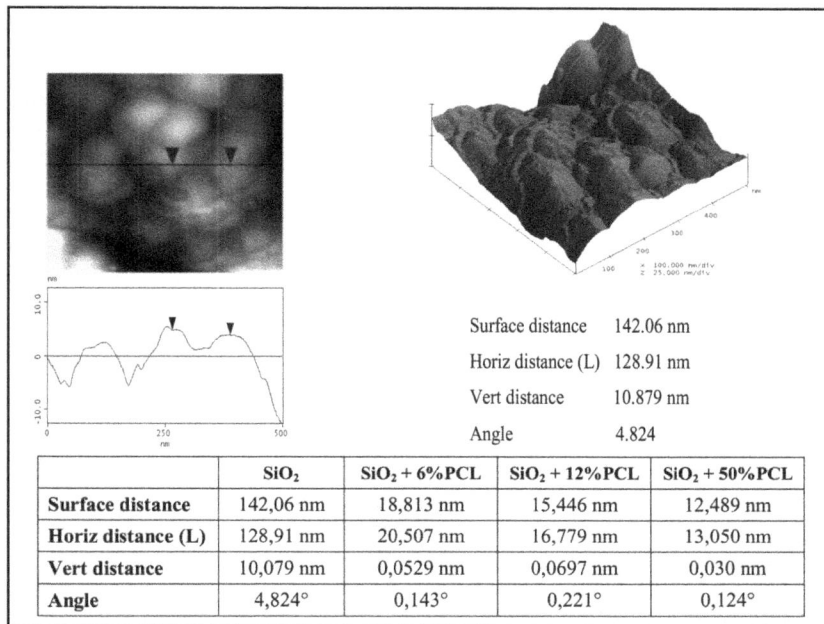

Surface distance	142.06 nm
Horiz distance (L)	128.91 nm
Vert distance	10.879 nm
Angle	4.824

	SiO$_2$	SiO$_2$ + 6%PCL	SiO$_2$ + 12%PCL	SiO$_2$ + 50%PCL
Surface distance	142,06 nm	18,813 nm	15,446 nm	12,489 nm
Horiz distance (L)	128,91 nm	20,507 nm	16,779 nm	13,050 nm
Vert distance	10,079 nm	0,0529 nm	0,0697 nm	0,030 nm
Angle	4,824°	0,143°	0,221°	0,124°

Fig. 13. AFM image showing the microstructure of SiO$_2$ gel .

The AFM topographic images of SiO_2 and SiO_2/PCL (0, 6, 12, 50 wt%) gel samples are shown in Figs. 13 and 14. It can be observed that the average domain size is less than 130 nm. This result confirms that the synthesized PCL/SiO_2 gels can be considered organic-inorganic hybrid materials as suggested by literature data (Hench & Clark, 1978).

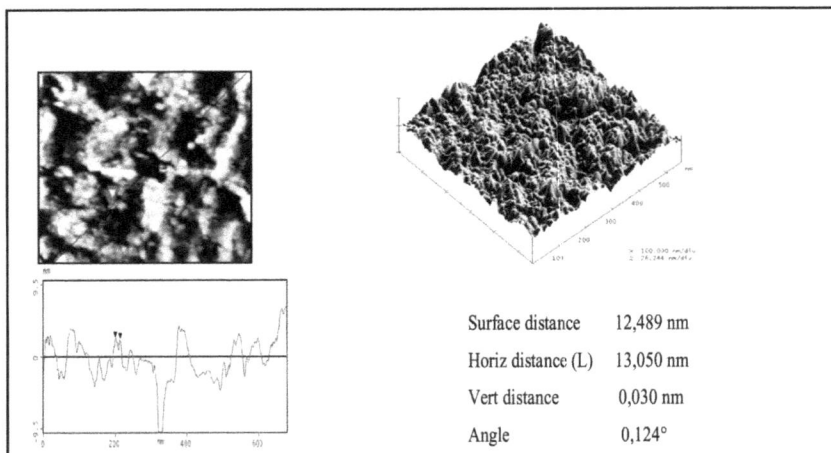

Surface distance	12,489 nm
Horiz distance (L)	13,050 nm
Vert distance	0,030 nm
Angle	0,124°

Fig. 14. AFM image showing the microstructure of SiO_2/PCL gel.

4.2 Biological characterisation

The hybrid materials were soaked in SBF, as indicated by Ohtsuki et al. (1992), for in vitro bioactivity tests. The FTIR spectra after several exposures to SBF, 7, 14 and 21 days are shown in Fig. 15b, 15c and 15d. Evidence of the formation of an hydroxyapatite layer is given by the appearance of the 1116 and 1035 cm^{-1} bands, usually assigned to P-O stretching (Teoli et al., 2006) and of the 580 cm^{-1} band usually assigned to the P-O bending mode (Teoli et al., 2006). The splitting, already after a 7 day soaking, of the 580 cm^{-1} band into two others at 610 and 570 cm^{-1} can be attributed to formation of crystalline hydroxyapatite (Ohtsuki et al.). Finally the band at 800 cm^{-1} can be assigned to the Si-O-Si band vibration between two adjacent tetrahedral, characteristic of silica gel (Teoli et al., 2006). These considerations support the hypothesis that a surface layer of silica gel forms as supposed in the mechanism proposed in the literature for hydroxyapatite deposition (Allen et al., 2000; Khor et al., 2002). Moreover an evaluation of the morphology of the apatite deposition and a qualitative elemental analysis were also carried out by electron microscopy observations on pelletized discs previously coated with a thin Au film. The EDS reported in Tab. 5 confirm that the surface layer observed in the SEM micrographs (Fig. 16) consists of calcium phosphate and which increases as the PCL.

Materials soaked in SBF for 21 days	Contents of Ca Atomic %	Contents of P Atomic %
SiO_2 + 0%PCL + Ketoprofen	2.65	1.64
SiO_2 + 6%PCL + Ketoprofen	3.13	1.94
SiO_2 + 12%PCL + Ketoprofen	5.04	3.23
SiO_2 + 50%PCL + Ketoprofen	7.82	4.86

Table 5. The EDS analyses of hybrid materials 21 days after immersion in SBF.

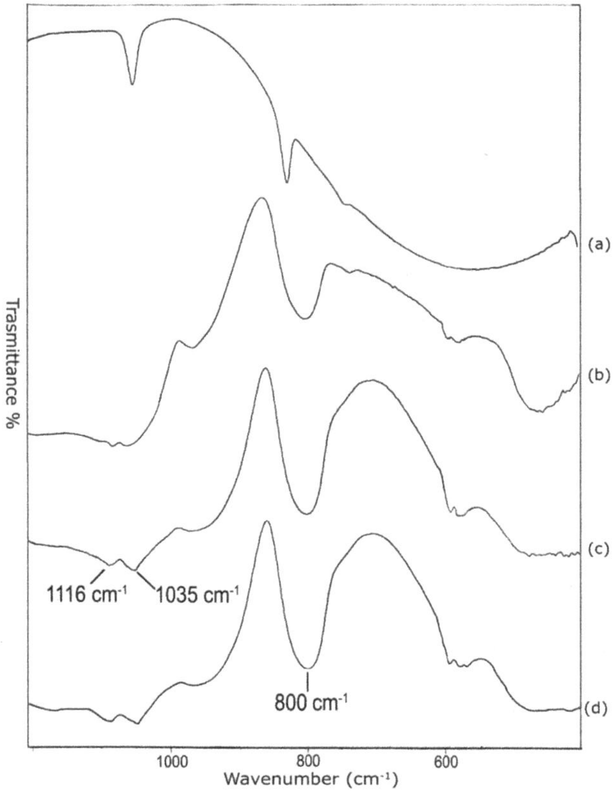

Fig. 15. FTIR spectra of SiO₂/PCL gel samples after different times of exposure to SBF: (a) not exposed; (b) 7 days exposed;(c) 14 days exposed; (d) 21 days exposed.

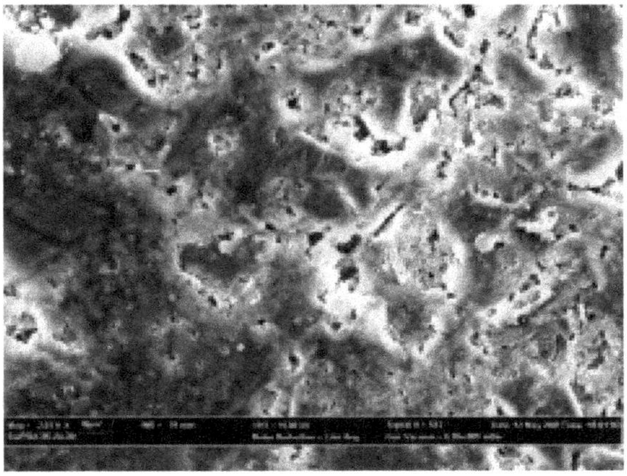

Fig. 16. SEM micrograph of SiO₂/PCL gel after being exposed to SBF 21 days.

4.3 Release kinetic characterization

Kinetic measurements of release from the studied materials were carried out in 15 ml of SBF incubated at 37±0.1°C and under continuous magnetic stirring at 150 rpm. Sink conditions were maintained throughout all studies. The discs used were obtained with particle size between 63-125μm compressed at 3 tons and aliquots of 600 μl were withdrawn at 1 h interval and replaced with an equal volume of release medium pre-equilibrated to temperature. Release was essayed by measuring the photometrical absorbance at 259.5 nm. In order to establish the relationship between the UV absorbance of at λ = 259.5 nm and the concentration of the solutions a calibration curve (r^2 =0.9907) was drawn for a standard solution with 4 levels of concentration: 0.0 mM, 1.0 mM, 2.0 mM and 3.0 mM (Fig. 17). All the standard solutions were prepared in SBF.

Fig. 18a, 19a, 20a and 18b, 19b, 20b show the drug release rates expressed as a percentage of the drug delivered, related to the drug-loading value, as a function of time. It was observed that from the SiO_2+PCL (0, 6, 12, 50 wt%)+ ketoprofen 5wt% gels about 60wt% of the drug was released in a relatively fast manner during the initial 2 hrs and it seems to be completed within 7 hrs without any evident difference in the time of release. For the SiO_2+PCL(0, 6, 12, 50 wt%)+ ketoprofen 10 and 15wt % gels about 60wt% of the drug was released during the initial 1 hr and 0,5 hr respectively and it is complete in about 3 hr and 4 hr respectively.

The differences observed in the release behaviour between SiO_2+ PCL (0, 6, 12, and 50 wt%) + ketoprofen might be due to the different networks of the four gels that are determined by the different content percentage of PCL. The two stage release observed in all cases suggests that the initial stage of release occurs mainly by dissolution and diffusion of the drug entrapped close to or at the surface of the samples. The second and slower release stage involves the diffusion of the drug entrapped within the inner part of the clusters. An interesting observation is the general presence of an early lag period, which indicates the need for the penetration of the solvent into the structure. Fig. 18b, 19b and 20b show this particular kinetic describing the changes of the release speed during the two stages.

Calibration curves

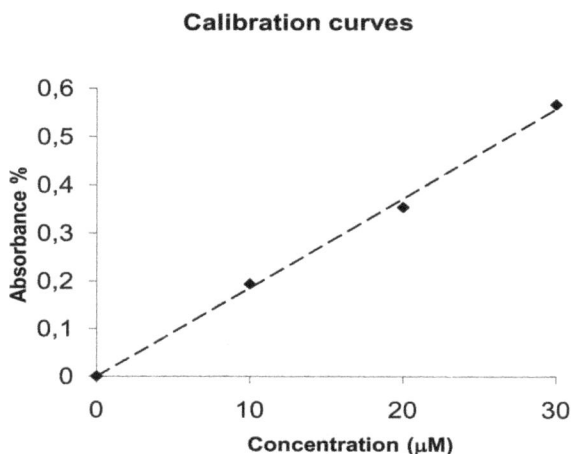

Fig. 17. Calibration curve (259.5 nm) depending on the concentration of Ketoprofen.

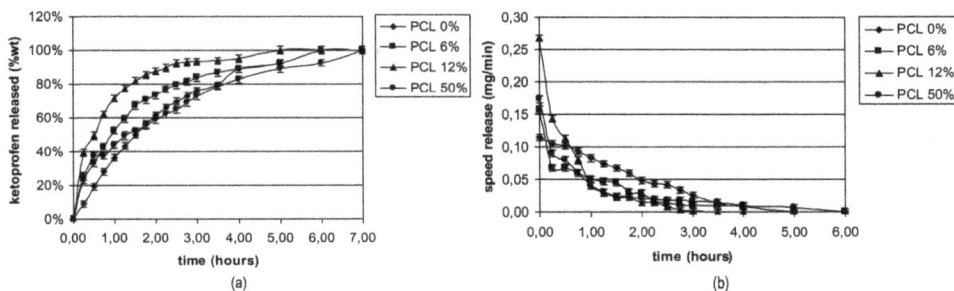

Fig. 18. (a) Time-dependent drug release plot for SiO_2 + PCL (0, 6, 12, 50%wt) + ketoprofen 5% at 37°C in SBF solution; (b) Time-dependent drug release rate plot for SiO_2 + PCL (0, 6, 12, 50%wt) + ketoprofen 5% at 37°C in SBF solution.

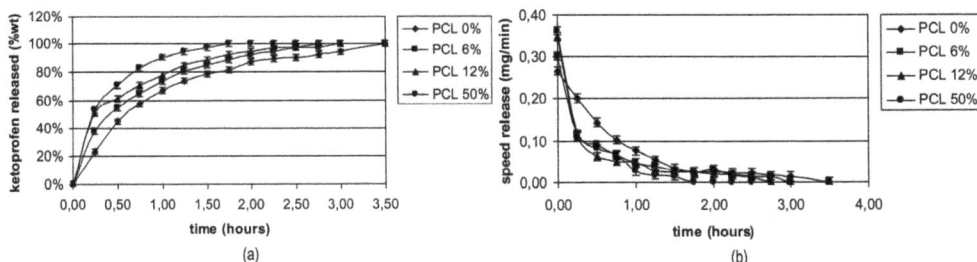

Fig. 19. (a) Time-dependent drug release plot for SiO_2 + PCL (0, 6, 12, 50%wt) + ketoprofen 10% at 37°C in SBF solution; (b) Time-dependent drug release rate plot for SiO_2 + PCL (0, 6, 12, 50%wt) + ketoprofen 10% at 37°C in SBF solution.

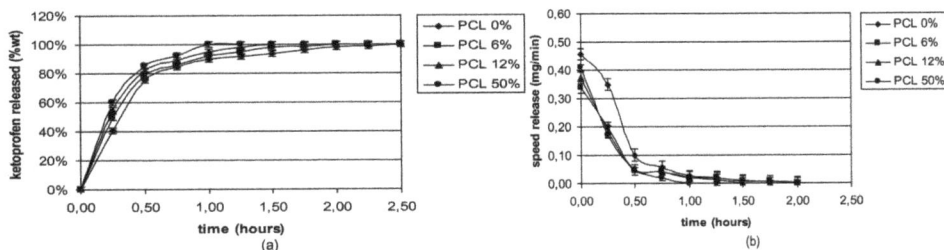

Fig. 20. (a) Time-dependent drug release plot for SiO_2 + PCL (0, 6, 12, 50%wt) + ketoprofen 15% at 37°C in SBF solution; (b) Time-dependent drug release rate plot for SiO_2 + PCL (0, 6, 12, 50%wt) + ketoprofen 15% at 37°C in SBF solution.

5. Applications

Applications for sol-gel processing derive from the various special shapes obtained directly from the gel state (e.g. monoliths, films, fibers, and monisized powders) combined with composition and microstructural control and low processing temperatures. Compared to conventional sources of ceramic raw materials, often minerals dug from the earth, synthetic chemical precursors are a uniform and reproducible source of raw materials than can be made extremely pure through various synthetic means. Low processing temperatures, which result from microstructural control (e.g. high surface areas and small pore sizes), expand glass-forming regions by avoiding crystallization or phase separation, making new materials available to the technologist. The advantages of the sol-gel process (for preparing glass) are shown in Tab. 6 (Brinker & Scherer 1990). The disadvantages of sol-gel processing include the cost of the raw materials, shrinkage that accompanies drying and sintering, and processing time, as it is shown in Tab 7.

1. Better homogeneity from raw materials.
2. Better purity from raw materials.
3.Low temperature of preparation:
a. Saving energy;
b. Minimizing evaporation losses;
c. Minimizing air pollution ;
d. No reactions with containers, thus purity;
e. Bypassing crystallization.
4. New noncrystalline solids outside the range of normal glass formation.
5. New crystalline phases from new noncrystalline solids.
6. Better glass products from special properties of gel.
7. Special products such as film.

Table 6. Some advantages of the Sol-Gel Methods over Conventional Melting for Glass

1. High cost of raw materials.
2. Large shrinkage during processing.
3. Residual fine pores.
4. Residual hydroxyl.
5. Residual carbon
6. Health hazards of organic solution.
7. Long process time

Table 7. Some disadvantages of the Sol-Gel Methods

There are books (Klein, 1988) and a number of review papers (Dislich, 1986; Johnson, 1985; Klein & Garvey, 1982; Mackenzie, 1988; Uhlmann et al., 1984; Ulrich, 1988a) which discuss this topic in detail and whose primary purpose is to provide a source of references to current technology, and at the same time to analyze critical issues associated with the various classes of applications that must be addressed in order to advance the sol-gel technology; for example, a short outline can be as follows:

1. Thin films and coatings, which can be applied to optical, electronic, protective, and porous thin films or coatings. That represents the earliest commercial application of sol-gel technology.
2. Monoliths, i.e. applications for cast bulk shapes dried without cracking in such areas as optical components, transparent superinsulation, and ultralow-expansion glasses.
3. Powders, which can be used as ceramic precursors or abrasive grains and applications of dense or hollow ceramic or glass spheres.
4. Fibers, which are drawn directly from viscous sols and are used primarily for reinforcement or fabrication of refractory textiles.

5. Gels can also be used as matrices for fiber-, whisker-, or particle-reinforced composites and as host for organic, ceramic, or metallic phases.
6. Porous Gel; many applications exist which result from the ability to tailor the porosity of thin free-standing membranes, as well as bulk xerogels or aerogels.

In recent years interest in bioactive and biocompatibility of surface-active, biomaterials has grown.

Biomaterials have been used extensively in medical, personal care and food applications, with many similar polymers being used across disciplines. This perspective will emphasize hybrid materials used in medicine and specifically those designed as scaffolds for use in tissue engineering and regenerative medicine. The areas of active research in tissue engineering include: biomaterials design (incorporation of the appropriate chemical, physical, and mechanical/structural properties to guide cell and tissue organization); cell/scaffold integration (inclusion into the biomaterial scaffold of either cells for transplantation or biomolecules to attract cells, including stem cells, from the host to promote integration with the tissue after implantation); and biomolecule delivery (inclusion of growth factors and/or small molecules or peptides that promote cell survival and tissue regeneration). While a significant and growing area of regenerative medicine involves the stimulation of endogenous stem cells, this perspective will emphasize hybrid materials scaffolds used for delivery of cells and biomolecules. The challenges and solutions pursued in designing polymeric biomaterial scaffolds with the appropriate 3-dimensional structure are curently studied.

Ceramics and hybrid dioxide-based materials for the repairing of muscle–skeletal tissues are being increasingly applied over the last half century (Hench, 1991; Li, et al., 1996). Orthopaedic and maxillo–facial prosthesis provide evidence for the enhanced biomechanical performance of titanium and its alloys among metallic prosthetic components (Kitsugi et al., 1996). TiO_2-based bioactive ceramic suggests that bone grafting is achieved by supporting the precipitation of calcium (Ca) and phosphorus (P) into a structure similar to the mineral phase of bone. Accordingly, titanium is very promising to develop biomedical materials and devices designed as hard tissue substitutes with improved interface properties (Coreno & Coreno, et al. 2005; Hench, 1991; Li, et al., 1996; Kitsugi et al., 1996).

6. Conclusions

Sol-gel processing has attracted much attention, for the possibility that the method offers to new materials. We define sol-gel rather broadly as the preparation of glass, glass-ceramic and hybrid materials by a sol, its gelation and removal of the solvent. The sol-gel chemistry is based on the hydrolysis and polycondensation of molecular precursors such as metal alkoxides $M(OR)x$, where M = Si, Sn, Ti, Zr, Al, Mo, V, W, Ce and so forth.

There are many potential applications of sol-gel derived materials in the form of films, fibers, monoliths, powders, composites, and porous media. The most successful applications are those that utilize the potential advantages of sol-gel processing such as purity, homogeneity, and controlled porosity combined with the ability to form shaped objects at low temperatures, avoiding inherent disadvantages such as costs of raw materials, slow processing times, and high shrinkage.

The SiO_2 + PCL (0, 6, 12 and 50 %wt) materials, prepared via sol-gel process, were found to be organic - inorganic hybrid materials. The polymer (PCL) can be incorporated into the network by hydrogen bonds between the carboxylic groups of organic polymer and the hydroxyl groups of inorganic matrix. The release kinetics demonstrates that the investigated materials supply high doses of the anti-inflammatory during the first hours when soaked in SBF and then a slower drug release allows a maintenance dose until the end of the experiment.

7. References

Aelion, R., Loebel, A. & Eirich, F. (1950). Hydrolysis and polycondensation of tetraalkoxysilanes, *Recueil des Travaux Chimiques des Pays-Bas et de la Belgique*, Vol. 69, pp. 61-75, ISSN 0370-7539

Aelion, R., Loebel, A. & Eirich, F. (1950). Hydrolysis of ethyl silicate, *Journal of the American Chemical Society*, Vol. 72, pp. 5705-5712, ISSN 0002-7863

Allen, C., Han, J., Yu, Y., Maysinger, D. & Eisenberg, A. (2000). Polycaprolactone-*b*-poly(ethylene oxide) copolymer micelles as a delivery vehicle for dihydrotestosterone. *Journal of Controlled Release*, Vol. 63, No 3, pp. 275-286, ISSN 0168-3659

Arcos, D., Ragel, C.V. & Vallet-Regí, M. (2001). Bioactivity in glass/PMMA composites used as drug delivery system, *Biomaterials*, Vol. 22, No. 7, pp. 701-708, ISSN 0142-9612

Avnir, D. & Kaufman, V.R. (1987). Alcohol is an unnecessary additive in the silicon alkoxide sol-gel process, *Journal of Non-Crystalline Solids*, Vol. 92, No. 1, pp. 180-182, ISSN 0022-3093

Black, J. & Hastings, G. (1998). *Handbook of Biomaterial Properties*, Chapman & Hall, ISBN 0412603306, New York, U.S.A.

Brinker, C.J., Keefer, K.D., Schaefer, D.W. & Ashley, C.S. (1982). Sol-gel transition in simple silicates, *Journal of Non-Crystalline Solids*, Vol. 48, No. 1, pp. 47-64, ISSN 0022-3093

Brinker, C.J. (1988). Hydrolysis and condensation of silicates: effects on structure, *Journal of Non-Crystalline Solids*, Vol. 100, No. 1-3, pp. 31-50, ISSN 0022-3093

Ceaglske, N.H. & Hougen, O.A. (1937). Drying granular solids, *Journal of Industrial and Engineering Chemistry (Washington, D. C.)*, Vol. 29, pp. 805-813, ISSN 0095-9014

Cogan, H.D. & Setterstrom, C.A. (1946). Properties of ethyl silicate, *Chemical & Engineering News*, Vol. 24, pp. 2499-2501, ISSN 0009-2347

Coreno, J. & Coreno, O. (2005). Evaluation of calcium titanate as apatite growth promoter. *Journal of Biomedical Materials Research*, Vol. 75A, No. 2, pp. 478–84, ISSN 478-484

Dislich, H. (1986). Sol-gel: science, processes and products, *Journal of Non-Crystalline Solids*, Vol. 80, No 1-3, pp.115-121, ISSN 0022-3093

Dwivedi, R.K. (1986). Drying behavior of alumina gels, *Journal of Materials Science Letters*, Vol. 5, No. 4, pp. 373-376, ISSN 0261-8028

Flory P.J. (1953). *Principles of Polymer Chemistry*, Cornell University Press, ISBN 0-8014-0134-8, Ithaca, New York

Flory, P.J. (1974). Gels and gelling process, *Faraday Discussions of the Chemical Society*, Vol. 57, pp. 7-18, ISSN 0301-7249

Fortes, M. & Okos, M. (1980). Drying Theories: Their Bases and Limitations as Applied to Foods and Grains, In: *Advances in Drying Vol 1*, A. Mujumdar (Ed.), 119-154, Hemisphere Publishing, ISBN 0070439753, New York, U.S.A.

Gigant, K., Posset, U. & Schottner, G. (2002). FT-Raman spectroscopic study of the structural evolution in binary UV-curable vinyltriethoxysilane/tetraethoxysilane mixtures from the sol to the xerogel state, *Applied Spectroscopy*, Vol. 56, No. 6, pp. 762-769, ISSN 0003-7028

Hench, L.L. & Clark, D.E. (1978). Physical chemistry of glass surfaces, *Journal of Non-Crystalline Solids*, Vol. 28, No. 1, pp. 83-105, ISSN 0022-3093

Hench, L.L., Orcel, G. & Nogues, J.L. (1986). The role of chemical additives in sol-gel processing, *Materials Research Society Symposium Proceedings*, Vol. 73, No. Better Ceramic Chemistry 2, pp. 35-47, ISSN 0272-9172

Hench, L.L. & West, J.K. (1990). The sol-gel process, *Chemical Reviews*, Vol. 90, No. 1, pp. 33-72, ISSN 0009-2665

Hench, LL. (1991). Bioceramics: from concept to clinic, *Journal of American Ceramic Society*, Vol. 74, No. 7, pp. 1487–1510, ISSN 1551-2916.

Hsiue, G.H., Kuo, J.K., Jeng, R.J., Chen, J.I., Jiang, X.L., Marturunkakul, S., Kumar, J. & Tripathy, S.K. (1994). Stable second-order nonlinear optical polymer network based on an organosoluble polyimide, *Chemistry of Materials*, Vol. 6, No. 7, pp. 884-887, ISSN 0897-4756

Hutmacher, D.W., Schantz, T., Zein, I., Woei N.K., Hin T.S. & Cheng T.K. (2001). Mechanical properties and cell cultural response of polycaprolactone scaffolds designed and fabricated via fused deposition modeling, *Journal of Biomedical Materials Research*, Vol. 55, No. 2, pp. 203-216 ISSN 0021-9304

Johnson, D.W.J. (1985). Sol-gel processing of ceramics and glass, *American Ceramic Society Bulletin*, Vol. 64, No. 12, pp. 1597-1602, ISSN 0002-7812

Joshua, D.Y., Damron, M., Tang, G., Zheng, H., Chu, C.-J. & Osborne, J.H. (2001). Inorganic/organic hybrid coatings for aircraft aluminum alloy substrates. *Progress in Organic Coatings*, Vol. 41, No. 4, pp. 226-232, ISSN 0300-9440

Judeinstein, P. & Sanchez, C. (1996). Hybrid organic-inorganic materials: a land of multidisciplinarity, *Journal of Materials Chemistry*, Vol. 6, No. 4, pp. 511-525, ISSN 0959-9428

Keefer, K.D. (1984). The effect of hydrolysis conditions on the structure and growth of silicate polymers, *Materials Research Society Symposium Proceedings*, Vol. 32, No. Better Ceramic Chemistry, pp. 15-24, ISSN 0272-9172

Khor, H.L., Ng, K.W., Schantz, J.T., Phan, T.-T., Lim, T.C., Teoh, S.H. & Hutmacher, D.W. (2002). Poly(e-caprolactone) films as a potential substrate for tissue engineering an epidermal equivalent. *Materials Science and Engineering: C*, Vol. 20, No. 1-2, pp. 71-75, ISSN 0928-4931

Kitsugi, T., Nakamura, T., Oka, M., Yan, W.Q., Goto, T., Shibuya, T., Kokubo, T. & Miyaji, S. (1996). Bone bonding behavior of titanium and its alloys when coated with titanium oxide (TiO_2) and titanium silicate (Ti_5Si_3), *Journal of Biomedical Materials Reserch*, Vol. 32, No. 2, pp. 149–56, ISSN 1097-4636

Klein, L.C. & Garvey, G.J. (1982). Silicon alkoxides in glass technology, *ACS Symposium Series*, Vol. 194, No. Soluble Silicon, pp. 293-304, ISSN 0097-6156

Klein, L.C. (1988). *Sol-Gel Technology for Thin Films, Fibers, Preforms, Electronics and Specialty Shapes*, Noyes, ISBN 0-8155-1154-X, Park Ridge, New Jersey, U.S.A.

Klukowska, A., Posset, U., Schottner, G., Wis, M.L., Salemi-Delvaux, C. & Malatesta, V. (2002). Photochromic hybrid sol-gel coatings: preparation, properties, and applications, *Materials Science*, Vol. 20, No. 1, pp. 95-104, ISSN 0137-1339

Li, P., de Groot, K. & Kokubo, T. (1996). Bioactive $Ca_{10}(PO_4)_6(OH)_2$–TiO_2 composite coating prepared by sol–gel process. *Journal of Sol–gel Science and Technology*. Vol. 7, No. 1, pp. 27–34, ISSN 0928-0707

Macey, H.H. (1942). Clay-water relationships and the internal mechanism of drying, *Transactions of the British Ceramic Society*, Vol. 41, pp. 73-120, ISSN 0371-5469

Mackenzie, J.D. (1988). Applications of the sol-gel process. *Journal of Non-Crystalline Solids*, Vol. 100, No. 1-3, pp. 162-168, ISSN 0022-3093

Mackenzie, J.D. & Bescher, E.P. (1998). Structures, properties and potential applications of Ormosils, *Journal of Sol-Gel Science and Technology*, Vol. 13, No. 1/2/3, pp. 371-377, ISSN 0928-070

Matsuura, Y., Matsukawa, K., Kawabata, R., Higashi, N., Niwa, M. & Inoue, H. (2001). Synthesis of polysilane-acrylamide copolymers by photopolymerization and their application to polysilane-silica hybrid thin films, *Polymer*, Vol. 43, No. 4, pp. 1549-1553, ISSN 0032-3861

Moore, F. (1961). Mechanism of moisture movement in clays with particular reference to drying-concise review, *Transactions of the British Ceramic Society*, Vol. 60, pp. 517-539, ISSN 0371-5469

Morrison, R.T. & Boyd R.N. (1966). *Organic Chemistry*, Allyn & Bsacon, ISBN 0136436773, Boston, U.S.A.

Novak, B.M. (1993). Hybrid nanocomposite materials - between inorganic glasses and organic polymers, *Advanced Materials*, Vol. 5, No. 6, pp. 422-433, ISSN 1521-4095

Ohtsuki, C., Kokubo, T. & Yamamuro, T. (1992). Mechanism of apatite formation on $CaOSiO_2P_2O_5$ glasses in a simulated body fluid. *Journal of Non-Crystalline Solids*, Vol. 143, No. 1, pp. 84-92, ISSN 0022-3093

Paul, A. (1990). *Chemistry of Glasses*, Chapman and Hall, ISBN 0412230208, New York, U.S.A.

Ragel, C.V. & Vallet-Regí, M. (2000). In vitro bioactivity and gentamicin release from glass-polymer-antibiotic composites, *Journal of Biomedical Materials Research*, Vol. 51, No. 3, pp. 424-429, ISSN 1552-4965

Sanchez, C. & Ribot, F. (1994). Design of hybrid organic-inorganic materials synthesized via sol-gel chemistry. *New Journal of Chemistry*, Vol. 18, No. 10, pp. 1007-1047, ISSN 1144-0546

Schmidt, H., Scholze, H. & Kaiser, A. (1984). Principles of hydrolysis and condensation reaction of alkoxysilanes, *Journal of Non-Crystalline Solids*, Vol. 63, No. 1-2, pp. 1-11, ISSN 0022-3093

Shih, W.Y., Alkay, I.A. & Kikuchi, R. (1987), Phase diagrams of charged colloidal particles, *Journal of Chemistry and Physics*, Vol. 86, No. 9, pp. 5127-5132, ISSN 0021-9606

Spanhel, L., Popall, G. & Muller, G. (1995). Spectroscopic properties of sol-gel derived nanoscaled hybrid materials, *Journal of chemical sciences*, Vol. 107, No. 6, pp. 637-644, ISSN 0974-3626

Stauffer D., Coniglio A. & Adam M. (1982). Gelation and critical phenomena. *Advances in Polymer Science*, Vol. 44, No. Polymer Networks, pp. 103-158, ISSN 0065-3195

Stockmayer W.H. (1945). *Advancing Fronts in Chemistry. Vol. 1. High Polymers*, Reinhold Publishing Company, New York, U.S.A.

Stoeber, W., Fink, A. & Bohn, E. Controlled growth of monodisperse silica spheres in the micron size range, *Journal of Colloid and Interface Science*, Vol. 26, No. 1, pp. 62-69, ISSN 0021-9797

Teoli, D., Parisi, L., Realdon, N., Guglielmi, M., Rosato, A. & Morpurgo, M. (2006). Wet sol-gel derived silica for controlled release of proteins, *Journal of Controlled Release*, Vol. 116, No. 3, pp. 295-303, ISSN 0168-3659

Uhlmann, D.R., Zelinski, B.J.J. & Wnek, G.E. (1984). The ceramist as chemist - opportunities for new materials, *Materials Research Society Symposium Proceedings*, Vol. 32, No. Better Ceramics Through Chemistry, pp. 59-70, ISSN 0272-9172

Ulrich, D.R. (1988). Prospects of sol-gel processes, *Journal of Non-Crystalline Solids*, Vol. 100, No. 1-3, pp. 174-193, ISSN 0022-3093

Vallet-Regì, M., Ramila, A., del Real, R.P. & Perez-Pariente, J. (2000). A new property of MCM-41: drug delivery system, *Chemistry of Materials*, Vol. 13, No. 2, pp. 308-311, ISSN 0897-4756

Vallet-Regí, M. (2001). Ceramics for medical applications, *Journal of the Chemical Society Dalton Transactions*, No. 2, pp. 97-108, ISSN 1472-7773

Vallet-Regí, M. & Arcos, D. (2006). Nanostructured hybrid materials for bone tissue regeneration, *Current Nanoscience*, Vol. 2, No. 3, pp. 179-189

Vallet-Regí, M. (2006). Bone repair and regeneration: possibilities, *Materialwissenschaft und Werkstofftechnik*, Vol. 37, No. 6, pp. 478-484, ISSN 1521-4052

Vallet-Regí, M. (2006). Ordered mesoporous materials in the context of drug delivery systems and bone tissue engineering, *Chemistry - A European Journal*, Vol. 12, No. 23, pp. 5934-5943, ISSN 1521-3765

Voronkov, M.G., Mileshkevich, V.P. & Yuzhelevskii, Y.A. (1978). *Studies in Soviet Science. The Siloxane Bond: Physical Properties and Chemical Transformations*, Consultants Bureau, ISBN 0306109409, New York, U.S.A.

Wang, D.S. & Pantano, C.G. (1992). Surface chemistry of multicomponent silicate gels, *Journal of Non-Crystalline Solids Advanced Materials from Gels*, Vol. 147-148, pp. 115-122, ISSN 0022-3093

Wei, Y., Xu, J., Dong, H., Dong, J.H., Qiu, K. & Jansen-Varnum, S.A. (1999). Preparation and physisorption characterization of d-glucose-templated mesoporous silica sol-gel materials, *Chemistry of Materials*, Vol. 11, No. 8, pp. 2023-2029, ISSN 0897-4756

Zallen R. (1983). *The Physics of Amorphous Solids,* Wiley, ISBN 9783527602797, New York, U.S.A.

Young, S.K., Gemeinhardt, G.C., Sherman, J.W., Storey, R.F., Mauritz, K.A., Schiraldi, D.A., Polyakova, A., Hiltner, A. & Baer, E. (2002). Covalent and non-covalently coupled polyester-inorganic composite materials, *Polymer,* Vol. 43, No. 23, pp. 6101-6114, ISSN . 0032-3861

Part 2

Biomedical Engineering

8

Domain-Specific Software Engineering Design for Diabetes Mellitus Study Through Gene and Retinopathy Analysis

Hua Cao, Deyin Lu and Bahram Khoobehi
Louisiana State University, University of Mississippi Medical Center, LSU Eye Center,
USA

1. Introduction

Software engineering designs and practices differ widely among various application domains. This chapter is concentrating on high performance software engineering design for bioinformatics and more specifically for diabetes mellitus study through gene and retinopathy analysis. Complex gene interaction study offers an effective control of blood glucose, blood pressure and lipids. Early detection of retinopathy is effective in minimizing the risk of irreversible vision loss and other long-term consequence associated with diabetes mellitus.

Type 2 diabetes mellitus is a disorder of glucose homeostasis involving complex gene and environmental interactions that are incompletely understood. Mammalian homologs of nematode sex determination genes have recently been implicated in glucose homeostasis and type 2 diabetes mellitus. The Fem1b knockout (Fem1b-KO) mice have been developed, with targeted inactivation of Fem1b, a homolog of the nematode fem-1 sex determination gene. It shows that the Fem1b-KO mice display abnormal glucose tolerance and that this is due predominantly to defective glucose-stimulated insulin secretion. Arginine-stimulated insulin secretion is also affected. These data implicate Fem1b in pancreatic islet function and insulin secretion, strengthening evidence that a genetic pathway homologous to nematode sex determination may be involved in glucose homeostasis and suggesting novel genes and processes as potential candidates in the pathogenesis of diabetes mellitus. In addition, this chapter is going to introduce basic gene analysis approaches that can be applied on diabetes mellitus study. These approaches include searching Genbank online database using BLAST, mapping DNA, locating genes, aligning different DNA or protein sequences, determining genotypes, and comparing nucleotide or amino acid sequences using global and local alignment algorithms. Fem1b gene, as an example, is going to be discussed with these basic gene analysis approaches.

Diabetic retinopathy is the leading cause of new cases of blindness among Americans aged 20 to 64 in both predominantly white and black populations [1]. Despite the recommendation for yearly eye examinations and efforts to achieve this, of the approximately 17 million Americans with diabetes, about 6 million nationwide remain undiagnosed and untreated, or not receiving annual eye examinations, which can lead to diabetic retinopathy [2].

Early indications of retinal blood vessel abnormalities and complications provide important indicators for clinical timely diagnosis and treatment of diabetes mellitus and eye disorders. The software engineering design tool facilitates increasing the number of annual diabetic screening eye examinations, thereby reducing the long time wait for diabetic patients to receive eye examinations. Common activities in software engineering approach for retinopathy include single or multi-modality retinal image registration, fusion, vessel pattern recognition, arteries & veins identification, and vessel diameter measurement. These methods play a major role in the development of better methods of diagnosing and treating diabetic retinopathy. Fusing the multi-modality retinal images, which usually requires intensive computational effort, is a very challenging problem because of the possible vast content change and non-uniform distributed intensities of the involved images.

This chapter is going to present a novel approach of retinal image fusion. Control points are detected at the vessel bifurcations using adaptive exploratory algorithm. Mutual-Pixel-Count (MPC) maximization based heuristic optimization adjusts the control points at the sub-pixel level. The iteration stops either when MPC reaches the maximum value, or when the maximum allowable loop count is reached. A refinement of the parameter set is obtained at the end of each loop, and finally an optimal fused image is generated at the end of the iteration. By locking the multi-modality retinal images into one single volume, the algorithm allows ophthalmologists to match the same eye over time to get a sense of disease progress and pinpoint surgical tools to increase accuracy and speed of the surgery. The new algorithm can be easily expanded to human or animals' 3D eye, brain, or body image registration and fusion.

2. Diabetes mellitus type 2 and fem1b gene

2.1 Diabetes mellitus type 2 occurrence and its diagnosis

Diabetes mellitus type 2 is a metabolic disorder that is characterized by high blood glucose in the context of insulin resistance and relative insulin deficiency [3]. The pathophysiology of Type 2 diabetes mellitus involves impaired insulin secretion, and impaired insulin action in regulating glucose and fatty acid metabolism in the liver, skeletal muscle, and adipose tissue. Many individuals with Type 2 diabetes mellitus have hypertension and perturbations of lipoprotein metabolism, as well as other manifestations of the insulin resistant syndrome. In addition to the risk for development of diabetes - specific complications of retinopathy, Type 2 diabetes mellitus is recognized as a substantial risk factor for cardiovascular disease [4].

It is recommended by the National Diabetes Data Group that diagnosis of diabetes mellitus be based on [5]

1. Two fasting plasma glucose levels of 126 mg/dL (7.0 mmol/L) or higher;
2. Two two-hour postprandial plasma glucose (2hrPPG) readings of 200 mg/dL (11.1 mmol/L) or higher after a glucose load of 75 g;
3. Two casual glucose readings of 200 mg per dL (11.1 mmol per L) or higher.

Fasting plasma glucose was selected as the primary diagnostic test because it predicts adverse outcomes (e.g., retinopathy) and is easy to perform in a clinical setting.

A mammalian Fem1 gene family, encoding homologs of fem-1, has been characterized and consists of at least three members in the mouse, designated Fem1a, Fem1b, and Fem1c; these have highly conserved homologs in humans, designated FEM1A, FEM1B, and FEM1C, respectively. Mammalian homologs of two other nematode sex determination genes, tra-2 and tra-3, have recently been implicated in glucose homeostasis and type 2 diabetes mellitus. In producing susceptibility to type 2 diabetes mellitus, NIDDM1 is known to interact with a gene, whose identity is unknown, on human chromosome 15 near the CYP19 locus at 15q21.3 [6]. This is near 15q22, where FEM1B, the human homolog of mouse Fem1b, localizes [7].

2.2 Glucose and insulin measurements

Blood glucose from tail vein was measured using an OneTouch FastTake Glucometer. Insulin was measured from plasma, tissue extracts, or cell supernatants using the Rat (Mouse) Sensitive Insulin radioimmunoassay (RIA) kit and the manufacturer's instructions. For the intraperitoneal glucose tolerance test (iPGTT), intraperitoneal insulin tolerance test (iP-ITT), and acute-phase glucosestimulated insulin secretion (A-GSIS) test, there were 12 animals in each group (12 male homozygous Fem1b-KO, 12 male wild-type controls, 12 female homozygous Fem1b-KO, and 12 female wild-type controls), aged 3 to 4 months. The arginine-stimulated insulin secretion test compared eight Fem1b-KO homozygous males with eight wild-type males, aged 6 months. D-Glucose (200 mg/ml) was administered at 2 mg/g body weight by intraperitoneal injection [8]. Tail vein blood was sampled for blood glucose determination from nonsedated animals before and at 15, 30, 60, and 120 min after glucose administration.

2.3 Fem1b-KO mice development

In this study, the gene targeting by homologous recombination has been used to generate Fem1b knockout (Fem1b-KO) mice with inactivation of the Fem1b gene. It was performed with a deletion of Fem1b coding exon 1, which contains the translation initiation codon and the first two ankyrin repeats [9]. The results show that these mice display abnormal glucose homeostasis, with abnormal glucose tolerance tests and defective glucose-stimulated insulin secretion. These findings indicate that Fem1b is involved in pancreatic islet β-cell function and provide further evidence for involvement of a pathway resembling nematode sex determination in mammalian glucose homeostasis. This approach utilized standard methodology (Figure 1) and the basic elements of the targeting vector and screening strategy by Southern blot and PCR genotyping. Figure 2 was generated by Zeiss AvioVision with the immunohistochemical analysis demonstrates a loss of specific Fem1b staining in islets of Fem1b-KO homozygotes.

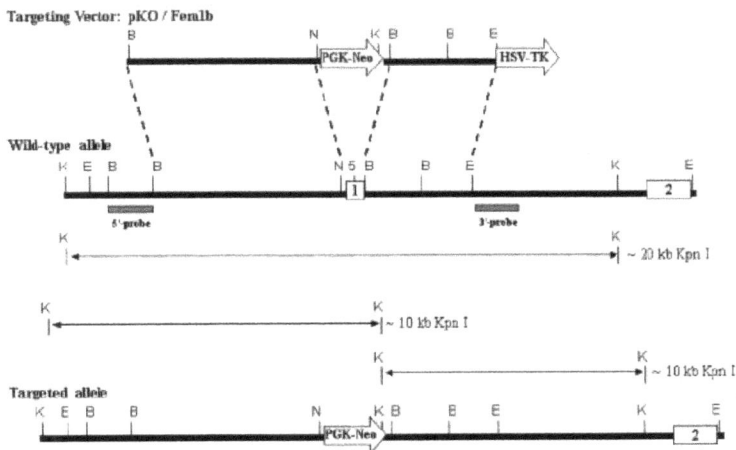

Fig. 1. General strategy, with expected KpnI digestion products along with 5_ and 3_ probes for Southern blot. The boxes labeled 1 and 2 represent exons 1 and 2, respectively. Exon 1 is replaced by the PGK-Neo gene (labeled arrow) in the targeted allele.

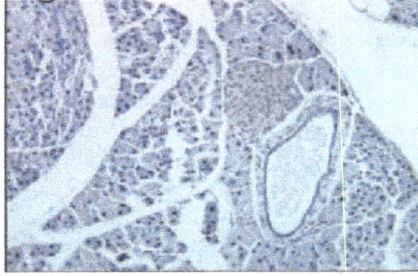

Fig. 2. Immunostaining of homozygous Fem1b-KO pancreas with anti-Fem1b C-terminus antibody Li-51, demonstrating the absence of specific staining for Fem1b.

2.4 Glucose homeostasis in Fem1b-KO mice

As noted above, mammalian homologues of nematode sex determination genes have recently been shown to be involved in glucose homeostasis and type 2 diabetes mellitus. Based on this logic, glucose homeostasis was evaluated in the Fem1b-KO mice by using established experimental methods. As a first-line screen, these mice were between 3 and 4 months of age. The iP-ITT showed minimally abnormal results (Figure 3), suggesting that insulin resistance is not the primary defect in homozygotes, although it could be contributing.

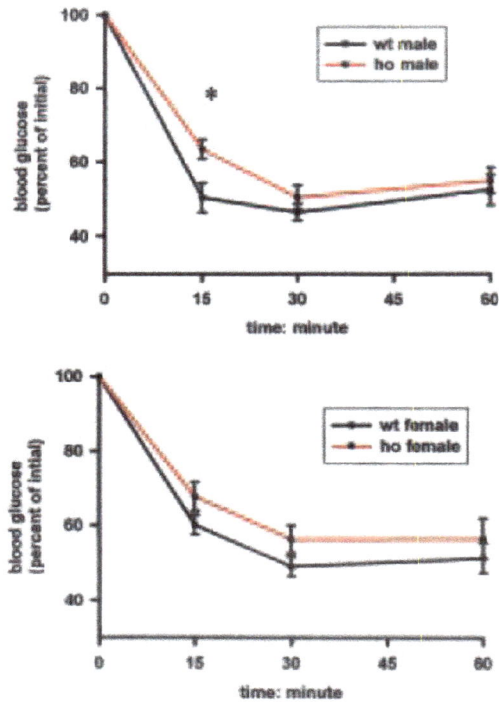

Fig. 3. Intraperitoneal insulin tolerance test (males and females).

To evaluate whether the defective acute-phase insulin secretion is related to a defect in secretion per se as opposed to a defect in insulin production, the insulin content was measured o in these mice (Figure 4), which demonstrates that Fem1b-KO homozygotes have increased insulin content compared to that of wild-type controls.

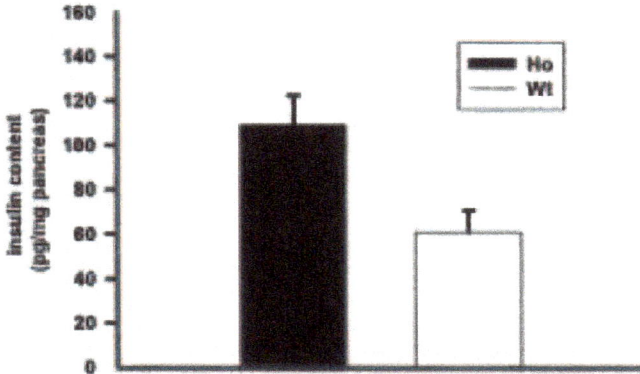

Fig. 4. Pancreatic insulin content in fasted mice (four Fem1b-KO homozygous mice and four wild-type mice).

2.5 Immunostaining of Fem1b in pancreatic islets
In humans, FEM1B has been shown to be expressed within whole pancreas [10], but cell type distribution within this organ was unknown. Immunostaining of wild-type pancreas with immunoaffinity-purified antibody shows that Fem1b protein is expressed in pancreatic islets (Figure 5).

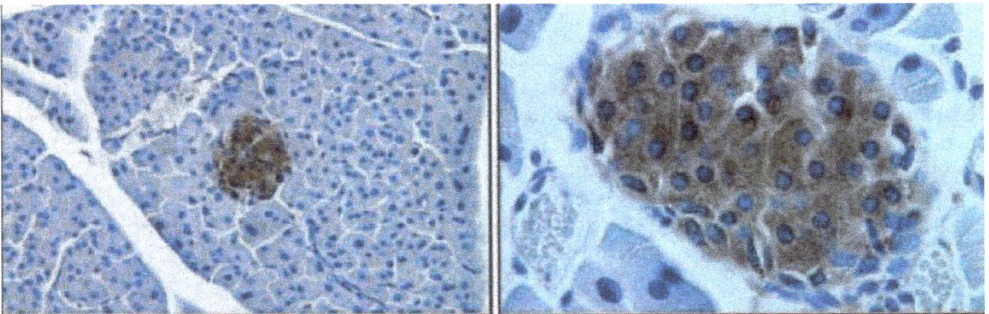

Fig. 5. Lower and higher magnification, respectively, of immunoperoxidase staining with antibody Li-51 (against the C terminus of Fem1b) counterstained with hematoxylin.

Immunostaining of the pancreas with a commercially available goat polyclonal antibody against Fem1b demonstrates the same islet staining pattern, with an absence of specific staining in the Fem1b-KO homozygotes. Coimmunostaining with antibodies to insulin, a β-cell marker, demonstrates that Fem1b is expressed in virtually all β cells (Figure 6).

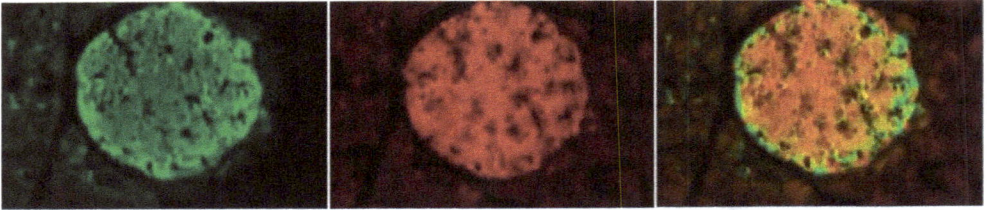

Fig. 6. Immunofluorescence staining of insulin (green), Fem1b (red), and merged image demonstrating that Fem1b is expressed not only in insulin-positive β cells but also in insulin-negative non-β cells.

The coimmunostaining with antibodies to glucagon and somatostatin, markers for α cells and δ cells, respectively, demonstrates that the Fem1b protein is also expressed in these non-β cells (Figure 7).

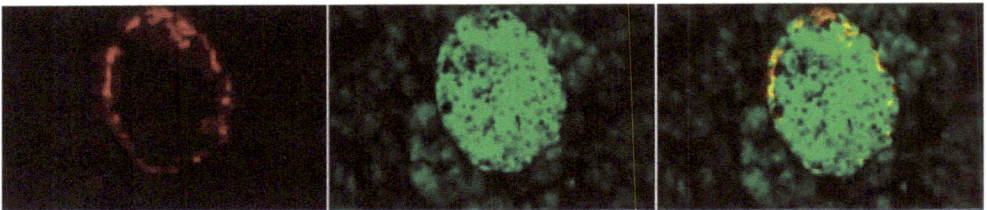

Fig. 7. Immunofluorescence staining with a combination of antibodies to glucagon and somatostatin (red) and Fem1b (green) and a merged image verifying expression of Fem1b within non- β cells.

3. Fem1b gene search and alignment

3.1 BLAST search

Basic Local Alignment Search Tool (BLAST) is an algorithm for comparing amino-acid sequences or the nucleotides. By performing a BLAST search, one is able to compare an unknown sequence with a library or database of known sequences, and identify library sequences that resemble the unknown sequence above a certain score percentage [11] (usually 40%). This chapter is going to give an example that follows the discovery of a previously unknown fem1b gene in the mouse and performs a BLAST search of the human genome to see if humans carry a similar fem1b gene. BLAST identifies sequences in the human fem1b genome that resemble the mouse fem1b gene based on similarity of sequence. Given a sequence of one fragment of mouse gene (Figure 8), BLAST software is going to search all human gene banks and find similar genes. http://blast.ncbi.nlm.nih.gov/Blast.cgi is the official BLAST website.

```
CAGAAGTCGG ATTGAAGCCT TGGAGCTCTT GGGTGCCTCC TTTGCAAATG ATCGTGAGAA
CTATGACATC ATGAAGACAT ACCACTATTT ATATTTAGCT ATGTTGGAGA GATTTCAGGA
TGGTGACAAC ATTCTTGAAA AAGAGGTTCT CCCACCCATC CATGCTTATG GGAACAGAAC
TGAGTGTAGG AACCCACAGG AATTGGAGGC TATTCGGCAA GACAGAGATG CTCTTCACAT
GGAGGGCCTT ATAGTTCGGG AACGGATTTT AGGTGCTGAC AACATTGATG TTTCCCACCC
CATCATTTAC AGAGGGGCTG TCTATGCTGA TAACATGGAG TTCGAGCAGT GCATCAAATT
```

Fig. 8. Unknown mouse gene.

When the results page appears, click the identifier with the highest score and you will see the following information. Here the highest score is 481. The score was calculated on the match quality and the length of the most-similar segments that occur between the unknown mouse gene and the target human fem1b gene.

Human - Sore: 481 Query Coverage 98%

```
LOCUS        NM_015322              2583 bp    mRNA    linear   PRI 26-DEC-2010
DEFINITION   Homo sapiens fem-1 homolog b (C. elegans) (FEM1B), mRNA.
ACCESSION    NM_015322
VERSION      NM_015322.3  GI:52851431
SOURCE       Homo sapiens (human)
ORGANISM     Homo sapiens
             Eukaryota; Metazoa; Chordata; Craniata; Vertebrata; Euteleostomi;
             Mammalia; Eutheria; Euarchontoglires; Primates; Haplorrhini;
             Catarrhini; Hominidae; Homo.
REFERENCE    1  (bases 1 to 2583)
  AUTHORS    Ewens,K.G., Stewart,D.R., Ankener,W., Urbanek,M., McAllister,J.M.,
             Chen,C., Baig,K.M., Parker,S.C., Margulies,E.H., Legro,R.S.,
             Dunaif,A., Strauss,J.F. III and Spielman,R.S.
  TITLE      Family-based analysis of candidate genes for polycystic ovary
             syndrome
  JOURNAL    J. Clin. Endocrinol. Metab. 95 (5), 2306-2315 (2010)
  PUBMED     20200332
  REMARK     GeneRIF: Observational study of gene-disease association. (HuGE
             Navigator)
REFERENCE    2  (bases 1 to 2583)
  AUTHORS    Subauste,M.C., Sansom,O.J., Porecha,N., Raich,N., Du,L. and
             Maher,J.F.
  TITLE      Fem1b, a proapoptotic protein, mediates proteasome
             inhibitor-induced apoptosis of human colon cancer cells
  JOURNAL    Mol. Carcinog. 49 (2), 105-113 (2010)
  PUBMED     19908242
```

When you scroll down the page, you see reach a long list of the human fem1b nucleotide sequence starting with

```
ATGGAAGGACTTGCGGGGTATGTTTACAAAGCAGCGAGCGAGGGTAAAGTGCTGACCCTGG
CTGCCCTATTATTAAATCGCTCGGAATCCGATATACGATACCTTCTTGGGTACGTTAGTCA
GCAAGGAGGCCAGCGGAGTACCCCCTTGATTATAGCCGCACGTAACGGCCACGCGAAGGTC
GTGAGACTCCTACTCGAACATTATCGGGTACAAACGCAGCAAACAGGAACCGTACGGTTCG
ACGGATACGTTATAGATGGCGCGACAGCTTTATGGTGTGCCGCAGGCGCCGGTCACTTCGA
AGTAGTC
```

3.2 Sequence statistics analysis

Sections of a nucleotide sequence with a certain percentage of A+T or C+G usually indicates intergenic parts of the sequence. Figure 9 is a plot of monomer densities and combined monomer densities. One can use such statistic plot to determine if the sequence has the characteristics of a protein-coding region.

Figure 10 is the visualization of the nucleotide distribution. Figure 11 is the codon distribution showing a high amount of GAA, GAT and AAC. The amino acids for GAA, GAT and AAC are Glutamate, Aspartate, and Asparagine respectively. The corresponding bar chart distribution is displayed at figure 12. It is noticeable that it contains high volume of leucine, alanine, and valine; low volume of tryptophan, methionine, and proline.

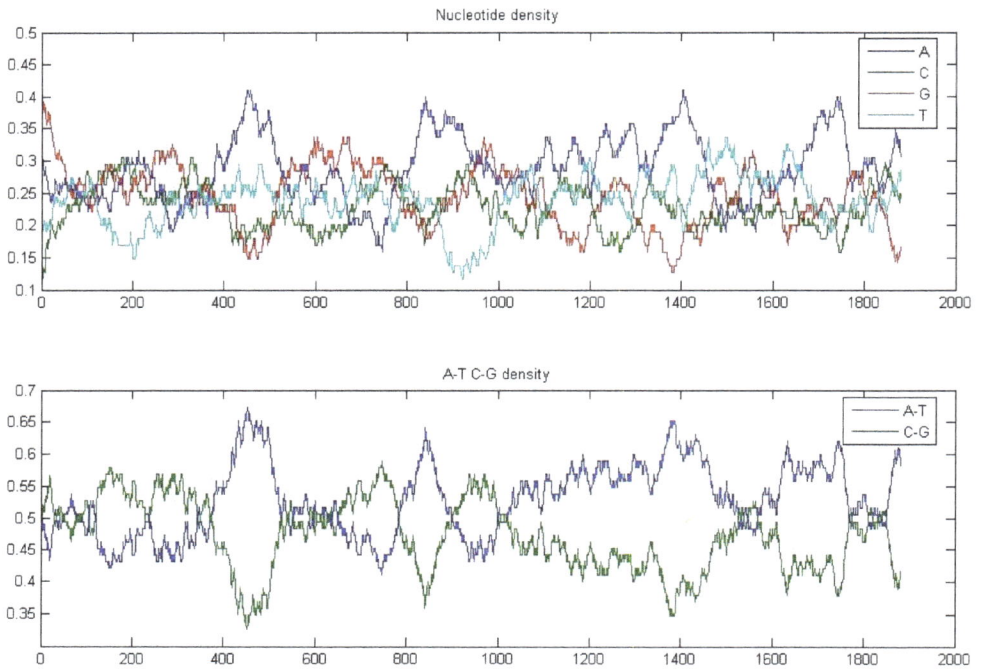

Fig. 9. Human fem1b gene's monomer densities and A-T &C-G combined monomer densities.

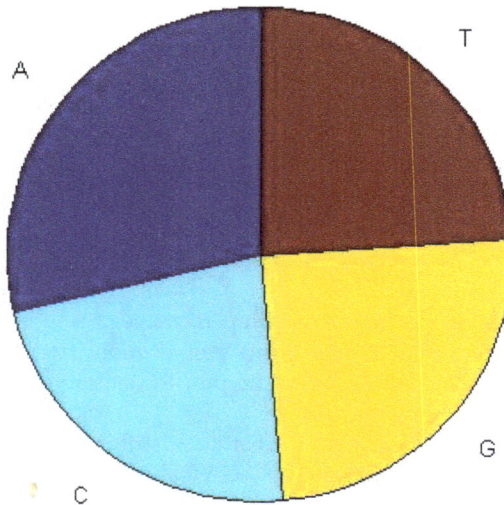

Fig. 10. Human fem1b gene's nucleotide distribution (A: 542, C: 426, G: 462, T: 451).

```
AAA  -  16        AAC  -  22        AAG  -  13        AAT  -  19
ACA  -   9        ACC  -   7        ACG  -   8        ACT  -  10
AGA  -   6        AGC  -   3        AGG  -   3        AGT  -   8
ATA  -  13        ATC  -  16        ATG  -  12        ATT  -  12
CAA  -   9        CAC  -   9        CAG  -  14        CAT  -  18
CCA  -   3        CCC  -   3        CCG  -   7        CCT  -   2
CGA  -   5        CGC  -   8        CGG  -   8        CGT  -   5
CTA  -  11        CTC  -  12        CTG  -  14        CTT  -  14
GAA  -  24        GAC  -  15        GAG  -  14        GAT  -  23
GCA  -  15        GCC  -  15        GCG  -  17        GCT  -  12
GGA  -   7        GGC  -   7        GGG  -  12        GGT  -   6
GTA  -  17        GTC  -   9        GTG  -  10        GTT  -   9
TAA  -   0        TAC  -  10        TAG  -   0        TAT  -  13
TCA  -   0        TCC  -   5        TCG  -   9        TCT  -   4
TGA  -   0        TGC  -  10        TGG  -   3        TGT  -   7
TTA  -  11        TTC  -   7        TTG  -   8        TTT  -   9
```

Fig. 11. Human fem1b gene codon distribution.

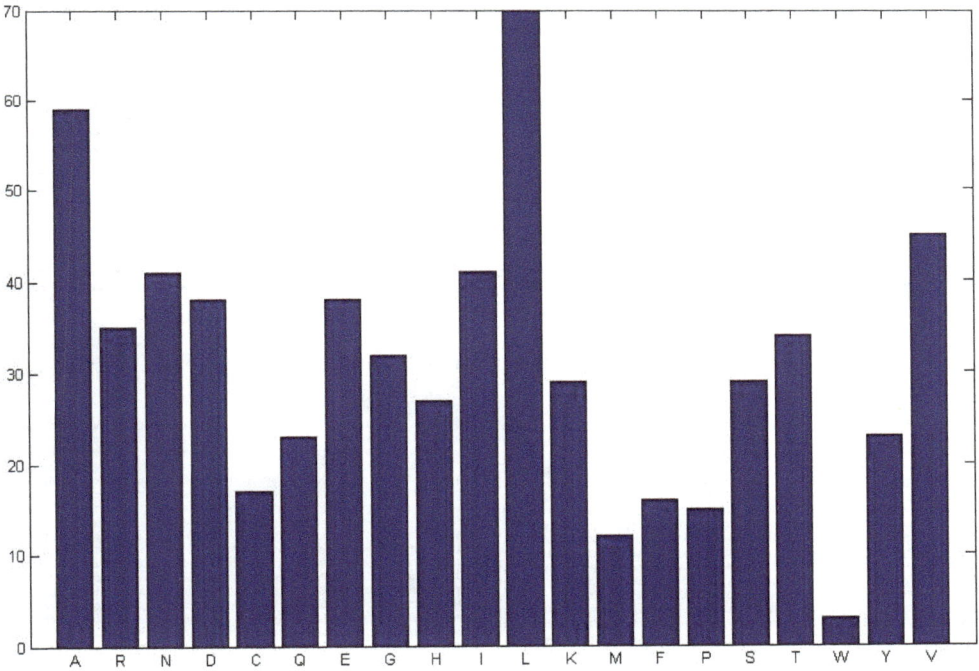

Fig. 12. Amino acids distribution of human fem1b gene.

A: 59 R: 35 N: 41 D: 38 C: 17 Q: 23 E: 38 G: 32 H: 27 I: 41 L: 70 K: 29 M: 12 F: 16 P: 15 S: 29 T: 34 W: 3 Y: 23 V: 45

3.3 Open reading frame of Fem1b gene from both human and mouse

An open reading frame (ORF) is a nucleotide sequence without having a stop codon in a given reading frame. ORFs can be identified by examining each of the three possible reading frames on each strand. A DNA sequence must contain a translation start codon and it is usually "AGT". Possible stop codons are "TAA", "TAG and "TGA" [11]. Identifying the start and stop codons for translation determines the ORF in a given nucleotide sequence. Once an ORF is located for a gene or mRNA, a nucleotide sequence can be translated into its corresponding amino acid sequence. Figure 13 – 15 display three reading frames for human's and mouse's fem1b gene sequences. Both genes show the longest ORF on the first reading frame.

Dot plots are one of the easiest ways to look for similarity between two sequences. The diagonal line shown in figure 16 indicates a good alignment between the human's and mouse's fem1b gene.

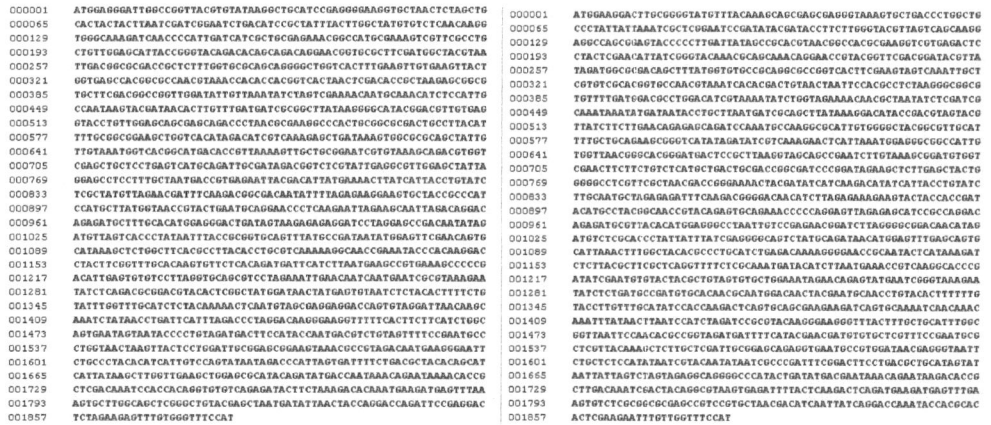

Fig. 13. First ORF of fem1b gene (left – mouse; right – human).

Fig. 14. Second ORF of fem1b gene (left – mouse; right – human).

```
000001  ATGGAAGGACTTGCGGGGTATGTTTACAAAGCAGCGAGCGAGGGTAAAGTGCTGACCCTGGCTG     000001  ATGGAAGGACTTGCGGGGTATGTTTACAAAGCAGCGAGCGAGGGTAAAGTGCTGACCCTGGCTG
000065  CCCTATTATTAAATCGCTCGGAATCCGATATACGATACCTTCTTGGGTACGTTAGTCAGCAAGG     000065  CCCTATTATTAAATCGCTCGGAATCCGATATACGATACCTTCTTGGGTACGTTAGTCAGCAAGG
000129  AGGCCAGCGGAGTACCCCCTTGATTATAGCCGCACGTAACGGCCACGCGAAGGTCGTGAGACTC     000129  AGGCCAGCGGAGTACCCCCTTGATTATAGCCGCACGTAACGGCCACGCGAAGGTCGTGAGACTC
000193  CTACTCGAACATTATCGGGTACAAACGCAGCAAACAGGAACCGTACGGTTCGACGGATACGTTA     000193  CTACTCGAACATTATCGGGTACAAACGCAGCAAACAGGAACCGTACGGTTCGACGGATACGTTA
000257  TAGATGGCGCGACAGCTTTATGGTGTGCCGCAGGCGCCGGTCACTTCGAAGTAGTCAAATTGCT     000257  TAGATGGCGCGACAGCTTTATGGTGTGCCGCAGGCGCCGGTCACTTCGAAGTAGTCAAATTGCT
000321  CGTGTCGCACGGTGCCAACGTAAATCACACGACTGTAACTAATTCCACGCCTCTAAGGGCGGCG     000321  CGTGTCGCACGGTGCCAACGTAAATCACACGACTGTAACTAATTCCACGCCTCTAAGGGCGGCG
000385  TGTTTTGATGGACGCCTGGACATCGTAAAATATCTGGTAGAAAACAACGCTAATATCTCGATCG     000385  TGTTTTGATGGACGCCTGGACATCGTAAAATATCTGGTAGAAAACAACGCTAATATCTCGATCG
000449  CAAATAAATATGATAATACCTGCTTAATGATCGCAGCTTATAAAGGACATACCGACGTAGTACG     000449  CAAATAAATATGATAATACCTGCTTAATGATCGCAGCTTATAAAGGACATACCGACGTAGTACG
000513  TTATCTTCTTGAACAGAGAGCAGATCCAAATGCCAAGGCGCATTGTGGGGCTACGGCGTTGCAT     000513  TTATCTTCTTGAACAGAGAGCAGATCCAAATGCCAAGGCGCATTGTGGGGCTACGGCGTTGCAT
000577  TTTGCTGCAGAAGCGGGTCATATAGATATCGTCAAAGAACTCATTAAATGGAGGGCGGCCATTG     000577  TTTGCTGCAGAAGCGGGTCATATAGATATCGTCAAAGAACTCATTAAATGGAGGGCGGCCATTG
000641  TGGTTAACGGGCACGGGATGACTCCGCTTAAGGTAGCAGCCGAATCTTGTAAAGCGGATGTGGT     000641  TGGTTAACGGGCACGGGATGACTCCGCTTAAGGTAGCAGCCGAATCTTGTAAAGCGGATGTGGT
000705  CGAACTTCTTCTGTCTCATGCTGACTGCGACCGGCGATCCCGGATAGAAGCTCTTGAGCTACTG     000705  CGAACTTCTTCTGTCTCATGCTGACTGCGACCGGCGATCCCGGATAGAAGCTCTTGAGCTACTG
000769  GGGGCCTCGTTCGCTAACGACCGGGAAAACTACGATATCATCAAGACATATCATTACCTGTATC     000769  GGGGCCTCGTTCGCTAACGACCGGGAAAACTACGATATCATCAAGACATATCATTACCTGTATC
000833  TTGCAATGCTAGAGAGATTTCAAGACGGGGACAACATCTTAGAGAAAGAAGTACTACCACCGAT     000833  TTGCAATGCTAGAGAGATTTCAAGACGGGGACAACATCTTAGAGAAAGAAGTACTACCACCGAT
000897  ACATGCCTACGGCAACCGTACAGAGTGCAGAAACCCCCAGGAGTTAGAGAGCATCCGCCAGGAC     000897  ACATGCCTACGGCAACCGTACAGAGTGCAGAAACCCCCAGGAGTTAGAGAGCATCCGCCAGGAC
000961  AGAGATGCGTTACACATGGAGGGCCTAATTGTCCGAGAACGGATCTTAGGGGCGGACAACATAG     000961  AGAGATGCGTTACACATGGAGGGCCTAATTGTCCGAGAACGGATCTTAGGGGCGGACAACATAG
001025  ATGTCTCGCACCCTATTATTTATCGAGGGGCAGTCTATGCAGATAACATGGAGTTTGAGCAGTG     001025  ATGTCTCGCACCCTATTATTTATCGAGGGGCAGTCTATGCAGATAACATGGAGTTTGAGCAGTG
001089  CATTAAACTTTGGCTACACGCCCTGCATCTGAGACAAAAGGGGAACCGCAATACTCATAAAGAT     001089  CATTAAACTTTGGCTACACGCCCTGCATCTGAGACAAAAGGGGAACCGCAATACTCATAAAGAT
001153  CTCTTACGCTTCGCTCAGGTTTTCTCGCAAATGATACATCTTAATGAAACCGTCAAGGCACCCG     001153  CTCTTACGCTTCGCTCAGGTTTTCTCGCAAATGATACATCTTAATGAAACCGTCAAGGCACCCG
001217  ATATCGAATGTGTACTACGCTGTAGTGTGCTGGAAATAGAACAGAGTATGAATCGGGTAAAGAA     001217  ATATCGAATGTGTACTACGCTGTAGTGTGCTGGAAATAGAACAGAGTATGAATCGGGTAAAGAA
001281  TATCTCTGATGCCGATGTGCACAACGCAATGGACAACTACGAATGCAACCTGTACACTTTTTTG     001281  TATCTCTGATGCCGATGTGCACAACGCAATGGACAACTACGAATGCAACCTGTACACTTTTTTG
001345  TACCTTGTTTGCATATCCACCAAGACTCAGTGCAGCGAAAGAAGATCAGTGCAAAATCAACAAAC   001345  TACCTTGTTTGCATATCCACCAAGACTCAGTGCAGCGAAAGAAGATCAGTGCAAAATCAACAAAC
001409  AAAYYTATAACTTAATCCATCTAGATCCGCGTACAAGGGAAGGGTTTACTTTGCTGCATTTGGC    001409  AAATTTATAACTTAATCCATCTAGATCCGCGTACAAGGGAAGGGTTTACTTTGCTGCATTTGGC
001473  GGTTAATTCCAACACGCCGGTAGATGATTTTCATACGAACGATGTGTGCTCGTTTCCGAATGCG     001473  GGTTAATTCCAACACGCCGGTAGATGATTTTCATACGAACGATGTGTGCTCGTTTCCGAATGCG
001537  CTCGTTACAAAGCTCTTGCTCGATTGCGGAGCAGAGGTGAATGCCGTGGATAACGAGGGTAATT    001537  CTCGTTACAAAGCTCTTGCTCGATTGCGGAGCAGAGGTGAATGCCGTGGATAACGAGGGTAATT
001601  CTGCTCTCCATATAATCGTACAATATAATCGCCCGATTTCGGACTTCCTGACGCTGCATAGTAT    001601  CTGCTCTCCATATAATCGTACAATATAATCGCCCGATTTCGGACTTCCTGACGCTGCATAGTAT
001665  AATTATTAGTCTAGTAGAGGCAGGGGCCCATACTGATATGACGAATAAACAGAATAAGACACCG    001665  AATTATTAGTCTAGTAGAGGCAGGGGCCCATACTGATATGACGAATAAACAGAATAAGACACCG
001729  CTTGACAAATCGACTACAGGCGTAAGTGAGATTTTACTCAAGACTCAGATGAAGATGAGTTTGA    001729  CTTGACAAATCGACTACAGGCGTAAGTGAGATTTTACTCAAGACTCAGATGAAGATGAGTTTGA
001793  AGTGTCTCGCGGCGCGAGCCGTCCGTGCTAACGACATCAATTATCAGGACCAAATACCACGCAC    001793  AGTGTCTCGCGGCGCGAGCCGTCCGTGCTAACGACATCAATTATCAGGACCAAATACCACGCAC
001857  ACTCGAAGAATTTGTTGGTTTCCAT                                          001857  ACTCGAAGAATTTGTTGGTTTCCAT
```

Fig. 15. Third ORF of fem1b gene (left – mouse; right – human).

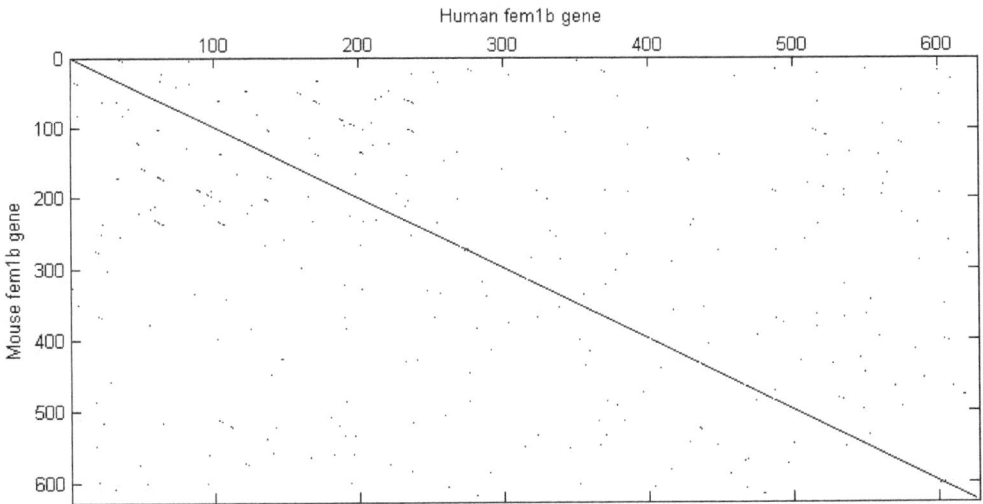

Fig. 16. Dot plot comparing the human and mouse amino acid sequences.

3.4 Sequence alignment
3.4.1 Global alignment

The Needleman-Wunsch algorithm, which was first published by Saul Needleman and Christian Wunsch in 1970 [12], performs a global alignment on two amino acid or nucleotide sequences. Such algorithm was the first application of dynamic programming to molecular sequence comparison. The following output was performed on two nucleotide sequences of mouse's and human's by the Needleman-Wunsch algorithm

```
Identities = 1430/1890 (76%), Positives = 1591/1890 (84%)

0001   ATGGAGGGATTGGCCGGTTACGTGTATAAGGCTGCATCCGAGGGGAAGGTGCTAACTCTAGCTG
       |||||:||| |  || || || ||  || ||:||:||:: |||||| ||:|||||:|| |||||
0001   ATGGAAGGACTTGCGGGGTATGTTTACAAAGCAGCGAGCGAGGGTAAAGTGCTGACCCTGGCTG
0065   CACTACTACTTAATCGATCGGAATCTGACATCCGCTATTTACTTGGCTATGTGTCTCAACAAGG
       | ||| || |:||||| |||||||| || || || ||  |:|||||| || || : |||:|||||
0065   CCCTATTATTAAATCGCTCGGAATCCGATATACGATACCTTCTTGGGTACGTTAGTCAGCAAGG

0129   TGGGCAAAGATCAACCCCATTGATCATCGCTGCGAGAAACGGCCATGCGAAAGTCGTTCGCCTG
       :|| ||: |:: :|||||| |||||| || || || ||: |:|||||||| |||||:|||||| | ||
0129   AGGCCAGCGGAGTACCCCCTTGATTATAGCCGCACGTAACGGCCACGCGAAGGTCGTGAGACTC
```

3.4.2 Local alignment
The Smith-Waterman algorithm was first published by Temple Smith and Michael Waterman in 1981 [13]. It is a well-known dynamic programming algorithm for local amino acid or nucleotide sequence alignment. Unlike the global alignment, the Smith-Waterman algorithm performs comparison among segments of all lengths and optimizes the similarity. It is guaranteed to find the optimal local alignment with respect to the scoring method. However, the Smith-Waterman algorithm requires $O(mn)$ (m and n are the length of two input sequences) . In practical use, it has been replaced by the heuristic BLAST algorithm, which is much more efficient although not guaranteed to find the optimal alignments. The following output was from local alignment of the amino acid sequences of mouse's and human's using the Smith-Waterman algorithm.

```
Identities = 621/627 (99%), Positives = 627/627 (100%)

001    MEGLAGYVYKAASEGKVLTLAALLLNRSESDIRYLLGYVSQQGGQRSTPLIIAARNGHAKVVRL
       |||||||||||||||||||||||||||||||||||||||||||||||||||||||||||||||
001    MEGLAGYVYKAASEGKVLTLAALLLNRSESDIRYLLGYVSQQGGQRSTPLIIAARNGHAKVVRL

065    LLEHYRVQTQQTGTVRFDGYVIDGATALWCAAGAGHFEVVKLLVSHGANVNHTTVTNSTPLRAA
       |||||||||||||||||||||||||||||||||||||||||||||||||||||||||||||||
065    LLEHYRVQTQQTGTVRFDGYVIDGATALWCAAGAGHFEVVKLLVSHGANVNHTTVTNSTPLRAA
```

4. Diabetic retinopathy study through retinal image fusion

Hypoxia of the retina is believed to be a factor in the development of diabetic retinopathy, the leading cause of blindness worldwide. Retina image fusion provides a practical way for determination of the oxygenation status of the ocular fundus. Such method would be a valuable medical diagnostic tool for diabetic retinopathy [14], age-related macular degeneration, glaucoma [15], retinopathy of prematurity, and central retinal vein occlusion [16].

4.1 Acquisition of retinal images
Retinal images presented in this chapter were taken by a modified Topcon TRC-50EX fundus camera, with a lens and a c-mount through the vertical path of the camera. Hyperspectral images were taken through the vertical viewing port by an imaging

spectrograph and digital camera (model VNIR 100; Photon Industries Inc., Mississippi Stennis Space Center, USA) across the fundus image (Figure 17).

The subjects of the retinal images were Cynomolgus monkeys of 4 to 4.5 years of age and 2.5 to 3 kg body weight with normal eyes [17]. The use of animals for taking retinal images was approved by Louisiana State University Health Sciences Center Institutional Animal Care and Use Committee [18]. This animal usage is also conformed to the ARVO Statement for the Use of Animals in Ophthalmic and Vision Research. The monkeys were housed in an air conditioned room with normal temperature and humidity with a 12 hour light-dark diurnal cycle.

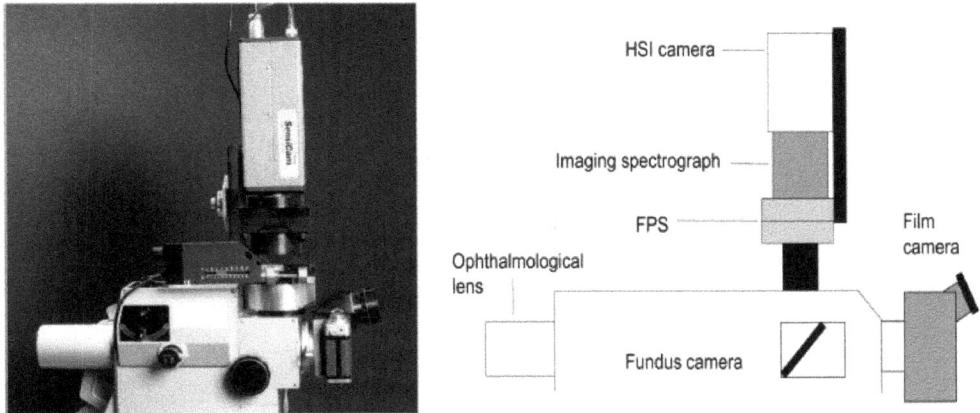

Fig. 17. Hyperspectral imaging system in relation to the fundus camera. The image is redirected upward by a mirror. The imaging system is translated over the camera port by a linear actuator mounted below the imaging spectrograph and CCD camera [17].

4.2 Retinal image fusion

There are five major steps involved in image fusion (Diagram 1):

1. The first step is the image segmentation. The segmentation subdivides an input image into its constituent regions or objects and extract/detect salient features/structure for the automated procedure.

2. The second step feature extraction is going to detect the salient structures on the target images for the feature-based approach.

3. The third step is the feature matching. The purpose of feature matching is to bringing together the information that represents same features detected at different images. The first three steps will provide the initial guess of the features for the fusion algorithm.

4. The fourth step is the optimization of the initial guess. The previously detected features will be adjusted in this step through a certain objective function.

5. The final step will transforms the images from single or different modalities into spatial alignment [19] through a certain mathematical model and then display combined view of the involved images.

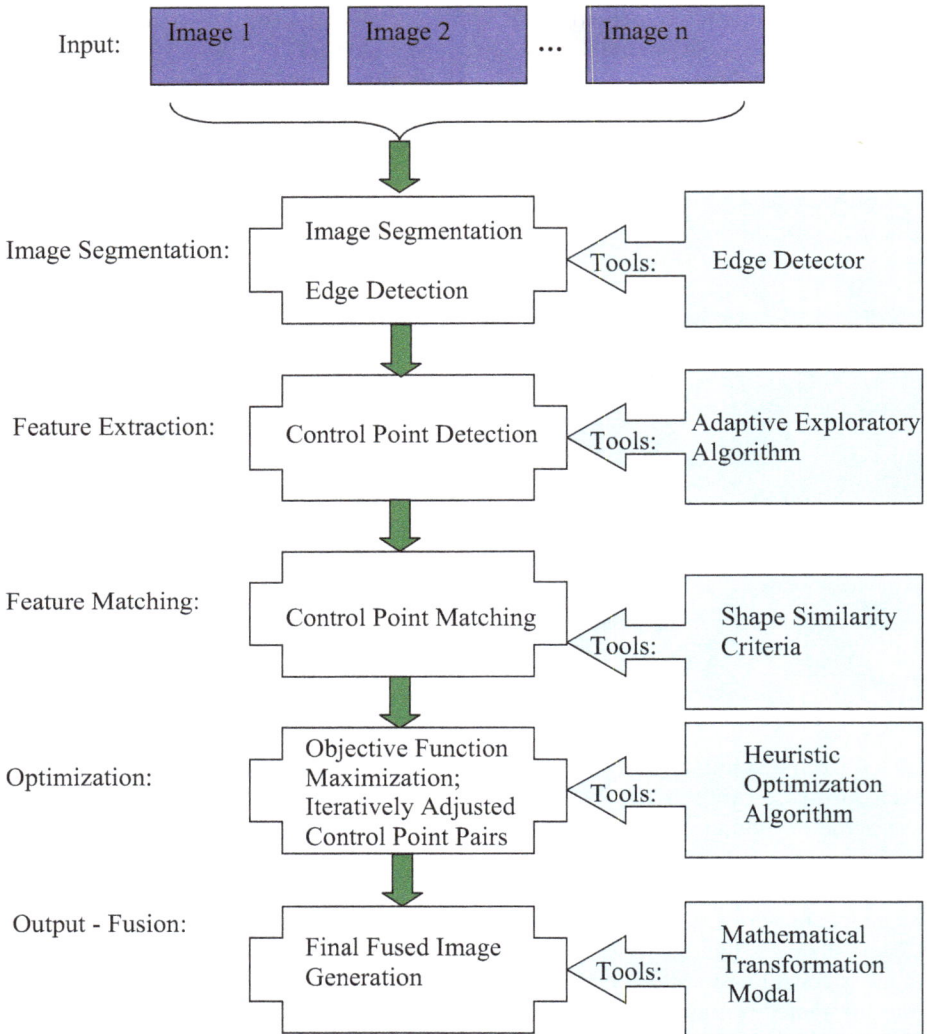

Diagram 1. Image fusion procedure.

4.2.1 Retinal vasculature extraction using canny edge detector

The Canny operator [20] is less likely than the others to be "fooled" by noise, and more likely to detect true weak edges. The Canny operator preserves most edges among all other edge detectors. Therefore, the Canny edge detector is employed in this research to extract the retinal vasculature edges. There are two criteria used in the Canny Operator to locate the rapidly changed intensity pixels. They are:

1. Pixels where the 1st derivative of the intensity is larger in magnitude than a certain threshold;

2. Pixels where the 2nd derivative of the intensity has a zero crossing.
3. Canny's method detects edges at the zero-crossings of the second directional derivative of the image. It performs zero-crossings of

$$\frac{d^2(G \times I)}{dn^2} = \frac{d\left(\left(\frac{dG}{dn}\right) \times I\right)}{dn} \qquad (1)$$

where, n is the direction of the gradient of the image; G is the edge signal; I is the image intensity. The zero-crossings of Canny's method correspond to the first directional-derivative's maxima and minima in the direction of the gradient. Edges will be identified as the maxima in magnitude. Each pixel's edge gradient is computed and compared with the gradients of its neighbors along the gradient direction. If the central pixel is smaller, mark the current edge's intensity as 0; if largest among all neighbors, keep the original intensity. Based on the nine-pixel neighborhood, the normal to the edge direction has two u_x and u_y. In order to estimate the gradient on the discrete sampling, two pixels closest to u are selected. A plane can be identified by the gradient magnitudes of three pixels. By using this plane, the gradient magnitude and the intensity at each pixel on the line can be locally estimated. The gradient magnitude at $P_{x+1,\ y+1}$ and $P_{x-1,\ y-1}$ (Figure 18) can be calculated as:

$P_{x+1,\ y+1}$: $\qquad G(P_{x+1,y+1}) = \frac{u}{u_y}G(x+1,y+1) + \frac{u_y-u_x}{u_y}G(x,y+1)$ $\qquad (2)$

$P_{x-1,\ y-1}$: $\qquad G(P_{x-1,y-1}) = \frac{u}{u_y}G(x-1,y-1) + \frac{u_y-u_x}{u_y}G(x,y-1)$ $\qquad (3)$

If the gradient at $P_{x,y}$ is greater than both of $G(P_{x+1,y+1})$ and $G(P_{x-1,y-1})$, $P_{x,y}$ will be identified as a maximum.

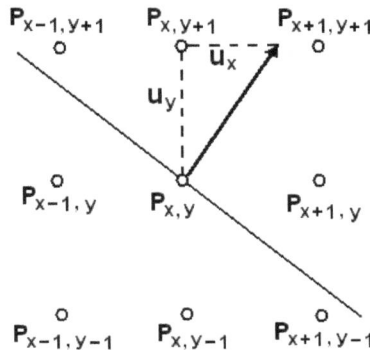

Fig. 18. Canny Edge Detection – Localization of Maxima [34].

In order to make the localization of magnitude maxima accurate, Canny defined a filter by optimizing a performance index which enhances real positive and real negative. The filter is used to minimize the probability of non-detected edge points and false detection.

$$SNR = \frac{\left| \int_{-w}^{w} G(-x)f(x)dx \right|}{n_0 \sqrt{\int_{-w}^{w} f^2(x)dx}} \qquad (4)$$

where, SNR stands for Signal-to-Noise Ratio; f is the filter; denominator is the RMSE response to noise $n(x)$. The identification of the real edge localization is defined as:

$$Localization = \frac{1}{\sqrt{E[x_0^2]}} = \frac{\left| \int_{-w}^{w} G'(-x)f'(x)dx \right|}{n_0 \sqrt{\int_{-w}^{w} f'^2(x)dx}} \qquad (5)$$

Two adaptive thresholds are used in Canny's method. They are high threshold and low threshold. The high threshold is used to find the start point of strong edges. Any points that meet the high threshold will be selected as the edge point. These start points are growing into different directions as long as there is no edge strength falling below the low threshold.

Figure 19 is the 3D shaded surface plots of the original retinal angiogram image. The X-Y axis corresponds to the original image size. The height Z axis is a single-valued function defined over a geometrically rectangular grid. Z specifies the color data as well as surface height, so color is proportional to surface height with range of [0, 1] of each pixel on the image. All the retinal salient features are preserved in the Canny edges. Those salient features are the retinal vessel bifurcations, from which the control points will be selected using the Adaptive Exploratory Algorithm. Figure 20 and 21 show the retinal vessel edges detected by the Canny operator.

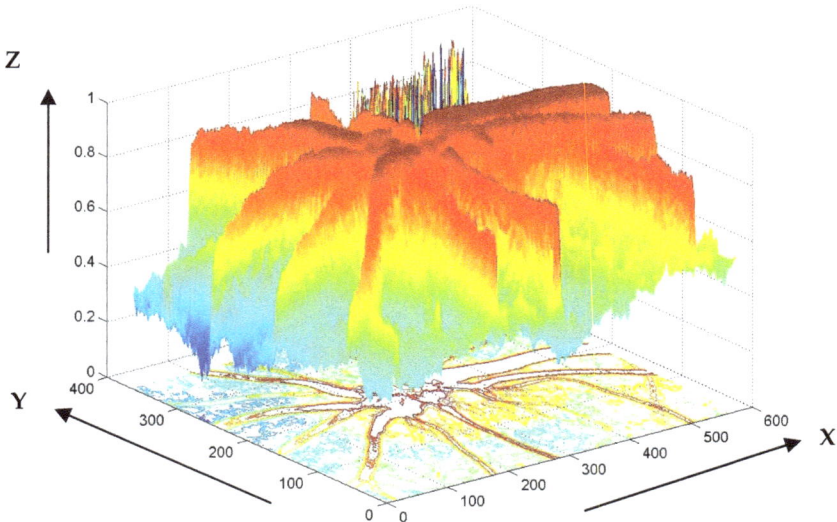

Fig. 19. 3D shaded surface plot of the angiogram image.

4.2.2 Control point detection

A good-guess of the initial control point selection ensures fused image generated at an efficient computational time. Bad control point selection will significantly increase the computation cost, or even cause the image fusion fail. Vessels or some particular abnormalities make images not necessarily matching the retina structures. Even when structure and function correspond, the abnormality still happens sometimes if inconsistence exists between structural and functional changes. Further more, angiogram images usually have higher resolution and are rich in information, whereas fundus images have lower resolution and are indeed abstract with some details or even missing some small vessels. Practically, those situations are unavoidable and will create difficulties in extracting the control points because the delineation of the vein boundaries may not be precise. In this study, control points are detected using the adaptive exploratory algorithm (Figure 20 and 21) [21].

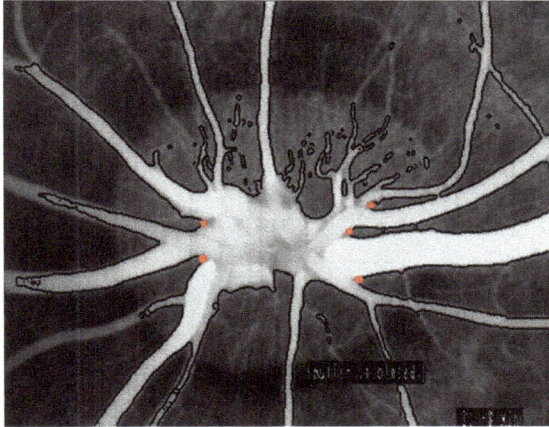

Fig. 20. Angiogram grayscale reference image's control point selection.

Fig. 21. Fundus true color input image's control point selection.

4.2.3 Heuristic optimization algorithm

An optimization procedure is required to adjust the initial good-guess control points in order to achieve the optimal result. The process can be formulated as a heuristic problem of optimizing an objective function that maximizes the Mutual-Pixel-Count between the reference and input images. The algorithm finds the optimal solution by refining the transformation parameters in an ordered way. By maximizing the objective function, one image's vessels are supposed to be well overlaid onto those of the other image (Figure 23). Mutual-Pixel-Count measures the optic nerve head vasculature overlapping for corresponding pixels in both images. It is assumed that the retinal vessels are represented by 0 (black pixel) and background is represented by 1 (white pixels) in the binary 2D map. When the vasculature pixel's transformed (u, v) coordinates on the input image correspond to the vasculature pixel's coordinates on the reference image, the MPC is incremented by 1 (Figure 22). MPC is assumed be maximized when the image pair is perfectly geometrically aligned by the transformation. After pre-processing, the binary images of the reference and input images are obtained, i.e. I_{ref} and I_{input}. Only black pixels from both images contribute to MPC. The ideal case is that all zero pixels of the input image are mapped onto zero pixels of the reference image. The problem can be mathematically formulated as the maximization of the following objective function:

$$f_{mpc}(x,y,u,v) = \sum_{I_{ref}(x,y)=1 \ and \ I_{input}(u,v)=1}^{u,v \in ROI} I_{input}(T_x(u,v),T_y(u,v)) \tag{6}$$

where f_{mpc} denotes the value of the Mutual-Pixel-Count. T_x and T_y are the transformations for u and v coordinates of the input image. The ROI (Region-of-Interest) is the vasculature region where the MPC is calculated on.

Coordinates' adjustment is iteratively implemented until one of the following convergence criteria is reached: (1). Predefined maximum number of loops is reached; or (2). the updated f_{MPC} is smaller than ε, i.e.

$$\left| f_{MPC^{n+1}}(x,y,u,v) - f_{MPC^n}(x,y,u,v) \right| < \varepsilon \tag{7}$$

where ε is a very small non-negative threshold.

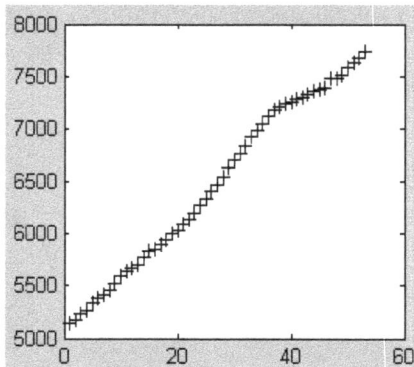

Fig. 22. f_{MPC} increasing during the iteration (Y-axis is f_{MPC}; X-axis is the loop count).

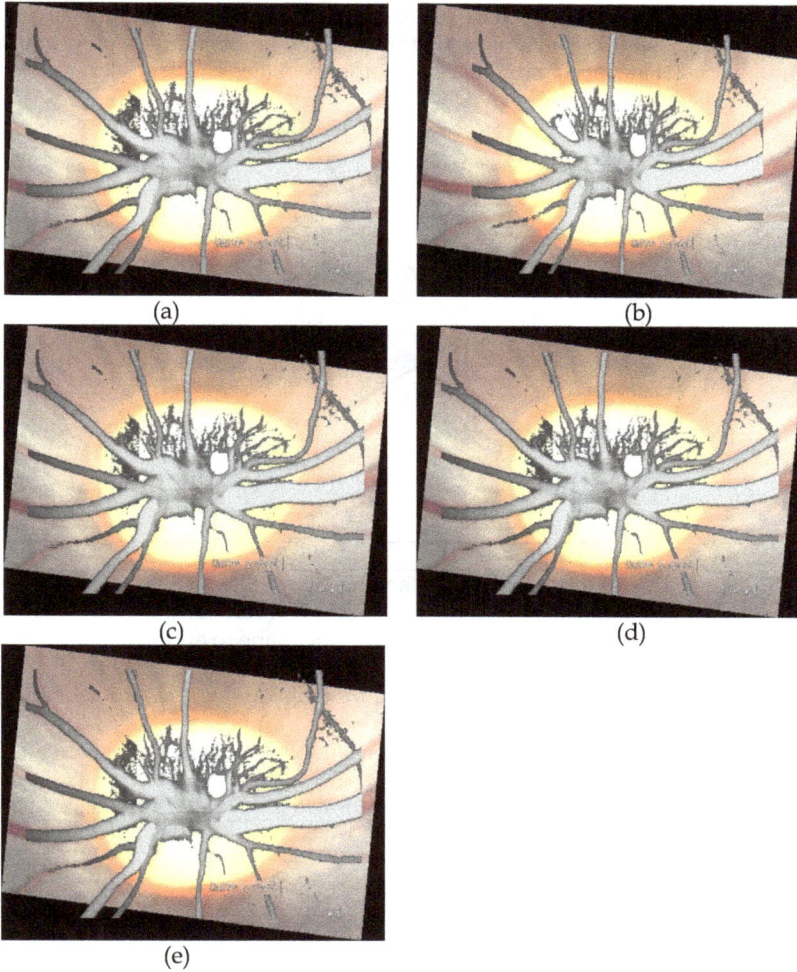

(a)   (b)

(c)   (d)

(e)

Fig. 23. Fused image improvement during the iteration. From (a)-(e) fMPC = 5144, 7396, 7484, 7681, 7732.

4.2.4 Affine transformation model

A mathematical model is the tool for transforming the target images and fusing them into one single volume. Affine model is a basic geometrical transformation in image processing and is defined as Eq. 8 and 9. The DOF of the affine model is 6 because it has six parameters, i.e. a_1, a_2, b_1, b_2, a_3, and a_4.

$$\begin{pmatrix} U \\ V \\ 1 \end{pmatrix} = \begin{pmatrix} a_1 & a_2 & b_1 \\ a_3 & a_4 & b_2 \\ 0 & 0 & 1 \end{pmatrix} \begin{pmatrix} x \\ y \\ 1 \end{pmatrix} \qquad \begin{aligned} U(x, y) &= a_1 x + a_2 y + b_1 \\ V(x, y) &= a_3 x + a_4 y + b_2 \end{aligned}$$

$$(8)$$
$$(9)$$

Affine model's advantage lies in that it can measure lost information such as skew, translation, rotation, shearing and scaling that maps finite points to finite points and parallel lines to parallel lines (Figure 24 and Table 1). Its drawback lies in the strict requirement that at least six pairs of control points are needed [19].

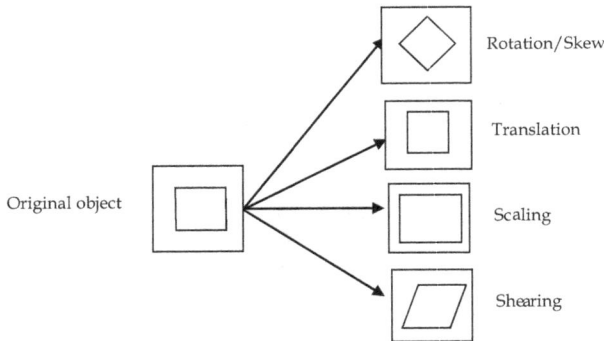

Fig. 24. Transformation Graphs.

Transformation	Description
Rotation/Skew	Points are rotated by an angle θ.
Translation	A linear shift in the position of the vertical and horizontal coordinates of the image in one plane to another set in the same spatial domain.
Scaling	A transformation of the horizontal and vertical coordinate points characterized by a certain scale factor.
Shearing	A transformation in which all points along a line remain fixed while other points are shifted parallel to the line by a certain distance proportional to their perpendicular distance from the line [22].

Table 1. Transformation Descriptions.

5. Conclusion and future directions

Fem1b functions in vivo to regulate insulin secretion and plasma glucose levels. Fem1b-KO mice do not have fasting hyperglycemia but rather have defective acutephase GSIS. Such defective acute-phase GSIS is the earliest detectable defect in humans destined to develop diabetes and may represent the primary genetic risk factor predisposing to diabetes [23]. With aging and superimposed insulin resistance, fasting hyperglycemia and overt diabetes later develop. Therefore, the Fem1b-KO mouse model is a key component of the complex pathogenesis of type 2 diabetes mellitus. Both male and female homozygotes display abnormal glucose tolerance. The role of Fem1b in pancreatic islet insulin secretion strengthens evidence that a genetic pathway homologous to nematode sex determination may be involved in mammalian glucose homeostasis. This novel pathway could be involved in the β-cell dysfunction seen in type 2 diabetes mellitus. Since Calpain-10/NIDDM1 is known to interact with a gene that is near where human FEM1B localizes, to increase susceptibility to type 2 diabetes[24], whether FEM1B could be the responsible interacting gene becomes a pertinent question Although the mechanism of this regulation by Fem1b remains to be established, this finding strengthens evidence that a genetic pathway homologous to nematode sex determination may be involved in mammalian glucose homeostasis and promises to offer insight into novel genes and processes as potential candidates in the pathogenesis of diabetes mellitus.

Multi-modality analysis has been emerging as a major trend in the remote sensing, computer visualization, and biomedical image fusion. Fusing biomedical images is a very challenging problem because of the possible vast content change and non-uniform distributed intensities of the involved images. The new algorithm presented in this chapter, which consists of the Adaptive Exploratory Algorithm for the control point detection and heuristic optimization fusion, is reliable and time efficient. The new approach has achieved an excellent result by giving the visualization of fundus image with a complete angiogram overlay. By locking the multi-sensor images in one place, the algorithm allows ophthalmologists to match the same eye over time to get a sense of disease progress and pinpoint surgical tools to increase accuracy and speed of the surgery. The new algorithm can be easily expanded to human or animals' 3D eye, brain, or body image feature extraction, registration, and fusion. Many biomedical registration and fusion methods are still primarily used for research activity [19]. Very few of them have been developed into the integrated user-friendly computer software. The eventual aim is to developing and distributing the advanced and easy-to-use software which is suitable for various clinical environments. This plan requires intensive user interface developing work, which allows users adjusting a few threshold parameters if necessary. The user interface must be stable, simple and informative for every day clinical routine, because most of the end-users do not have much knowledge about algorithms and computer programs. Working closely with clinicians is extremely important when applying the new methods to practical clinic applications. As computation speed has been dramatically increased, real-time live ophthalmic image processing [25] will be used to handle larger and larger volumes of data in short periods. The data transmission rate, image size, higher resolution pixels, and many other issues will inevitably stress such live imaging fusion systems. The algorithms presented in this book have potential ability to handle those challenges. The presented method is a promising step towards useful clinical tools for retinopathy diagnosis, and thus forms a good foundation for further development.

6. References

[1] Klein R, Klein BE, Moss SE, Davis MD, DeMets DL. "The Wisconsin Epidemiologic Study of Diabetic Retiopathy, II: prevalence and risk of diabetic retinopathy when age at diagnosis is less than 30 years." *Arch Ophthalmol;* 102: 520-526. 1984

[2] Ryan, M; "Diabetes The Leading Cause of New Cases of Blindness"; Ohio Department of Health, Public Affairs; November 2003;

[3] Brissova, M., M. Shiota, W. E. Nicholson, M. Gannon, S. M. Knobel, D. W. Piston, C. V. Wright, and A. C. Powers. "Reduction in pancreatic transcription factor PDX-1 impairs glucose-stimulated insulin secretion"; *J. Biol. Chem.* 277:P11225–11232. 2002

[4] KELLEY, D. E., and B. H. GOODPASTER. Effects of exercise on glucose homeostasis in Type 2 diabetes mellitus. *Med. Sci. Sports Exerc.*, Vol. 33, No. 6, Suppl., p S495–S501.

[5] J. MAYFIELD, "Diagnosis and Classification of Diabetes Mellitus: New Criteria"; Bowen Research Center, Indiana University, Indianapolis, Indiana

[6] Cox, N. J., M. Frigge, D. L. Nicolae, P. Concannon, C. L. Hanis, G. I. Bell, and A. Kong. "Loci on chromosomes 2 (NIDDM1) and 15 interact to increase susceptibility to diabetes in Mexican Americans"; *Nat. Genet.* 21: 213–215.1999

[7] Ventura-Holman, T., N. B. Hander, and J. F. Maher. "The human FEM1B gene maps to chromosome 15q22 and is excluded as the gene for Bardet-Biedl syndrome, type 4": *Am. J. Med. Sci.* 319:268–270. 2000

[8] D. Lu, Tereza Ventura-Holman, J. Li, Robert W. McMurray, Jose S. Subauste, and Joseph F. Maher. "Abnormal Glucose Homeostasis and Pancreatic Islet Function in Mice with Inactivation of the Fem1b Gene". *Molecular and Cellular Biology.* 25(15): 6570–6577;2005

[9] Ventura-Holman, T., M. F. Seldin, W. Li, and J. F. Maher. "The murine Fem1 gene family: homologs of the Caenorhabditis elegans sex-determination protein FEM-1"; *Genomics* 54:221–230. 1998

[10] Chan, S. L., K. O. Tan, L. Zhang, K. S. Yee, F. Ronca, M. Y. Chan, and V. C. Yu. "F1Aα, a death receptor-binding protein homologous to the Caenorhabditis elegans sex-determining protein, FEM-1, is a caspase substrate that mediates apoptosis". *J. Biol. Chem.* 274:32461–32468. 1999

[11] J. Claverie; C. notredame; "Bioinformatics For Dummies, 2nd Edition"; ISBN: 978-0-470-08985-9; Wiley Publishing, Inc; 2007

[12] Needleman, Saul B.; and Wunsch, Christian D. "A general method applicable to the search for similarities in the amino acid sequence of two proteins"; *Journal of Molecular Biology*; V48, I3: P443–53; 1970.

[13] Smith, Temple; and Waterman, Michael. "Identification of Common Molecular Subsequences"; *Journal of Molecular Biology*; V147: P195–197. 1981

[14] Denninghoff KR, Smith MH, Hillman L. "Retinal imaging techniques in diabetes"; *Diabetes Technol Ther.* V2:111-113. 2002

[15] Schweitzer D, Hammer M, Kraft J, Thamm E, Konigsdorffer E, Strobel J. In vivo measurement of the oxygen saturation of retinal vessels in healthy volunteers. Transactions on Biomdeical Engineering. 46: 1454-1465. 1999

[16] Yoneya A, Saito T, Nishiyama Y, et al. "Retinal oxygen saturation levels in patients with central retinal vein occlusion". *Ophthalmology*;109:1521-1526. 2002

[17] J. Beach, J. Ning, B. Khoobehi; "Oxygen Saturation in Optic Neve Head Structures by Hyperspectral Image Analysis"; Current Eye Research; V32, P161-170, 2007.

[18] B. Khoobehi, J. Beach, H. Kawano; "Hyperspectral imaging for measurement of oxygen saturation in the optic nerve head"; Invest Ophthalmol Vis Sci. V45, P1464–1472; 2004

[19] H. Cao; "Automated Fusion of Multi-Modality Biomedical Images"; ISBN: 363906622;VDM, Verlag Dr. Muller Publisher, August 2008.

[20] J.F. Canny; "A computational approach to edge detection"; IEEE Transactions on Pattern Analysis and Machine Intelligence; V8, P 679-698; 1986.

[21] H. Cao, B. Khoobehi, S. S. Iyengar; "Automated Optic Nerve Head Image Fusion of Nonhuman Primate Eyes Using Heuristic Optimization Algorithm"; 5th IEEE Symposium on Computational Intelligence in Bioinformatics and Computational Biology (CIBCB 2008); Sun Valley, Idaho, USA; September 15-17, 2008; P 228-232

[22] Z. Millwala; "A dual-state approach to dental image registration"; Masters Thesis; *Lane Department of Computer Science and Electrical Engineering; West Virginia University*; 2004

[23] Denninghoff KR, Smith MH, Hillman L. "Retinal imaging techniques in diabetes"; *Diabetes Technol Ther.* V2:111-113. 2002

[24] Schweitzer D, Hammer M, Kraft J, Thamm E, Konigsdorffer E, Strobel J. In vivo measurement of the oxygen saturation of retinal vessels in healthy volunteers. Transactions on Biomdeical Engineering. 46: 1454-1465. 1999

[25] Yoneya A, Saito T, Nishiyama Y, et al. "Retinal oxygen saturation levels in patients with central retinal vein occlusion". *Ophthalmology*;V109: P1521-1526. 2002

9

Diabetes Mechanisms, Detection and Complications Monitoring

Dhanjoo N. Ghista[1], U. Rajendra Acharya[2], Kamlakar D. Desai[3],
Sarma Dittakavi[4], Adejuwon A. Adeneye[5] and Loh Kah Meng[6]

[1]*Department of Graduate and Continuing Education,*
Framingham State University, Framingham, Massachusetts,
[2]*School of Engineering, Division of ECE, Ngee Ann Polytechnic,*
[3]*Mukesh Patel School of Technology Management & Engineering,*
[4]*Biomedical Engineering department, Osmania University,*
[5]*Department of Pharmacology, Faculty of Basic Medical Sciences,*
Lagos State University College of Medicine, Ikeja,
Department of Pharmaceutical Sciences,
College of Pharmacy,
University of Kentucky, Kentucky,
[6]*VicWell Biomedical Private Limited,*
[1,5]*USA*
[2,6]*Singapore*
[3,4]*India*
[5]*Lagos State, Nigeria*

1. Introduction

Historically, the word *diabetes* was coined from the Greek word meaning a *siphon* by the 2nd century Greek physician, Aretus the Cappadocian. He used the word to connote a condition of passing water (urine) like a *siphon*. Later the Latin description *mellitus* meaning *sweetened or honey-like* was added. Put together, the term *diabetes mellitus* was literarily used to denote a disease condition which was associated with *the persistent passage of sweetened urine* (Krall & Braser, 1999).

In 1999, the World Health Organization described diabetes mellitus as a metabolic disorder of multiple aetiology characterized by chronic hyperglycaemia (the fasting blood glucose level equal or above 200 mg/dl taken at least twice, on different occasions) with disturbances of carbohydrate, fat and protein metabolism resulting from defects in insulin secretion, insulin action, or both. In other words, diabetes mellitus is a chronic disease with insidious onset in which the fasting blood glucose is persistently raised above the normal range values, the normal range being between 60 to 120 mg/dl of blood [Krall & Braser, 1999]. It occurs either because of a lack of insulin (the hormone responsible for glucose metabolism), or due to the presence of certain factors opposing the action of insulin on the body tissues that are involved in glucose metabolism, particularly, the liver and the skeletal muscles.

The consequence of insufficient insulin action is hyperglycaemia which may be associated with many associated metabolic abnormalities notably the development of hyperketonaemia resulting from disordered protein metabolism, and derangements in fatty acid and lipids metabolism. If the fasting blood glucose lies between 100 to 130 mg/dl, it is referred to as *Prediabetes* which is associated with an increased tendency or potential of developing *frank* diabetes. A fasting blood glucose of 140 mg/dl or higher is consistent with either type of diabetes mellitus, particularly, when accompanied by classic symptoms of diabetes [Diabetes Control and Complication Trial Research Group, 1997].

2. Diabetes mechanisms

Defects in glucose metabolizing machinery (such as defective insulin secretion, insulin action due to de-expression of insulin receptors or insensitivity of expressed insulin receptors and glucose transporters, decreased peripheral glucose utilization and defective glucose metabolizing enzymes, *etc.*) and consistent efforts of the physiological system to correct the imbalance in glucose metabolism or maintain glucose homeostasis (such as increased insulin secretion, lipolysis, gluconeogenesis, glycogenolysis, *etc.*) place an over exertion on the endocrine system, resulting in hyperglycaemia. The persistent chronic exposure of pancreatic β-cells to the supraphysiological glucose concentrations (hyperglycaemia) results in non-physiological and potentially irreversible β-cell damage, a term known as glucose toxicity which is a gradual, time-related onset of irreversible lesion to pancreatic β-cellular components of insulin content and secretion.

Multiple biochemical pathways and cellular mechanisms for glucose toxicity have been identified and these include glucose autoxidation (resulting from oxidative stress in the presence of chronic hyperglycaemia), protein kinase C (PKC) activation, increased flux through the hexosamine biosynthesis pathway (HBP), formation of advanced glycation end-products (AGEs), altered polyol pathway flux and altered gene expression. However, all these pathways share in common the formation of highly reactive oxygen intermediates (ROIs) or reactive oxygen species (ROS) which in excess amount and on prolonged exposure induce chronic oxidative stress on the pancreatic β-cell population, which in turn causes defective insulin gene expression and insulin secretion as well as increase pancreatic β-cell death.

Hyperglycaemia leads to the production of ROS which modulates various biological functions by stimulating transduction signals, some of which are involved in the pathogenesis of diabetes mellitus. Thus, redox-sensitive signalling pathways have been shown to play a pivotal role in the development, progression, and damaging effect on β-cells population within the pancreatic islet of Langerhans. In the pancreatic tissues, as hyperglycaemia worsens, the redox-sensitive signalling pathways mediating insulin synthesis, storage and release from the pancreatic β-cells becomes compromised progressively. In addition, the oxidative stress induced by chronic hyperglycaemia promotes pancreatic β-cells apoptosis which ultimately resulting in an overt reduction in the insulin secreting pancreatic β-cells population. The hallmarks of these molecular events are pancreatic β-cells failure and hypoinsulinaemia, which constitute the major pathogenic factors in type 1 diabetes mellitus.

Similarly, chronic hyperglycaemia-induced oxidative stress (the presence of an excess amount of reactive oxygen intermediates, due to an imbalance between their formation and degradation as a result of chronic hyperglycaemia) has been considered a proximate cause and common pathogenic factor for tissue/systemic complications of diabetes such as endothelial cells (micro- and macro-angiopathies), nerve cells (neuropathy), proximal renal epithelial cell (nephropathy), pancreatic β-cells (pancreatic β-cell failure) through lipid peroxidation and glycation mechanisms in these organs. Hyperglycaemia has been shown to result in glycation (a non-enzymatic conjugation of glucose to proteins leading to the formation of advanced glycation (glycosylation) end-products (AGEs) and tissue damage. Increased glycation and build-up of tissue AGEs have been implicated in the aetiology of diabetes mellitus, its complications and progression because they alter glucose metabolizing enzyme activity, decrease ligand binding, modify protein half-life and alter immunogenicity.

One mechanism by which the effects of glucose toxicity result in chronic hyperglycaemia are thought to be mediated is oxidative stress [Baynes, 1991; Evans *et al.*, 2002], and hyperglycaemia is known to be one of the main causes of oxidative stress in type 2 diabetes mellitus [Bonnefont-Rousselot, 2002; Robertson *et al.*, 2003]. Oxidative stress is a state of imbalance between free radical generation and mopping up.

Oxidative stress is known to play a pivotal role in the pathogenesis of insulin resistance which is itself is thought to be mediated via its contribution to glucose toxicity, particularly, in insulin target tissues including the pancreatic β-cells [Gleason *et al.*, 2000; Fantus, 2004]. Tissues such as the mesangial cells (in the kidneys), retinal cells and pancreatic islets are least endowed with intrinsic antioxidant enzyme expression, including *superoxidases-1* and -*2, catalase* and *glutathione peroxidase* [Hayden & Tyagi, 2002; Robertson, 2004]. Prolonged exposure of pancreatic β-cell to hyperglycaemia, as in diabetes, results in decreased expression of the antioxidant gene *γ-glutamylcysteine ligase* (*γ-GCL*) and down-regulation of the rate-limiting enzyme for glutathione synthesis [Robertson, 2004]. The *γ-GCL* catalyses the rate-limiting step in the synthesis of γ-glutamyl cysteine from cysteine, which forms the substrate for the second enzyme regulating glutathione synthesis [Yoshida *et al.*, 1995; Tanaka *et al.*, 2002]. Reduced gluthathione plasma and tissue concentrations, as marked by elevated levels of ceruloplasmin, promote free radical generation, production of advanced glycation products (AGEs) and acute flunctuations in glucose concentrations.

In addition, oxidative stress promotes the onset and development of diabetes mellitus by directly decreasing insulin sensitivity and causing direct cytotoxicity to the pancreatic insulin-producing β-cells [Maiese *et al.*, 2007]. The generated ROS penetrates through the cell membranes and reacts with the membrane phospholipids through the process of lipid peroxidation as well as reacts with the mitochondrial DNA to distrupt the mitochondrial respiratory machinery (mitochondrial electron transport) which is regulated by NADPH ubiquinone oxidoreductase and ubiquinone-cytochrome c reductase systems [Maiese *et al.*, 2007].

Oxidative stress is known to depress the mitochondrial oxidoreductase and citrate synthase activities resulting in significant reductions in mitochondrial oxidative and phosphorylation activities as well as reduces the levels of mitochondrial proteins and mitochondrial DNA in adipocytes, particularly in type 2 diabetes mellitus (Petersen et al., 2003). Oxidative stress has been shown to trigger the opening of the mitochondrial membrane permeability transition pore which results in a significant depletion of mitochondrial NAD+ stores and subsequently apoptotic cell injury (Maiese et al., 2007). In the pancreatic tissues, these cellular events result in depletion of the β-cells population, insulin deficiency while in the skeletal muscle, it manifests as insulin resistance.

Oxidative stress is also known to modify a number of cellular signalling pathways that can results in insulin resistance. For example, a significant increase in muscle protein carbonyl content (often used as a reliable biological marker of oxidative stress) and elevated levels of malondialdehyde and 4-hydrononenal (as reliable indicators of lipid peroxidation) have been implicated in the aetiology of insulin resistance diabetes mellitus [Haber *et al.*, 2003].

3. Glucose-insulin regulatory system modeling and simulation of OGTT blood glucose concentration dynamics to obtain indices for diabetes risk and detection

This section deals with the bioengineering modelling of the glucose-insulin regulatory system and the OGTT blood glucose dynamics data, for more reliable detection of diabetes as well as designation of risk to diabetes.

The conventional way of diagnosing diabetes is based on designation of specific values of fasting plasma glucose equal or greater than 126 mg/dl (7.0 mmol/l), and (ii) 2-hour plasma glucose concentration equal or greater than 200 mg/dl (11.1 mmol/l) during OGTT. Instead of this rigid approach, we are proposing that for more reliable monitoring and diagnosis of diabetes, it is more relevant to mathematically characterise the trend of blood glucose concentration rise and decline after an oral intake of 75 g glucose load in OGTT. Hence, we provide the bioengineering analysis of the Glucose-insulin regulatory system and glucose response data, leading to the formulation of a novel nondimensional diabetes index for diagnosis of diabetic patients as well as of those who are at risk of becoming diabetic.

So, in this section, we present the Glucose-Insulin Regulatory System (GIRS) modeling in the form of governing differential equations, and converge to the equation representing blood glucose response to glucose infusion rate. This equation forms the basis of modeling of the Oral Glucose Tolerance Test (OGTT). We then demonstrate how this OGTT model equation's solutions can simulate the OGTT data, to evaluate the model parameters distinguishing diabetes subjects from normal subjects. The climax to this section is the formulation of the Non-dimensional Diabetes Index (DBI), involving combination of the model parameters into just "one number" by which we can reliably detect diabetes. In fact, by determining the range of values of DBI for a big patient population, we can even detect "patients at risk of being diabetic".

3.1 Differential equation model of the glucose-insulin system

With reference to the Blood Glucose-Insulin Control System (depicted in Fig. 1), the corresponding first-order differential equations of the insulin and glucose regulatory sub-systems are given by equations (1) and (2) [Dittakavi et al., 2001].

$$x' = p - ax - \beta y \tag{1}$$

$$y' = q - \gamma x - \delta y \tag{2}$$

where x' and y' denote the first time-derivatives of x and y, x: insulin output, y: glucose output, p: insulin input, q: glucose input, for unit blood-glucose compartment volume (V). In these equations, the glucose-insulin model system parameters (regulatory coefficients) are α, β, γ, δ.

These coefficients, when multiplied by the blood-glucose compartment volume V (which is proportional to the body mass) denote, respectively,

• the sensitivity of insulinase activity to elevated insulin concentration (αV),

- the sensitivity of pancreatic insulin output to elevated glucose concentration (βV),
- the combined sensitivity of liver glycogen storage and tissue glucose utilization to elevated insulin concentration (γV), and
- the combined sensitivity of liver glycogen storage and tissue glucose utilisation to elevated glucose concentration (δV).

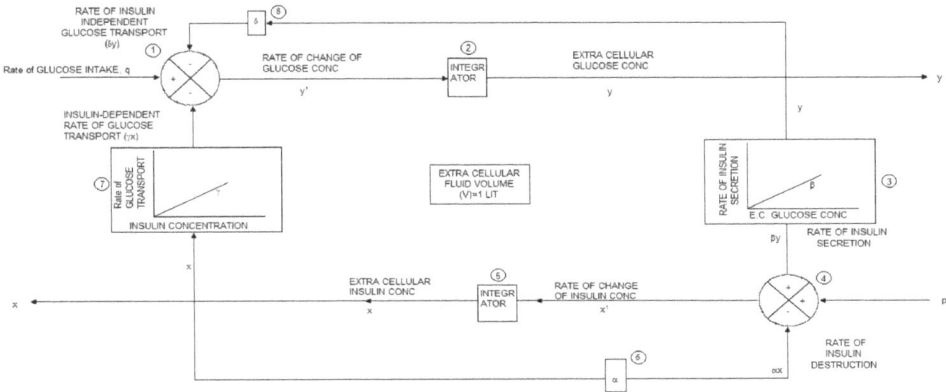

Fig. 1. Physiological model of the Blood Glucose Control system (represented by equations 1 and 2).

From equations (1) and (2), the differential equation model in glucose concentration (y) for insulin infusion rate ($p = 0$) and glucose in flow rate (q), is obtained as

$$y'' + y'(\alpha + \beta) + y(\alpha\delta + \beta\gamma) = q' + \alpha q \tag{3}$$

where y' and y'' denote first and second time derivatives of y.
The transfer-function corresponding to Eqn. (3) is obtained by taking the Laplace transforms on both sides (assuming the initial conditions to be zero). Thereby, we obtain (for glucose response)

$$Y(s)/Q(s) = \frac{(s+\alpha)}{s^2 + s(\alpha+\delta) + (\alpha\delta + \beta\gamma)} = G(s) \tag{4}$$

3.2 Model analysis to simulate Oral Glucose Tolerance Test (OGTT)

The OGTT model-simulation response curve is considered to be the result of giving an impulse glucose dose (of 4 gm of glucose/liter of blood-pool volume) to the combined system consisting of GI tract and blood glucose concentration (BGCS). Now, we can put down the transfer-function (TF) of the gastro-intestinal (GI) tract to be 1/ ($s + \alpha$), because the intestinal glucose-concentration variation is an exponential decay, and the exponential parameter value is close to that of the parameter α. When we multiply this GI tract TF [1/($s + \alpha$)] by the TF of the blood-pool glucose-metabolism given by Eqn. (4), and put Q(S) = 'G' gm of glucose per litre of blood-pool volume per hour, we get

$$Y(s) = G / \left\{ s^2 + s(\alpha + \delta) + (\alpha\delta + \beta\gamma) \right\} \tag{5}$$

The corresponding governing differential equation is now:

$$y'' + 2Ay' + \omega_n^2 y = G\delta(t)$$
$$\text{or}$$
$$y'' + \lambda T_d y' + \lambda y = G\delta(t) \tag{6}$$

wherein ωn ($= \lambda^{1/2}$) is the natural frequency of the system, A is the attenuation or damping constant of the system, $\lambda = 2A / Td = \omega_n^2$, and $\omega = (\omega_n^2 - A^2)^{1/2}$ is the angular frequency of damped oscillation of the system.

The solution of Eq. (6), for an under-damped response (corresponding to that of normal subjects, represented by the lower curve in Fig. 2) is given by

$$y(t) = (G/\omega)e^{-At} \sin \omega t, \tag{7}$$

where in ω (or ω_d) $= (\omega_n^2 - A^2)^{1/2}$.

The solution for over-damped response (corresponding to that of diabetic subjects, represented by the upper curve in Fig 2) is given by:

$$y(t) = (G/\omega)e^{-At} \sinh \omega t \tag{8}$$

where in ω (or ω_d) $= (A^2 - \omega_n^2)^{1/2}$

The solution for a critically-damped response (in which $A = \omega_n$), which applies to subjects at risk of becoming diabetic (whose blood glucose response curve would lie between the two curves of normal and diabetic subjects), is given by:

$$y(t) = G t e^{-At}; \tag{9}$$

for $\omega_n^2 = A^2 = \lambda$, and derivative-time period $T_d = \dfrac{2A}{\lambda} = \dfrac{2A}{\omega_n^2}$

These solutions are employed to simulate the clinical data, and to therefore evaluate the model-system parameters A and ω (or λ and T_d), to not only differentially-diagnose diabetes subjects as well as sbut also to characterize resistance-to-insulin.

Now, we can employ equations (7) and (8) to simulate the OGTT data shown in Fig. 2 to obtain the value of parameters: (i) $\lambda = 2.6\text{hr}^{-2}$, $Td = 1.08$ hr, for the normal subject, and (ii) $\lambda = 0.27\text{hr}^{-2}$ and $Td = 6.08$ hr, for the diabetic subject [Ghista, 2004].

We now formulate the Non-dimensional Diabetes Index (DBI), as

$$DBI = AT_d = \dfrac{2A^2}{\lambda} = \dfrac{2A^2}{\omega_n^2} \tag{10}$$

The value of DBI for the normal subject is 1.3, whereas for the diabetic subject it is 4.9. We have further found (in our initial clinical tests) that DBI for normal subjects is less than 1.6, while the DBI for diabetic patient is greater than 4.5. Hence a DBI value of 2-4 can suggest that the subject is at risk of becoming diabetic. This is a testimony of how well we have simulated the OGTT by our BME model and employed this DBI to diagnose diabetes.

$$y(t) = \frac{G}{\omega} e^{-ck} \sinh \omega t$$

$$(AT_d = 4.9)$$

$$(A = 0.808 \ hr^{-1}, \lambda = 0.2657 \ hr^{-2}$$

$$T_d = 6.08 \ hr,$$

$$G = 2.9464 \ gL^{-1}hr^{-1})$$

$$y(t) = \frac{G}{\omega} e^{-ck} \sinh \omega t$$

$$(AT_d = 1.3)$$

$$(A = 1.4 \ hr^{-1}, \lambda = 2.6 \ hr^{-2}$$

$$T_d = 1.08 \ hr,$$

$$G = 1.04 \ gL^{-1}hr^{-1})$$

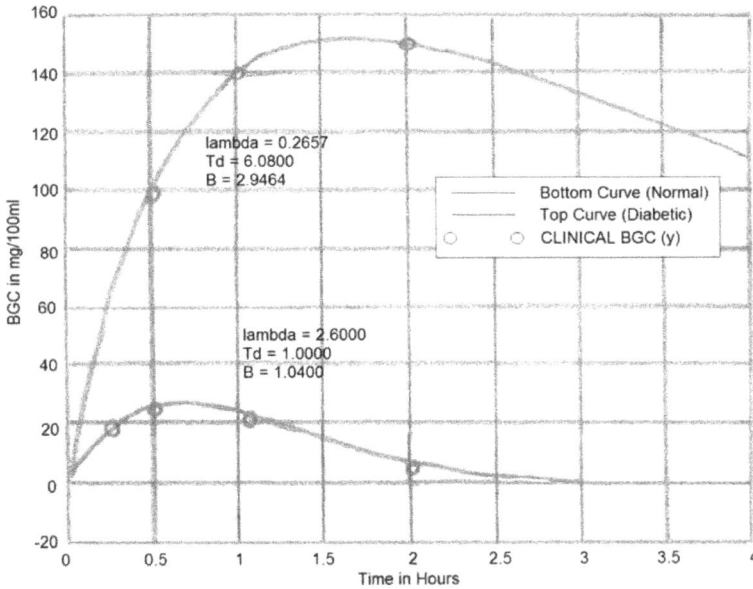

Fig. 2. OGTT Response Curve [Ghista, 2004], showing the glucose concentration responses of normal and diabetic subject.

4. Biomedical signal processing and image processing techniques for diabetes analysis

This section presents different signal and image processing methods that are used to evaluate the effect of diabetes on different organs.

4.1 Analysis of the heart rate variability signal

Heart rate variability (HRV) decreases in patients with diabetes [Acharya et al., 2006; Acharya et al., 2011b; Faust et al., 2011]. This variability can be analyzed in the time domain, frequency domain, and by using non-linear methods. Fig. 3 shows typical HRV signals of normal and diabetes subjects. Visually, it is difficult to notice the variability in these two signals. Hence, analysis in time domain and frequency domain with the use of non-linear methods is necessary. These methods are explained in this section.

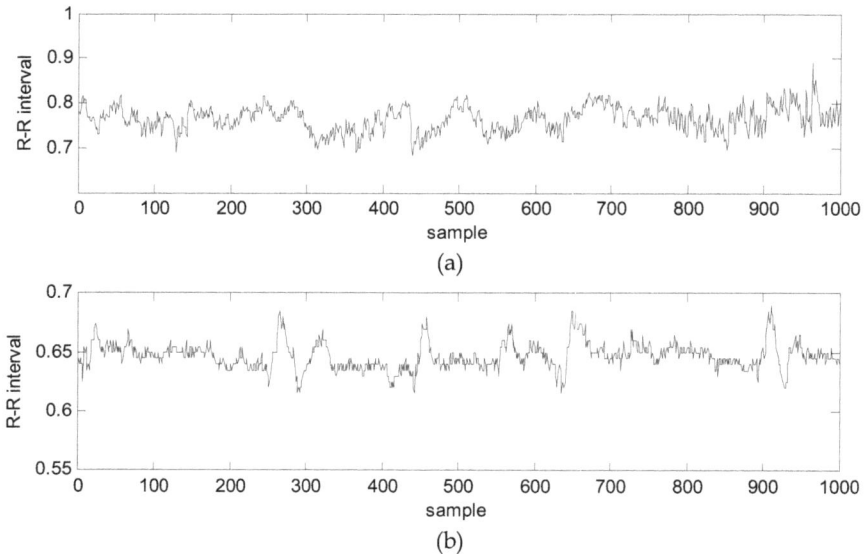

Fig. 3. Typical heart rate signals; (a) normal (b) diabetes.

4.1.1 Time domain analysis

The time-and frequency-domain measures of HRV were analyzed by the Task Force of the European Society of Cardiology [Task Force, 1996]. Several time domain parameters are calculated from the original R-R interval: mean R-R interval, standard deviation of the NN intervals (SDNN), standard deviation of differences between adjacent RR (NN) intervals (SDSD), Standard Error, or Standard Error of the Mean (SENN), which is an estimate of the standard deviation of the sampling distribution of means based on the data, number of successive difference of intervals which differ by more than 50 msec expressed as a percentage of the total number of ECG cycles analyzed (pNN50%).

The HRV triangular index (TINN) is the integral of the density distribution (i.e. the number of all NN intervals) divided by the maximum of the density distribution. Thus, six standard measures namely Mean RR, SDNN, SENN, SDSD, pNN50% and TINN were studied.

4.1.2 Frequency domain analysis

Spectral analysis of HRV signal results in three main components: high frequency (HF) component, low frequency (LF) component, and very low frequency (VLF) component [Task Force, 1996]. The influence of the vagus nerve in modulating the sinoatrial node is indicated by the HF component (0.15Hz -40Hz) of the spectrum. The LF component (0.04Hz-.155 Hz) indicates the sympathetic effects on the heart. The VLF component (0.003Hz -.04 Hz) explains many details of the heart, chemoreceptors, thermareceptors, and renin-angiotensin system [Task Force, 1996; Kamath et al., 1987; Van der Akker et al., 1983].

Fig. 4 shows a typical power spectral density (PSD) distribution of the heart rate signals obtained from a normal subject (Fig. 4-a) and a diabetes patient (Fig. 4-b). The beat to beat variation is greater in the normal heart rate signal compared to the diabetes heart rate signal. Hence, the power spectral density is more predominant in HF in the normal subject[Faust et al., 2011].

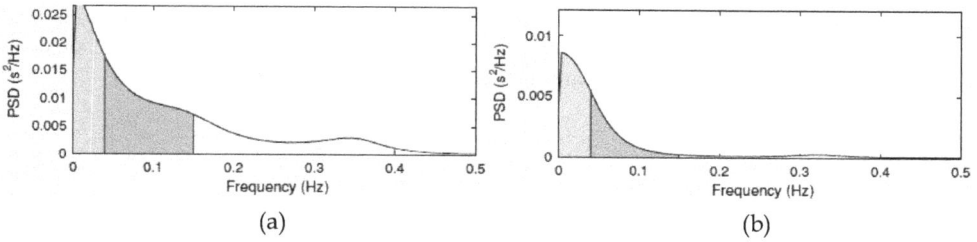

(a)   (b)

Fig. 4. Typical power spectral density of heart rate signal (a) normal (b) diabetes subject. The PSD of normal heart rate signal has LF, HF components. The diabetic heart rate signal, however, does not have HF components due to lower variability in the heart rate signal [Acharya et al., 2011b].

4.1.3 Non-linear parametric analysis of heart rate signals

Various non-linear parameters can be used to analyze the diabetes heart rate signals. They are Approximate Entropy (ApEn), Correlation Dimension (CD), Largest Lyapunov Exponent (LLE), The Hurst exponent (H), Recurrence plot (RP), and Fractal Dimension (FD).

The Approximate Entropy *ApEn* measures regularity of the time series. The method proposed by Pincus et al can be used to evaluate the ApEn [Pincus, 1991]. For the data points $x(1), x(2), ..., x(N)$, with an embedding dimension m, the ApEn or *APEN* is given by:

$$APEN(m, r, N) = \frac{1}{N - m + 1} \sum_{i=1}^{N - m + 1} \log C_i^m(r) - \frac{1}{N - m} \sum_{i=1}^{N - m} \log C_i^{m+1}(r) \quad (11)$$

where $C_i^m(r) = \frac{1}{N - m + 1} \sum_{j=1}^{N-m+1} \Theta(r - \|\mathbf{x}_i - \mathbf{x}_j\|)$ is the correlation integral. For this study, m is

set to 2, and r is chosen as 0.15 times the standard deviation of the original data sequence, and N is the total number of data points.

The *Correlation dimension (CD)* is a quantitative measure of the informational complexity of the heart rate signal [Grassberger, 1983]. Some unique ranges of CD for different cardiac diseases have been proposed by Acharya et al. [2007]. The formula for CD involves the correlation function $C(r)$, which is the probability that two arbitrary points on the orbit are closer together than r. This is done by calculating the separation between every pair of N data points and sorting them into bins of width dr proportionate to r. The correlation dimension can be calculated by using the distances between each pair of points in the set of N number of points, $a(i, j) = |X_i - X_j|$

$$C(r) = \frac{1}{N^2} \times \left(Number\ of\ pairs\ of\ (i, j)\ with\ a(i, j) < r \right) \quad (12)$$

Correlation dimension (CD) is given by:

$$CD = \lim_{r \to 0} \frac{\log(C(r))}{\log(r)} \quad (13)$$

The Largest Lyapunov Exponent (LLE) measures the predictability of the system and determines sensitivity of the system to initial conditions [Rosenstien et al., 1993]. A positive LLE indicates chaos. The LLE is estimated by using a least squares fit to "average "line, and is given by:

$$y(n) = \frac{1}{\Delta t}\langle \ln(d_i(n))\rangle \tag{14}$$

where $d_i(n)$ is the distance between i^{th} phase-space point and its nearest neighbor at n^{th} time step, and $\langle \cdot \rangle$ denotes the average overall phase space points.

The Hurst Exponent (HE) indicates the self-similarity and correlation properties of heart rate signal. The *HE* has been defined and proposed by Dangel et al [Dangel et al., 1999]. Unique range of H values has been proposed by Acharya et al, for various cardiac states [Acharya et al., 2007].

$$H = \log(R/S)/\log(T) \tag{15}$$

where T is the duration of the sample of data and R/S is the corresponding value of rescaled range. An *HE* value of 0.5 indicates the presence of a random walk, *HE* < 0.5 depicts anti persistence, and *HE* > 0.5 indicates the persistence in the signal.

The *Recurrence plot (RP)* can be used to unearth the non-stationarity in the heart rate signals [Acharya et al., 2006], and was originally introduced by Eckmann et al. [Eckmann et al., 1987].

A *Fractal* is a set of points which, when looked at smaller scales, looks similar to the whole group [Madelbrot, 1983]. The Fractal Dimension (FD) determines the complexity of the time series. FD has been used in heart rate analysis to recognise and differentiate specific states of physiologic functions [Acharya et al., 2007].

The heart rate signal is a non-linear and non-stationary signal. The hidden intricacies of the signal can be easily extracted using non-linear analysis methods. The heart rate variation is more random in normal subjects as compared to the diabetes subjects. Hence, most of these non-linear parameters may show distinct values for normal and diabetes subjects. These clinically significant non-linear parameters can be fed into the classifiers as features for automatic classification. Moreover, these non-linear parameters can be combined in the form of an integrated index [Ghista, 2004; 2009a; 2009b]. Such an index may have unique range of values for normal and diabetes classes. Hence, one can diagnose normal and diabetes subjects by just using one index value without the need for automatic classifiers.

4.2 Image processing of digital fundus images in diabetic retinopathy

Diabetic retinopathy is an important complication of diabetes. As the diabetes retinopathy progresses, the number of blood vessels varies, and the exudates appear in the advanced DR stages [Yun et al., 2008; Acharya et al., 2011a]. Different image processing techniques have been used to extract blood vessels and exudates in DR subjects, and these techniques are explained in this section. Moreover, techniques for plantar pressure images analysis, which have proved to be useful in detecting diabetic neuropathy conditions, are also been presented in this section.

4.2.1 Retinal blood vessels detection

The detailed steps involved in the blood vessel detection are shown in Fig. 5 [Nayak et al., 2008; Acharya et al., 2011a; Acharya et al. 2009; Acharya et al., 2011b]. The green component of the RGB (Red, Green Blue) blood vessel image is considered for this study. The border of the image is obtained by applying an edge detection algorithm on the inverted green component of the image. Morphological operation is performed by using a disk shaped structuring element (SE) for blood vessels detection. Adaptive histogram equalization is then performed on these images to enhance the image, and subsequently, morphological opening operation is performed using a ball structuring element. Thresholding is carried out on the resulting image followed by the median filtering to obtain the boundary of the image. The small holes are then filled and the boundary is removed. Finally, the image with only blood vessels is obtained (Fig. 7) [Acharya et al., 2011b]. It can be seen from Fig. 7(a) that the number of blood vessels is different in the normal and the proliferative diabetes retinopathy (PDR) classes.

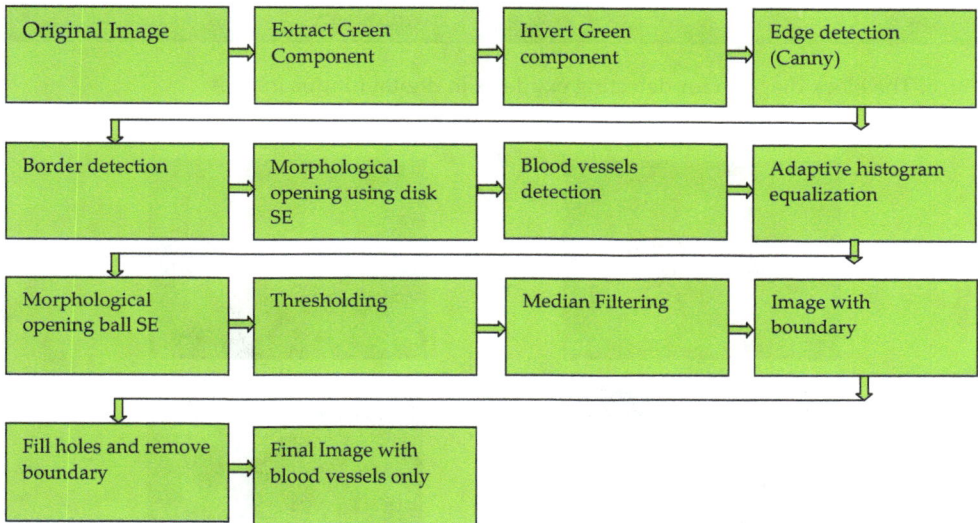

Fig. 5. The block diagram for detecting retinal blood vessels.

4.2.2 Exudates detection in digital fundus images

Fig. 6 shows the block diagram of the exudates extraction in digital fundus images [Acharya et al., 2008; Nayak et al., 2008; Acharya et al., 2011a; Acharya et al., 2011b]. The green component of the original image is extracted and subjected to the morphological closing operation by using octagonal shaped structuring element. Then, the resulting image is subjected to thresholding, and morphological closing operation is carried out by using disk shaped SE.

The edges are detected by using the Canny method. Subsequently, an 80x80 region of interest (ROI) is considered to remove the optic disc, and then the border of the image is also removed. Finally, by performing morphological erosion operation with disk shaped SE of size 3, the final image with only exudates is obtained (Fig. 7) [Acharya et al., 2011b]. It can be seen from the Fig. 7(b) that there are no exudates in the normal image, while the PDR image has exudates.

Original Image	→	Extract Green Component	→	Morphological closing using octagon shaped SE	→	Column wise neighborhood operation

Thresholding	→	Morphological closing using disk SE	→	Edge detection (Canny)	→	Region of interest (ROI)

Remove optic disc	→	Remove border	→	Morphological operation using disk using SE	→	Final Image with only exudates

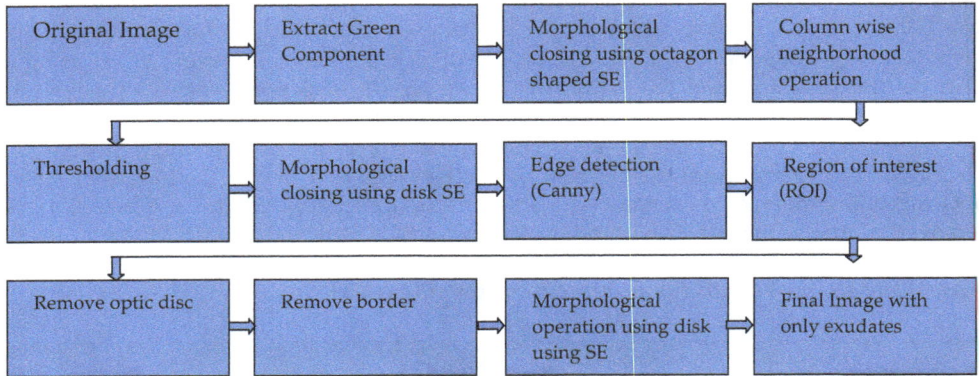

Fig. 6. The block diagram for detecting exudates in digital fundus images.

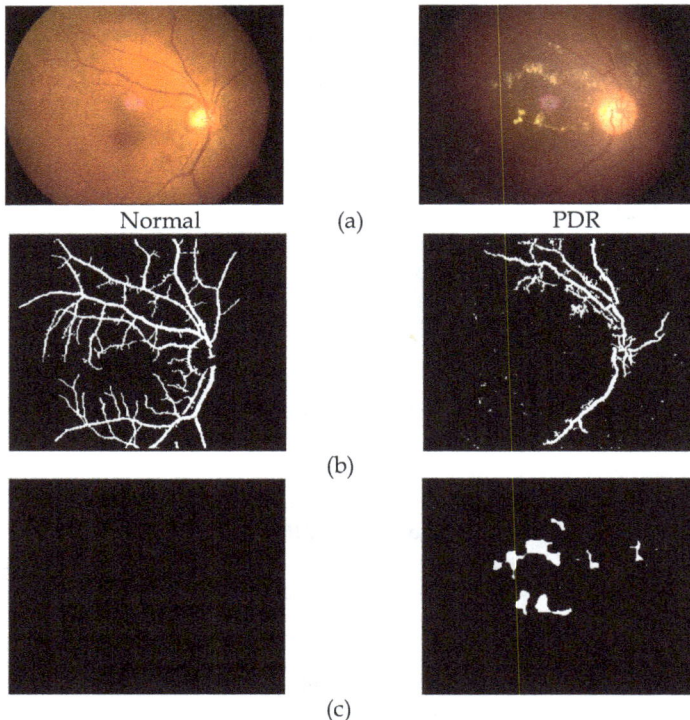

Normal   (a)   PDR

(b)

(c)

Fig. 7. Results of blood vessel detection and exudate detection from normal and PDR images. (a) Original normal and PDR images (b) Results of blood vessel detection (c) Results of exudate detection. The number of blood vessels are different for normal and PDR images, and exudates are absent in the normal fundus image.

4.3 Plantar pressure distribution image analysis

Fig. 8 shows the plantar pressure distribution images of normal subjects, and subjects with diabetes type II without and with neuropathy. It can be seen from the figure that the pressure distribution is different for normal, diabetes without and with neuropathy subjects [Acharya et al., 2008; Acharya et al., 2011b]. This difference can be further analyzed using Fourier transform and discrete wavelet transform (DWT).

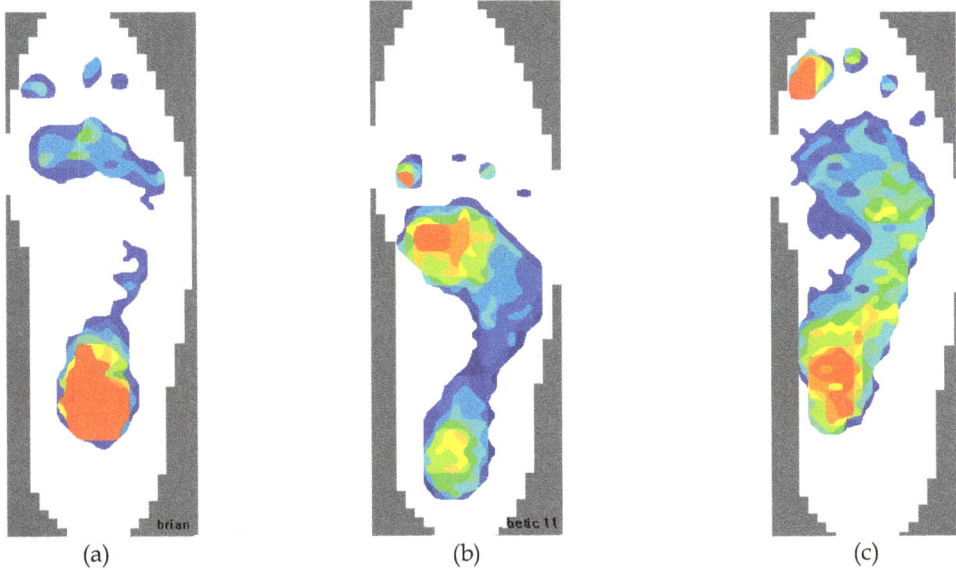

(a)   (b)   (c)

Fig. 8. Static pedobarograph images of (a) the normal foot, (b) a diabetic foot with neuropathy, and (c) a diabetic foot without neuropathy.

The important feature used to diagnose the normal, diabetes type II with and without neuropathy classes is the power ratio (PR) that is obtained using the Fourier transform [Rahman et al., 2006]. This method is clearly explained below.

Fourier domain analysis: The Fourier spectrum $F(u,v)$ of each region of the image can be obtained by using the below equation (16) [Cavanagh et al., 1991]. In this equation, M and N represent the numbers of rows and columns of the image. The power ratio (PR) is the ratio of the high frequency power (HFP) to the total power (TP). The Fourier spectrum is given by

$$F(u,v) = \frac{1}{MN} \sum_{x=0}^{M-1} \sum_{y=0}^{N-1} f(x,y).e^{-j2\Pi\left(\frac{ux}{M}+\frac{vy}{N}\right)} \tag{16}$$

where x, y, u and v are the variables.

$F(0,0)$ is the DC component of the image in the frequency domain and is the sum of all the pixels of an image in spatial domain [Cavanagh et al., 1991]. The *total power (TP)* of the image is given by

$$TP = \left\{ \sum\sum |F(u,v)|^2 \right\} - |F(0,0)|^2 \tag{17}$$

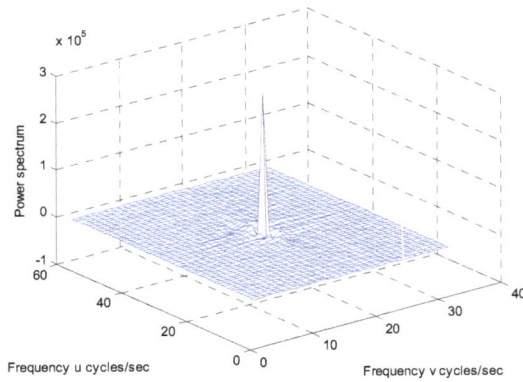

(a)

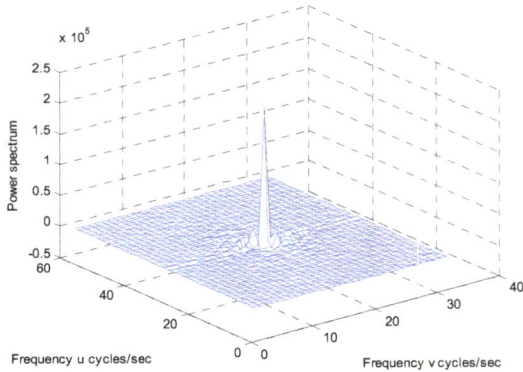

(b)

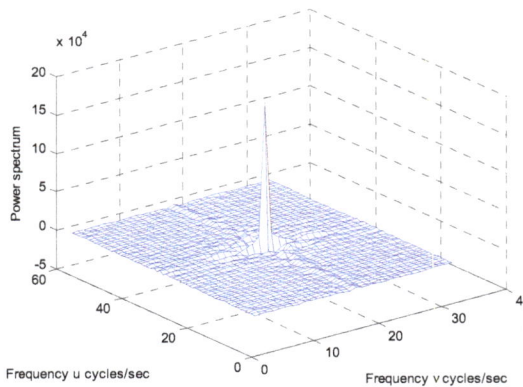

(c)

Fig. 9. Typical power spectra after deleting the DC component from region 6 of the left foot for (a) normal subject (b) diabetes subject without neuropathy (c) diabetes subject with neuropathy[Acharya et al., 2011b].

The low frequency and high frequency components are separated by S_0, which is given by

$$
S_o = \begin{cases} \dfrac{M}{4} & if\ M \leq N \\ \dfrac{N}{4} & if\ M < N \end{cases}
\tag{18}
$$

$$
LFP = \left\{ \sum_{S(u,v)=0}^{S_o} |F(u,v)|^2 \right\} - |F(0,0)|^2
\tag{19}
$$

$$
HFP = TP - LFP
\tag{20}
$$

$$
PR = \left(\frac{HFP}{TP} \right) x100
\tag{21}
$$

where LFP, HFP, and PR, denote the low frequency power, high frequency power, and the power ratio, respectively.

Fig. 9 shows the typical power spectra obtained for a normal subject, having diabetes without neuropathy, and subject having diabetes with neuropathy. It is a 3D figure, with u and v frequencies corresponding to row and column. The Y-axis indicates the power. The power spectrum of normal class has a peak in the centre and very small peaks around it. In the case of diabetes without neuropathy, the adjacent peaks are slightly larger; in the case of diabetes with neuropathy, there are dominating peaks on four sides. These plots are unique and depict variation of power spectrum. The PR values extracted from various regions of the plantar image are shown in Table 1[Acharya et al., 2011b].

Type	Control subjects (CS)	Diabetic control (DC)	Neuropathic (N)	p-value
Region 1	12.80 ± 3.49	9.562 ± 2.25	17.657 ± 3.27	<0.0001
Region 2	11.865 ± 2.13	9.678 ± 2.58	14.453 ± 2.31	<0.0001
Region 5	13.769 ± 3.31	9.512 ± 2.530	14.542 ± 2.69	<0.0001
Region 6	10.179 ± 2.09	9.697 ± 1.23	12.35 ± 2.19	<0.0001
Region 7	9.28 ± 6.03	8.67 ± 3.30	11.56 ± 1.45	<0.0001

Table 1. Power ratio values for the various regions of the plantar pressure images obtained from the three classes.

The PR is the ratio of HF power to the total power. This value is higher for diabetes subjects with neuropathy when compared to the normal and diabetes without neuropathy subjects for regions 1, 2, 5, 6, and 7 (Table 1). These ranges are unique and clinically significant (p<0.0001). These PR features can be used to diagnose the three classes automatically using classifiers.

Likewise, DWT coefficients have also been used to identify the normal, diabetes type II with and without neuropathy classes [Acharya et al., 2008; Acharya et al., 2011b].

5. Diabetic autonomic neuropathy diagnosis from HRV power spectrum plots

The RR interval files are processed to get HRV and HRVPS [Desai, K.D et al., 2011]. The sampling frequency used to get HRV form RR file is 2Hz. The power spectrum plots depict power in (BPM)^2 versus Frequency (in Hertz). The auto regression statistics gives display of the following parameters:

Power under Low frequency range: frequency range from 000 to 0.04Hz

Power under Mid frequency range: frequency range from 0.04 to 0.15Hz

Power under High frequency range: frequency range from 0.15 to 0.40Hz

Sympatho/Vagal balance ratio: ratio of mid to high frequency powers

The Sympatho-Vagal ratio is found in the different frequency characteristics of the parasympathetic and sympathetic influences on heart rate. The HRVPS plots (for the supine, standing and deep breathing modes) are plotted with time-scale up to 150 seconds and heart rate scale in the range of 40 bpm to 140 bpm.

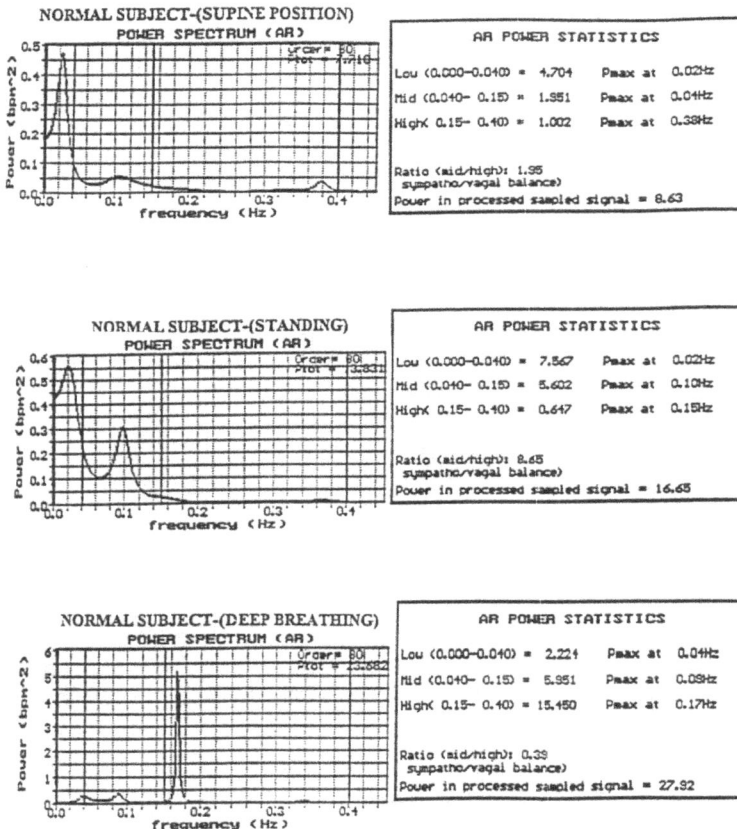

Fig. 10. HRVPS plots of a normal subject in supine, standing and deep breathing modes. The power statistics on the right side show the power in low, medium and high frequency bands. There is an increase in the mid frequency power in standing position and in high frequency power in deep breathing mode.

Figure 10 displays the HRVPS of a typical normal subject in supine, standing and deep breathing modes. In this figure, the power statistics show the power in low, medium and high frequency bands. It can be noted that there is an increase in the mid-frequency power in standing position and in the high-frequency power in deep breathing mode. Figure 11 depicts the HRVPS plot of a typical diabetic subject in supine, standing and deep-breathing modes. Now, it can be seen that there is a decrease in mid-frequency power and in high-frequency power in deep-breathing mode compared to corresponding power levels of a normal subject (in Figure 10) [Desai, K.D et al., 2011].

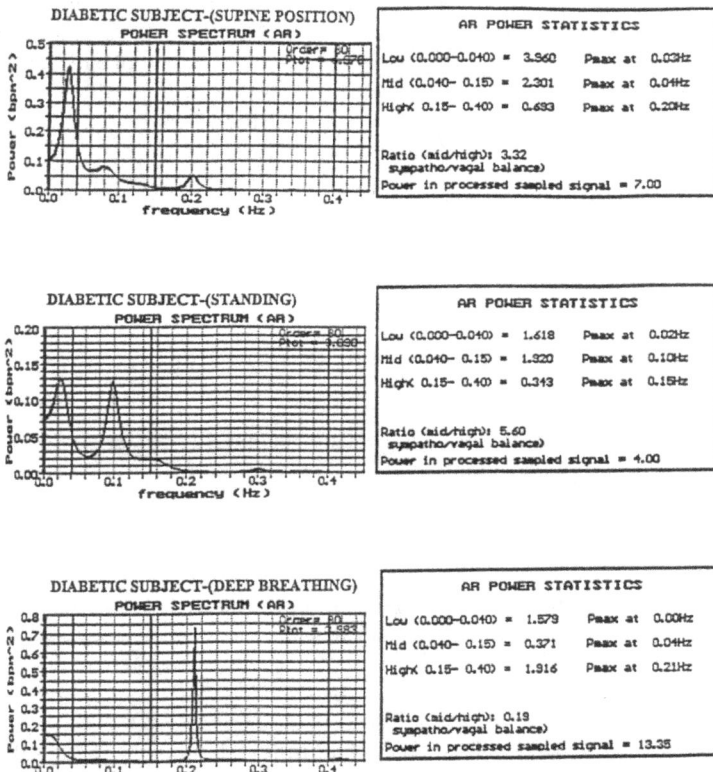

DIABETIC SUBJECT-(SUPINE POSITION)
POWER SPECTRUM (AR)

AR POWER STATISTICS

Low (0.000-0.040) = 3.560   Pmax at 0.03Hz
Mid (0.040- 0.15) = 2.301   Pmax at 0.04Hz
High 0.15- 0.40) = 0.693   Pmax at 0.20Hz

Ratio (mid/high): 3.32
sympatho/vagal balance)
Power in processed sampled signal = 7.00

DIABETIC SUBJECT-(STANDING)
POWER SPECTRUM (AR)

AR POWER STATISTICS

Low (0.000-0.040) = 1.618   Pmax at 0.02Hz
Mid (0.040- 0.15) = 1.920   Pmax at 0.10Hz
High 0.15- 0.40) = 0.343   Pmax at 0.15Hz

Ratio (mid/high): 5.60
sympatho/vagal balance)
Power in processed sampled signal = 4.00

DIABETIC SUBJECT-(DEEP BREATHING)
POWER SPECTRUM (AR)

AR POWER STATISTICS

Low (0.000-0.040) = 1.579   Pmax at 0.00Hz
Mid (0.040- 0.15) = 0.371   Pmax at 0.04Hz
High 0.15- 0.40) = 1.916   Pmax at 0.21Hz

Ratio (mid/high): 0.19
sympatho/vagal balance)
Power in processed sampled signal = 13.35

Fig. 11. HRVPS plots of a diabetic subject in supine, standing and deep breathing modes. The power statistics on the right side show the power in low, medium and high frequency bands. There is a decrease in the mid frequency power in standing position and in high frequency power in deep breathing mode compared to corresponding power levels of normal subject as shown in Fig 10a[Desai, K.D et al., 2011].

5.1 Diagnostic indices (based on HRVPS)

The analysis of HRV power spectra is commonly focused on the power in different frequency bands. In particular, the power in the high-frequency range reflects the fast parasympathetic never activity [Fallen et al., 1985], and the power in the mid-frequency range reflects both parasympathetic and sympathetic never activity [Akselrod, S., et al., 1981].

The ratio of the mid-frequency range power to the high-frequency range power is sometimes used as a relative index of the sympatho/vagal balance [Bianchi et al., 1990]. The high frequency range power ratio between supine and standing position is used as a parasympathetic index [Fallen et al., 1985]. The same ratio is used to study sympathetic function in standing position in the mid frequency range [Fallen et al., 1985]. Sympathetic vasomotor nerve function is quantified by the baro receptor oscillation frequency (i.e., the mid-peak frequency) in the HRVPS [Kamath et al., 1987].

In our study, autonomic function indices are defined in terms of spectral power indices and HRV period (or frequency) shift indices. The mid and high-frequency ranges are considered for defining indices. The diagnostic indices are based on parameters measured from the HRVPS. The values of diagnostic indices for the three groups of subjects have shown significant difference and can provide rational basis for selecting prognostic therapy before a diabetic patient develops cardiac arrhythmic complications.

The **diagnosis indices** are defined as follows:

$$I_1 = \text{Relative sympathetic} - \text{to} - \text{vagal balance index} = P_2 / P3 , \tag{22}$$

where
(P_2)=area under HRVPS spectral plot between 0.04 Hz and 0.15 Hz
(P3)=area under HRVPS spectral plot between 0.15 and 0.4 Hz.

$$I_2 = \text{Orthostatic Stress Index} = (P2sta - P2sup) / P2sup , \tag{23}$$

where
(P2)=area under HRVPS spectral plot between 0.04 Hz and 0.15 Hz, and subscripts 'sta' and 'sup' refer to standing supine positions.

$$I_3 = \text{Sympatho} - \text{Vagal Integrity Index} = \sum (\text{Hrmax} - \text{Hrmin}) / n , \tag{24}$$

where
HRmax = Local maximum heart rate (beats per minute) during one breathing cycle.
HRmin = Local minimum heart rate (beats per minute) prior to local maximum in the same breathing cycle.
n=number of breathing cycles.

$$I_{4std} = \text{Sympathetic HRV} - \text{Spectral Frequency Shift Index(standing)} = (F2_{std} - 0.1) / 0.1, \tag{25}$$
where F_2 = Frequency of the Baroreceptor reflex peak

$$I_{5sup} = \text{Sympathetic HRV} - \text{Spectral Frequency Shift Index(supine)} = (F2_{supine} - 0.1) / 0.1 \tag{26}$$

$$I_6 = \text{Respiratory Stress Index} = \left(P_{3db} - P_{3supine} \right), \tag{27}$$

where

$$\left(P_{2supine} \right) = \text{area under HRVPS spectral plot}$$
$$\text{frequency 0.04 Hz and 0.15 Hz in supine position .} \tag{28}$$

$$\left(P_{3db}\right) = \text{area under HRVPS spectral plot between} \tag{29}$$
frequency 0.15 Hz and 0.4 Hz in deep breathing .

$$\left(P_{3supine}\right) = \text{area under HRVPS spectral plot between} \tag{30}$$
frequency 0.15 Hz and 0.4 Hz in supine position .

The following Table 2 shows the calculated indices, for a sample normal subject, obtained from the HRVPS parameters.

Autonomic	Index Formula	Index Value
Relative Sympathetic-to-Vagal Balance Index	$I_1 = P_2/P_3$	$I_{1(sup)} = 2.06$ $I_{1(st)} = 7.82$ $I_{1(db)} = 0.32$
Orthostatic Stress Index	$I_2 = (P_{2sta} - P_{2sup})/P_{2sup}$	$I_2 = 1.64$
Sympatho-Vagal Integrity	$I_3 = \sum (HR_{max} - HR_{min}) / n$	$I_3 = 6.62$
HRV-Freq-Shift Index (standing)	$I_4 = (F_{2std} - 0.1) / 0.1$	$I_4 = -0.05$
HRV-Freq-Shift Index (supine)	$I_{5sup} = (F_{2supine} - 0.1) / 0.1$	$I_5 = -0.10$
Respirator Stress Index	$I_6 = (P_{3db} - P_{3supine}) / P_{3supine}$	$I_6 = 16.70$

Table 2. Computed Indices for a typical normal subject.

In this Table 2, $I_{1(sp)} = P_2/P_3$ (equation 22) in supine position; $I_{2(st)} = P_2/P_3$ (equation 22) in standing position; $I_{2(db)} = P_2/P_3$ (equation 22) in deep – breathing mode.

Now, diagnosis based on six indices makes it somewhat difficult to track in a patient as regards how much each index varies from its normal value, for making an appropriate diagnosis. So now we will adopt the novel approach, as in Ghista [Ghsta, 2004; 2009a], of formulating an index by combining the parameters in such a way that the index values are distinctly different for normal subjects, diabetics, and diabetics with ischemic heart disease. Hence, we are proposing that, from a diagnostic and classification viewpoint, it would be more convenient to formulate a DAN Integrated Index (DAN-IID) [Desai, K.D et al., 2011], as :

$$DAN - IID = \left[\left(I1, st\right) + \left(I1, db\right) + \left(I2\right) + \left(I3\right) + \left(I6\right) \right] - - \left[\left(I4\right) + \left(I5\right) \right] \tag{31}$$

5.2 Results and analysis: HRVPS of normal subjects, diabetic subjects, and diabetic subjects with ischemic heart disease

The instantaneous heart rate average (IHRav), average of difference between maximum and minimum heart rate over a cycle (ΔHrav), power and frequency measurements (P,F) measured from HRVPS are determined. There from the diagnostic indices are computed (as per equation 22-27).

Descriptive Statistics of Indices of the Three Groups

The computed indices for the three categories of subjects are displayed in the following Tables
Table 3 for normal subject group.
Table 4 for diabetic subject group.

Table 5 for IHF subject group.
Then, using the values in Tables (2), (3) and (4), the mean and standard deviation values of three groups are calculated and presented in the Table 6.

Name	$I_{1sp}(N)$	$I_{1,st}(N)$	$I_{1,db}(N)$	$I_2(N)$	$I_3(N)$	$I_4(N)$	$I_5(N)$	$I_6(N)$	DAN-IID
Ahamidm	2.08	14.57	0.51	3.78	3.26	-0.1	-0.07	16.26	38.55
Awmeah	2.07	7.83	0.33	1.64	6.62	-0.05	-0.1	16.7	33.27
Fahmia	1.88	2.67	0.36	2.04	8.56	0.13	-0.1	17.81	31.41
Fatimah	0.47	7.46	0.84	1.08	4.9	-0.57	0.07	0.24	15.02
Gitakr	1.93	3.93	2.49	2.66	9.51	0.03	-0.07	2.8	21.43
Indvai	1.63	8.25	0.26	2.01	3.86	-0.4	-0.07	16.14	30.85
Kploga	1.4	2.31	0.3	2.32	7.66	-0.12	0.5	10.67	22.88
Mattarh	1.21	9.17	2.02	2.2	3.59	-0.52	-0.07	1.53	19.1
Mohdsae	3.09	1.47	0.3	-0.65	4.43	-0.08	0.13	1.81	7.31
Mohsed	5.78	9.95	0.75	1.36	6.63	-0.02	0.1	21.87	40.48
Ramial	8.95	3.53	0.18	0.92	11.73	0.3	0.1	8.69	24.65
Sekarm	2.15	5.19	0.25	0.02	3.29	-0.05	0.3	9.88	18.38
Average	2.72 ±2.36	6.36 ±3.87	0.71 ±0.75	1.61 ±1.19	6.17 ±2.76	-0.12 ±0.25	6.0E-02 ±0.18	10.36 ±7.45	25.277 ±9.88

Table 3. Results of Indices for normal subject group.

Name	$I_{1sp}(D)$	$I_{1,st}(D)$	$I_{1,db}(D)$	$I_2(D)$	$I_3(D)$	$I_4(D)$	$I_5(D)$	$I_6(D)$	DAN-IID
Ahmedn	3.54	7.09	0.19	-0.06	2.83	-0.08	-0.57	1.88	12.58
Altmoh	2.07	4.95	0.31	-0.32	2.47	-0.57	-0.57	5.40	13.95
Aminaha	1.34	2.11	0.70	-0.28	2.94	-0.57	-0.57	1.08	7.69
Bakmh	4.25	2.45	0.52	-0.35	1.66	-0.57	-0.57	25.00	30.42
Elmamol	0.51	0.49	0.44	-0.41	2.25	-0.57	-0.57	-0.47	3.44
Fikria	3.57	12.72	0.16	0.78	1.51	-0.57	-0.57	16.89	33.2
Ghyarh	3.78	7.75	1.56	-0.36	2.81	-0.52	-0.57	0.20	13.05
Humoya	3.58	3.78	0.68	2.80	3.68	-0.35	-0.57	4.59	16.45
Kmilmo	2.84	5.30	0.59	0.11	2.38	-0.30	-0.57	1.15	10.4
Krshpr	0.85	0.59	0.13	-0.36	1.86	-0.57	-0.57	20.42	23.78
Kurubrl	1.55	1.74	3.08	0.03	1.44	-0.57	-0.57	3.78	11.21
Mahabs	1.54	5.78	1.06	0.51	6.91	-0.57	-0.57	1.28	16.68
Mohdosb	4.39	1.92	0.32	-0.76	2.01	0.03	0.01	1.32	4.77
Mohikat	2.41	0.55	0.29	-0.29	1.87	-0.57	-0.23	13.25	16.47
Muisdr	0.86	1.08	3.30	-0.30	2.98	-0.10	-0.57	0.34	8.07
Nasah	2.59	26.95	1.19	1.83	2.09	-0.32	0.57	0.44	32.25
Naya	0.75	3.51	3.92	-0.71	1.36	-0.57	-0.57	-0.58	8.64
Salmm	0.35	1.89	0.54	-0.30	0.78	-0.57	-0.57	-0.59	3.43
Average	2.26 ±1.36	5.03 ±6.31	1.05 ±1.17	8.66E-02±0.9	2.43 ±1.32	0.43 ±0.2	-.455 ±0.29	5.29 ±7.94	14.804 ±9.43

Table 4. Results of Indices for diabetic subject group.

Name	$I_{1,sp}(H)$	$I_{1,st}(H)$	$I_{1,db}(H)$	$I_2(H)$	$I_3(H)$	$I_4(H)$	$I_5(H)$	$I_6(H)$	DAN-IID
Aminase	0.45	0.74	0.99	-0.81	1.45	-0.57	-0.57	1.74	5.25
Hamamak	6.22	1.73	0.52	-0.63	1.77	-0.57	-0.57	13.10	17.63
Mayara	2.03	3.80	1.00	0.52	2.90	-0.23	-0.08	5.53	14.06
Mdshr	1.33	1.05	0.11	0.21	3.41	-0.57	-0.05	9.62	15.02
Mohmust	2.14	2.45	1.34	0.10	2.22	-0.57	-0.28	0.69	7.63
Omarsh	2.95	2.60	0.22	-0.49	2.17	-0.20	-0.02	1.79	6.51
Shamsa	0.66	1.10	2.57	-0.16	1.78	-0.57	-0.57	0.25	6.68
Tamebr	2.79	7.10	2.09	2.50	1.58	-0.57	-0.15	9.60	23.59
Average	2.32 ±1.82	2.57 ±2.09	1.1 ±0.87	0.155 ±1.05	2.16 ±0.68	-0.48 ±0.16	-.28 ±0.24	5.29 ±4.93	12.046 ±10.85

Table 5. Results of Indices for ischemic heart disease subject group.

Index	Normal (N)		Diabetic (D)		Diabetic + IHD (H)	
	Mean	Sd	Mean	Sd	Mean	Sd
I_1 (supine)	2.719	2.357	2.266	1.356	2.32	1.821
I_1 (standing)	6.361	3.864	5.036	6.312	0.155	1.49
I_1 (deep breathing)	0.715	0.755	1.053	1.166	1.107	0.869
I_2 (orthostatic stress)	1.614	1.185	0.085	0.908	0.155	1.049
I_3 (sympatho-vagal integrity)	6.195	2.736	2.435	1.323	2.16	0.681
I_4 (sym HRVPS freq shift by standing)	-0.121	0.257	-0.439	0.203	-0.481	0.165
I_{5sup} (sym HRVPS freq shift in sup)	0.06	0.185	-0.519	0.152	-0.286	0.248
I_6 (resp stress index)	10.366	7.447	5.261	7.969	5.29	4.92
		9.88	14.804	9.43	12.046	
DAN-IID	25.277					6.57

Table 6. Descriptive Statistics of indices of the three groups.

Diagnostically significant indices

In order to demonstrate the effectiveness of the diagnostic indices (I_1 to I_6) to distinguish the three groups, the diagnostically significant indices are calculated using Mann Whitney Wilicoxon Rank test (Non-Parametric Tests), and the p values(<0.05)are tabulated in Table 7 below[Desai, K.D et al., 2011].

Index	Significance Between Two Groups	P-value(<0.05)
I_1 (standing)	N & H	0.0109
I_2	N & H	0.0253
I_3	N & H	0.0004
I_4	N & H	0.0025
I_5	N & H	0.0083
I_2	N & D	0.0020
I_3	N & D	0.0000
I_4	N & D	0.0015
I_5	N & D	0.0000
I_6	N & D	0.0422
I_5	H & D	0.0105

Table 7. Diagnostically Significant Indices.

5.3 Physiological relevance of the computed indices

The computed indices reflect the sympatho-vagal interactions that modulate cardiovascular function. The low-frequency component (in the 0.04Hz to 0.15Hz range) of the HRV power spectrum (F_2 peak) is an indicator of sympathetic modulation, and the high frequency component (in the 0.15Hz to 0.4Hz range) in the HRV power spectrum (F_3 peak) is a marker of vagal modulation.

The index I_1 (= P_2/P_3) represents relative sympathetic-to-vagal balance, I_1 is found to be reduced to a very low value, from 6.361 to 0.155 in standing position in the case of diabetics with ischemic heart disease. This indicates that diabetics with ischemic heart disease are not able to withstand orthostatic stress or load. Patients recovering from an acute myocardial infarction can be expected to have an increased I1 index during early convalescence, and a return to a normal value by 6 to 12 months

The orthostatic stress index I_2 shows significant reduction from a normal value of 1.614 to 0.085 in diabetics, and, to 0.155 in diabetics with ischemic heart disease. A similar trend is noted for **the sympatho-vagal integrity index I_3,** showing reduction in the index value from a normal value of 6.19 to 2.43 in the case of diabetics, and to 2.16 in diabetics with ischemic heart disease. This is indicative of damage to the sympathetic and parasympathetic systems controlling the SA node pacing activity

The sympathetic HRVRS frequency-shift Index in standing position (I_{4sd}) and Sympathetic HRVPS frequency-shift Index in supine position (I_{5sup}) are found to be decreased in diabetics as well as in diabetics with ischemic heart disease patients, compared to the normal subject group. This is indicative of the increased delay (of more than 10 seconds) in case of diabetics as well as diabetics with ischemic heart disease, due to demyelination of their nervous control system controlling the heart rate.

The Respiratory Stress Index I_6 denotes the effectiveness of vagal control on heart rate variation, and is found to be considerably reduced from a normal value of 10.36 to 5.26 in diabetics, and to 5.29 in diabetics with ischemic heart disease.

Thus the indices derived from the HRV power spectrum represent non-invasive signatures of the balance between sympathetic and parasympathetic components of the autonomic nervous system. These indices are shown to characterize diabetic autonomic neuropathy state, and to hence distinguish diabetics and diabetics with ischemic heart disease.

Integrated index composed of power-spectral indices

We have shown how well the HRVPS indices differentiate normal subjects from diabetics and diabetics with ischemic heart disease.

We now compute the values of this integrated Index (DAN-IID) for normal subjects (in Table 3), diabetic subjects (in Table 4), and diabetic patients with ischemic heart disease (in Table 5) . From these Index values, we compute its mean values and standard deviations, for normals, diabetics, and diabetics with ischemic heart disease (IHD). These values are tabulated in Table 6. It can be clearly seen, from this Table 6, that our integrated Index can be employed to effectively differentiate and diagnose diabetic subjects and diabetics with IHD. The Index can also be employed to assess the efficacy of diabetic medication and insulin administration.

We next make a distribution plot of this Integrated Index for normals, diabetics, and diabetics with IHD, in Figure 12. This plot graphically illustrates how well this integrated Index separates normal subjects, diabetic patients, and diabetic patients with ischemic heart disease[Desai, K.D et al., 2011].

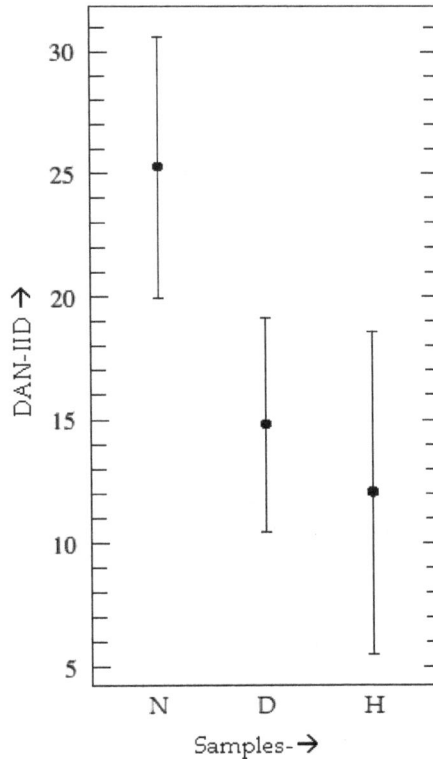

Fig. 12. Variation of DAN-IID for (N) normal subjects, (D) diabetic patients, and (H) diabetics with IHD. It can be noted that this DAN-IID clearly separates diabetics and diabetics with IHD from normal subjects.

6. Activity-based dynamic insulin infusion system

In section 3, we have introduced the glucose-insulin regulatory system and applied it to model the OGTT. We came up with our novel DBI, by means of which we can even detect supposedly normal subjects who are at risk of becoming diabetic. Now, we continue on the trail of this glucose-insulin regulatory system, by presenting its application to illustrate how for a diabetic patient the glucose level keeps going up after meal, and how it is regulated by automated infusion of insulin.

Herein, we demonstrate the operation of a Glucose activity-based Dynamic Insulin infusion (or release) system. The current insulin infusion systems are based on the diabetic patient's known activities history, in order to estimate the required insulin amount. These techniques do not allow the patients to deviate too much from their normal daily activities [Naylor et al., 1996]. Hence, our approach focuses on regular sampling of the diabetic patients' blood glucose concentration through a sensor, to compute the required amount of insulin to be released into the blood stream.

The amount of insulin infused to bring the blood glucose concentration down is regulated by a Closed-loop PD (Proportional-Derivative) Control system algorithm (Fig. 13). The closed loop system continuously monitors the blood glucose concentration at 0.5 h interval. Once the system detects that the blood glucose concentration exceeds a predetermined threshold e.g. 120mg/dl [International Diabetes Federation], the system is alarmed and 'calculates' the amount of insulin required [Loh, 2004] to bring the blood glucose concentration below the threshold.

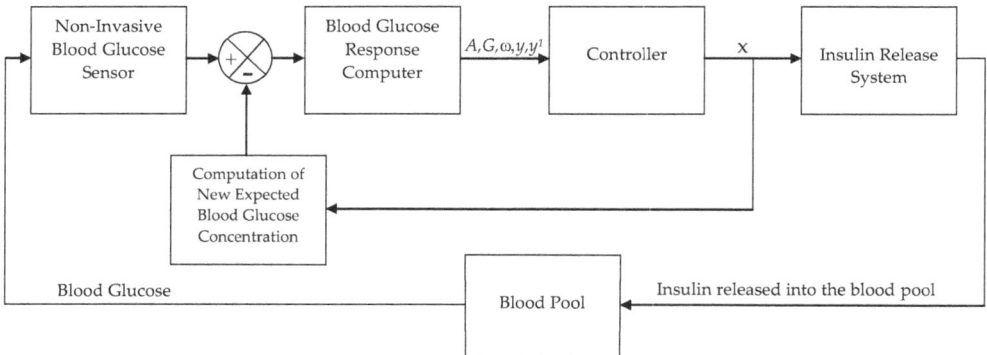

Fig. 13. Block diagram of the Glucose Regulating Insulin Release (GRIR) system: The glucose sensor monitors the existing blood glucose level. The error between the glucose sensor level and the computed expected blood glucose concentration is fed into the Closed-loop PD (Proportional-Derivative) Control system, and its algorithm computes the amount of insulin (x) to be released. Accordingly, the required amount of insulin is released into the blood. This now readjusts the blood glucose level, which is again monitored by the sensor.

Then, Fig. 14 shows the results of the application of the Insulin Infusion Release system of Fig 13. The diabetic subject D18's unaided glucose clinical data is fed into the system. On the Y axis, we have plotted blood-glucose concentration above the patient's glucose concentration of 120 mg/dl (or 1.2 g/l) at time 0 after meal. The insulin is released at 0.5 hour, 1 hour and 1.5 hours after meal. In figure 16, it is seen, how following insulin infusion, the blood glucose comes down. Once the blood glucose concentration drops below the threshold, the controller will stop releasing insulin into the blood stream.

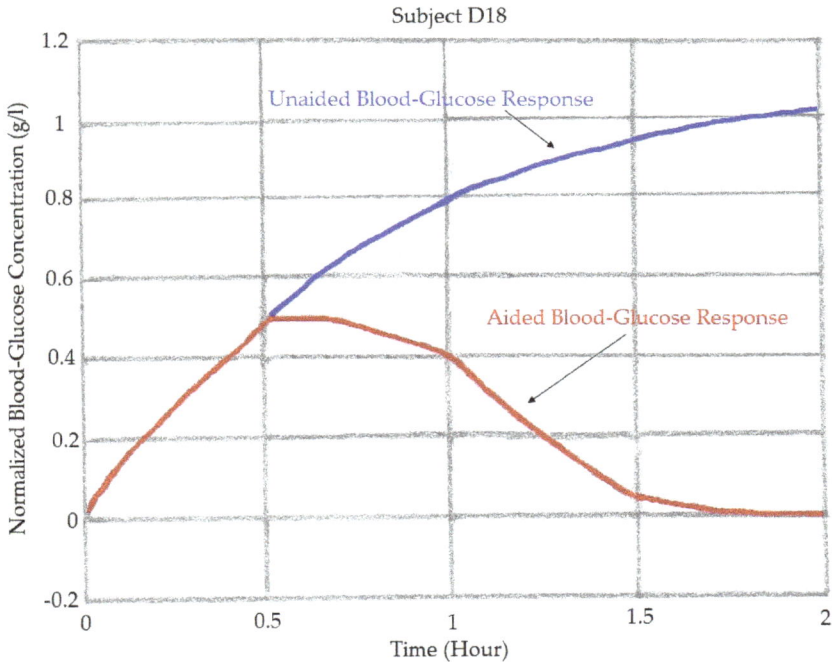

Fig. 14. The subject's unaided blood glucose concentration at time 0 is above 120mg/dl. The system is alarmed and samples the blood glucose concentration at 0.5h (170 mg/dl). The system sends a bolus of insulin 10mU/dl into the blood stream. The system keeps monitoring the resulting blood glucose concentration at 1.0h and 1.5 hour intervals, and infuses computed insulin bolus into the blood stream to bring the blood glucose concentration below the threshold value.

Thus, we have demonstrated the capability of the activity based adaptive dynamic real-time insulin release system. This system is able to protect the users from hyperglycemia.

7. Conclusion

This chapter is framed to provide useful insights into: (i) the mechanisms of diabetes; (ii) how the bioengineering analysis of the glucose regulatory system can be employed to diagnose diabetic patients and subjects at risk of becoming diabetic, based on an integrated index composed of parameters of the governing differential equation to simulate blood glucose concentration data of OGTT; (iii) parameters of time-and frequency-domain measures of HRV can be employed to differentiate diabetic subjects from normal subjects; (iii) processing of retinal digital fundus images to characterize retinopathy, and analysis of plantar pressure distribution images of normal subjects, and subjects with diabetes type II without and with neuropathy, (iv) diagnosis of diabetic autonomic neuropathy by means of a novel intregrated index composed of parameters of heartrate variability power-spectrum plots; (v) how we can apply the glucose-insulin regulatory system to illustrate how for a diabetic patient the glucose level keeps going up after meal, and how it is can be regulated by automated infusion of insulin.

8. Acknowledgments

Authors thank Journal of Mechanics in Biology and Medicine (published by World Scintific Publishing Co. Pte Ltd) for giving permission reproduce the portions of materials from their published papers, (i) U Rajendra Acharya, Jasper Tong, Vinitha Sree, Chua Kuang Chua, Tan Peck Ha, Dhanjoo N Ghista, Subhagata Chattopadhyay, Kwan-Hoong Ng, Jasjit S Suri, "Computer-Based Identification of Type 2 Diabetic Subjects With and Without Neuropathy using Dynamic Planter Pressure and Principal Component Analysis", Journal of Medical Systems, 2011 (In Press: DOI: 10.1007/s10916-011-9715-0); (ii) Desai, K.D, Dhanjoo N. Ghista, Issam Jaha El Mugamex, U Rajendra Acharya, Michael Towsey, Sultan Abdul Ali, Mohammed Saeed, M. Amin Fikri, *"Diabetic Autonomic Neuropathy Detection By Heart-Rate Variability Power-Spectral Analysis"*, Journal of Mechanics in Medicine and Biology, 2011 (In Press); (iii) Dhanjoo N Ghista, "Nondimensional Physiological Indices for Medical Assessment", Journal of Mechanics in Medicine and Biology, 9(4), 2009, 643-669.

9. References

Acharya, U.R.; Lim, C.M.; Ng, E.Y.K.; Chee, C. & Tamura, T. (2009). Computer-based detection of diabetes retinopathy stages using digital fundus images. *Proc Instn Mech Engrs. J Eng Med Part H,* Volume 223, No. 5, pp. 545-553.

Acharya, U.R.; Tan, P.H.; Subramaniam, T.; Tamura, T.; Chua, K.C.; Goh, S.C.E.; Lim, C.M.; Goh, S.Y.D.; Chung, K.R.C.& Law, C.(2008). Automated identification of diabetic type 2 subjects with and without neuropathy using wavelet transform on pedobarograph. *J Med Syst,* Volume 32, No. 1, pp.21-29.

Acharya, U.R.; Joseph, P.K.; Kannathal, N.; Lim, C.M. & Suri, J.S.(2006). Heart rate variability: A Review, *Med Biol Eng Comput,* Volume 44, No. 12, pp. 1031-1051.

Acharya, U.R.; Suri, J.S.; Spaan, J.A.E. & Krishnan, S.M.(2007). Advances in Cardiac Signal Processing. Springer Verlang GmbH. Berlin, Heidelberg.

Acharya, U.R.; Rahman, M.A.; Aziz, Z.; Tan, P.H.; Ng, E.Y.K. & Yu, W. (2008). Computer-based identification of plantar pressure in type 2 diabetes subjects with and without neuropathy. *Journal of Mechanics in Medicine and Biology,* Volume 8, No.3, pp. 363-375.

Acharya, U.R.; Faust, O.; Dua, S.; Hong, S.J.; Yang, T.S.; Lai, P.S. & Choo, K. (2011a). Computer-based diagnosis of diabetes retinopathy stages using digital fundus images. Dua S, Acharya UR, Ng EYK (Eds.), 'Computational image modeling of human eye with applications', *World Scientific Publishing Company.*

Acharya, U.R.; Ghista, D.N.; Nergui, M.; Chattopadhyay, S.; Ng, E.Y.K.; Sree, V.S.; Tong, J.W.K.; Tan, J.H.; Loh, K.M.& Suri, J.S. (2011b). Diabetes Mellitus: Enquiry into its Medical aspects and Bioengineering of its Monitoring and Regulation. *Journal of Mechanics in Medicine and Biology* (In Press).

Akselrod, S., et al (1981), Power spectrum analysis of heart rate fluctuation: A quantitative probe of beat-to-beat cardiovascular control, Science, Vol. 213, 10, pp. 220-222 .

Baynes, J.W. (1991). Role of oxidase stress in development of complications in diabetes (Review). *Diabetes,* Vol. 40, No. 4, pp. 405-412.

Bianchi, A., et al (1990) Spectral analysis of heart rate variability signal and respiration in diabetic subjects, Medical & Biological Engineering & Computing, Vol. 28, pp. 205-211.

Bonnefont-Rousselot, D. (2002). Glucose and reactive oxygen species. *Current Opinion in Clinical Nutrition and Metabolic Care,* Vol. 5, No. 5, pp. 561-568.

Cavanagh, P.R.; Sims, D.S.& Sanders, Jr. L.J. (1991). Body mass is a poor predictor of peak plantar pressure in diabetic men. *Diabetes Care*, Volume 14, pp.750-755.

Dangel, S.; Meier, P.F.; Moser, H.R.; Plibersek, S. & Shen, Y.(1999). Time series analysis of sleep EEG. *Computer Assisted Physics*, pp. 93-95.

Desai, K. D.; Ghista, D.N.; Mugamex, I. J. El.; Acharya, U.R.; Towsey, M.; Ali, S.A.; Saeed, M., Fikri, M. A. (2011). Diabetic Autonomic Neuropathy Detection By Heart-Rate Variability Power-Spectral Analysis, *Journal of Mechanics in Medicine and Biology*, 2011 (In Press);

Dittakavi, S.S. & Ghista, D.N.(2001). Glucose tolerance tests modeling & patient simulation for diagnosis. *J Mech Med Biol*, Volume 1, No.2, pp.193-223.

Eckmann, J.P.; Kamphorst, S.O. & Ruelle, D.(1987). Recurrence plots of dynamical systems. *Europhys Lett*,Volume 4, pp. 973-977.

Evans, J.L.; Golfine, I.D.; Maddux, B.A. & Grodsky, G.M. (2002). Oxidative stress and stress-activated signalling pathways: a unifying hypothesis of type 2 diabetes. *Endocrine Reviews*, Vol. 23, No. 5, pp. 599-622.

Fallen, E.L., Nandogopal, D., Connonlly, S., and Ghista, D.N. (1987) "How reproducible is the power spectrum of heart rate variability in health subjects?", Proceedings of International Symposium on Neural and Cardiovascular Mechanisms, Bologna, Italy May, 1985.

Fantus, G. (2011). Diabetes, glucose toxicity.
http://www.endotext.org/diabetes/diabetes12new/diabetes12.htm

Faust, O.; Acharya, U.R.; Molinari, F.; Chattopadhyay, S. & Tamura, T. (2011). Linear and Non-Linear Analysis of Cardiac Health in Diabetic Subjects. *Biomedical Signal Processing and Control*, 2011

Ghista, D.N.(2004). Physiological systems' numbers in medical diagnosis and hospital cost effective operation. *J Mech Med Biol.*, Volume 4, No. 4,pp. 401–418.

Ghista, D.N.(2009a). Nondimensional physiological indices for medical assessment. *J Mech Med Biol.*, Volume 9, No. 4, pp. 643-669.

Ghista, D.N. (2009b). Applied Biomedical Engineering Mechanics. *CRC Press*.

Gleason, C.E.; Gonzalez, M.; Harmon, J.S. & Robertson, R.P. (2000). Determinants of glucose toxicity and its reversibility in the pancreatic islet beta-cell line, HIT-T15. *American Journal of Physiology, Endocrinology and Metabolism*, Vol. 279, No. 5, pp. E997-E1002.

Grassberger, P. & Procassia, I.(1983). Measuring the strangeness of strange attractors. *Physica*, Volume D9, pp.189-208.

Haber, C.A.; Lam, T.K.; Yu, Z.; Gupta, N.; Goh, T.; Bogdanovic, E.; Giacca, A. & Fantus, I.G. (2003). N-acetylcysteine and taurine prevent hyperglycaemia-induced insulin resistance *in vivo*: possible role of oxidative stress. *American Journal of Physiology: Endocrinology and Metabolism*, Vol. 285, No. 4, pp. E744-E753.

Hayden, M.R. & Tyagi, S.C. (2002). Islet redox stress: the manifold toxicities of of insulin resistance, metabolic syndrome and amylin derived islet amyloid in type 2 diabetes mellitus. *Journal of Pharmacology*, Vol. 3, No. 4, pp. 86-108.

International Diabetes Federation. Website: http://www.idf.org/2000, last accessed in Dec 2010.

Kamath, M.V.; Ghista, D.N.; Fallen, E.L.; Fitchett, D.; Miller, D. & McKelvie, R.(1987). Heart rate variability power spectrogram as potential noninvasive signature of cardiac regulatory system response, mechanisms, and disorders. *Heart Vessels*, Volume 3, pp.33-41.

Krall, L.P. & Beaser, R.S. (1989). *Joslin Diabetes Manual* (12th edition), Lippincott Williams and Wilkins, ISBN 978012111200, London (United Kingdom).

Loh, K.C. (2004). Pharmacology of Oral Anti-hyperglycaemic Agents & Insulin (Invited Article). *Singapore Family Physician*, Volume 30, pp.16-20.

Maiese, K.; Chong, Z.Z. & Shang, Y.C. (2007). Mechanistic insights into diabetes mellitus and oxidative stress. *Current Medicinal Chemistry*, Vol. 16, No. 16, pp. 1729-1738.

Mandelbrot, B.B.(1983). Geometry of Nature. Freeman San Francisco.

Nayak, J.; Bhat, P.S.; Acharya, U.R.; Lim, C.M. & Gupta, M.(2008). Automated identification of different stages of diabetic retinopathy using digital fundus images. *J Med Syst*, Volume 32, No. 2, pp. 107-115.

Naylor, C.D.; Sermer, M.; Chen, E. & Sykora, K.(1996). Cesarean delivery in relation to birth weight and gestational glucose tolerance: pathophysiology or practice style? Toronto Tri-Hospital Gestational Diabetes Investigators. *JAMA*, Volume 275, pp.1165.

Petersen, K.F.; Befroy, D.; Dufour, S.; Dziura, J.; Ariyan, C.; Rothman, D.L.; DiPietro, L.; Cline, G.W. & Shulman, G.I. (2003). Mitochondrial dysfunction in the elderly: possible role in insulin resistance. *Science*, Vol. 300, No. 5622, pp. 1140-1142.

Pincus, S.M.(1991). Approximate entropy as a measure of system complexity. *Proc National Academic Science*, Volume 88, pp.2297-2301.

Rahman, M.A.; Aziz, Z.; Acharya, U.R.; Tan, P.H.; Natarajan, K.; Ng, E.Y.K.; Law, C.; Subramaniam, T. & Shuen, W.Y.(2006). Analysis of plantar pressure in diabetic Type 2 subjects with and without neuropathy. *Innov Technol Biol Med*, Volume 27, No. 2, pp.46-55.

Robertson, R.P. (2004). Chronic oxidative stress as central mechanism for glucose toxicity in pancreatic islet beta cells in diabetes. *Journal of Biological Chemistry*, Vol. 279, No. 41, pp. 42351-42354.

Robertson, R.P.; Harmon, J.; Tran, P.O.; Tanaka, Y. & Takahashi, H. (2003). Glucose toxicity in β-cells: Type 2 diabetes, good radicals gone bad, and glutathione connection. *Diabetes*, Vol. 52, No. 3, pp. 581-587.

Rosenstien, M.; Colins, J.J. & De Luca, C.J.(1993). A practical method for calculating largest Lyapunov exponents from small data sets. *Physica D*, Volume 65, pp. 117-134.

Tanaka, Y.; Tran, P.O.; Harmon, J. & Robertson, R.P. (2002). A role for glutathione peroxidase in protecting pancreatic beta cells against oxidative stress in a model of glucose toxicity. *Proceedings of National Academy of Science (USA)*, Vol. 99, No.19, pp. 12363-12368.

Task Force of the European Society of Cardiology and North American Society of Pacing and electrophysiology.(1996). Heart Rate Variability: Standards of measurement, physiological interpretation and clinical use. *Eur Heart J*, Volume 17, pp.354-381.

Van der Akker, T.J.; Koeleman, A.S.M.; Hogenhuis, L.A. & Rompelman, G.(1983). Heart-rate variability and blood pressure oscillations in diabetics with autonomic neuropathy. *Automedica*, Volume 4, pp.201-208.

Yoshida, K.; Hirokawa, J.; Tagami, S.; Kawakami, Y.; Urata, Y. & Kondo, T. (1995). Weakened cellular scavenging activity against oxidative stress in diabetes mellitus: regulation of glutathione synthesis and efflux. *Diabetologia*, Vol. 38, No. 2, pp. 201-210.

Yun, W.L.; Acharya, U.R.; Venkatesh, Y.V.; Chee, C., Lim C.M. & Ng, E.Y.K.(2008). Identification of different stages of diabetic retinopathy using retinal optical images. *Information Sciences*, Volume 178, No. 1, pp.106-121.

10

A Shape-Factor Method for Modeling Parallel and Axially-Varying Flow in Tubes and Channels of Complex Cross-Section Shapes

Mario F. Letelier and Juan S. Stockle
University of Santiago of Chile,
Chile

1. Introduction

In the study of some industrial, biological and natural fluidic systems it is often necessary to model fluid flow through tubes, channels or passages of complex geometries. The complexity may arise from the cross-sectional shape, or from longitudinal cross-section variation, or from both. Typical cases include flow of molten metals or plastics through dies and moulds, blood flow, microfluidic applications, and flow in porous media, among many others. Characteristics of these flows are laminar state, incompressibility, small rates of flow and varied time patterns. One field where pertinent applications are being developed at a fast rate is Microfluidics (Cetin and Li., 2008; Chen et al., 2008; Forte et al., 2008; Gebauer and Bocek, 2002 ; Mathies and Huang, 1992; Sommer et al., 2008; Srivastava et al., 2005; Woolley and Mathies, 1994; Yeger et al., 2006.) . In this specific field, present microchannel manufacturing techniques produce typically non-circular capillaries (Sommer et al., 2008). Also the introduction of electrical or magnetic field induce plastic behavior in the working fluid.

In particular, it is well known that blood is a biological fluid that behaves as a Newtonian fluid in arteries, veins and large capillaries, but becomes non-Newtonian in the smaller vessels, where the size of suspended particles is big as compared to the vessel´s diameter size (Pedley ,2008). A relevant problem in this field as to the method presented in the next sections is the analysis of diseased arteries and veins for quirurgical interventions. Specifically, stenosed arteries are blood conduits of irregular geometry in which cross-section geometry usually varies along the vessel length.

The above context implies that it is desirable, particularly for modeling and design purposes, to count with analytical techniques that can integrate variables such as the non-circular cross-section of conduits, axial variation of conduit geometry, and plastic flow in some cases.

In this chapter it is presented a method of analysis that allows to address in a general way the problem here outlined.

The standard analytical technique for tube flow problems is usually the search of specific solutions to the momentum equations with associated boundary and initial conditions (Batchelor, 2000). Otherwise numerical solutions are developed for some purposes (Xue et al.,1995).

The main aim of this chapter is, thus, to introduce and explore the potential use of a general analytical approach to irregular conduit flow, which makes it possible to determine velocity field, rate of flow, shear stress, recirculation regions and plug zones, this last when fluid plasticity is operant.

The method already referred to has been developed by the authors through specific applications mainly during the past decade. In this chapter some previous results are organized within a common analytical pattern, together with novel material.

This chapter includes sections for the *general model*, considering one velocity component and more than one velocity component versions, *applications* related to flow in straight tubes and to axially-varying flows, and a closing *conclusion* section.

2. The general model

The concept of "shape factor" herein used is applied to a function G of spatial coordinates, such as when $G = 0$, a series of closed curves are determined for a range of some parameters contained in G. One typical example is

$$G = 1 - r^2 + \varepsilon r^n \sin n\,\theta \tag{1}$$

In this (r, θ) are polar coorfinates, n is an integer number and ε is a parameter such as that for $\varepsilon = 0$ the curve described by (1) is a circle, and as ε increases, the shape evolves to some limiting shape, controlled by n. In all cases here considered, the maximum allowable value of ε is less than unity, and beyond that value, the curve is no longer a closed one. If $\varepsilon = \varepsilon_c$ is the critical, or maximum, allowable value of ε, then for the shape factor described by (1), ε_c is found to be

$$\varepsilon_c = \frac{2}{n} \left(\frac{n-2}{n}\right)^{(n-2)/2} \tag{2}$$

n	ε_c
3	0,385
4	0,250
5	0,186
6	0,148

A more general shape factor is

$$G = 1 - r^2 + \varepsilon_1\, r^{n_1} \sin n_1\,\theta + \varepsilon_2\, r^{n_2} \sin n_2\theta + \cdots \tag{3}$$

which leads to more complex shapes. Some instances of these shapes are shown in Fig. 1.

For the purposes of this presentation, a general shape factor in polar coordinates can be defined as

$$G = h_0(r) + \varepsilon_1 h_1(r, \theta) + \varepsilon_2 h_2\,(r, \theta) + \cdots \tag{4}$$

in which h_1, h_2 ... are boundary perturbation functions. For the case of channel flow, the structure of (4) may be the same, in which polar coordinated may be substituted by Cartesian coordinates. The specific characteristics of functions h_i are determined by the nature of the equations of motion and associated boundary conditions.

Two relevant cases can be highlighted, namely, flow with one velocity component, and flow with more than one velocity component.

A Shape-Factor Method for Modeling Parallel and Axially-Varying Flow in Tubes and Channels of Complex
Cross-Section Shapes

249

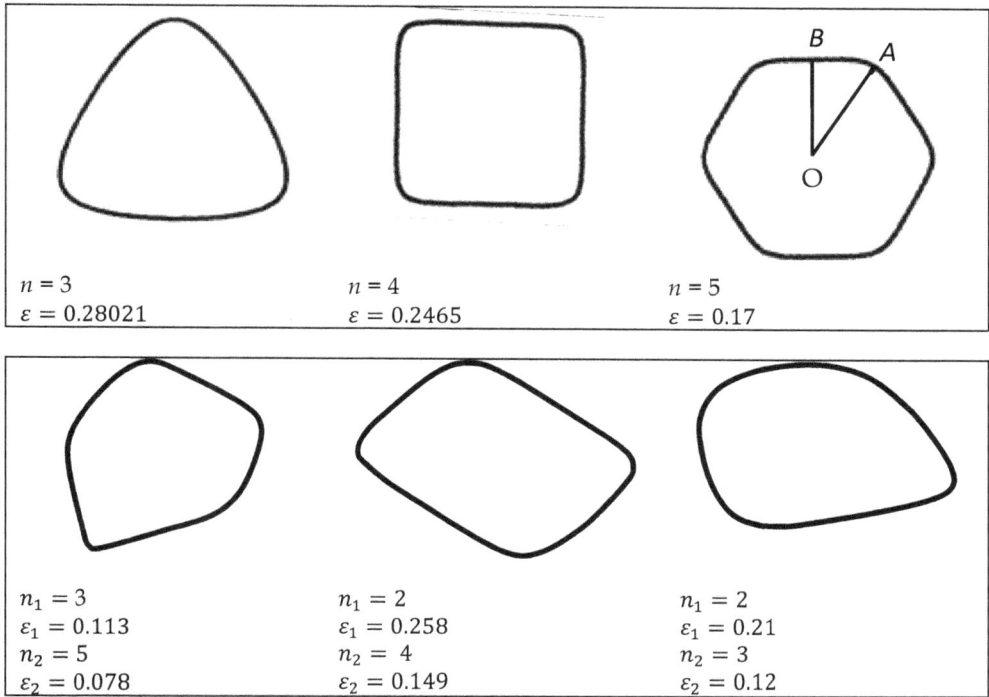

Fig. 1. Samples of tubes contours from shape factors (1) and (3).

2.1 Flow with one velocity component

These are flows in straight tubes of constant cross-section. In these cases the axial velocity w can be modeled as

$$w(r,\theta,t) = G(r,\theta)[\,f_0(r,t) + \varepsilon\,f_1(r,\theta,t) + \varepsilon^2 f_2(r,\theta,t) + \cdots\,] \tag{5}$$

where, for the sake of simplicity, only one boundary perturbation function has been considered. Functions f_i are to be determined from the equation of motion in terms of a standard regular perturbation scheme around the small parameter ε.

2.2 Flows with more than one velocity component

These are mainly flows with axial variation of tube or channel geometry. In these cases the solution procedure will usually involve the use of a stream function Ψ. In such problems both Ψ and the velocity components should be zero at the boundary, a condition that can be met by defining

$$\Psi = G^2[g_0(r,t) + \varepsilon\,g_1(r,\theta,t) + \varepsilon^2 g_2(r,\theta,t) + \cdots\,] \tag{6}$$

where again functions g_1 have to be determined from the equations of motion. The definition of Ψ is given for every specific application in the corresponding section.

In the following some specific applications of this method of analysis are presented.

3. Applications

3.1 Flow in straight tubes of constant non-circular cross-section
3.1.1 Newtonian unsteady flow

For incompressible, developed and isothermal flow, the equation of motion are the standard Navier-Stokes and continuity equations. In dimensionless variables they are

$$\Omega\frac{\partial w}{\partial t} - \left(\frac{1}{r}\frac{\partial w}{\partial r} + \frac{\partial^2 w}{\partial r^2} + \frac{1}{r^2}\frac{\partial^2 w}{\partial\theta^2}\right) = \phi(t) = -\frac{\partial P}{\partial z}(t) \tag{7}$$

$$\frac{\partial w}{\partial z} = 0 \tag{8}$$

In this

$$\Omega = \frac{\rho a^2}{\mu T_0} \tag{9}$$

is the so-called unsteadiness number of the flow, which measures the relative importance of a temporal inertia force against a steady viscous force, and where ρ =density, α =reference tube radius, μ =dynamic viscosity, and T_0 =reference time.

A convenient solution of (7) (Letelier et al., 1995) for round tubes (ie for $\partial/\partial\theta = 0$) can be worked out by postulating

$$w = A_2(1 - r^2) + A_4(1 - r^4) + A_6(1 - r^6) + \cdots \tag{10}$$

Where $A_{2n} = A_{2n}(t)$ for $n = 1,2,3 \ldots \infty$. Equation (10) meets the no-slip boundary condition $w(1,t) = 0$. After substituting (10) in (7) it is found that all functions A_{2n} can be expressed in terms of $A_2 = A$ so that the axial velocity takes the form

$$w_0 = (1 - r^2)\left\{A + \frac{\Omega}{4^2}\frac{dA}{dt}(1 + r^2) + \frac{\Omega^2}{4^2 6^2}\frac{d^2A}{dt^2}(1 + r^2 + r^4) + \cdots\right\} \tag{11}$$

where A is related to the forcing function $\phi(t)$ as follows

$$\frac{\phi}{4} = A + \frac{\Omega}{2^2}\frac{dA}{dt} + \frac{\Omega^2}{2^2 4^2}\frac{d^2A}{dt^2} + \cdots \tag{12}$$

In these expressions Ω can have any finite positive value.
According to (5), it is found

$$f_0 = A + \frac{\Omega}{4^2}\frac{dA}{dt}(1 + r^2) + \frac{\Omega^2}{4^2 6^2}\frac{d^2A}{dt^2}(1 + r^2 + r^4)\cdots \tag{13}$$

If

$$L = \Omega\frac{\partial}{\partial t} - \left(\frac{1}{r}\frac{\partial}{\partial r} + \frac{\partial^2}{\partial r^2} + \frac{1}{r^2}\frac{\partial^2}{\partial\theta^2}\right) \tag{14}$$

then, by collecting terms of order ε, it follows

$$L\{f_1(1 - r^2)\} = L\{f_0 r^n \sin(n\theta)\} \tag{15}$$

wherefrom

$$f_1 = \sin(n\theta)\sum_{i=1}^{\infty}\left\{\Omega^i\frac{d^iA}{dt^i}\right\}\sum_{s=0}^{2i-2}\{C_{is}r^{i+s}\} \tag{16}$$

A Shape-Factor Method for Modeling Parallel and Axially-Varying Flow in Tubes and Channels of Complex
Cross-Section Shapes

251

The constants C_{is} in these equations are obtained by putting the coefficients of all powers of
r, for any $d^i A/dt^i (i = 1,2, ...)$, equal to zero in (13). The result is

$$C_{10} = \frac{n-3}{16(n+1)} \tag{17}$$

$$C_{22} = \frac{72C_{10}+n-2.5}{576(n+2)} \tag{18}$$

$$C_{20} = \frac{(n+1)(576C_{22}+1)-144C_{10}-9}{576(n+1)} \tag{19}$$

and so on. Higher order terms in ε can be obtained in like fashion.

An example of velocity profiles is shown in figure 1 for $n = 6$, $\varepsilon = 0.148$ at two semi-axes (cf
Fig. 1).

In this case the tube contour is an approximate hexagon and $\phi = cost$, ie a purely oscillatory
flow is described.

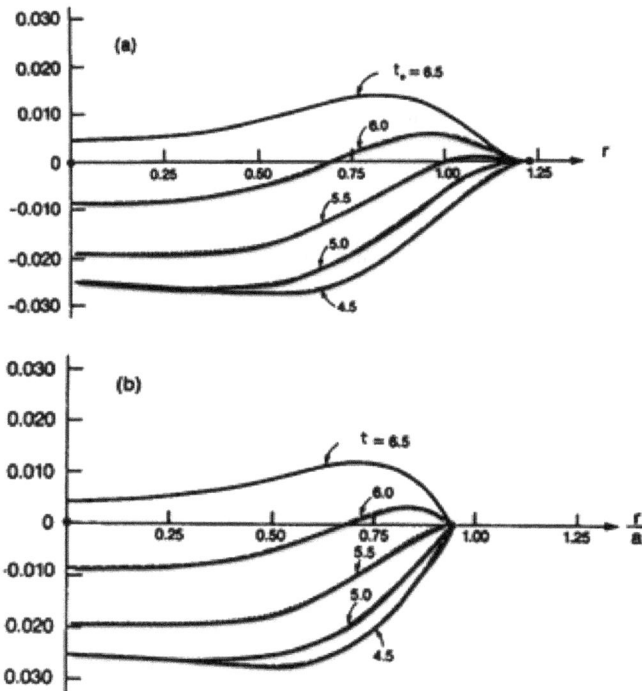

Fig. 2. Instantaneous velocity profiles for $\Omega = 40$. (a) along maximum semi-axis OA; (b)
along minimum semi-axis OB.

The structure of (4) makes it possible to apply a regular perturbation method of solution
around the dimensionless parameter ε. Since ε is bounded for a given value of n, and is
always less than unity, the solution becomes actually an exact one when enough terms are
obtained.

3.1.2 Steady plastic flow
In this application it is considered steady flow of a Bingham plastic. Here (5) is also applicable, and the equation of motion, in terms of shear stress, as defined below, is

$$\frac{\tau_{rz}}{r} + \frac{\partial \tau_{rz}}{\partial r} + \frac{1}{r}\frac{\partial \tau_{\theta z}}{\partial \theta} = \phi \tag{20}$$

$$\tau_{rz} = -\left(1 - \frac{N}{I}\right)\frac{\partial w}{\partial r} \tag{21}$$

$$\tau_{\theta z} = -\left(1 - \frac{N}{I}\right)\frac{1}{r}\frac{\partial w}{\partial \theta} \tag{22}$$

and

$$I = \sqrt{\left(\frac{\partial w}{\partial r}\right)^2 + \left(\frac{1}{r}\frac{\partial w}{\partial \theta}\right)^2} \tag{23}$$

is the second invariant of the rate of deformation tensor. The dimensionless yield stress is

$$N = \frac{\tau_0 a}{w_0 \eta_0} \tag{24}$$

The momentum equation (20) is the standard one for parallel steady flow. Its structure has been made consistent with (21-22) and with the standard mathematical ordering of terms. The constitutive expressions (21-22) come from the applicable form of the Bingham fluid model. Defining

$$w = w_0 + \varepsilon\, w_1 + \cdots \tag{25}$$

then, from (5) it follows

$$w_0 = (1 - r^2) f_0 \tag{26}$$

$$w_1 = (1 - r^2) f_1 + r^n f_0 \sin n\theta \tag{27}$$

and the following equations are found

$$\tau_{rz} = N - \frac{\partial w_0(r)}{\partial r} - \varepsilon \frac{\partial w_1(r,\theta)}{\partial r} \tag{28}$$

$$\tau_{\theta z} = \left(\frac{N}{\frac{\partial w_0(r)}{\partial r}} - 1\right)\frac{\varepsilon}{r}\frac{\partial w_1(r,\theta)}{\partial \theta} \tag{29}$$

$$-\frac{1}{r}\frac{\partial w_0(r)}{\partial r} - \frac{\partial^2 w_0(r)}{\partial r^2} = \phi - \frac{N}{r} \tag{30}$$

$$\frac{1}{r}\frac{\partial w_1(r,\theta)}{\partial r} + \frac{\partial^2 w_1(r,\theta)}{\partial r^2} + \left(1 - \frac{N}{\frac{\partial w_0(r)}{\partial r}}\right)\frac{1}{r^2}\frac{\partial^2 w_1(r,\theta)}{\partial \theta^2} = 0 \tag{31}$$

Equations (28-29) are the result of substituting (25) in (23) and of ordering terms in powers of ε through a linearization procedure. From (30-31) it is found

$$w_0(r) = N(r - 1) + \frac{\phi}{4}(1 - r^2) \tag{32}$$

$$w_1(r,\theta) = A_0\left(1 + n^2 \sum_{i=1}^{n} \frac{(-1)^i \phi^i (n^2-(i-1)^2)!}{2^i N^i (i^2)!} r^i\right) \cos(n\theta) \tag{33}$$

where

$$A_0 = \frac{2^n [n^2!] N^n}{(-1)^n \phi^n [n^2(n^2-1\ )(n^2-2^2)...(n^2-(n-1)^2)]} \tag{34}$$

Functions f_0, f, and following can be found equating terms in orders of ε in (5), ie

$$f_0 = \frac{w_0}{1-r^2} \tag{35}$$

$$f_1 = \frac{w_1 - r^n f_0 \sin n\,\theta}{1-r^2} \tag{36}$$

In this both functions are continuous for $r = 1$, and so can be built higher order functions.
Isovel plots and plug zones for selected instances of flow are shown in figure 3. A plug zone
is such that inside its limiting boundary the shear stress is less than the yield stress.

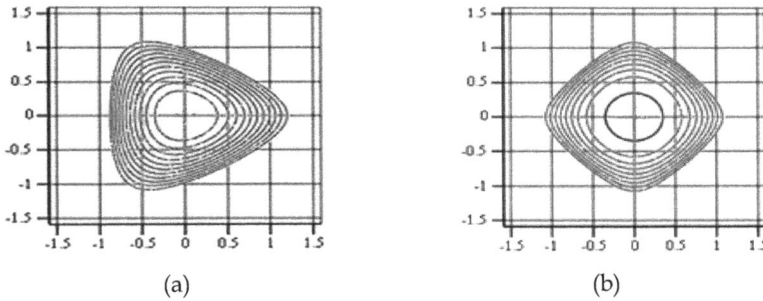

(a)   (b)

Fig. 3. Isovels from (25) for (a) $n = 3$, $\varepsilon = 0.3$, $N = 0.2$ and $\phi = 4$; (b) $n = 4$, $\varepsilon = 0.24$;
$N = 0.7$ and $\phi = 4$.

According to fig.3, in both cases therein depicted, the plug zone appears at the center and is
essentially circular.

3.2 Axially-varying flows in conduits
3.2.1 Newtonian flow in round tubes of arbitrarily axially-varying cross-section
A definition diagram is shown in fig.4

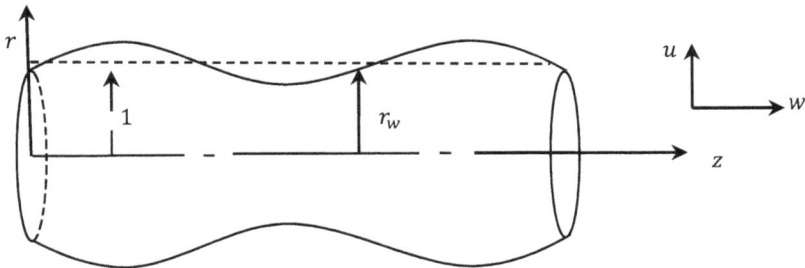

Fig. 4. Definitions diagram for flow in round tubes of axially-varying cross-section.

For this flow, the continuity and Navier-Stokes equations are

$$\frac{1}{r}\frac{\partial(ru)}{\partial r} + \frac{\partial w}{\partial z} = 0 \tag{37}$$

$$u\frac{\partial u}{\partial r} + w\frac{\partial u}{\partial z} = -\frac{2}{Re}\frac{\partial P}{\partial r} + \frac{2}{Re}\left(\frac{\partial}{\partial r}\left(\frac{1}{r}\frac{\partial(ru)}{\partial r}\right) + \frac{\partial^2 u}{\partial z^2}\right) \tag{38}$$

$$u\frac{\partial w}{\partial r} + w\frac{\partial w}{\partial z} = -\frac{2}{Re}\frac{\partial P}{\partial z} + \frac{2}{Re}\left(\frac{1}{r}\frac{\partial}{\partial r}\left(r\frac{\partial w}{\partial r}\right) + \frac{\partial^2 w}{\partial z^2}\right) \tag{39}$$

where Re is the Reynolds number. The velocity field is found after defining

$$u = \varepsilon u_1(r,z) \ldots \tag{40}$$

$$w = w_0(r) + \varepsilon w_1(r,z) \ldots \tag{41}$$

$$P = P_0(z) + \varepsilon P_1(r,z) \ldots \tag{42}$$

$$\Psi = \Psi_0(r) + \varepsilon \Psi_1(r,z) \ldots \tag{43}$$

The velocity is next expressed in terms of de stream function as follows

$$w = \frac{1}{r}\frac{\partial w}{\partial r} \tag{44}$$

$$u = -\frac{1}{r}\frac{\partial \Psi}{\partial z} \tag{45}$$

The wall radius is a function of the axial coordinate, that is here defined as

$$r_w = 1 + \varepsilon h(r)F(z) \tag{46}$$

in which ε is a small parameter. This algebraic structure allows to introduce a very large range of axial variation since h and F are arbitrary functions. Three cases will be considered. In the first case

$$h = r; \qquad F(z) = \sin \omega z \tag{47}$$

where is ω is an arbitrary frequency.
The stream function is modelled from (6) and (46), ie

$$\Psi = (r_w - r)^2[g_0(r) + \varepsilon\{g_{11}(r)\sin(\omega z) + g_{12}(r)\cos(\omega z)\} + 0(\varepsilon^2) + \cdots] \tag{48}$$

wherefrom

$$\Psi_0 = (1 - r)^2 g_0(r) = \frac{\phi}{16}(r^2 - 1)^2 \tag{49}$$

and thus

$$g_0(r) = \frac{\phi}{16} \tag{50}$$

The first order stream funtion is

$$\Psi_1 = (1 - r)^2\big(g_{11}(r)sin(\omega z) + g_{22}(r)cos(\omega z)\big) + 2(1 - r)\, r\, sin(\omega z)\frac{\phi}{16} \tag{51}$$

which is next written as

$$\Psi_1 = H_1(r) * \sin(\omega z) + H_2(r) * \cos(\omega z) \tag{52}$$

and where H_1 and H_2 are unknowns that are modelled as finite polynominals of even order terms, ie

$$H_1(r) = a_0 + a_2 r^2 + a_4 r^4 + a_6 r^6 + a_8 r^8 + a_{10} r^{10} + a_{12} r^{12} + a_{14} r^{14} + a_{16} r^{16} \tag{53}$$

$$H_2(r) = b_0 + b_2 r^2 + b_4 r^4 + b_6 r^6 + b_8 r^8 + b_{10} r^{10} + b_{12} r^{12} + b_{14} r^{14} + b_{16} r^{16} \tag{54}$$

The coefficients a_i and b_j are determined by substituting (53) and (54) in the equation for Ψ found from (38-39) once (44-45) are substituted in there. Examples of typical streamline and isovelocity patterns are shown in figures 5 and 6. Streamlines are plotted from (48) and isovelocity are curves where $u^2 + v^2$ is constant.

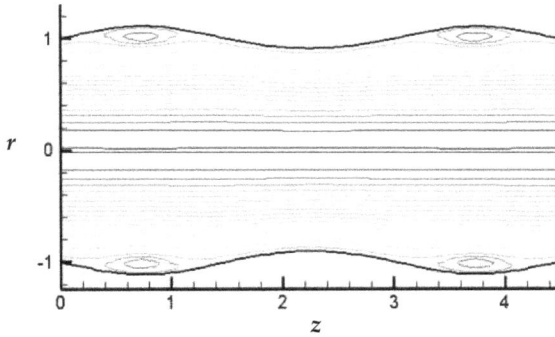

Fig. 5. Streamlines for Re=100 and ε =0.1.

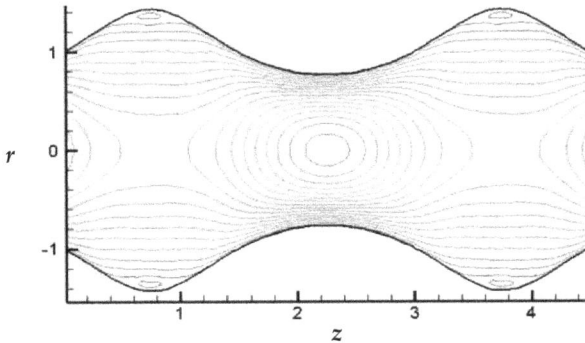

Fig. 6. Isovelocity lines for Re=1 and ε =0.3 from (48) and (44-45).

A second kind of contour is defined through the expression

$$F(z) = -0.028z^4 + 0.434z^3 - 2.156z^2 + 3.43z \tag{55}$$

which was transformed in a Fourier series in the range $0 \leq z \leq 3$ in terms of sine and cosine functions that allow a modeling similar, but more complex, to that already described. Examples of typical streamline and isovelocity patterns are shown in figure 7 and 8.

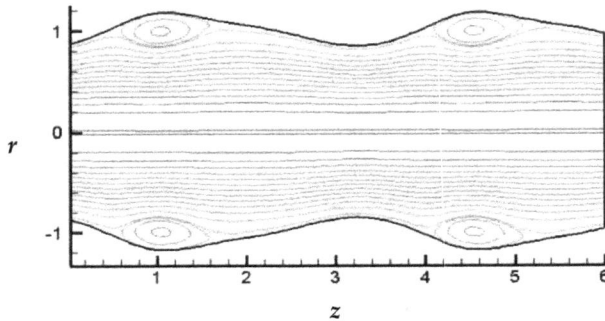

Fig. 7. Streamlines for Re=100 and ε =0.2.

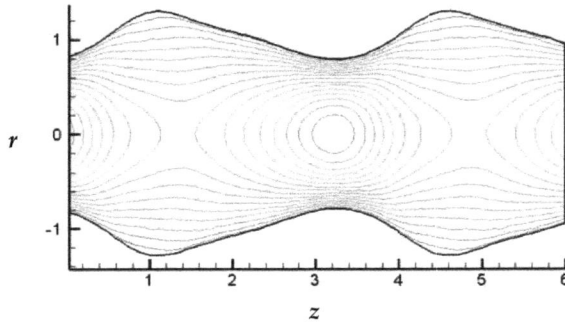

Fig. 8. Isovelocity lines for Re=1 and ε =0.3.

In similar fashion, the a third contour presented is defined by

$$F(z) = -0.03z^4 - 0.045z^3 + 0.405z^2 + 0.42z \qquad (56)$$

For this case, typical isovelocity and isobaric curves are shown in figures 9 and 10.

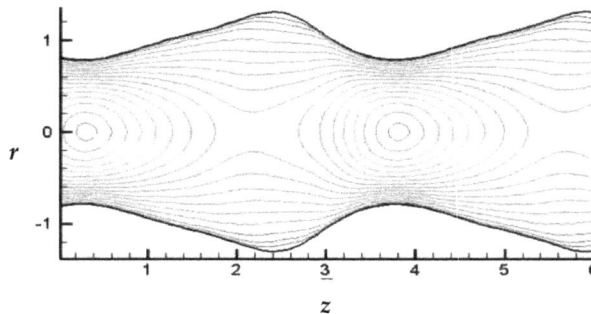

Fig. 9. Isovelocity lines for Re=1 and ε =0.3.

A Shape-Factor Method for Modeling Parallel and Axially-Varying Flow in Tubes and Channels of Complex
Cross-Section Shapes

257

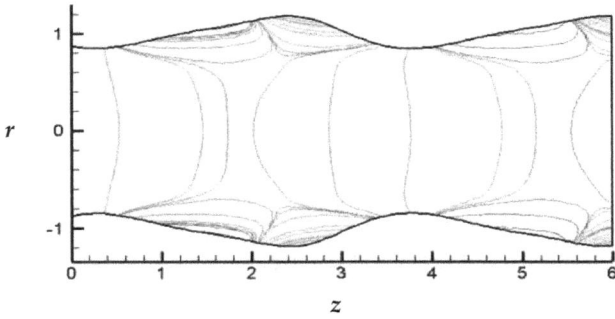

Fig. 10. Isobaric lines for Re=100 and ε =0.2 from (42).

3.2.2 Steady plastic flow in undulating channels
A definition diagram for this flow is shown in fig. 11.

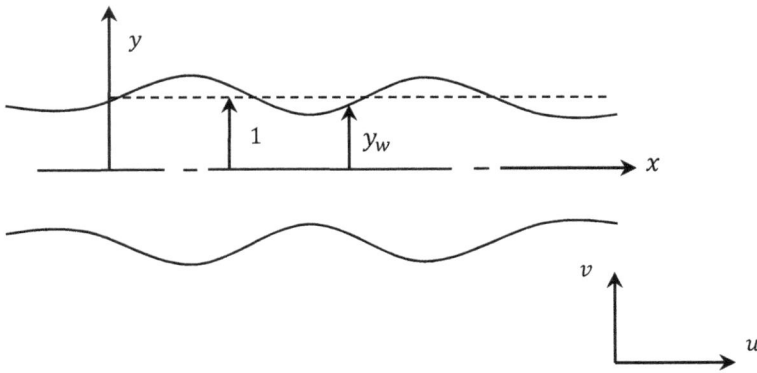

Fig. 11. Definition diagram for flow in undulating channels.

The Bingham constitutive equations are for this case are

$$\tau_{xx} = -\left(1 - \frac{N}{I}\right) 2 \left(\frac{\partial u}{\partial x}\right) \tag{57}$$

$$\tau_{yy} = -\left(1 - \frac{N}{I}\right) 2 \left(\frac{\partial v}{\partial y}\right) \tag{58}$$

$$\tau_{yx} = \tau_{xy} = -\left(1 - \frac{N}{I}\right) \left(\frac{\partial u}{\partial y} + \frac{\partial v}{\partial x}\right) \tag{59}$$

and the corresponding continuity and momentum equations are

$$\frac{\partial u}{\partial x} + \frac{\partial v}{\partial y} = 0 \tag{60}$$

$$u\frac{\partial u}{\partial x} + v\frac{\partial u}{\partial y} = -\frac{1}{Re}\left(\frac{\partial \tau_{xx}}{\partial x} + \frac{\partial \tau_{xy}}{\partial y}\right) - \frac{1}{Re}\frac{\partial P}{\partial x} \tag{61}$$

$$u\frac{\partial v}{\partial x} + v\frac{\partial v}{\partial y} = -\frac{1}{Re}\left(\frac{\partial \tau_{xy}}{\partial x} + \frac{\partial \tau_{yy}}{\partial y}\right) - \frac{1}{Re}\frac{\partial P}{\partial y} \tag{62}$$

Next, the velocity, pressure and stream function are expanded as

$$u(x, y) = u_0(y) + \varepsilon u_1(x, y) \ldots \tag{63}$$

$$v(x, y) = \varepsilon v_1(x, y) \ldots \tag{64}$$

$$P(x, y) = P_0(x) + \varepsilon P_1(x, y) \ldots \tag{65}$$

$$\psi(x, y) = \psi_0(y) + \varepsilon \psi_1(x, y) \ldots \tag{66}$$

$$u(x, y) = \frac{\partial}{\partial y}\psi(x, y) \tag{67}$$

$$v(x, y) = -\frac{\partial}{\partial x}\psi(x, y) \tag{68}$$

$$\psi = (y - y_w)(g_0(y) + \varepsilon g_1(x, y) + \varepsilon^2 g_2(x, y) + \cdots) \tag{69}$$

where the wall is described by

$$y_w = 1 + \varepsilon F(y)sin(\omega x) \tag{70}$$

Following a procedure similar to the one presented in section 3.2.1, it is found

$$\psi_0 = (y - 1)^2 g_0(y) \tag{71}$$

$$g_0(y) = -\frac{1}{6}(2y + 4 - 3N) \tag{72}$$

$$\psi_1 = (1 - y)^2\big(g_1(y)sin(\omega x) + g_2(y)cos(\omega x)\big) + 2(1 - y)F(y)sin(\omega z)f_0(y) \tag{73}$$

$F(y)$ is defined as $F = y$ and the first order stream function is modeled as

$$\psi_1 = A(y)sin(\omega x) + B(y)cos(\omega x) \tag{74}$$

The unknown functions $A(y)$ and $B(y)$ are modeled as finite polynomial, ie

$$A(y) = a_0 + a_1y + a_2y^2 + a_3y^3 + a_4y^4 + a_5y^5 + a_6y^6 + a_7y^7 + a_8y^8 + a_9y^9 \tag{75}$$

$$B(y) = b_0 + b_1y + b_2y^2 + b_3y^3 + b_4y^4 + b_5y^5 + b_6y^6 + b_7y^7 + b_8y^8 + b_9y^9 \tag{76}$$

The coefficients a_i and b_j are determined by substituting (75) and (76) in the equations for ψ found from the momentum equations. In the following figures are presented plots of streamlines (equation (69)), isovelocity lines ($u^2 + v^2 = const.$), plug zones and axial velocity profiles

In figure 13 the plug zones are shown as shaded areas, which were determined by putting the condition that the shear stress should be equal or less the yield stress. The quasi-plug zones are zones where only $\tau_{rz} \leq 0$.

A Shape-Factor Method for Modeling Parallel and Axially-Varying Flow in Tubes and Channels of Complex
Cross-Section Shapes

259

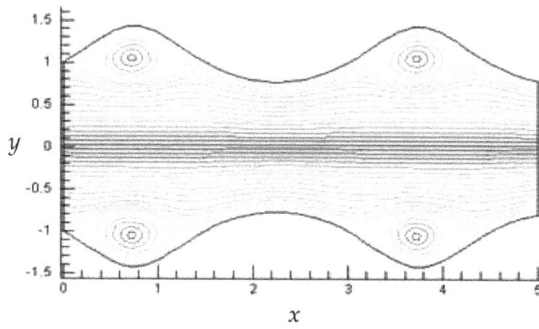

Fig. 12. Streamlines for Re=100, N=0.3 and ε = 0.3.

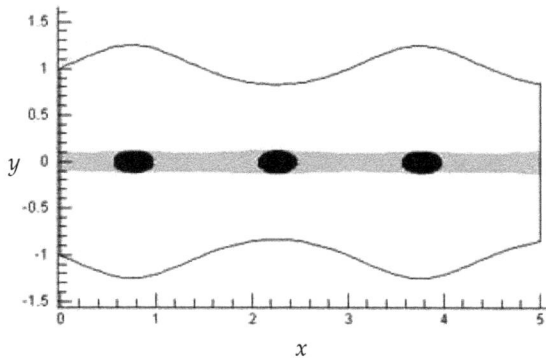

Fig. 13. Plug and quasi-plug zones for Re=1, N=02 and ε=0.2.

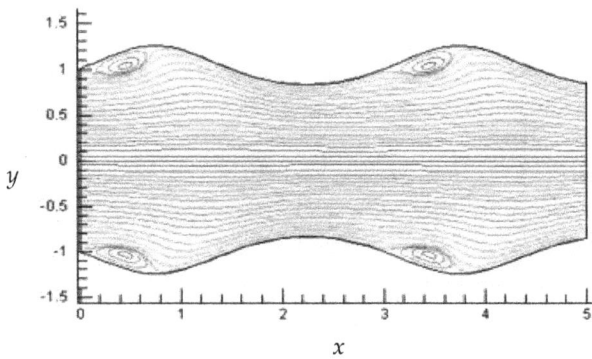

Fig. 14. Streamlines for Re=20, N=0.1 and ε=0.2.

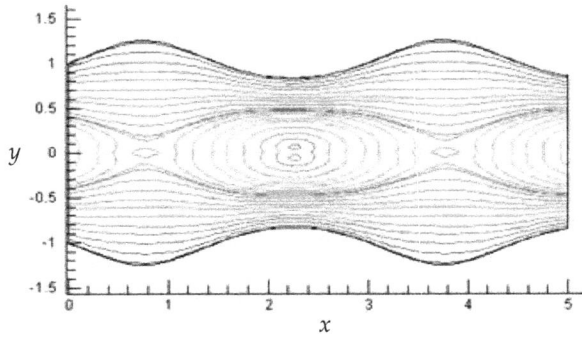

Fig. 15. Isovelocity lines for Re=1, N=0.1 and ε=0.2.

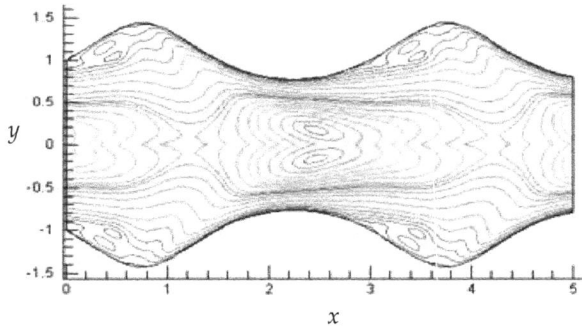

Fig. 16. Isovelocity lines for Re=20, N=0.2 and ε=0.3.

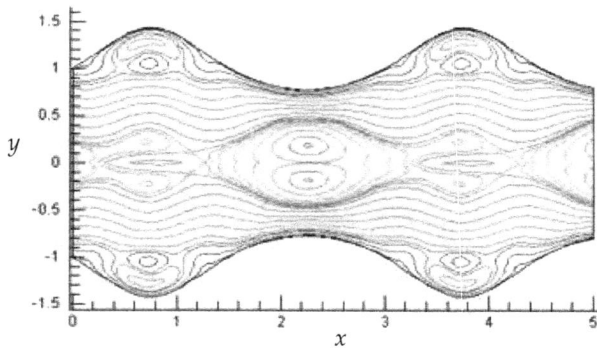

Fig. 17. Isovelocity lines for Re=100, N=0.1 and ε=0.3.

A Shape-Factor Method for Modeling Parallel and Axially-Varying Flow in Tubes and Channels of Complex Cross-Section Shapes

261

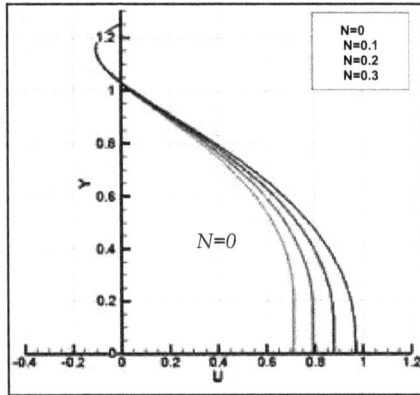

Fig. 18. Axial velocity profiles at x=0.75 for $\varepsilon = 0.2$ and Re =100.

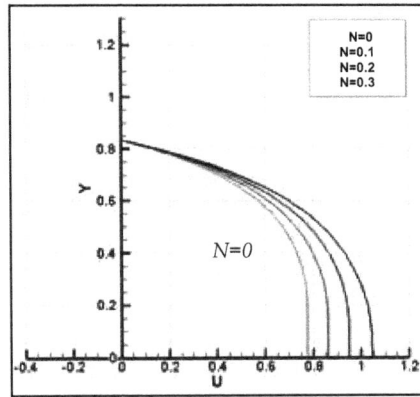

Fig. 19. Axial velocity profiles at x=2.25 for $\varepsilon = 0.2$ and Re =100.

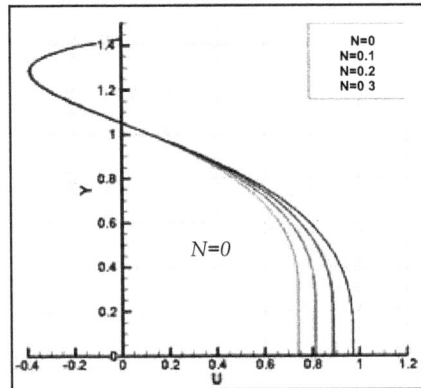

Fig. 20. Axial velocity profiles at x=0.75 for $\varepsilon = 0.3$ and Re =100.

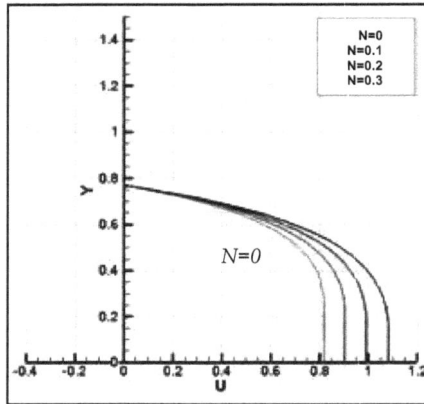

Fig. 21. Axial velocity profiles at x=2.25 for $\varepsilon = 0.3$ and Re =100.

4. Conclusion

The method here described can lead to very accurate solutions for the velocity field and related variables such as shear stress, rate of flow and pressure in a great variety of flows in tubes and channels. Symbolic software presently available, such Maple and MathCAD make it possible to obtain and compute higher order solutions that, in some cases, may have complex algebraic structures. The fact that for all cases here considered, ie cases where $n \geq 3$, ε is much less than unity (cf table 2), leads to a regular perturbation scheme that in most cases requires terms up to second order to achieve enough accuracy. The cases when $n = 1$ and $n = 2$ deserve special mention. For $n = 1$ the shape factor (1) describes an excentric circle, and for $n = 2$ an ellipse. In this last instance ε is not bounded and can take any finite value, which implies that the perturbation scheme would break down if $\varepsilon \geq 1$. So that, in this particular case, the method is limited to elliptical cross-sections of axes ratio close to unity. The method can be expanded to many more complex flow geometries. This possibility is implicit in the more general shape factor (3), which makes it necessary to develop a compound perturbation scheme, in terms of more than one perturbation parameter. The structure of the shape factor (1) determines that the analysis, especially for $n \geq 3$, is more sensible to the perturbation parameter for $r \geq 1$, ie close to the wall conduit. This requires a careful analysis of series convergency which should define the order of the higher order term considered. On the other hand, in the case of flow in straight tubes, in all cases studied, in a considerable region around the conduit axis, say for $r \leq 0.4$, the flow variables are independent of the boundary geometry and take the values of the corresponding flow in round tubes.

5. Acknowledgment

The authors acknowledge the financial support provided, at different stages of the work here presented, by FONDECYT-CONICYT and DICYT at the University of Santiago of Chile.

6. References

Batchelor G. K. (2000). *An Introduction to Fluid Dynamics*. Cambridge University Press.

Cetin B. and Li D. (2008). *Microfluidic continuous particle separation via ac- dielectrophoresis with 3d electrodes*. ASME International Mechanical Engineering Congress and Exposition.

Chen P.CH.; Wang H.; Park D.S.; Park S.; Nikitopoutos D.E.; Soper S. A. and Murphy M. C. (2008). *Protein adsorption in a continuous flow micro-channel environment*. ASME International Mechanical Engineering Congress and Exposition.

Forte J.A.; Sipahi R. and Ozturk A.(2008). *A novel device for nonmagnetic particle navigation using ferrofluids manipulated by magnetic fields*. ASME International Mechanical Engineering Congress and Exposition.

Gebauer P. and Bocek P.(2002). *Recent progress in capillary isotachophoresis*, pp. 3858-3864. Electrophoresis 23.

Letelier M.F. and Siginer D.A. (2003). *Secondary Flows of Viscoelastic Liquids in Straight Tubes*. pp. 5081-5095. International Journal of Solids and Structures, N°40.

Letelier, M.F. and Siginer, D.A. (2007). *On the Flow of a Class of Viscoinelastic-Viscoplastic Fluids in Tubes of Non-circular Contour*, pp. 873 – 881. International Journal of Engineering Science, Vol. 45, Issue 11.

Letelier, M.F.; Leutheusser, H.J.; Member, ASCE, and Marquez Z., (1995). *Laminar fluid transients in conduits of arbitrary cross section*, pp.1069-1074. Journal of Engineering Mechanics.

Letelier, M.F.; Siginer D.A. and Cáceres, C.G. (2002). *Pulsating Viscoelastic Flow in Tubes of Arbitrary Cross-Secction, Part I: Longitudinal Field*. International Journal of Non-Linear Mechanics.

Letelier, M.F.; Siginer, D.A.;. Stockle, J.S and Huilcan A. P. (2010). *Laminar flow in tubes of circular cross section with arbitrary axial variation. ASME* International Mechanical Engineering Congress and Exposition..

Mathies R.A. and Huang X.C.(1992). *Capillary array electrophoresis: an approach to high-speed, high-throughput DNA sequencing* pp. 167-169. Nature 359.

Pedley T. J. (2008). *The Fluid Mechanics of Large Blood Vessels*.   Cambridge University Press.

Sommer G.J.; Chang D. S.; Jain A.; Langelier S. M.; Park S.; Rhee M.; Wang F.; Zeitun R. I. and Burns M. A. (2008). *Introduction to microfluidics*, pp.1- 33. In microfluidics for biological applications. W-C, Tian and E. Fine hout, editor, Springer.

Srivastava N.; Davenport RD.; Burns MA. (2005). *Nanoliter viscometer for analyzing blood plasma and other liquid samples*, pp. 383 - 392. Anal Chem. 77.

Stevenson, B. (2010). *Análisis del campo de velocidades y de las condiciones de existencia y características de la zona tapón en flujos plásticos en ductos de secciones arbitrarias*. Mechanical Engineering thesis, Master degree, University of Santiago of Chile.

Svensson, P.J. (2010). *Análisis de la zona tapón en flujo plástico en canales ondulados*. Mechanical Engineering thesis, University of Santiago of Chile.

Woolley A. and Mathies R. (1994). *Ultra-high-speed and fragment separations using microfabricated capillary array electrophoresis chips*, pp. 11348 - 11352. Proceedings of the National Academy of Sciences 91.

Xue, S.C.; Phan-Thien N. and Tanner, R.I. (1995). *Numerical study of secondary flows of viscoelastic fluid in straight pipes by an implicit finite volume method*, pp. 191-213. Journal of Non-Newtonian Fluid Mechanics, 59.

Yager P.; Edwards T.; Fu E.; Helton K.; Nelson K.; Tam MR.; Weigl BH.(2006) *Microfluidic diagnostic technologies for global public health*, pp.412 - 418. Nature 442.

11

CSA – Clinical Stress Assessment

Sepp Porta et al.*
¹*Institute of Applied Stress Research, Judendorf – Strassengel,*
²*Institute of Pathophysiology, Medical University of Graz,*
³*Institute of Mathematics and Scientific Computing, KFU Graz,*
⁴*Rehabilitation Clinique of the AUVA, Tobelbad,*
⁵*Theresianische Militärakademie, Wiener Neustadt,*
⁶*St. Anna Hospital, Herne,*
[1,2,3,4,5]*Austria*
⁶*Germany*

1. Introduction

1.1 Theoretical background – Outlook to applications

What is stress?

Although stress may be defined seemingly differently by endocrinologists, physiologists, psychologists, a.s.o. prefering the tools of their own, specialized trades, the general denominator and most important fact to remember is, that stress never ever impacts somehow upon you or threatens you from somewhere. Stress is always and only your individual typical reaction to something beginning with a menace and ending perhaps with a slight challenge typical for everyday life. Your personal reaction to such provocations is called stress. The provocation itself is not at all a stress but is called "stressor". Those two technicalities are regrettably often confused by journalists, so that the word "stress" became an exceedingly wooly term.

But if we once agree upon the reactive nature of stress, further reasoning is simple: If your efforts to remove a provocation turn out to be too feeble, the provocation remains and your unsuccessful efforts become chronic – chronic stress ensues. Although perhaps a bit too feeble to remove the provocation, your stress efforts are still using up more than the portion of energy which you have allotted to the routine running of events. Thus your whole system needs more fuel over a longer time, which in its turn tends to exhaust your energy reserves. Your efforts grow feebler still – burnout threatens.

* Gertrud W. Desch, Harald Gell, Karl Pichlkastner, Reinhard Slanic, Josef Porta, Gerd Korisek, Martin Ecker and Klaus Kisters
1 Institute of Applied Stress Research, Judendorf – Strassengel, Austria,
2 Institute of Pathophysiology, Medical University of Graz, Austria,
3 Institute of Mathematics and Scientific Computing, KFU Graz, Austria,
4 Rehabilitation Clinique of the AUVA, Tobelbad, Austria,
5 Theresianische Militärakademie, Wiener Neustadt, Austria,
6 St. Anna Hospital, Herne, Germany.

It is interesting, that such reasoning describes on the one hand the well known development of chronically increased energy turnover into exhaustion and burnout. On the other hand it shows, that increasing exhaustion also increasingly curtails successful reactions to immanent provocations, meaning that those provocations cannot be fought with adequate reactions any more – an exhausted subject cannot mobilize enough reserves to fend off a challenge – there is not enough stress available to cope successfully. Thus one could appreciate the nonsense of statements repeated in journals over and over again, to "dismantle your stress" or "let your stress phase out". Far from getting rid of a personal reaction which may successfully release you from an impending menace, one has to fight the menace itself, which can be only done by successfully mobilizing ones reserves.

All those different reactions of the organism, due to differing workload intensities and different duration leading to the symptoms just described, can be quantified by a multiple parameter assessment called CSA.

Physiological aspects of stress and their utilization for stress assessment

Stress situations incite changes of Adrenaline and Noradrenaline which, in their turn effect variations in blood pH, CO_2, O_2, buffer parameters like BE or HCO_3, lactate and blood glucose as well as electrolytes like K, Na, Ca and Mg. Since we could show as far back as 1991 (1, 2), that those stress hormone effects do correlate highly significantly with adrenaline and noradrenaline changes themselves, the tedious, costly and time consuming catecholamine determination by HPLC could – at least for the purposes with which we tend to deal in this chapter – be abandoned in favour of a much quicker method for determination of stress induced metabolic changes. Estimation of those metabolic effects has the additional benefit, that the obstacle of the receptor situation, which influences hormonal effects and thus restricts the meaningfulness of catecholamine determination as a tool for assessing stress intensity is avoided, since all of our parameters depict post receptor effects.

Those non hormonal parameters therefore show interdependent stress hormone effects, which, when determined simultaneously, can be laid over an organism like a data net by especially designed online software which can be taken as the basis of an individually adaptable multi parameter stress index.

In the simplest case of a physical workload e.g., we find workload dependent change in blood glucose, increase in lactate, accompanied by adaptive changes in pCO_2 and/or HCO_3 and baseexcess, softening the lactate impact upon pH. Moreover, typical shifts in Ca, Mg and K provide us with information about the intensity and duration of stress. When e.g. more sensitive parameters like pCO_2 or less sensitive ones like HCO_3 and blood glucose, are determined at the same time in the same sample, they can give a good idea about the duration of the stress, depending upon the relative involvement of the said parameters.

Likewise, change of electrolytes can tell – together with either pH or pCO_2 about the momentary inclination to sportive performance, about the intensity of symapthoadrenal expectation situations or even about the individually felt efforts of competition and – in the long run – even help to diagnose and quantify a possible state of exhaustion. Chronic sympathoadrenal impact upon metabolism could also be detectable in psychopathological diseases like depressive disorders, where the question arises, whether the patients' mental exhaustion may not affect metabolic processes too.

Also, those interconnections of multiple metabolic effects can be used to uncover hitherto less well understood parameter interactions in metabolic diseases like the metabolic syndrome or even diabetes. There the quantity of electrolyte deficiency and its relation to the idiosyncratic behaviour of a chronically affected metabolism may open new aspects of diagnosis and therapy. Finally, diagnosis of mental and physical load leading to exhaustive stress can not only be used in managers and sports persons, but also to the purpose of being better able to judge upon correct treatment of livestock.

In most cases one has to take pains to collect pre- and post workload data.

The hardware used for such an assessment consists of well established determination systems, implemented in most ICUs all over the world. Due to the easy transportability, at least of those two examples shown below in fig.1 and fig.2., they have been used in assessment campaigns, ranging from determination of psychical workload of teachers in schools or managers in industrial plants to the evaluation of fitness of professional ski racing teams in mountain ranges and assessment of the impact of sleep deprivation in soldiers far away from human habitations.

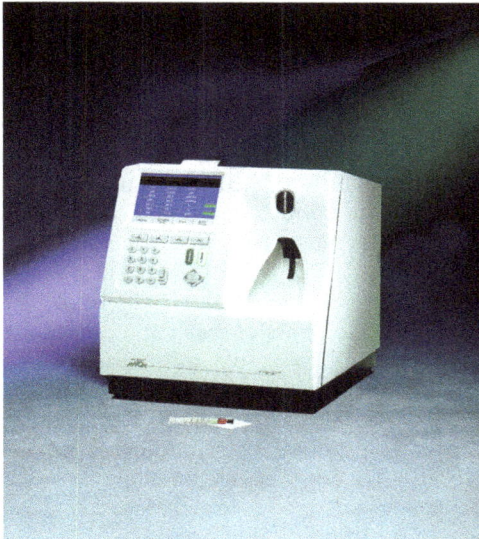

Fig. 1. NOVA CCX (Critical Care Express)   Fig. 2. NOVA Phox - M

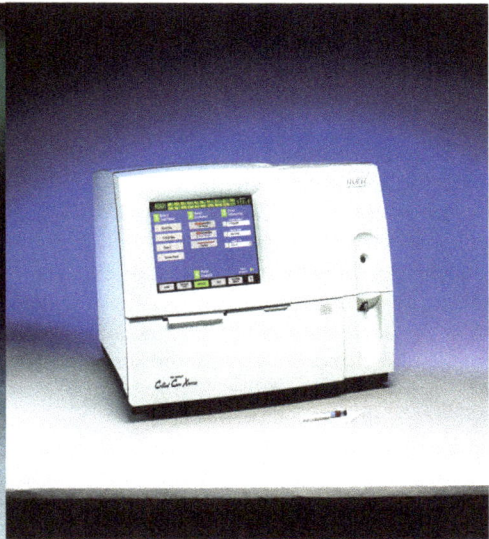

Fig. 1. and 2. Two types of transportable ICU analyzers (dimensions estimable by the syringe in the foreground)

1.2 Practical implementation

Sampling and sample determination – Single persons

Thus, a persons' or an animals' workload, stress compatibility, duration of stress and also the intensity and the kind of stress can be determined within 3 minutes by collecting about 100 microliters of capillary blood, usually from the finger tip. The sample is routinely analyzed for pH, pCO_2, pO_2, O_2saturation, ionized magnesium, ionized potassium, ionized calcium and ionized sodium, lactate, blood glucose, baseexcess and HCO_3 (optionally hemoglobine and hematocrit) using a CCX (Crital Care Express, fig.1)) analyzer (NOVA Biomedical) or a Phox – M (fig.2) of the same producer (NOVA Biomedical), with about the same functions but smaller and even easier transportable and CSA (Clinical Stress Assessment) software. Both devices are widely applied all over the world in Intensive Care Units (ICUs), the software for online data evaluation and interpretation however has been developed by an Austrian corporation (PLK, Judendorf - Strassengel).

Healthy persons are usually checked before and after a standardized ergometric workload, mainly 80 Watts during 8 minutes, or before and after absolving sporting activities, or before and after a standardized psychical load like a Shapiro – test or any kind of training routine the impact of which stands in question. Additionally, the effects upon a persons' metabolism by so called wellness activities - from steam bath to sauna baths, massages etc. - can be investigated, applying the same protocol. In other fields of application even single determinations can be useful e.g. in the course of daily glucose profiles from diabetic patients in rehabilitation hospitals.

Sampling and sample determination – Groups

Such determinations can therefore characterize the reaction of a single person but they also can do the same with a whole group of people. In the latter case the group reaction may serve as a mirror of the typical demands of a certain task on a sample of persons. Thus information about e.g. the usefulness of a bout of training for defined purposes can be collected. All data won from the 100 microliter sample are analysed simultaneously within two minutes, so that total measuring time from blood sampling until the printout of the online processed data takes no more than 3 minutes. Calculations of averages, standard errors of means (SEM), delta values between the groups and linear correlations between all sampled parameters with their regression coefficients are software immanent. This means, that basic group statistics are available immediately after the testing of the last group member. The automatic correlation of every single parameter with all others comes in useful in many ways as we will see during the progress of the chapter. It allows us to look behind the equalizing group averages, thus enabling us to quantify the position of every group member from the point of view of a certain parameter combination. Moreover, the automatically emphasized numbers of significant regression coefficients in the correlation table creates a quickly recognizable pattern of typical group behaviour, as explained in fig. 3. Seen for the first time, this coefficient tables seem somehow crowded with data, but after a short acclimatisation one appreciates the quick oversight over all relevant interconnections of the measured parameters under different conditions:

In the first table (correlations day 1 before load) there are 7 significant linear correlations between stress related parameters. In the second table, describing the correlative situation of the same parameters and the same group, but this time after workload (military obstacle run – HIB), the number of correlations more than doubles to 15. Especially the increase of correlations with lactate and pCO_2 after workload and the occasional difference in plus/minus signs are noteworthy. As we will endeavour to demonstrate, such correlative views can show at a glance, whether the present workload of a group still allows overcompensation (see below) or whether the group seems to be on the brink of exhaustion already. The usefulness of this tool in preventive medicine is obvious. The results of more than 2000 patients and experiments have been recorded in our data banks and published in about 60 papers and printed abstracts, thus providing comparative material for easier interpretation of results.

Data correlation before and after workload

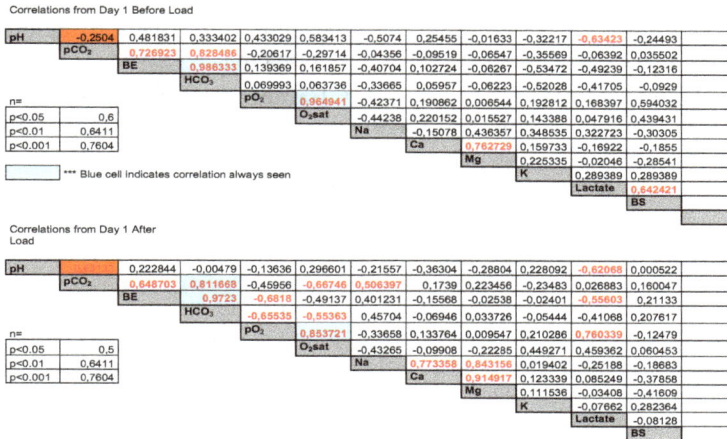

Correlations from Day 1 Before Load

	pCO₂	BE	HCO₃	pO₂	O₂sat	Na	Ca	Mg	K	Lactate	BS
pH	-0,2504	0,481831	0,333402	0,433029	0,583413	-0,5074	0,25455	-0,01633	-0,32217	-0,63423	-0,24493
pCO₂		0,726923	0,828486	-0,20617	-0,29714	-0,04356	-0,09519	-0,06547	-0,35569	-0,06392	0,035502
BE			0,986333	0,139369	0,161857	-0,40704	0,102724	-0,06267	-0,53472	-0,49239	-0,12316
HCO₃				0,069993	0,063736	-0,33665	0,05957	-0,06223	-0,52028	-0,41705	-0,0929
pO₂					0,964941	-0,42371	0,190862	0,006544	0,192812	0,168397	0,594032
O₂sat						-0,44238	0,220152	0,015527	0,143388	0,047916	0,439431
Na							-0,15078	0,436357	0,348535	0,322723	-0,30305
Ca								0,762729	0,159733	-0,16922	-0,1855
Mg									0,225335	-0,02046	-0,28541
K										0,289389	0,289389
Lactate											0,642421
BS											

n=	
p<0.05	0,6
p<0.01	0,6411
p<0.001	0,7604

*** Blue cell indicates correlation always seen

Correlations from Day 1 After Load

	pCO₂	BE	HCO₃	pO₂	O₂sat	Na	Ca	Mg	K	Lactate	BS
pH	0,222844	-0,00479	-0,13636	0,296601	-0,21557	-0,36304	-0,28804	0,228092	-0,62068	0,000522	
pCO₂		0,648703	0,811668	-0,45956	-0,66746	0,506397	0,1739	0,223456	-0,23483	0,026883	0,160047
BE			0,9723	-0,6818	-0,49137	0,401231	-0,15568	-0,02538	-0,02401	-0,55603	0,21133
HCO₃				-0,65535	-0,55363	0,45704	-0,06946	0,033726	-0,05444	-0,41068	0,207617
pO₂					0,853721	-0,33658	0,133764	0,009547	0,210286	0,760339	-0,12479
O₂sat						-0,43265	-0,09908	-0,22285	0,449271	0,459362	0,060453
Na							0,773358	0,843156	0,019402	-0,25188	-0,18683
Ca								0,914917	0,123339	0,085249	-0,37858
Mg									0,111536	-0,03408	-0,41609
K										-0,07662	0,282364
Lactate											-0,08128
BS											

n=	
p<0.05	0,5
p<0.01	0,6411
p<0.001	0,7604

Outprint of CSA – software immanent automatic correlations of all the parameters measured.
Upper triangle: before workload,
Lower triangle: after workload
Red numbers: significant correlation coefficients.

Fig. 3. Example of two tables of regression coefficients of a group before and after workload, useful for quick overall estimation of the changes of interparametric dynamics.

Ethical aspects

All participants in the investigations presented in this chapter consented to the anonymous use of their data after being carefully informed about the aim of the study according to the Helsinki Charter (http://www.wma.net/e/ethicsunit/helsinki.htm.). The most recent submission to and passing of our experimental intentions by the ethical commission of the Austrian Federal Ministry for Defence and Sports (BMfLVS) dated from Dec. 13th, 2010.

1.3 Application of CSA

Our own experience of CSA application ranges from investigations into the behaviour of teachers, students, triathletes, aircraft pilots, bungee jumpers, military combat groups, special police units, rescue teams, sleep deprivation, free radical research, patient's stress in rehabilitation clinics to the impact of wellness treatments, even to the exhaustion of farmers providing holidays for tourists additionally to their usual tasks.

The following four applications that will be discussed in the course of the chapter form a concentrated substrate, presented to make the reader think about own, customized applications:

1. Determination of the impact of sport training or training of so called "first responders", like military training units, special police groups, fire fighters and others. We maintain that it is not only possible to link changes of blood parameters with sportive success but also to predict success chances before competition or deployment.
2. Determination of the quantity of mental stress.
3. In the field of internal medicine mainly idiosyncrasies of diabetic metabolism, especially those due to the newly found importance of mineral deficiencies in type2 diabetics

which, by CSA application, we were able to link up with the deterioration of other metabolic disturbances.

4. Qualification and quantifications as well as predictions of success chances in competing animals like horses or camels and also prevention of cruelty to animals by stress documentations are one our next step of development.

ad 1: Determination of the impact of sport training or training of so called "first responders", like military training units, special police groups, fire fighters and others

Up to now the decision about a persons' fitness for a certain competition mainly rests upon the trainers' subjective adjudication, bolstered by lactate tests or even more demanding workout procedures and a more or less profound experience. Thus the availability of a hardly molesting, objective, in depth assessment of competitors and first responder personnel concerning their momentary ability to perform a certain task (3, 4, 5) comes in useful. Investigations in that area revealed not unexpectedly, that the metabolic situation of a person before a contest contributes decisively to the degree of the later success. Therefore it could be advantageous if we would be able to quantify the metabolic turnover before the contest in each case. Because, by quantifying pre contest metabolic situations, both the individual position within a group of contesters could be determined and eventually significant deviations of possible group outsiders could be marked down, understood and subsequently discussed with the person in question. That e.g. exhausted soldiers or sportsmen are no more able to perform satisfactorily is a truism. However, there are unsatisfactory performances which are less well explainable. The most common reasons are mostly privately known to the performer but not eagerly revealed to the trainer or group commander like lack of sleep due to entertainments during the previous evening.

But also unexpected bouts of good or bad performances occur, unexplainable to both performer and trainer. Moreover, the very same scores obtained easily by performer A could have been demanding for performer B, so that equal scoring may not mean equal potential at all. Even slight influences during the pre contest situation which are hardly felt and therefore frequently ignored, like temperature differences or even the changing of a routine can be reasons for a significant shift in interdependencies of electrolytes and metabolic parameters with measurable impact upon the performance to follow. Especially correlative changes between Mg, K, the Ca/Mg quotient or, to a lesser extent Ca alone with H^+ donors like lactate and the consecutive pH and blood buffer situation react rather sensibly to changes of the sympathoadrenal situation in man and even in horses (3,4,5,6). Since in our experience the demand for such determination focuses upon comparatively small groups and their individual members, we tried to offset the comparatively small number of experiments by software immanent statistics from diverse points of view to keep results controlled as strictly as possible.

Since our assessment provides us with at least 24 interconnected data per person (from determinations before and after workload), a more comprehensive description than that by the widely used lactate test or by catecholamine assay alone is possible.

It is remarkable, that nearly the same small amount of capillary blood which is still routinely used for lactate tests could easily yield eleven times more information instead of the single lactate determination. As an example we would like to show an investigation of 14 Ensigns of the Theresianische Militärakademie in Wiener Neustadt, Austria before and after a demanding military obstacle race of 3 – 5 minutes duration. It turned out that the metabolic changes in the experiments due to that military obstacle race were considerable, although the overall duration did not exceed five minutes:

Changes in group averages before and after the obstacle race (*military steeplechase, Hindernisbahn - HIB*) are shown in table 1: An expected significant change in stress dependent parameters like pH, pCO2, BE, HCO3 and K, due to severe exertions was visible. The averages of Mg and Ca however did not change significantly.

	pH	pCO$_2$	BE	HCO$_3$	pO$_2$	O$_2$sat	Na	Ca	Mg	K	Latate	BS
Averages before HIB	7,433	34,714	-1,114	23,400	68,293	94,164	145,886	1,135	0,514	4,339	1,793	102,357
+ / - SEM	0,005059	0,568578	0,419295	0,383162	1,5218	0,522737	0,429906	0,008238	0,008228	0,077598	0,11117	1,99892
Averages after HIB	7,257214	26,14286	-15,3643	12,01429	92,59286	96,29286	146,3714	1,140714	0,515	3,805	12,98462	151,2857
+ / -SEM	0,020643	0,53171	1,048002	0,72323	2,565461	0,215464	0,39103	0,009167	0,009931	0,093631	0,864019	6,807743

Average group values before and after sports +/- standard errors of means (SEM). Changes in pH, pCO2, BE, HCO3, pO2, O2sat., K, Ca, lactate and blood glucose are highly significant (two sided t- tests (p<0,01), (7))

Table 1. CSA outprint, example of the change of group averages due to a military obstacle race (HIB).

Beyond the information which the average group values provide, we were able to uncover four different facts by correlating appropriate parameters of the pre- contest situation.

a. The phenomenon of "overcompensation" in the pre contest phase can be detected and its effect upon the assessment of a persons' fitness and sportive compatibility can be discussed.

b. The impact of pre contest sympathoadrenal arousal upon the metabolism of each group member can be quantified. If outliers are present, they can be characterized and marked down for interviews

c. Connections between the now quantifiable position of each group member in the pre contest phase and the definite scoring during the contest can be established. Success/effort relationship can be characterized.

In other words, the answers to those four points provide the trainer with comprehensive information about the individual pre contest situation of each group member, whether the adrenergic arousal before contest remains within beneficiary boundaries or already uses up too much energy which would be more profitable employed later during the contest. They detect outsiders and even provide the trainer with scoring - chance predictions for the contest to follow. How is it done? First of all we have to become familiar with the usual state of affairs before a contest, the knowledge of which seems to be not usual at all. We call it "overcompensation". Different group members tend to express it in differing quantities.

a. Overcompensation

Slight to moderate mental or physical load frequently results in metabolic overcompensation. Simplified, the well known respiratory compensation of metabolic acidity can become over efficient, so that persons with increased lactate or other H$^+$ donors end up with a more alkaline pH brought about by a disproportionally increased breathing frequency, connected of course with an equally disproportional loss of CO2. The benefit of this seemingly wasteful behaviour is a kind of run up into higher pH regions to premeditate a later fall into dangerous acidity by an eventually more demanding workload in the immediate future which the organism seems to prudently forestall. Accumulation of O2 in the blood because of the more alkaline conditions points the same way. A good example of such a reaction is the pre contest situation before the military steeplechase concerning pH/pCO2 relationship. (fig. 4). This figure deals with and quantifies an "over successful" removal of acidity from the blood by increased breathing frequency.

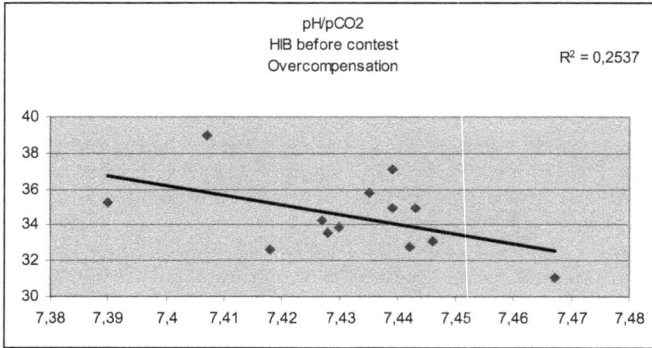

Abscissa: pH
Ordinate: pCO2 in mmHg
P< 0,05, significant
Situation before the military steeplechase contest.

Fig. 4. Situation before contest: Increased breathing frequency leads to diminishing pCO2 and consecutive pH increase.

The highest breathing frequency, borne out by the most pronounced loss of CO2 leads to the most alkaline pH. This means, that here the most pronounced metabolic activity, expressed by the highest breathing frequency, paradoxically yields the highest pH values (fig. 4) The picture is completely contrary to the familiar concomitant fall in pH and pCO2 during pronounced physical action (contest), which is shown in the next graph to underline the striking differences between the "warming up"(fig.4) and the real contest situation (fig.5): pH / pCO2 after workload

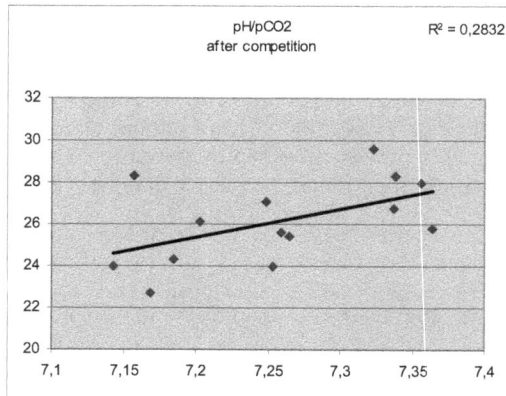

Abscissa: pH
Ordinate: pCO2 in mmHg
P< 0,01, highly significant
Situation after the military steeplechase contest.

Fig. 5. During the demanding military exercise pH/pCO2 relation turns around, CO2 release is now unable to prohibit the fall in pH (note the extremely low pH and pCO2 values due to the heavy workload).

The adrenaline induced slight, individually different lactate increase in the pre contest situation should lead to a concomitant decrease in pH. But since at the same time H ions are indirectly got rid of by increased breathing, proportional pH decrease along with lactate increase is counteracted. Consequently, in the pre contest situation, pH and lactate do not correlate positively any more – as they do during the contest – due to the increased loss of CO_2 during the said overcompensation, as shown in fig.6.

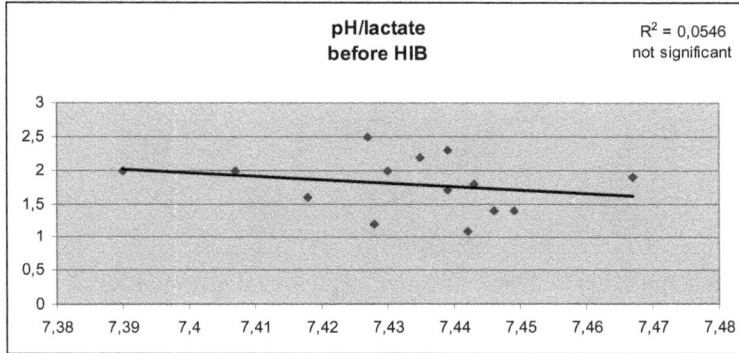

Abscissa: pH
Ordinate: lactate in mM/l
p>0,05, not significant
Situation before the military steeplechase contest.

Fig. 6. Due to successful CO_2 control of pH before contest, pH/lactate relationship vanishes.

Consequently again, all our following correlative graphs of pre contest situations dealing at least with either pCO_2 or pH have to be interpreted as situations, when the highest pH and the lowest pCO_2 occur in the person with the most clearly increased metabolism. Having ourselves used those automatic correlative evaluations regularly, we came across overcompensation surprisingly frequently, mostly in pre contest situations or other moments of sympathoadrenal arousal. Therefore we would like to forward the supposition that overcompensation is a general feat of adaptation to probable demands in the future and thereby possibly an important part of evolutionary survival strategy.

b. Quantification of pre- contest conditions and characterization of outliers

Following up our suppositions of a possible impact of the pre- contest situation upon the contest proper, we checked as a first step the K/pH proportions of the experimentees, because a possible sympatho – adrenal arousal in expectation of the contest may well lead to an individual increase of lactate values (the average increase being only a slight one, see tab.1), consequently to increased H ions, which would be exchanged with K ions from the tissue in a rate presumably proportional to the H ion production and therefore proportional to catecholamine impact. Indeed, a highly significant, but negative correlation between pH and Ionized K ensues, positioning the most pronounced K loss along with the highest metabolic turnover, characterized in this overcompensating situation by the highest pH and lowest K values. A combination of electrolyte- and metabolic parameters therefore are seemingly able to characterize typical group idiosyncrasies (overcompensation in this case) as well as the individual position of the participants within the group (fig.7):

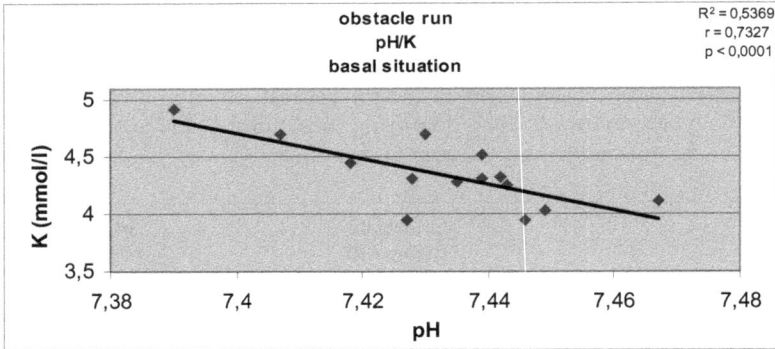

Abscissa: pH
Ordinate: K in mM/l,
p<0,001, highly significant
Correlation between pH and potassium before sports.

Fig. 7. Sympathoadrenal arousal before contest incites overcompensation, also shown by significant inverse pH/K relationship.

In such an adrenergic state of expectation, a further correlation can be expected, namely a proportional behaviour of pH and ionized magnesium, because adrenaline increase changes of pH via the mechanism mentioned above and can also increase ionized Mg in blood (8,9).

However, no significant correlation could be found. This could have been due to two outliers which are marked by an oval inclusion in the graph below (fig.8):

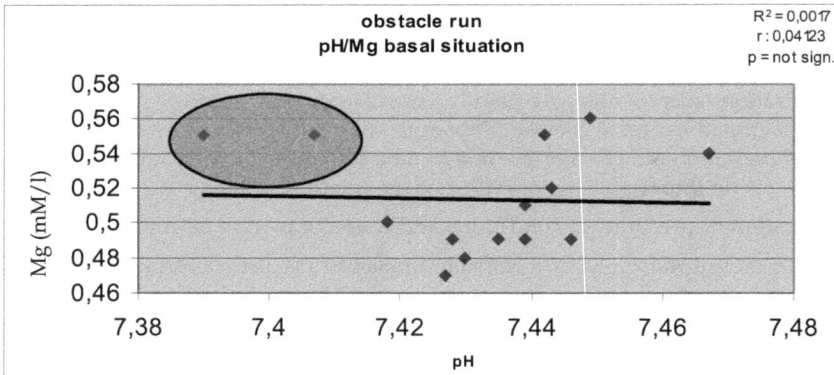

Abscissa: pH
Ordinate: Mg in mM/l,
p>0,05, not significant
points within the elliptic figure: presumptive outliers

Fig. 8. Positive pH/Mg correlation is disturbed by two outliers.

When the outliers are removed, a highly significant positive correlation between ionized Mg and pH in the blood of the pre-contest experimentees evolved (fig.9):

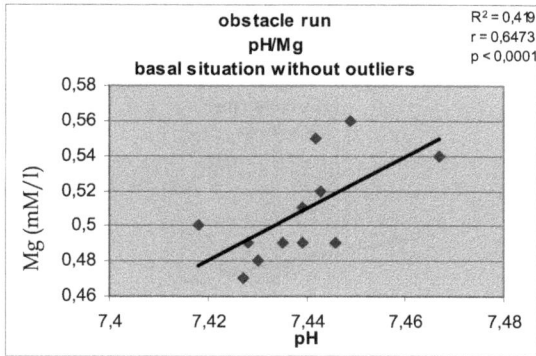

Abscissa: pH
Ordinate: Mg in mM/l,
p<0,001
Correlation between pH and magnesium before sports without outliers.

Fig. 9. Removal of the outliers leads to restored pH/Mg correlation.

Similar as in fig.8, the highest Mg turnover (here Mg increase) goes along with the most pronounced metabolic turnover, again characterized by pH increase, due to overcompensation, provided the outliers have been removed. However, to characterize and/or remove the outliers out of purely statistical reasons is not correct, although it seems obvious, that they are not part of the sample. They both show high Mg values at concomitantly low pH which does not fit the group behaviour at all. On the contrary, this combination of parameters points towards an already most active metabolism before sports, which may not be able any more to meet the subsequently further increased energy turnover, necessary for high scoring. It follows, that they do not develop any sign of overcompensation, as does the rest of the group, since their high Mg values exist concomitantly with low pH, a feat that does nowhere occur in the rest of the overcompensating group. The pre – contest diagnosis of a prematurely increased metabolism of the outliers is consistent with some of the outliers' values *after* sports, forming a different multi parameter pattern:
1.   very high BE values after sports
2.   very high pCO2 values after sports
3.   very high HCO3 values after sports
4.   low scoring (ranking on 12th respectively 13th position within a group of 14).
According to this additional information, we felt justified to separate those two participants from the other members of the group, which – without those two - forms the mentioned well definable, highly significant pH/Mg relationship (fig.3). At least for practical reasons the information of their outlying position - which is only shown by the correlation analysis and would never have come forth by group average calculations alone - should on no account be discarded and at least used for closer observation and extended interviews with the two experimentees.

c. Prediction of success chances by quantitative evaluation of pre- contest conditions

Investigations into interdependencies between basal K and awarded scores (fig.10).For the first time we introduce non CSA values in our correlative interpretations – the awarded scores for the obstacle run, which are nearly identical with ranking of the period of time needed for its absolvation.

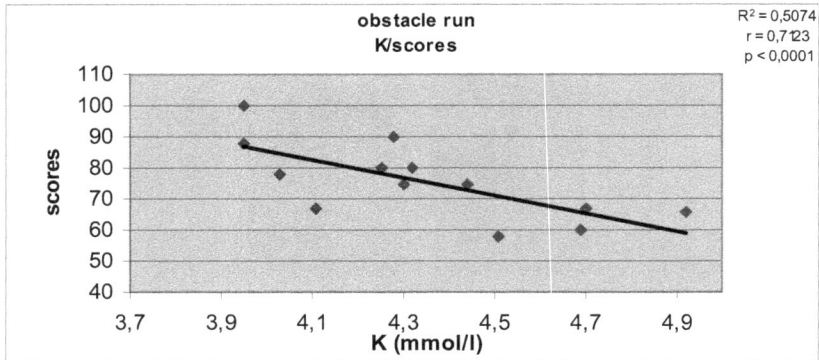

Abscissa: Basal K in mM/l
Ordinate: awarded scores
p<0,001

Fig. 10. Blood K levels before contest possibly predict chances of scoring at the contest proper.

Pre contest K concentrations and the scores awarded after steeplechase correlated negatively in a highly significant manner, meaning that persons with lowest pre- contest K values cherish the best chances for high scoring in the subsequent contest. Lowest K concentrations on the other hand coincide with highest pH levels in the overcompensating group (fig.1), which – under those circumstances – mark high metabolic activity. Within reason therefore, those who have been most successfully mentally "warming up" themselves, stand to be rewarded with better scoring chances. It is important, not to be led astray by the high scoring of contestants with an alkaline starting position, Slightly alkaline pH in this context is definitely not a sign of low metabolic turnover, but, as we have already been able to demonstrate, a feat of increased energy turnover by sympathoadrenal arousal, what we just called "mental warming up". Summing up the information of the graphs hitherto presented, the pre contest metabolic pattern of a presumable high scorer seems to be:
1.   Low potassium (fig. 7)
2.   High pH (fig 1))
3.   High Mg (fig. 9)
4.   Low pCO2 (fig. 4)
This multi parameter pattern shows, that increased metabolism characterized by the most pronounced loss of CO2, increases Mg clearance from the tissue and also increase blood pH. Since pre- contest pH is already increased by excessive CO2 loss, there is no urgent need for cation exchange, which is underlined by the low blood K values. Roughly spoken, K seemingly is allowed to stay put within the tissue, regardless of increased metabolism. Such a tissue reserve of readily exchangeable K ions however, could facilitate a more successful H ion removal into tissues later, during the demanding contest expected. That lactate values did not enter our multi parameter pattern more prominently can be explained by the failing

correlation between lactate and pH. Increased breathing frequency obviously buffers direct lactate impact. Also – during the predominantly mental stress before contest – a participation of free fatty acids from catecholamine induced beta oxidation is to be expected and can indeed be roughly calculated by subtracting lactate values from baseexcess (both in mM/l)

d. Characterization of success/effort relationships

The role of certain electrolytes in blood, changed by the psychic arousal of the pre- contest situation could be shown already by correlating them with metabolic parameters like pH or pCO2. But they play a further role – again together with metabolic parameters - as indicators of competition success, part of which we have already exemplarily shown by the predictive power of K changes before competition. During our investigations of blood samples before and after the military obstacle run (HIB), however, we found connections of lactate and Mg changes with running times and scoring which contributed substantially to a better understanding of a contestants' attitude to the task. This may come in useful for basic research about the role of Mg in energy turnover but also has practical importance in such cases when one wants to check effort and performance of tasks where other means of objectivation cannot be applied or one is simply not present.

Therefore let us have a look into the Mg/lactate relationship, which we purport to be especially apt to reveal the individual attitude towards the contest: Although individual Mg+ changes were rather pronounced, no corresponding *average* Mg+ increase or decrease was visible because Mg+ changes often pointed in opposite directions. This individual behaviour of the Mg+ changes however correlates highly significantly with lactate changes. Fig.11 shows, that sometimes more information can be gained by using the – also automatically compiled – delta values, the changes between the values before and after workload and not the measured values themselves. Thus clear proportionalities between effort and electrolyte turnover could be shown.

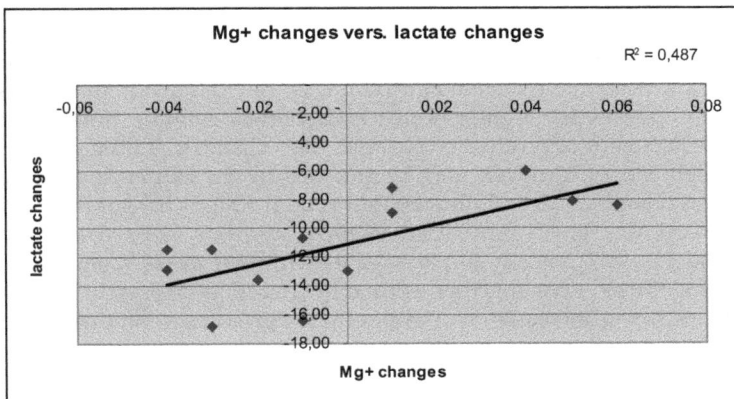

Abscissa: Mg+ changes (delta values: Mg+ before exercise minus Mg+ after exercise)
Ordinate: Lactate changes (delta values: lactate before exercise minus lactate after exercise)
$P < 0,001$, highly significant

Fig. 11. Documentation of delta values sometimes are better able to reveal parameter interactions than absolute concentrations.

Higher lactate increases beyond 11 mM/l correspond with Mg+ increase, lactate changes below that value corresponds with Mg+ decrease in blood. That lactate changes do not correlate with both awarded scores and running time is not surprising, since it is common knowledge that subjects with better fitness may score higher with less lactate increase than less well trained subjects with higher lactate increase (3). But unexpectedly, Mg+ changes (automatically computed delta values) did correlate with awarded scores and running times as well, highly significantly in a polynomial curve of the 2nd order. Fig. 12 shows, that Mg changes not only correlate with lactate changes (effort) but also with awarded scores, but now in a different, polynomial manner.

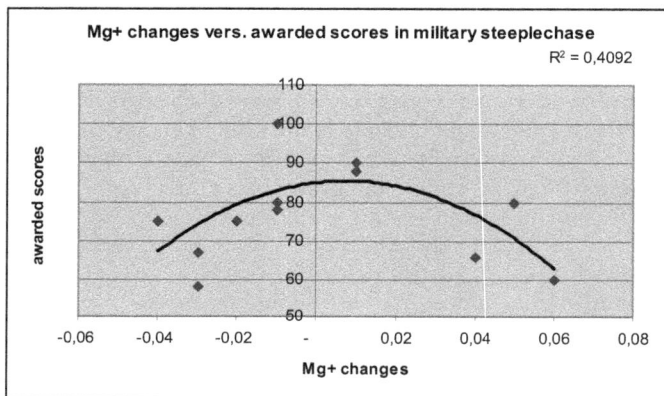

Abscissa: Mg+ changes (delta values)
Ordinate: scores awarded
P< 0,001, highly significant

Fig. 12. Mg delta values this time plotted against scores, once more prove their usefulness.

Since changes in blood Mg+ are to be considered as subtractions of Mg+ influx from tissue and Mg+ redistribution (10) by clearance from the blood, people with the best balanced Mg+ in- and efflux show nearly zero deviation. And it is exactly this group that has been shown to have the best chances for highest scores and shortest running times. Moreover, the supposed unpredictability of positive or negative Mg+ changes, at least during short term exercise with high energy turnover, may be qualified by the existence of this polynomial correlation between both scores and running times and the Mg+ changes in question.

Distinctly increased Mg+ blood concentrations therefore, went along with the high lactate increases beyond 11 mM/l, mostly in participants with low scores. Contrarily, low scores and long running times went together with pronounced decreases in Mg+ along with comparatively low lactate increases otherwise seen in high scorers, but then with hardly any Mg+ changes.

Thus, a typical combination of lactate and the obviously quicker Mg changes is characteristic not only for a certain score, but also provides more information about the reason for low scoring, when one combines the information of fig.4 and fig.5: High Mg+ increase is mostly associated with pronounced lactate increase and pronounced Mg+ decrease with a relatively small lactate increase for this demanding kind of exercise. All the high lactate increases therefore are on the side of Mg+ increase and vice versa. Moderate lactate increase, along with moderate Mg+ change seem to characterize the reaction of a well trained subject.

One is tempted to deduce, that in our short and demanding exercise the mark of an average delta value of 11mM/l of lactate may be the turning point between a preponderance of Mg+ clearance from the blood during the first stages of the exercise and a consecutively increased Mg+ influx into the blood, overcoming the clearance rate, because of considerably increased demand upon muscular tissue.

This turning point should be demonstrable by a polynomial curve, which indeed it has been. It furthermore may even serve as a kind of standardization mark, beyond which additional lactate increase (effort) does correspond less and less with effectiveness. A possible practical application of the combined information of fig.4 and fig.5 could be i.a. an objectivation of subjective adjudications of any persons' effort and success. Hitherto, mostly increases of Mg+ blood concentrations during short term exercise have been advocated, while Mg+ decrease has been mainly associated with longer lasting workloads (11, 12, 13) Biphasic Mg+ changes, at least during short term exercises have not been described up to now. As mentioned above, the results of such an investigation allow basic considerations about new aspects of the role of electrolytes as well as pave the way towards some practical progress in adjudication of success and effort relationship.

ad.2: Metabolic changes due to mental stress in depressive patients

Having been able to show that the predominantly psychically induced change in metabolic parameters before a contest can be measured and thus quantification of psychical arousal by metabolic determinations can be attempted at least proportionally, we would like to show an application of this idea at psychiatric in - patients. Our clinical study included 19 patients (17 females and 2 males) with a mean age of 44 years (range from 24 to 65, with a median of 44). All of them were suffering from major depressive disorders (Hamilton depression scale from 18 to 33). We compared them with a group of .46 subjects (35 males and 11 females, nearly equally aged) Before and after a slight ergometric effort (60 watts for 6 minutes) capillary blood samples were drawn as described above and the resulting group averages

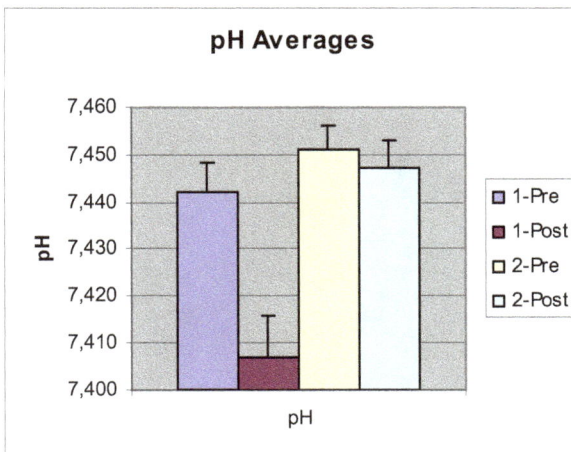

Abscissa: determination situations (see text)
Ordinate: pH

Fig. 13.

for pH, pCO2, Basexcess and Magnesium are shown in the next four automatically generated graphs. The first two columns of each graph show the group averages before (col.1) and after workload (col.2) of the psychiatric patients, columns 3 and 4 show the group averages of the equally treated healthy group

Fig. 13: Figs. 13, 14, 15 and 16 deal with the average changes of pH (fig.13), pCO2 (fig.14), base excess (fig.15) and magnesium (fig.16) of psychiatric in – patients (blue and red column) and a matched control group from our data banks (yellow and green columns). Remark the much more sensible reaction of the patients to workload.

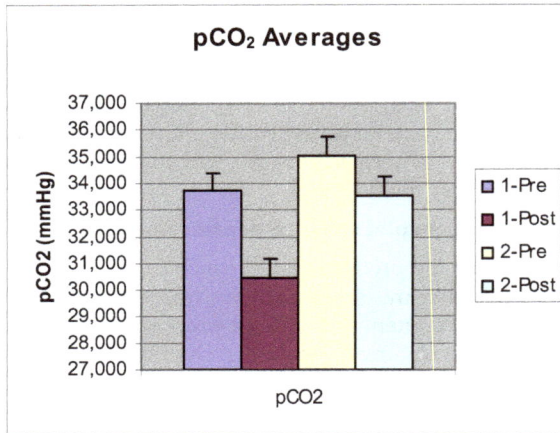

Abscissa: determination situations (see text)
Ordinate: pCO2 in mmHg

Fig. 14.

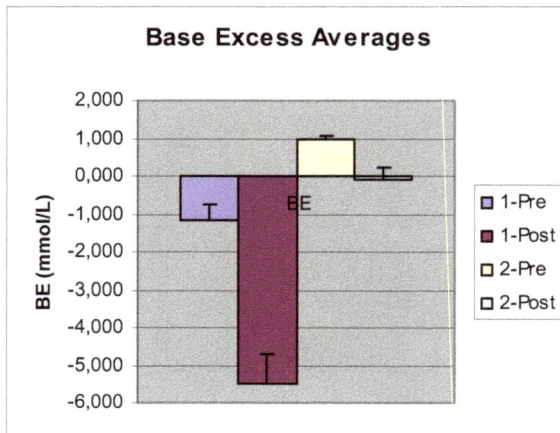

Abscissa: determination situations (see text)
Ordinate: Baseexcess in mM/l

Fig. 15.

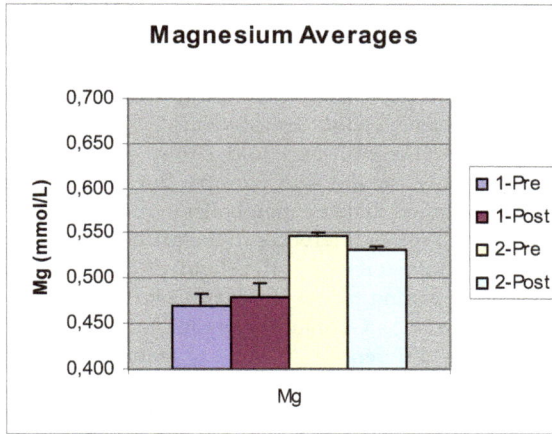

Abscissa: determination situations (see text)
Ordinate: Ionized Mg in mM

Fig. 16.

With the exception of the baseexcess there were no significant differences in the basal situation between the depressive and the healthy group. Since the significantly lower buffer capacity, shown by the baseexcess values of the depressive group (fig. 15, col.1 and col. 3) cannot be due to acutely increased breathing, (fig. 14, col.1 and col.3) it has to be acknowledged as a chronic buffer diminishment of longer standing, developed in the course of the illness. This is underlined by the significantly and clearly more intensive reaction to the slight workload by the depressive patients. Their pCO2, their pH and their baseexcess react disproportionately sensitive to the moderate workload. Such an expected accumulation of over sensitive reactions to daily demands may well have been the reason for a chronic decrease of their total buffer capacity in the course of the illness. Accordingly, our investigations into the differences of metabolic reaction between depressive and healthy people yield – just by glancing at the automatically generated graphs and average statistics - two general results:

- The well known unwillingness of depressive patients or burnout patients to perform bodily feats is obviously not only due to depressive moods but also to a handicapped metabolism, demonstrable mainly by a disproportional reaction to provocation. Careful attempts to increase bodily fitness of the patients may be rewarded.
- Although metabolic tests can never be used diagnostically in mental illnesses, repeated checks of the metabolic situation could be helpful for documentation and quantification e.g. of the progress of medication.

ad 3: Idiosyncrasies of the diabetic metabolism, especially those due to the newly found importance of mineral deficiencies in type2 diabetics uncovered by CSA diagnosis

The blood glucose status of diabetic in - patients is routinely checked by a daily glucose profile, which consists of glucose determination from capillary blood at three different times. With nearly the same small amount of blood and the same effort we could determine not only glucose but also 11 additional parameters. The results of those investigations have been published in some full papers and several congress abstracts. We could show i.a. that not only type1 diabetics but also about 36% of the nearly tenfold higher number of type2 diabetics suffer from severe loss of electrolytes, especially from hypomagnesemia. However, the magnesium state of type 2 diabetics has not been considered to be crucially important for the patients' wellbeing up to now, since only easily treatable cramps were thought to ensue from magnesium deficiency. But by correlatively combining of some of our simultaneously determined parameters, we could show, that diabetic hypomagnesemia seems to be responsible not only for the said cramps, but for a whole series of negative influences upon the already strained diabetic metabolism:

Let us direct, e.g. our attention to some differences in metabolic behaviour in patients with Mg levels below and above the hypomagnesemic threshold of 0,45mM/l ionized Mg in blood (hypomagnesemic threshold according to the Austrian Consensus Conference as well as the Deutsche Gesellschaft für Ernährung – German Society for Nutrition) and exemplarily look at some facts accompanying those differences: As already mentioned above, severe deficits in ionized blood magnesium became increasingly conspicuous during investigations into interactions of blood glucose, buffers and electrolytes during daily glucose profiles of type2 diabetic patients, since we had the opportunity of magnesium determination with ion sensitive electrodes (NOVA CCX, CSA). This fraction, according to our knowledge, has not been compared yet with blood glucose metabolism in type2 diabetic patients to any larger extent.

Similarly, investigations about the behaviour of ionized calcium in type2 diabetics seem at least to be rare. Its average values in our patients are, like those of magnesium, very low. Also remarkably low were the base excesses of the patients, though lactate concentrations in blood did not exceed normal elevations found on moderately busy days.

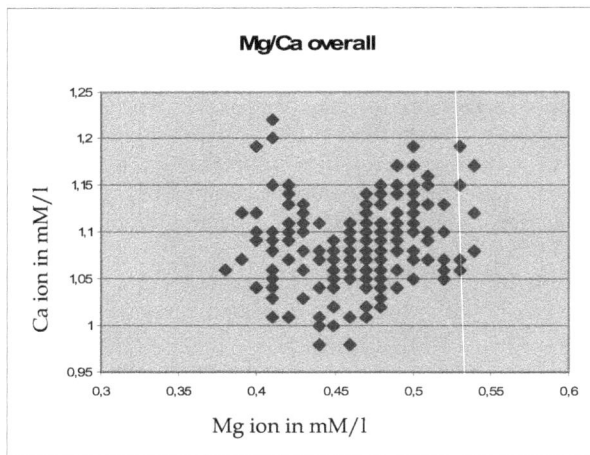

Abscissa. Ionized Ca in mM/l
Ordinate: ionized Mg in mM/l

Fig. 17. Remarkable change of relationship of Mg and Ca in the blood of diabetics nearly exactly at the point of the hypomagnesiemic threshold.

Magnesium and calcium averages give the impression to be inversely proportional to the concomitant blood glucose values, a feat that has been already mentioned by others together with magnesium and blood glucose or insulin sensitivity (17). But when we put together all single values of all our patients, regardless of sampling times, there was no significant inverse correlation between ionized magnesium and blood glucose. Still, the pattern of the individual points in the Ca/Mg graph was exceptional (fig.6). Exactly at the Mg value of 0,45 mM/l (the agreed hypomagnesemic threshold) the Ca/Mg relationship seemed to switch directions. Fig.17 shows the turnaround of the relationship of ionized Ca with ionized Mg. Obviously, only from the hypomagnesemic threshold (0,45mM/l) upwards a significant, positive correlation has developed.

Even without indrawn regression lines one can observe, that the Ca/Mg ratio takes an opposite course nearly exactly above and below the hypomagnesemic threshold of 0,45 mM/l. We acknowledged this ambivalent behaviour by splitting the sample along this threshold of 0,45 mM/l into a high- and a low Mg subgroup. Consequently, we found a highly significant positive correlation between Mg and Ca in the higher Mg subgroup but no correlation at all in the lower subgroup. This finding encouraged us to look for more correlations, not within the overall sample but again within the subgroups above and below the hypomagnesemic threshold, trying to find at least some hints for the reason of the very low Mg values in our diabetics, where 36 % had Mg concentrations of 0,45mM/l and lower, since the mechanism of low magnesium values in NIDDMs seems to be unclear, Some authors (18,19) discuss a recurrent metabolic acidosis, along with episodes of osmotic diuresis as possibilities among others for magnesium diminishment in the diabetic patients, while stating that this diabetic hypomagnesemia seems to merit poor attention by physicians anyway. Shaffie et al (20) observed a lowering of bicarbonate along with low tissue pCO_2 and hyperventilation.

Indeed, pCO_2 values in our patient sample are low. Additionally, correlations between pH and pCO_2 overall and in the subgroups showed a significantly inverse behaviour, most clearly expressed in the low Mg subgroup, with a slope more than double as steep (y=59x) as in the higher Mg subgroup (y=23,6x) , pointing to an increasing need for respiratory compensation along with diminishing Mg concentrations. Thus, pH seems to be kept at an average of 7, 43 by constant loss of CO2, obviously slightly overcompensating a steady input of anions.

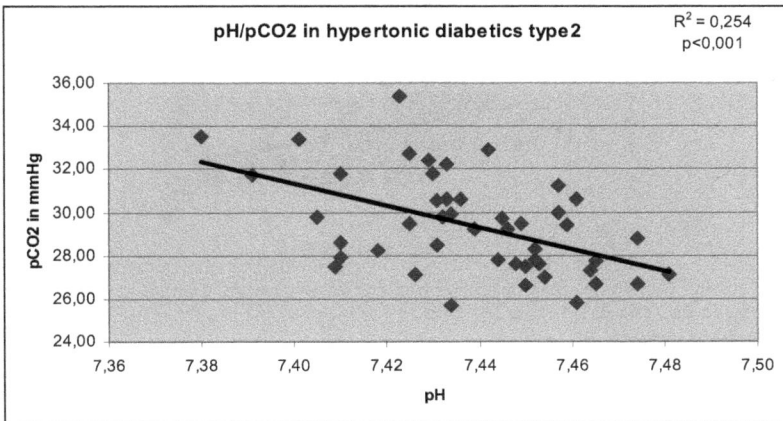

Abscissa. pH
Ordinate: pCO_2 in mmHg
P<0,001 highly significant

Fig. 18. Increasingly lower pCO_2 creates more and more alkaline pH in hypertonic diabetics.

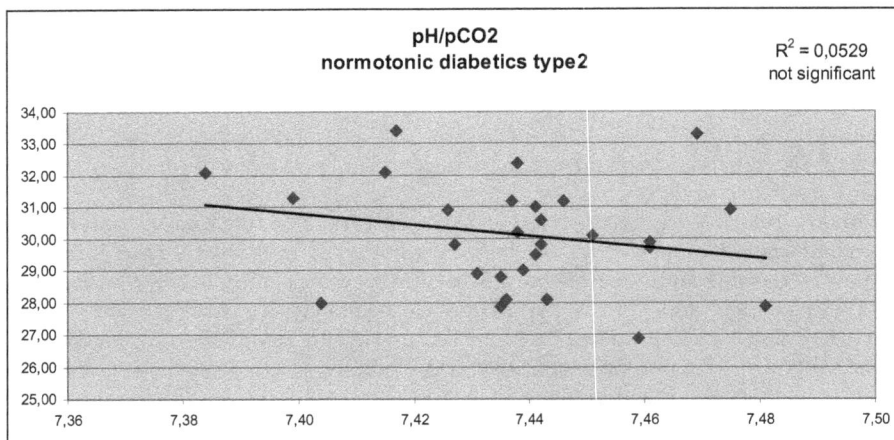

Abscissa. pH
Ordinate: pCO2 in mmHg
P>0,05 not significant

Fig. 19. Normotonic diabetics do not show any relationship between pH and pCO2 (see fig. 18) at all.

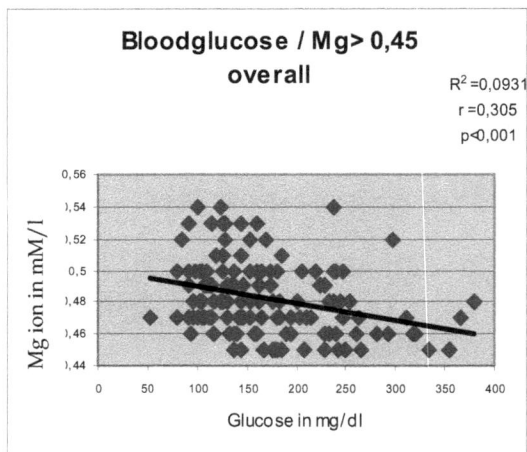

Abscissa. Glucose in mg/dl
Ordinate: ionized Mg in mM/l
P<0,001 highly significant

Fig. 20. Above the hypomagnesiemic threshold blood glucose increases along with lower Mg levels – a good argument for substitution.

Such chronic processes accompanied by a moderately increased breathing frequency may slowly but successfully waste the magnesium (and calcium) resources of the patient. The interesting observation, that patients with Mg concentrations below 0,45mM/l do not seem to show correlations with blood glucose any more, may indeed point towards a certain exhaustion, for which low Mg concentrations are usually characteristic. But low Mg in those patients may be not only a marker of increasing metabolic exhaustion, but could actively contribute to wasteful anaerobic glycolysis by limiting ATP – ADP turnover. At least in non diabetic patients, we could show that low magnesium concentrations before extensive liver surgery deteriorate the prognosis about the final outcome significantly (21). The most important result concerning type2 diabetics is unquestionably the highly significant negative correlation between Mg levels and blood glucose above the hypomagnesemic threshold.

It means that diabetic patients with lower Mg levels are prone to higher blood glucose values.

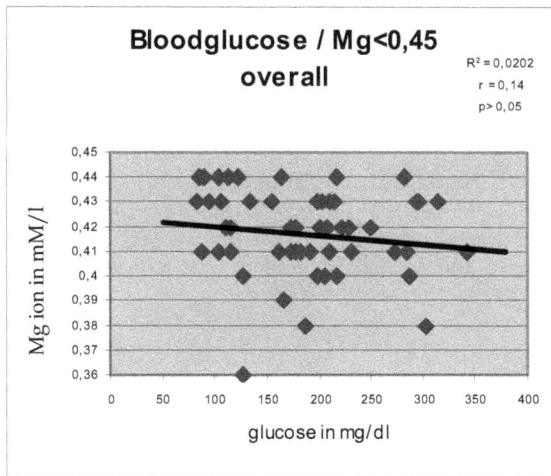

Abscissa. Glucose in mg/dl
Ordinate: ionized Mg in mM/l
P 0,05 not significant

Fig. 21. Below the hypomagnesiemic threshold the calculable Blood glucose/ Mg relationship vanishes, but the number of patients with high glucose values is remarkable.

Therefore we think that Mg determination, especially, that of the more active ionized fraction, should be included into the monitoring at least of hospitalized diabetic patients. When interpreted together with blood glucose levels and other CSA parameters, it reveals a much deeper insight into the metabolic state of the patient. In our opinion, increased knowledge of physicians about the impact of Mg deficiency upon the diabetic (and also non–diabetic) metabolism would increase the demand for magnesium determination and also for magnesium medication.

Consequently, we can see, that without increased effort, just by substituting a more up to date measuring device coupled with appropriate software, the daily glucose profile, a routine diagnostic method of very long standing, could be changed into a much more sensitive investigative tool, capable of quickly unearthing new knowledge about metabolic dynamics for the benefit of the patients.

ad 4: CSA application in animals (outlook)

About 15 years ago we investigated the catecholamine state of immobilized rats and of pigs in abattoirs. Immobilisation of animals led to dramatic increase in catecholamines and to vastly diminished stress compatibility (22, 23) Catecholamine levels of pigs before slaughtering in abattoirs have been found to be an incredible hundred thousand fold higher than normal.

Both, immobilisation and the pre- slaughtering situation are widely common in the treatment of pigs, since mother sows are often kept practically immobilized in very small cages. Objectivation of the effects of the demonstrated catecholamine increase upon the metabolic parameters provided by simple and scarcely molesting CSA testing could reveal at least electrolyte changes in blood, most probably vastly increased electrolyte input into muscle tissue. Low quality, watery meat could well be the outcome. In immobilized mother sows the metabolic effects of catecholamine elevation may also have a whole bunch of negative effects, easily imaginable (24). At the very least CSA metabolic investigation may provide prove, that cruelty to animals does not pay, in fact actually decreases profits of husbandry.

Concerning the application of CSA tests in the training of e.g. racing animals like horses, dogs or camels, it seems easy to adapt our methods and results from our investigations in humans.

It is obviously an asset for both animals and trainers to be able to adjudge the pre contest condition, thereby the contest chances and the metabolic changes during a given contest of a specific animal. The familiarity of a good trainer with animals in his care, by which up to now subjective judgement has been delivered, could be supplemented by objective testing. Hitherto surprising reactions during the contest may thus be more successfully avoided, momentary fitness state of the animal more correctly adjudicated, latent illnesses better anticipated as benefits for animal, trainer and owner.

2. General outlook

By providing therapists, trainers or commanders of first responder units with a multi – parameter pattern of stress hormone effects and their statistics within minutes, an objective support of decisions which range from selection of specific treatments, adaptation of training efforts or educated guesses about success chances of teams to detection of team *outliers* is provided.

Still, the decision making responsibility of the physician, the trainer or the commander remains untouched. On the one hand, the intricate pattern of the interwoven parameters provides a rather sensitive metabolic picture, which can be broadened by additional correlations with sociological and psychological scores (for which free spaces in our correlation tables are already provided). But the same multi parametric intricacy prohibits – at least at the moment – a mathematical overall calculation including all changing parameters according to their momentary importance. It turned out, that pure multiple regressions of the bulk of all data are not stable enough, so that factor analyses have to be applied in advance. Some smaller problems could be already solved in this way, but generally the best way to use our tool at the moment is a one days course in interpretation with lots of practical applications and some theoretical background, which in nearly all cases enables the user to proceed successfully on her or his own.

3. Summing up

Stress hormone changes correlate significantly with their also changing effects like lactate, blood gases, buffer capacities and electrolyte concentrations. About twelve of those parameters can be determined simultaneously by easily transportable ICU equipment from 100 microliter of capillary blood within three minutes and the data be transferred on line by CSA (Clinical Stress Assessment) software into data banks. Individual statistics or group statistics are automatically calculated, so hat a kind of data net is spread over the proband or a whole group of probands. Changing of sensitive parameters like pCO2 or Ca are indicative of acute effects, while changing of less sensitive values like HO3 or blood glucose point towards more chronic developments.

Application of CSA has been found fruitful i.a. to detect and quantify:

- The pre – contest stress situation of sports persons or pre deployment situation of first responders like soldiers or fire fighters, thereby being able to select outliers and calculate scoring changes preliminarily. A persons' success/effort relationship can be quantified.
- Since mental stress imprints its quantity in nearly all cases upon metabolic changes, those changes and their idiosyncrasies can be quickly measured and taken as sensitive pointers for e.g. a gradual effectiveness check of medical treatment.
- Inclusion of CSA application in diabetic check ups can provide the physicians with a more then 10 – fold increased information about blood glucose changes within the individual metabolic ambient of a patient, out of nearly the same small amount of blood within 3 minutes. It enabled us at least to find hitherto hardly suspected deficiencies and interactions of electrolytes and glucose in type2 diabetics, with a significant further deterioration in hypertonic diabetics.
- Concerning the predictive and in depth assessing capabilities of CSA concerning sports and first responders, it hits the eye that those very capabilities could be applied successfully in training of competing animals as well as in the control of unnecessary stress of husbandry situations.

4. Acknowledgment

The authors would like to thank G. Porta and T. Porta for their valuable assistance.

5. References

[1] Porta S., Emsenhuber W., Petek W., Pürstner P, Vogel W., Schwaberger G., Slawitsch G., Korsatko Detection and Evaluation of Persisting Stress Induces Hormonal Disturbances by a Post Stress Provocation Test in Humans. W. Life Sciences, Vol.53,pp.1583-1589, 1991).

[2] W. Emsenhuber, S. Porta, W. Petek, P. Pürstner, W. Vogel, P. Felsner, G. Schwaberger, P. Slawitsch, Retrospective Stress Measurement By Standardized Post-Stress Provocation, in „Stress: Neuroendocrine and Molecular Approaches" Edited by R. Kvetnansky, R. McCarty and J. Axelrod Gordon and Breach Science Publishers S.A., New York, USA, 1992

[3] Desch W., Schappacher W., Schappacher G., Wintersteiger R., Ecker M., Köhler U., Korisek G., Porta S. An attempt to quantify the influence of some IC parameters upon the levels of ionized Mg in blood. Trace Elem. Elec. 26; 2/2009 89 – 94

[4] S. Porta, K. Pichlkastner, H. Gell, W. Desch, J. Porta, and M. M. Bratu, Interdependencies of electrolyte and metabolic parameters can characterize handicaps and predict success in sports. Trace. Elem. Electr. 27, 3, 2010, 103 – 109

[5] S. Porta and K. Kisters, Electrolytes and sports – special role of magnesium, Editorial, Trace. Elem. Electr. 27, 3, 2010, 103 – 109

[6] JH Foreman, JK Waldsmith and RB Lalum. Physical, acid–base and electrolyte changes in horses competing in training. Preliminary and intermediate horse trials, Equine and Comparative Exercise Physiology (2004), 1:99-105 Cambridge University Press.

[7] Porta S, Gell H, Pichlkastner K, Cichocki G, Desch W, Schappacher W, Korisek G, Grieshofer P and Stelbrink U. Direct correlation between Mg++ changes and awarded scores in military steeplechase (HIB). Trace Elem.26,4/2009 177 – 180.

[8] Thomas L. "Labor und Diagnose", TH – Books Verlagsgesellschaft Frankfurt/Main 1998.

[9] S. Porta, W. Desch, G. Korisek, K. Kisters, J. Porta, H. Gell und M. M. Bratu. Eine neue Möglichkeit zur Erfassung von Stoffwechselbesonderheiten bei hypertonen Typ2 Diabetikern. Nieren und Hochdruckkrankht. 39, 5/2010, 220 – 230.

[10] Kisters K, Hoffmann O, Hausberg M, Quang Nguyen M, Micke O, Gremmler B. Plasma magnesium status and pulse pressure in essential hypertension. Trace Elem Elec 22; 2005; 67 – 70

[12] Westmoreland D, Porta S, Leitner G, Knapp M, Spencer K, Merback J, Leitner T. The effect of magnesium supplementation on exercise-induced plasma magnesium shifts and lactic acid accumulation in female youths. Trace Elem Elec. 2004a; 21: 95-98.

[13] Westmoreland D, Orland U, Porta S. The diurnal rhythm of plasma magnesium and its probable influence on the exercise-induced plasma magnesium shift. Trace Elem Elec. 2004b; 21: 185-189.

[15] S. Porta, G. Korisek, B. Machan, H. Gaggl, W. Desch,∙ J Porta, G. Schappacher, K. Kisters∙ Changes in electrolyte- and metabolic interactions below the hypomagnesaemic threshold in Type 2 Diabetics. Trace elements and electrolytes 27/3 119 – 124, 2010-.

[16] S. Porta, K. Kisters, G. Korisek, G. Desch, J. Porta and H. Gell, Differences in electrolyte mismanagement between normotonic and hypertonic Type2 diabetics detectable by correlative capillary blood evaluation, Trace. Elem. Electr. 28/1 2011, 31 – 36.

[17] Barbagallo, M, Dominguez, LJ. Magnesium metabolism in Type2 diabetes mellitus, metabolic syndrome and insulin resistance. Arch. Biochem. Biophys. 2007, 458: 40 - 47

[18] Sales CH, Pedrosa LdeF: Magnesium and diabetes mellitus: their relation. Clin. Nutr. 25, 2006, 554 – 5.

[19] Phuong – Chi T. Pham, Phuong – Mai T. Pham, Son V. Pham, Miller J-M, Phuong – Thu T. Pham. Hypomagnesaemia in patients with Type2 diabetes. Clin J Am Soc Nephrol 2: 2007 366-373.

[20] Shaffiee M.A, Kamel S. Kamel, Halperin M.L: A conceptual approach to the patient with metabolic acidosis. Nephron (Suppl.1): 46 – 55 92; 2002.

[21] Bacher H, Mischinger HJ, Cerwenka H, Werkgartner G, El-Shabrawi A, Supancic A, Porta S. Liver ischemia, catecholamines and preoperative condition influencing postoperative tachycardia in liver surgery. Life Sciences 1999; 66(1):11-18.

[22] Porta S., Kvetnansky R., Oprsalova O., Emsenhuber W., Leitner G, Plasma Epinephrine Elevation Increases Stress induced Catecholamine Increase Even After an Interval of 24 Hours. 7th International Catecholamine Symposium, Amsterdam June 22-26, 1992.

[23] Porta S., Leitner G., Kvetnansky R., Kaciuba Uscilko H., Nazar K., Teter A., Emsenhuber W. Cumulative Effects of Repeated Stress Detectable Even After Very Long Intervals. International Symposium On Stress And Adaption, Amsterdam June 19, 1992.

[24] Porta S., Ehrenberg A., Helbig J., Classen H.G. ,Egger G., Weger M., Zimmermann P., Weiss U., Time and Dose Dependent Influence of Magnesium-Aspartate-Hydrochloride Treatment Upon Hormonal and Enzymatic Changes as well as Alterations in Meat Quality Due to Slaughtering Stress in Pigs.Mg. Bull. 12/3, 1993, 21 – 24.

12

Neurotechnology and Psychiatric Biomarkers

William J. Bosl

Harvard Medical School,
Children's Hospital Boston Informatics Program,
Boston University School of Medicine Behavioral Neuroscience Program,
USA

1. Introduction

Neuroscience as a scientific discipline has enjoyed enormous growth and success in the past decade. Some have called the early 20th century the golden age of physics, the latter half of the 20th century a period when the genomic revolution blossomed and predict that the early 21st century will be a period when brain sciences achieve remarkable success. While understanding the basic mechanisms of the brain and how they relate to thought and behavior may be the foundation for applications to medicine, there is a great need for technological innovation if more than academic results are to be achieved. The need for neurotechnology is already great. Traumatic brain injury and damaged limbs require prosthetic devices that can be controlled in some way by willful volition. Ideally, direct connections between thought and action are desirable to restore natural functions. Mental health is a branch of medicine that has long been relegated to a secondary status within medicine. The reasons for this may be many, but they certainly include the difficulty of understanding and measuring brain activity in a quantitative way and relating those measurements to behavior and cognitive activity. As healthcare costs continue to spiral out of control in both developed and developing regions of the world, the need for engineers to become involved in neuroscience and neurotechnology research and development has never been greater. Innovative engineering ideas, with a view toward practical application and affordable cost have much to contribute to clinical applications of brain science. A key contribution of neuroengineering will be innovative methods for quantitative measurement of brain activity and mapping of those measurements to behavior and thought. The term psychiatric biomarkers will be used here in this broad sense to indicate quantitative measurements of the brain and the algorithms necessary to interpret them in psychiatric or psychological descriptions or diagnoses.

One important way in which psychiatric biomarkers differ from other physiological biomarkers is in that the mapping from biomarker to symptom or disease is much more complex. A biomarker for a specific cancer, for example, may be a gene mutation that is in some way directly involved in the disease progression itself. The relationship between biomarker and the manifestation of interest – cancer, in this case – is rather simple and direct. That is not to say that the gene or the gene expression patterns are simple to find, but only that the conceptual relationship between the marker and disease is simple to understand. In the case of psychiatric biomarkers, the phenomena of interest, thought and behavior, are complex, emergent phenomena of brain neural activity. The relationship between neural firing patterns and the communication deficits that are clearly evident in a person with a mental disorder is not at all clear, even if we posit that all thought is indeed dependent upon neural activity. The relationship in this case is somewhat like the relationship between letters of the alphabet and a metaphor in great literature. Certainly metaphors depend on spelling and grammar, but the concept is much more than spelling and grammar. Similarly, the complex patterns of neural activity that distinguish the way a person with autism responds to someone speaking directly to them from someone considered "normal" are quite complex.

Normal and abnormal behavior are differentiated by subtle, complex patterns of activity that a trained expert observes or discovers through systematic diagnostic tests. If brain function and behavior are mirrors of each other, as is commonly accepted (Cowan and Kandel, 2001; Hyman, 2007; Kandel, 1998; Singh and Rose, 2009), then biomarkers of mental disorders may be hidden in subtle, complex patterns of neurobiological data. There is a growing realization that the neurophysiological mechanisms that underlie brain function cannot be understood by pure reduction to physiological causes (Stam, 2005; Ward, 2003). The dynamics of the brain is inherently nonlinear, exhibiting emergent dynamics such as chaotic and transiently synchronized phenomena that may be central to understanding the mind-brain relationship (Varela et al., 2001) or the 'dynamic core' (Le Van Quyen, 2003). The behaviors and thoughts that characterize mental dysfunction may be emergent phenomena or complex patterns of physiological processes, especially neural processes. For example, major depression or the communication deficits present in a child with autism are emergent phenomena that reflect complex patterns in brain function that differ from some socially-defined norm. The task of the neuroengineer is to create new technology to measure and interpret the patterns of brain activity that connect brain measurements to observed behavioral patterns.

A key challenge in cognitive neuropsychiatry is to discover the neural correlates underlying behavior. To be clinically useful, these discoveries must be accompanied by technology that enables brain activity to be measured and interpreted safely, inexpensively and easily. The explosive growth of neuroimaging studies that link functional brain activity to behavior promises exciting opportunities for measuring nonlinear brain activity that may indicate abnormalities or allow response to therapy to be monitored. While several imaging modalities are available for neuroscience *research*, most have significant limitations that prohibit their use as routine *clinical* tools. Cost and ease of use are essential qualities for clinically useful tools, which may not be as important or relevant in a scientific research context. Neuroengineers must be cognizant of these constraints when considering the intended use of the technology.

Measurements of brain electrical activity with electroencephalography (EEG) have long been a valuable source of information for neuroscience research, yet this rich resource may be under-utilized for clinical applications in neurology and psychiatry (Niedermeyer, 2003; Niedermeyer and Lopes da Silva, 2005). To fully exploit this data, methods for discovering subtle nonlinear patterns and deeper understanding of the relationship between emergent signal features, neurophysiology and behavior are needed. Near infrared spectroscopy (NIRS) has recently been introduced as a safe, portable alternative for measuring blood oxygen level dependent (BOLD) response in infants (Irani et al., 2007; Muehlemann et al., 2008). One of the primary advantages of NIRS, like EEG, is that it is safe for all ages, relatively inexpensive and portable. As a new brain-imaging tool, much remains to be discovered about the value and limitations of NIRS as a clinical instrument. In addition, coupling EEG and NIRS may have some advantages for clinical use and remains to be explored by researchers. Many of the advances in non-invasive functional brain measurement are being driven by the brain computer interface community, where mobility and cost requirements limit the technologies that can be adopted (Dornhege, 2007). Neuropsychology and cognitive neuropsychiatry can learn from this community, while adapting the methods to the particular needs of behavioral, affective and cognitive assessment.

In this chapter some relevant information concerning our current understanding of complex network organization and implications for finding neural correlates of behavior will be reviewed with goal of motivating engineers to consider contributing their skills to developing new neurotechnology. Considerable attention will be given to EEG measurements as one of the most promising technologies for clinical application to neuropsychiatry. Novel methods for extracting information from EEG signals are beginning to appear, taking advantage of advances in the physics of nonlinear systems and signals, complex network theory and machine learning algorithms. The need for innovative neurotechnology to meet the need for mental and neurological healthcare in developing regions of the world is great, but the promise is even greater. The primary goal of this chapter is to provide information to enable researchers interested in brain disorders and mental health to become involved in creating innovative neurotechnology for clinical use.

2. The brain as a complex system

2.1 Complex networks

The human brain contains on the order of 10^{11} neurons and more than 10^{14} synaptic connections (Kandel et al., 2000). Although sparsely connected, each neuron is within a few synaptic connections of any other neuron (Buzsáki, 2006). This remarkable connectivity is achieved by a kind of hierarchical organization that is not fully understood in the brain, but is ubiquitous in nature, called scale-free or complex networks (Barabasi, 2009; Bassett and Bullmore, 2006; Ravasz and Barabasi, 2003). Complex networks are characterized by dense local connectivity and sparser long-range connectivity (Barabasi, 2009) that is fractal or self-similar at all scales.

Many brain disorders appear to be associated with abnormal brain connectivity that may vary between different regions and different scales (Bassett and Bullmore, 2009; Craddock et al., 2009; Noonan et al., 2009). Examples include autism (Belmonte et al., 2004; Noonan et al., 2009), schizophrenia (Raghavendra et al., 2009; Uhlhaas et al., 2008; Whittington, 2008), depression (Li et al., 2008; Sheline et al., 2009) and epilepsy (Douw et al., 2010; Percha et al., 2005). Methods for estimating neural connectivity or changes in neural connectivity might be effective diagnostic biomarkers for abnormal connectivity development that is associated with brain dysfunction. The electrical signals produced by neural networks are believed to contain information about the neural network structure on several scales in the vicinity of the EEG sensor (Raghavendra et al., 2009; Stam, 2005; Zavaglia et al., 2008).

Novel analysis methods from nonlinear systems theory are able to extract information from EEG signals that reflect the underlying network organization. System invariants will be encoded in the time series produced and measured by EEG sensors (Fuchs et al., 2007; Gao and Jin, 2009). Multiscale entropy (MSE) is one invariant measure of system dynamics that has been shown to particularly sensitive to changes in physiological systems (Costa et al., 2005b; Hu et al., 2009b; Takahashi et al., 2010), including mental disorders such as

Alzheimer's Disease (Abasolo et al., 2006), schizophrenia (Takahashi et al., 2010), the effect of antipsychotic drugs (Takahashi et al., 2010) normal aging (Bruce et al., 2009) and autism spectrum disorders (Bosl et al., 2011).

A comparison of functional network properties using fMRI showed that children and young-adults' brains have similar "small-world" organization at the global level, but differ significantly in hierarchical organization and interregional connectivity (Supekar et al., 2009). The networks measured with fMRI in this study are those that are formed among correlated voxels, which may represent a different spatial scale from that measured by single EEG sensors. Transient or sustained generalized synchronization between EEG sensors is another measurement of functional connectivity in the brain. A simultaneous study of EEG and fMRI signals in patients with a degenerative type of epilepsy showed that in the non-myoclonus state, subtle abnormalities that were detected in EEG signals did not affect fMRI, suggesting that EEG measurements of connectivity may measure different connectivity or may be more sensitive to temporal synchronization that occurs on a time scale less than that of fMRI (on the order of one second).

Neuroscience has made great progress using linear methods of spectral (Fourier) analysis, it is likely that much more information is contained in the complex patterns of brain activity (see, for example, (Bruce et al., 2009). EEG may be under-utilized for clinical applications in neurology and psychiatry (Niedermeyer, 2003; Niedermeyer and Lopes da Silva, 2005) particularly now as new developments in complex systems and multivariate nonlinear time series analysis may allow previously unexplored information to be extracted from EEG signals (Kulisek et al., 2008; Mizuhara et al., 2005; Stam, 2005; Varela et al., 2001). Sensors such as EEGs measure the coordinated electrical response of many neurons to produce time series that reflect the dynamics of this complex system. In order to extract salient information from this data, methods appropriate for analyzing nonlinear time series are required. Although many useful techniques for nonlinear time series and system analysis have been developed in other disciplines (Bosl, 2000; Braha et al., 2006; Elnashaie and Grace, 2007; Holland, 1995; Stauffer, 2006), it is not immediately clear which are most appropriate for neuroscience research. Nevertheless, recent research results are quite promising.

The complexity of EEG signals was found in one study to be associated with the ability to attend to a task and adapt to new cognitive tasks; a significant difference in complexity was found between normal subjects and those with diagnosed schizophrenia (Li et al., 2008). Schizophrenic patients were found to have lower complexity than normal controls in some EEG channels and significantly higher interhemispheric and intrahemispheric cross mutual information values (another measure of signal complexity) than normal controls (Na et al., 2002).

Methods for chaotic signal and phase synchronization analysis arose from a need to rigorously describe physical phenomena that exhibited what was formerly thought to be purely stochastic behavior but was then discovered to represent complex, aperiodic self-organized dynamics (Pikovsky et al., 2001). The analysis of signal complexity and interaction between signals leading to transient synchronization may reveal information about local neural complexity and long-range communication between brain regions (Buzsáki, 2006; Stam, 2005; Varela et al., 2001).

The synchronization patterns of complex networks have been shown to be closely related to the topology of the network (Arenas et al., 2006) and are related to brain connectivity (Sakkalis et al., 2008). EEG signals are believed to derive from pyramidal cells aligned in parallel in the cerebral cortex and hippocampus (Sörnmo and Laguna, 2005), which act as many interacting nonlinear oscillators (Nunez and Srinivasan, 2006). As a consequence of the scale-free network organization of neurons, EEG signals exhibit complex system characteristics reflecting the underlying network topology, including transient synchronization between frequencies, short and long range correlations and cross-modulation of amplitudes and frequencies (Gans et al., 2009). A great deal of information about interrelationships in the nervous system likely remains hidden because the linear analysis techniques currently in use fail even to detect them (Drongelen, 2007).

2.2 Mental processes as emergent phenomena

Much of modern scientific medicine is reductionist, involving a search for ultimate basic causes of disease. This paradigm for scientific research follows naturally from the extraordinary success of physics in the last century, with its search for the fundamental laws of the universe. But that model is giving way to a new vision of the universe as a complex dynamical system, one in which fundamental laws may in fact be emergent properties of the system (Laughlin, 2005).

Emergent properties are those that result from the organization of individual parts and do not exist apart from the organizational whole. The saying that the whole is more than the sum of the parts is a description of an emergent property. The process of cell division, for example, is an emergent process that cannot be explained or studied using quantum mechanics – even though quantum mechanics is a good description of how atoms interact, and a cell is made up of many atoms, each obeying the fundamental physics of quantum theory. Similarly, the difference between a well-written high school essay for a college admissions committee and a Pulitzer Prize winning novel is not to be found in grammar and spelling, even though proper spelling and grammar are essential to the meaning of literature. Literature, genre and metaphor are emergent phenomena, more than words and sentences.

Developing neurotechnology, including devices and analysis methods, may be the most challenging subfield of biomedical engineering because the phenomenon of interest, a mental state or a complex set of behaviors that may indicate a diagnosis of a mental disorder, is an emergent phenomenon. Human behavior is controlled by the brain, which is ultimately a complex network of neurons that transmit electrochemical signals. Thought and behavior cannot be understood or measured by studying neurons (or genes) alone. Psychiatric biomarkers that focus on complex system properties may be the most informative measurements for assessing mental state. We now present a survey of complex system properties that can be computed from time series of brain electrical activity. Neural activity is electrochemical activity. Taking into account degradation due to the skull and scalp and the introduction of noise by the electronic sensors, EEG may be the most direct measurement of brain function that is possible. Thought might be considered an emergent phenomenon of neural electrical activity.

3. Methods

3.1 Univariate measures

Many different methods for computing the complexity of time series have been defined and used successfully to analyze EEG data (Chen et al., 2009; Kuusela et al., 2002). Sample entropy, a measure of nonlinear time series complexity, was significantly higher in certain regions of the right hemisphere in pre-term neonates that received skin-to-skin contact than in those that did not, indicating faster brain maturation (Scher et al., 2009). Sample entropy has also been used as a marker of brain maturation in neonates (de la Cruz et al., 2007) and was found to increase prenatally until maturation at about 42 weeks, then decreased after newborns reached full term (Zhang et al., 2009). A study of the correlation dimension (another measure of signal complexity) of EEG signals in healthy subjects showed an increase with aging, interpreted as an increase in the number of independent synchronous networks in the brain (Stam, 2005).

Intuitively, complexity is associated with structural richness, depth, patterns upon patterns, incorporating correlations over multiple spatio-temporal scales (Costa et al., 2005b). There is no consensus on a definition of complexity, but algorithms have been developed to attempt to give meaning to complexity. In the context of time series analysis, the concept of entropy is relevant. The use of entropy measures to describe the information content of time series began with the publication of Shannon's Mathematical Theory of Communication (Shannon and Weaver, 1949). In an intuitive sense, information is a measure of the difference in uncertainty before and after a measurement. In the context of time series, information is related to the predictability of the series. Entropy is a mathematical function of the probability that the next point in a sequence or time series will be a certain values, given the previous (Baddeley, 2000). Several different entropy measures can be defined algorithmically, including Shannon entropy, spectral entropy, approximate entropy, Lempel-Ziv complexity and sample entropy (Sabeti et al., 2009), each with certain advantages for particular time series characteristics (length, amount of noise, for example). The sample entropy has been used for a number of investigations of physiological signals. Changes in sample entropy appear to correlate with aging and pathological conditions in the context of cardiac health (Bruce et al., 2009; Costa et al., 2008; Norris et al., 2008) and for normal brain development (Zhang et al., 2009) and to distinguish certain mental disorders such as schizophrenia (Sabeti et al., 2009; Takahashi et al., 2010).

The multiscale entropy (MSE) analysis is one method for computing the complexity of a time series that builds on the sample entropy and expands the concept. It has been used to analyze a number of physiological processes (Costa et al., 2005b; Hornero et al., 2009; Norris et al., 2008; Takahashi et al., 2010). The multiscale entropy algorithm incorporates two steps. The first is a coarse-graining procedure that uses successive averaging of a time series to create new coarse-grained time series. For a window size τ, τ = 1, 2, ... j, the jth coarse-grain series, y^{τ}_j, is computed by averaging non-overlapping windows:

$$y^{\tau}_j = \frac{1}{\tau} \sum_{i=(j-1)\tau+1}^{j\tau} x_i \tag{1}$$

where x_i is the original time series of length N and τ is the scale factor satisfying $1 < \tau < N/\tau$. A schematic illustration of this process is shown in equation 2.

Scale 1 :	s_1 s_2 s_3 s_4 s_5 s_6 s_7 s_8 s_9 s_{10} s_{11} s_{12} s_{13} s_{14} s_{15} s_{16}						original time series	
Scale 2 :	x_1	x_2	x_3	x_4	x_5	x_6	x_7 x_8	$x_k = (s_i + s_{i+1}) / 2$
Scale 3 :	y_1		y_2	y_3	y_4	y_5		$y_k = (s_i + s_{i+1} + s_{i+2}) / 3$
Scale 4 :	z_1		z_2		z_3	z_4		$z_k = (s_i + s_{i+1} + s_{i+2} + s_{i+3}) / 4$

$$\tag{2}$$

The coarse-graining method for extracting signal variability on different scales used by (Costa et al., 2005b) seems to be a heuristic procedure without any solid theoretical foundation. Other procedures can be substituted that may be justified on similar grounds. Perhaps the most immediate alternative would be to use the median rather than the mean value in each coarse graining step. This would have the effect of emphasizing the variability of the original signal rather than smoothing out such variability. Another procedure would be to select every kth point from the original series, where k is the desired scale, and use a pre-selected window size to compute an average value at the kth point. A systematic discussion and computational experiments have yet to be done for the coarse graining procedure that is central to the multiscale entropy algorithm.

The second step is to then compute the entropy of each of the coarse-grain time series y^τ_j, using some entropy measure. The sample entropy is the most common entropy formulation to be used for analyzing physiological signals (Costa et al., 2005b). A useful variation to the original multiscale entropy algorithm uses the modified sample entropy defined in (Xie et al., 2008). The practical effect of using the modified sample entropy is the computed entropy values are more robust to noise and results are more consistent with short time series. In brief, the similarity functions A^m and B^m defined by equations (7) and (9) in (Xie et al., 2008) are computed for each coarse-grained time series defined in equation 1. The modified multiscale entropy (mMSE) is then defined as the series of modified sample entropy values at each of the coarse grain scales. This method was used for complexity analysis of EEG time series as a biomarker for autism risk (Bosl, et al. 2011).

An alternative to the MSE is the scale dependent Lyapunov exponent (SDLE) algorithm described in (Gao, 2007; Gao et al., 2006). This measure of complexity is stable for short, noisy time series and reportedly is able to distinguish a number of different types of chaotic motion, including noise-induced chaos, stochastic oscillations and others, which entropy measures are not able to do. SDLE has not yet been used to analyze EEG signals in young children or infants. The SDLE algorithm is based on following the time evolution of all pairs of vectors in phase space that satisfy a given embedding restriction. This results in a rather straightforward algorithm for computing the SDLE. The SDLE is reportedly better at distinguishing noise from chaotic dynamics in time series. SDLE was shown to be a more effective measure of heart disease than sample entropy and MSE (Hu et al., 2009a). Similarly, SDLE was shown to be more effective in retrospectively identifying changes in EEG signal complexity just prior to the onset of epileptic seizures than MSE, but few other studies of SDLE with EEG time series have been done. This is a potentially promising measure to be investigated further.

3.2 Detecting nonlinearity in time series

Living systems exhibit a fundamental propensity to move forward in time. This property also describes physical systems that are far from an equilibrium state. For example, heat moves in only one direction, from hot to cold areas. In thermodynamics, this property is related to the requirement that all systems must move in the direction of higher entropy. Time irreversibility was found to be a characteristic of healthy human heart electrocardiogram (EKG) recordings and was shown to be a reliable way to distinguish between actual EKG recordings and model EKG simulations (Costa et al., 2008). EKG signals from patients with congestive heart disease were found to have lower time irreversibility indices than healthy patients (Costa et al., 2005a). Interestingly, time irreversibility of EEG signals has been associated with epileptic regions of the brain and this measure has been proposed as a biomarker for seizure foci (Gautama et al., 2003). Time irreversibility may be used as a practical test for nonlinearity in a time series.

As an illustration, a time irreversibility index (t_{rev}) was computed for different resolutions of the EEG time series using the algorithm of (Costa et al., 2008). The third column of Figure 1 shows t_{rev} for several different linear and nonlinear time series. Of particular note is that only the sine wave time series and both random time series have nearly zero irreversibility indices, while the index for the nonlinear logistic series and the representative EEG signal are both nonzero on all scales shown.

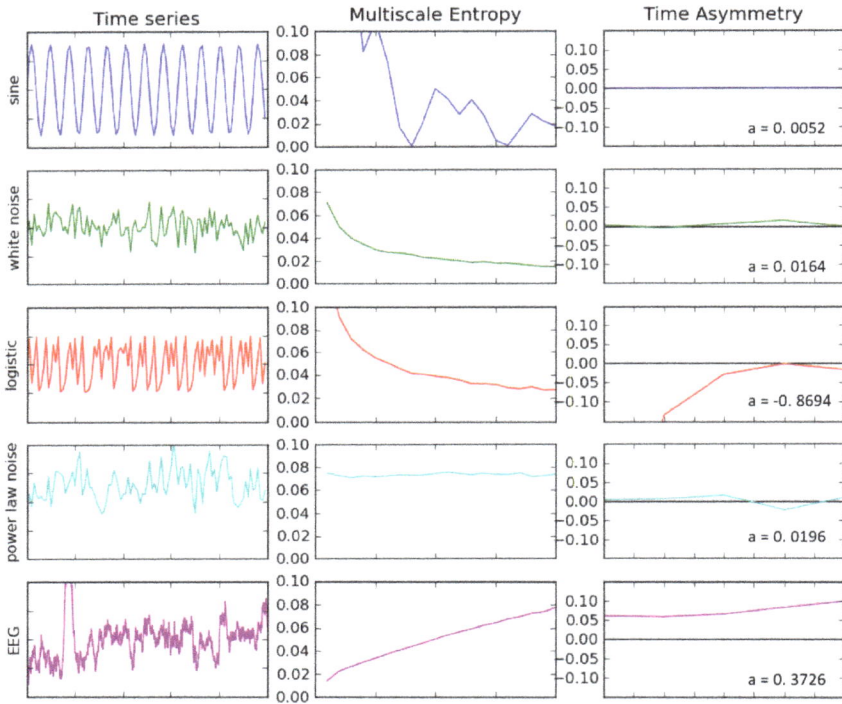

Fig. 1. Characteristics of five different time series are shown. Column 1 shows the time series amplitudes. Column two is the multiscale entropy, where the horizontal axis is the coarse graining scale, from 1 to 20. Column 3 is the multiscale time asymmetry value. The value of a in the lower right corner of the time asymmetry plot is the value of the time asymmetry index summed over scales 1 to 5. A non-zero time asymmetry value is a sufficient condition for nonlinearity of a time series.

After computing multiple resolutions of the EEG time series using the multiscale algorithm shown in equation 2, an estimate of the time irreversibility for each resolution is computed by noting that a symmetric function or time series will have the same number of increments as decrements. That is, the number times $|x_{i+1} - x_i| > 0$ will be approximately the same as the number of times $|x_{i+1} - x_i| < 0$. Thus, an estimate of the time series symmetry (or reversibility) was found by summing increments and decrements and dividing by the length of the series. A reversible time series will have a value of zero. For a series of 5000 points, as used in figure 1, $t_{rev} > 0.1$ is a significant indicator of irreversibility and thus of nonlinearity (Schreiber and Schmitz, 1997). This information is used only to indicate that nonlinear information is contained in the EEG time series that is not used in linear analysis methods, suggesting that the MSE may contain more diagnostically useful information than power spectra analysis alone.

Additional methods for characterizing nonlinear signals may be derived from recurrence plot analysis, to be discussed separately below. Communications and electrical engineers may be especially well suited to research in analyzing brain activity and applying methods from communication signals analysis to find information that can be correlated to behavioral and cognitive assessment data. Integration of new results from both nonlinear time series analysis and complex network research may prove to be a fruitful approach for engineers interested in finding patterns in neural activity that are correlated to complex behavioral patterns that psychiatrists and psychologists use to characterize mental health.

3.3 Synchronization

While signal complexity is a property of a single time series or EEG channel, transient synchronized activity is a measure of the interaction between different channels and an indication of communication and coordination between different brain regions. Synchronization may be used as a marker for diagnosing underlying mental disorders that involve aberrant long-range connectivity in the brain and may also reveal causal mechanisms (Whittington, 2008). The complexity of synchronization patterns appears to change during network development and reflects different neural wiring schemes and levels of cluster organization (Fuchs et al., 2007).

The synchronization patterns of complex networks have been shown to be closely related to the topology of the network (Arenas et al., 2006) and are related to brain connectivity (Sakkalis et al., 2008). Synchronization between sensors is an indicator of connectivity between brain regions on a scale commensurate with the sensor spacing. EEG signals are believed to derive from pyramidal cells aligned in parallel in the cerebral cortex and hippocampus (Sörnmo and Laguna, 2005), which act as many interacting nonlinear oscillators (Nunez and Srinivasan, 2006). Synchronization between gamma activity (typically defined to be 30-50 Hz range) is believed to be involved in long-range communication between brain regions. A possible link between gamma activity and the hemodynamic response measured by fMRI was found in a study of auditory response. Distinct activations in the gamma frequency range were found in subcortical structures, including the anterior cingulate cortex (ACC) and thalamus (Mulert et al., 2007).

Synchronized oscillations are transiently stable, thus form and decay rapidly. Synchrony can result from a common input oscillator, such as in cardiac synchronization. It can also be an emergent, self-organized phenomenon that is related to the network structure itself. The latter is of particular relevance to the search for psychiatric biomarkers that are associated with complex behaviors. The complexity of synchronization patterns appears to change during network development and reflects different neural wiring schemes and levels of cluster organization (Fuchs et al., 2007). It is thus reasonable to suppose that the developing brain will show different but characteristic synchronization patterns at different developmental stages. While the fundamental neurophysiological correlates of these patterns may be difficult to ascertain, they nevertheless may serve as a marker for normal and abnormal brain functional development. The emergence of a "social brain network" during early childhood was found in a study of evoked response potentials (ERPs) in 3-, 4-, and 12-month-old infants viewing faces of different orientation and direction of eye gaze (Johnson et al., 2005), suggesting a particular pattern of brain connectivity that develops in early childhood. Default mode networks (DMNs) found in adults and (negatively) associated with particular cognitive and sensorimotor activities were lacking in a study of (premature) 3 month old infants (Fransson et al., 2007). However, resting state networks have recently been shown to emerge in the first year of life, suggesting development of brain networks and their potential disruption in neurodevelopmental disorders (Supekar et al.; Uddin et al., 2010).

There is no consensus at this time on the best methods for determining nonlinear synchronization in neurological data and a number of different algorithms have been proposed (Kreuz et al., 2007; Sakkalis et al., 2009). Although strong signal synchronization would likely be detected by any of several nonlinear synchronization measures, two measures that are based on different algorithmic approaches are chosen here (see (Sakkalis et al., 2009) for more thorough discussion of some of the relative merits of each).

Two methods outlined here for computing bivariate synchronization matrix are: (1) the synchronization likelihood (SL) method (Montez et al., 2006); and (2) the instantaneous cross modulation from the circular phase Hilbert transform-based synchronization index (HI), which is robust to signal noise and short time windows (Gans et al., 2009).

Synchronization indices may be searched for correlation in each frequency bands using centered moving averages (Bashan et al., 2008). This approach will find weak or strong correlations with time lags. For each pair, the relative phase index can be computed and stored in a correlation matrix. At each time, some channels may be synchronized. A clustering algorithm can be applied to all channels at a single (averaged) time segment. The result is analogous to gene expression profile clustering (Ramoni et al., 2002). Statistical significance of clusters can be determined by assigning a numerical label to each channel involved in a cluster and the fractional overlap of clusters in different individuals computed. Synchronized clusters may also exhibit very low frequency oscillations, with frequencies of 1 to 0.01 hertz. These have been found in fMRI studies of default mode networks (Broyd et al., 2009; Greicius et al., 2008; Uddin et al., 2010).

Synchronization likelihood (SL) is a method based on the assumption that neurons are highly nonlinear devices, hence methods from chaotic dynamical systems may effectively capture the relevant dynamics of the system (Sakkalis et al., 2009). It is an unbiased generalized synchronization method that relies on detection of simultaneously occurring patterns that may differ in two time series. A method for automatically computing all but two user parameters for the SL algorithm has been developed and will be used here (Montez et al., 2006).

Instantaneous cross modulation (synchronization) of EEG channels can be computed using the Hilbert transform method (Gans et al., 2009). This method is robust to noise and detects synchronization across all frequency bands. The n:m cyclic relative phase index $y_{1,2}$ between two signals, $f_1(t)$ and $f_2(t)$, at a specific time t is computed over a time interval using a sliding window as:

$$\Psi_{1,2}^{n,m}(t) = \left| n\phi_1(t) - m\phi_2(t) \right|, \quad \text{mod } 1 \tag{3}$$

where f(t)=arctan(H(y)/y) and H(y) is the Hilbert transform of the time series y. This approach is stable for nonstationary data (Gans et al., 2009), which is appropriate for our data. The mod 1 term ensures that significant phase differences will be detected even in the presence of noise-induced phase jumps. In most cases n=m=1 is commonly assumed, though cross correlation of signals with n not equal m will be used here. Two signals are defined to be synchronized when $\Psi_{1,2}$ is less than a specified constant. This algorithm is stable for nonstationary data and will detect synchronization without the need to distinguish between noise and chaos (Gans et al., 2009). A sliding window will be used to compute sync over 5 minutes.

A number of methods have been used for determining synchronization in neurological data and a number of different algorithms have been proposed (Kreuz et al., 2007; Sakkalis et al., 2009). Although useful, many of these methods have difficulties with nonstationary, nonlinear signals and either fail to find true synchrony or introduce spurious synchronization (Fine et al., 2010). Spurious synchronization due to volume conduction effects can be removed by applying a spatial algorithm to ICA decomposition (Hironaga and Ioannides, 2007). ICA eliminates volume conduction effects while maintaining the same time resolution, thus still allowing generalized synchronization to be computed.

As Fourier spectrum can only give meaningful interpretation to linear and stationary processes, its application to data from nonlinear and nonstationary processes is problematical. A relatively new method for extracting *instantaneous* phase and frequency information from both linear and nonlinear, chaotic signals is the Huang-Hilbert transform (Huang and Wu, 2005; Huang et al., 2009). Determination of instantaneous phase and frequency is usually accomplished using the Hilbert transform method (Kreuz et al., 2007). However, this is only appropriate for monofrequency analytic signals that have a single center of rotation in the complex plane. The Empirical Mode Decomposition (EMD) introduced by Huang makes no assumptions about linearity. The EMF decomposes a nonlinear, nonstationary time series into adaptively determined characteristic time scales of each of the components (Huang et al., 2009). These component functions are termed intrinsic mode functions (IMF) and are analogous to Fourier components in a traditional linear decomposition. The IMFs computed using the empirical mode transform (EMF) have the property of a single center of rotation in the complex plane, ideally satisfying the requirements for the application of the Hilbert transform to determine instantaneous phase and frequency (Fine et al., 2010; Huang et al., 2009).

After computation of IMFs for each EEG channel, the IMF components with the highest power will be used to compute an instantaneous phase coherence matrix, R, using a sliding window. Following (Bialonski and Lehnertz, 2006), R is computed:

$$R_{ij} = \left(\frac{1}{w} \sum_{t=0}^{w-1} e^{i(\phi_{it} - \phi jt)} \right), \text{mod } 1 \qquad (4)$$

where w is the number of time samples in the time series segment or window, i and j designate the channel number (or the IMF component of the channel) and f(t)=arctan(H(y)/y) and H(y) is the Hilbert transform of the IMF component. The mod 1 term ensures that significant phase differences will be detected even in the presence of noise-induced phase jumps. The Hilbert transform obtains the best fit of a sinusoid to each IMF at every point in time, identifying an *instantaneous frequency* (IF), along with its associated *instantaneous amplitude* (IA). The IF and IA provide a time-frequency decomposition of the data that is highly effective at resolving non-linear and transient features. This algorithm is stable for nonstationary data and will detect synchronization without the need to distinguish between noise and chaos (Gans et al., 2009). An example of bivariate synchronization between two EEG sensors in the right medical parietal region is shown in figure 2. The synchronization likelihood in this example was computed using the only the first three IMFs from each sensor, without searching for cross band synchronization.

To identify synchronized clusters of EEG channels, a method based on an eigenvector space method, using eigenvalues of R, can be used, following the algorithm developed and applied in (Allefeld and Bialonski, 2007; Bialonski and Lehnertz, 2006; Fine et al., 2010). The outcome of this algorithm will be synchronized clusters of EEG channels. These may be mapped onto scalp plots and the identified clusters compared to default mode networks that have been identified in young children (Sauseng and Klimesch, 2008; Supekar et al., 2009; Supekar et al., 2010). It will be of particular interest to determine if synchronization clusters are significantly correlated with functional networks in the brain and are biomarkers of abnormalities in brain network function (Assaf et al., 2010; Kennedy and Courchesne, 2008). To date, most research on functional brain networks, including the default mode network, has relied on functional MRI. Networks determined by fMRI reflect only the hemodynamic or metabolic response of neurons (Power et al., 2010). This can be considered a kind of amplitude correlation but not true synchronization of brain regions.

If synchronization of electrical activity can be shown to be an alternative measure of brain network activity it would open up much more exploration of the role of brain networks in cognitive activity, brain computer interfaces and neuropsychiatric disorders. Aberrations to default mode networks have been implicated in a number of brain disorders (Broyd et al., 2009) including post traumatic stress syndrome (Daniels et al., 2010), social phobias (Gentili et al., 2009), depression (Sheline et al., 2009), ADHD (Uddin et al., 2008), autism (Di Martino et al., 2009), and schizophrenia (Lagioia et al., 2010). fMRI is far too expensive to be used routinely as a clinical screening and monitoring tool. Yet the apparent widespread role of synchronized brain networks in many neuropsychiatric disorders suggests that a less expensive and easy to administer technology for analyzing brain networks would be widely useful in clinical practice.

3.3 Recurrence quantitative analysis

Several univariate measures of time series complexity and a number of approaches for computing the degree of synchronization between signals have been used to analyze EEG data. Applications of these methods to psychiatric care and mental health continue to show promise. A more general framework for characterizing the dynamics of complex systems may be to construct recurrence plots (Marwan et al., 2007) and compute quantitative properties. The idea to use recurrence plots as a representation of complex system dynamics was first proposed by (Eckmann et al., 1987) in the late 1980s. The original tool presented a graphical means for visualizing differences in system dynamics. Methods for quantifying the small scale structures in recurrence plots were devised and shown to be capable of revealing system parameters and transitions that are not easily obtained by other methods (Marwan et al., 2002; Zbilut et al., 2002). Some dynamical parameters, such as K2 entropy and mutual information can also be derived from recurrence plots without RQA methods by computing the distribution of line lengths in the recurrence plot (Marwan et al., 2007). Readers are referred to the references for reviews of this unifying approach to nonlinear systems analysis.

(Schinkel et al., 2007) demonstrated that a single measure from RQA analysis could detect a change in the N400 response in single trials when subjects were presented with an oddball task, suggesting that RQA may be a sensitive measure of transient brain states. Few studies have been done using RQA to determine more stable brain functional characteristics. This may be a promising new field for research on EEG biomarkers of psychiatric disorders.

4. Clinical applications

4.1 Infant brain development and autism spectrum disorders

Autism spectrum disorder (ASD) constitutes a heterogeneous developmental syndrome that is characterized by a triad of impairments that affect social interaction, communication skills, and a restricted range of interests and activities (APA, 2000), with highly variable outcomes. Studies have consistently shown that early intervention leads to better long-term outcomes. But early intervention is predicated on early detection. Behavioral measures have thus far proven ineffective in diagnosing autism before about 18 months of age, in part because the behavioral repertoire of infants is so limited. Neural development may precede overt behavioral observations and thus provide an earlier marker for emerging autistic behaviors. Yet, measuring functional brain development is difficult because few noninvasive methods are available for infants and it is not clear what features to measure that are biomarkers of normal development. As discussed above, multiscale entropy computed from EEG time series has been shown to be a particularly informative analysis tool for physiological signals.

Complex mental disorders such as autism are associated with abnormal brain connectivity that may vary between different regions and different scales (Noonan et al., 2009). Estimation of changes in neural connectivity might be an effective diagnostic biomarker for abnormal connectivity development that leads to ASD behaviors. The electrical signals produced by neural networks are believed to contain information about the network structure (Raghavendra et al., 2009; Stam, 2005; Zavaglia et al., 2008). The physics of complex networks suggests that system invariants will be encoded in the time series produced and measured by EEG sensors (Fuchs et al., 2007; Gao and Jin, 2009). Multiscale entropy (MSE) discussed above is one invariant measure of system dynamics that has been shown to particularly sensitive to changes in physiological systems (Costa et al., 2005b; Hu et al., 2009b; Takahashi et al., 2010), including mental disorders such as Alzheimer's Disease (Abasolo et al., 2006), schizophrenia (Takahashi et al., 2010), the effect of antipsychotic drugs (Takahashi et al., 2010) and normal aging (Bruce et al., 2009). Preliminary results suggest that MSE may be a biomarker for autism endophenotypes (Bosl et al., 2011) and may provide an earlier diagnosis than behavioral assessments.

MSE values were computed for 79 different infants: 46 at high risk for ASD (hereafter referred to as HRA) based on having an older sibling with a confirmed diagnosis of ASD and 33 controls, defined on the basis of a typically developing older sibling and no family history of neurodevelopmental disorders. The study participants were part of an on-going longitudinal study and for this analysis visits were evaluated at regular intervals. Infants were seated on their mothers' laps in a dimly lit room while a research assistant engaged their attention by blowing bubbles. This procedure was followed to limit the amount of head movement made by the infant that would interfere with the recording process. Continuous EEG was recorded with a 64-channel Sensor Net System. The data were amplified, filtered (band pass 0.1-100.0 Hz), and sampled at a frequency of 250 Hz. Data were manually reviewed to remove sections with obvious artifacts and continuous, clean 20 second segments were identified to compute alert, resting state MSE values.

Results shown in figure 2 were computing by averaging all channels in a given region, including left/right hemispheres, left/right frontal and total scalp MSE values. The plots are derived by averaging regional MSE values for all infants at a given age. These reveal a significant difference between high risk and typically developing infants (Bosl et al., 2011). The data suggest that there are not only significant differences between the high risk group and typically developing controls, but MSE follows characteristic trajectories during development. To our knowledge, these preliminary results are the first demonstration of a complex systems analysis of EEG data for biomarkers in infants at risk for a complex neurodevelopmental disorder. More details about this study and the use of machine learning to discover diagnostically significant patterns in MSE data can be found in (Bosl et al., 2011).

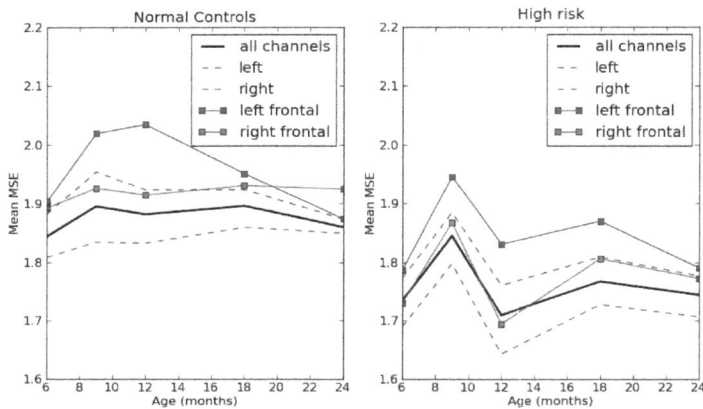

Fig. 2. The change in mean MSE over all channels is shown for each age. Averaging over all channels reveals that in general MSE is higher than the normal controls than in the high risk group, but regional differences cannot be seen.

The trajectory of the curves between 6 and 12 months in figure 2 appear to be as informative as information at any specific age. This is a period of important changes in brain function that are foundational for the emergence of higher level social and communicative skills that are at the heart of ASD. A key characteristic of autism is lack of social perception; autistic spectral disorders are also associated with abnormal brain connectivity (Belmonte et al., 2004; Kleinhans et al., 2008). A number of major cognitive milestones typically occur beginning at around the age 9 months and perhaps earlier in girls. These include joint perception (Behne et al., 2005) and loss of the ability to discriminate certain categories of faces (Pascalis et al., 2002). The latter developments are especially significant because they reveal how *socially-grounded experiences* influence changes in the neurocognitive mechanisms that underlie speech and face processing. In a prospective study, (Ozonoff et al., 2010) found that social communicative behaviors in infants that later developed ASD declined dramatically between 6 and 18 months when compared to typically developing infants. The results shown here may indicate that MSE is sensitive to neural correlates underlying the observed social and cognitive developments.

Currently there are no objective medical tests for diagnosing autism. According to the American Academy of Pediatrics (AAP), data strongly suggest better outcomes in children whose conditions are diagnosed early and participate in early intervention programs. Children who are diagnosed early and given intensive behavioral intervention can often be mainstreamed and live normal, productive lives. The AAP recommends that *all* children be screened for autism at 18 and again at 24 months. Unfortunately, behavioral assessment is time consuming and requires specialized training, both of which pose a problem for routine screening. Because atypical brain development is likely to precede abnormal behavior by months or even years, a critical developmental window for early intervention may be missed if diagnosis is based entirely on a behavioral phenotype. If reliable methods can be developed to lower the age of diagnosis, and if insight is gained into the biological mechanisms that underlie the disorder, it may be possible to develop intervention strategies that can be implemented during the first year of life. Complex systems analysis of EEG may enable a relatively inexpensive and reliable way to estimate the risk of developmental disorders within the first year of life.

4.2 Schizophrenia

Early studies of signal synchronization in patients diagnosed with schizophrenia used a *linear* measure of complexity based on global correlation of all channels (Wackermann et al., 1993) and synchronization based on linear correlation between amplitudes (Saito et al., 1998). Although these methods will necessarily miss nonlinear signal characteristics, they found significant differences between patients and normal controls that were interpreted to reflect decreased coordination between different brain regions, a characteristic of schizophrenia (Uhlhaas et al., 2008; Whittington, 2008).

One interpretation of biological complexity is that it reflects a systems' ability to quickly adapt and function in a changing environment (Costa et al., 2005b). The complexity of EEG signals was found in one study to be associated with the ability to attend to a task and adapt to new cognitive tasks; a significant difference in complexity was found between normal subjects and those with diagnosed schizophrenia (Li et al., 2008). Schizophrenic patients were found to have lower complexity than normal controls in some EEG channels and significantly higher interhemispheric and intrahemispheric cross mutual information values (another measure of signal complexity) than the normal controls (Na et al., 2002).

While signal complexity is a property of a single time series or EEG channel, transient synchronized activity is a measure of the interaction between different channels and an indication of communication and coordination between different brain regions. Synchronization may be used as a marker for diagnosing underlying mental disorders such as schizophrenia and may also reveal causal mechanisms (Whittington, 2008). The complexity of synchronization patterns changes during network development and reflects different neural wiring schemes and levels of cluster organization (Fuchs et al., 2007).

Abnormalities in phase synchronization between multiple bands have been found to be sensitive biomarkers for mental dysfunction in schizophrenic patients (Uhlhaas et al., 2008; Whittington, 2008). Unfortunately, similar abnormalities in synchronous activity have been found associated with a number of other mental disorders (Uhlhaas et al., 2008), so further research is required to discover if more refined patterns of synchrony exist for discriminating different disorders or subtypes. A developmental perspective may be useful here. For example, while many attempts to correlate cortical thickness with intelligence have failed, recent research demonstrated that specific characteristic growth *trajectories* of cortical thickness from infancy to early teen years were highly correlated with above or below average intelligence (Shaw et al., 2006), suggesting that growth curves of brain function may contain more information than any combination of measurements at one specific age. This would require routine brain measurements become part of the medical record and algorithms that recognize abnormal trends would need to be used to interpret data after regular checkups.

5. Neurotechnology for global mental health

Mental healthcare in developing nations has long been overlooked by organizations concerned with global health needs. A number of reasons can be cited for this, but one significant factor is the difficulty of accurately diagnosing and classifying mental disorders and the relatively intensive need for human expertise to administer assessments and care. Precision healthcare is a term coined by Clayton Christensen to describe objective medical diagnosis and prescription that is enabled by new technologies. Innovative neurotechnologies that enable basic mental diagnosis to be administered in a cost effective manner in clinics that may be staffed by technicians is urgently needed in developing nations. Even if the diagnostic precision is not as high as might be possible in big city hospital in the developed world, low-cost basic care would serve to alleviate a great deal of personal suffering as well as the economic burden imposed.

The need for neurotechnology to enable precision mental healthcare has never been greater. The developed world is spending greater and greater amounts of money on healthcare and costs threaten their economic vitality. Developing nations will benefit tremendously from even basic neurological and mental healthcare.

While considerable resources have been devoted to finding cures for infectious diseases such as HIV and malaria, relatively little attention has been given to the neurological and cognitive effects of these infections on young patients that recover from the acute infection. A majority of HIV infected patients manifest HIV-associated neurocognitive disorders despite receiving highly active antiretroviral therapy (Van Rie et al., 2009). The developing brains of children who have acquired HIV through vertical infection are particularly vulnerable to central nervous system insults either directly, through secondary infections or from side effects of antiretroviral therapies (LLorente et al., 2009). For example, school age children with perinatally acquired HIV infections receiving antiretroviral therapy displayed lower cognitive function than either HIV affected or HIV unaffected children. Although the antiretroviral therapy reduced HIV symptoms, the cognitive deficits were not improved (Puthanakit et al., 2010).

These clinical observations inform the need to obtain a better understanding of how and when HIV affects cognitive function and to develop novel therapeutic drug candidates to prevent or interfere with progression of HIV associated neurocognitive impairment (Robertson et al., 2010). Some treatments for the primary or secondary infections, may be more harmful to vulnerable brain tissue in children that have been compromised by HIV infection than in adults (LLorente et al., 2009). Research on the effects of HIV infection on the developing brain is a topic of increasing importance as new antiretroviral treatments become more widely available causing many more HIV-1 infections to be treated as a chronic condition to be managed rather than as a fatal disease (Abubakar et al., 2008). Few longitudinal studies have been done of HIV infected children in developing regions who have been receiving antiretroviral therapy. The methods for measuring and analyzing brain electrical activity developed here may help to fill the need for objective, cross-cultural measures of cognitive function to estimate the effects of HIV on brain development in children.

It is hoped that neuroengineers will adopt methods outlined in this chapter to discover biomarkers for monitoring the progression of cognitive development and demonstration their utility in the context of HIV associated neurocognitive disability (HAND). Protocols are needed for early assessment and diagnosis of mental and cognitive impairment due to HIV infection, secondary infections and drug therapy. These may also be used as a clinical tool for widespread screening of HIV infected or exposed infants that may enable developmental cognitive impairments to be diagnosed before other HIV symptoms are apparent. Such tools would represent a major step forward in the effort to understand the long-term impact of chronic HIV infection and other infections on cognitive development in children and to provide a new approach for monitoring the cognitive effects of long term management of HIV as a chronic condition in developing regions of the world.

6. Conclusion

Neuroscience has been called the next frontier science after the breakthroughs achieved by physics in the mid-twentieth century and genetics and biophysics at the end of the 20th century and early 21st century. The brain is of great scientific interest because of the emergent phenomenon of thought that arises from the vast complexity of this organ. While the question of how the brain thinks is likely to remain a challenging scientific and philosophical topic for a long time, the need for practical innovations to meet the healthcare needs of the world is more urgent. Two issues drive this need at the present time. First, of course, is the tremendous amount of suffering that is the result of brain injury and mental disease. Mental disorders in particular are a difficult challenge for medicine because in most cases there is not a single cause that can easily be diagnosed and "cured". Rather, the perspective throughout this chapter has been that mental disorders are complex emergent phenomena and must therefore be analyzed as such.

This is an exciting and promising time for clinical neuroscience research due to the convergent development of several technologies at this time in history. Although EEG equipment has been around since the 1920s, the creation of inexpensive, wireless, high quality EEG headsets by innovative companies such as Emotiv Systems and Neurosky opens up research possibilities with EEG equipment for many more neuroscientists. As the cost comes down and quality goes up, these tools become practical for routine clinical use. The rapid development of the physics of complex networks and systems, as well as mathematical methods for analyzing complex (chaotic) time series that are produced by complex networks enables new information from the signals measured with inexpensive EEG devices to be extracted and studied as markers for neural correlates of behavior. Finally, the continuing advance of pattern recognition algorithms from the artificial intelligence community enables more subtle pattern-markers to be discovered in complex EEG signal features.

The high cost of healthcare threatens the economies of developed nations and prevents many people in developing regions from obtaining the psychiatric care that they need. On both of these fronts, innovative neurotechnologies for early detection and monitoring of mental disorders are urgently needed. The ready availability of mobile communication devices introduces a platform for making information technology based mental healthcare available to many people that have no access at this time. Integration of EEG devices and complex systems methods with clinical decision support tools and mobile device-based health records promises both greatly improved neuropsychiatric healthcare and lower costs. It is hoped that the methods introduced in this chapter will inspire a new generation of bioengineers concerned with mental health to create new tools for clinical neuroscience while closing the gap between neurology and psychiatry.

7. Acknowledgements

This research was supported by grants from the National Institute of Mental Health and the Simons Foundation.

8. References

Abasolo, D., Hornero, R., Espino, P., Alvarez, D., and Poza, J. (2006). Entropy analysis of the EEG background activity in Alzheimer's disease patients. Physiol Meas 27, 241-253.

Abubakar, A., Van Baar, A., Van de Vijver, F.J., Holding, P., and Newton, C.R. (2008). Paediatric HIV and neurodevelopment in sub-Saharan Africa: a systematic review. Trop Med Int Health 13, 880-887.

Allefeld, C., and Bialonski, S. (2007). Detecting synchronization clusters in multivariate time series via coarse-graining of Markov chains. Phys Rev E Stat Nonlin Soft Matter Phys 76, 066207.

Arenas, A., Diaz-Guilera, A., and Perez-Vicente, C.J. (2006). Synchronization reveals topological scales in complex networks. Phys Rev Lett 96, 114102.

Assaf, M., Jagannathan, K., Calhoun, V.D., Miller, L., Stevens, M.C., Sahl, R., O'Boyle, J.G., Schultz, R.T., and Pearlson, G.D. (2010). Abnormal functional connectivity of default mode sub-networks in autism spectrum disorder patients. Neuroimage 53, 247-256.

Baddeley, R. (2000). Information Theory and the Brain (New York, Cambridge University Press).

Barabasi, A.L. (2009). Scale-free networks: a decade and beyond. Science 325, 412-413.

Bashan, A., Bartsch, R., Kantelhardt, J.W., and Havlin, S. (2008). Comparison of detrending methods for fluctuation analysis. Physica A 387, 5080-5090.

Bassett, D.S., and Bullmore, E. (2006). Small-world brain networks. Neuroscientist 12, 512-523.

Bassett, D.S., and Bullmore, E.T. (2009). Human brain networks in health and disease. Curr Opin Neurol 22, 340-347.

Behne, T., Carpenter, M., Call, J., and Tomasello, M. (2005). Unwilling versus unable: infants' understanding of intentional action. Dev Psychol 41, 328-337.

Belmonte, M.K., Allen, G., Beckel-Mitchener, A., Boulanger, L.M., Carper, R.A., and Webb, S.J. (2004). Autism and abnormal development of brain connectivity. J Neurosci 24, 9228-9231.

Bialonski, S., and Lehnertz, K. (2006). Identifying phase synchronization clusters in spatially extended dynamical systems. Phys Rev E Stat Nonlin Soft Matter Phys 74, 051909.

Bosl, W.J. (2000). Modeling Complex Crustal Processes. In GeoComplexity and the Physics of Earthquakes, J.R. Rundle, D.L. Turcotte, and W. Klein, eds. (Washington, D.C., American Geophysical Union).

Bosl, W.J., Tager-Flusberg, H., and Nelson, C.A. (2011). EEG Complexity as a Biomarker for Autism Spectrum Disorder. BMC Medicine 9.

Braha, D., Minai, A., and Bar-Yam, Y. (2006). Complex engineered systems : science meets technology (Berlin ; New York, Springer).

Broyd, S.J., Demanuele, C., Debener, S., Helps, S.K., James, C.J., and Sonuga-Barke, E.J. (2009). Default-mode brain dysfunction in mental disorders: a systematic review. Neurosci Biobehav Rev 33, 279-296.

Bruce, E.N., Bruce, M.C., and Vennelaganti, S. (2009). Sample entropy tracks changes in electroencephalogram power spectrum with sleep state and aging. J Clin Neurophysiol 26, 257-266.

Buzsáki, G. (2006). Rhythms of the brain (Oxford ; New York, Oxford University Press).

Chen, W., Zhuang, J., Yu, W., and Wang, Z. (2009). Measuring complexity using FuzzyEn, ApEn, and SampEn. Med Eng Phys 31, 61-68.

Costa, M., Goldberger, A.L., and Peng, C.K. (2005a). Broken asymmetry of the human heartbeat: loss of time irreversibility in aging and disease. Phys Rev Lett 95, 198102.

Costa, M., Goldberger, A.L., and Peng, C.K. (2005b). Multiscale entropy analysis of biological signals. Phys Rev E Stat Nonlin Soft Matter Phys 71, 021906.

Costa, M.D., Peng, C.K., and Goldberger, A.L. (2008). Multiscale analysis of heart rate dynamics: entropy and time irreversibility measures. Cardiovasc Eng 8, 88-93.

Cowan, W.M., and Kandel, E.R. (2001). Prospects for neurology and psychiatry. Jama 285, 594-600.

Craddock, R.C., Holtzheimer, P.E., 3rd, Hu, X.P., and Mayberg, H.S. (2009). Disease state prediction from resting state functional connectivity. Magn Reson Med 62, 1619-1628.

Daniels, J.K., McFarlane, A.C., Bluhm, R.L., Moores, K.A., Clark, C.R., Shaw, M.E., Williamson, P.C., Densmore, M., and Lanius, R.A. (2010). Switching between executive and default mode networks in posttraumatic stress disorder: alterations in functional connectivity. J Psychiatry Neurosci 35, 258-266.

de la Cruz, D.M., Manas, S., Pereda, E., Garrido, J.M., Lopez, S., De Vera, L., and Gonzalez, J.J. (2007). Maturational changes in the interdependencies between cortical brain areas of neonates during sleep. Cereb Cortex 17, 583-590.

Di Martino, A., Ross, K., Uddin, L.Q., Sklar, A.B., Castellanos, F.X., and Milham, M.P. (2009). Functional brain correlates of social and nonsocial processes in autism spectrum disorders: an activation likelihood estimation meta-analysis. Biol Psychiatry 65, 63-74.

Dornhege, G. (2007). Toward brain-computer interfacing (Cambridge, Mass., MIT Press).

Douw, L., de Groot, M., van Dellen, E., Heimans, J.J., Ronner, H.E., Stam, C.J., and Reijneveld, J.C. (2010). 'Functional connectivity' is a sensitive predictor of epilepsy diagnosis after the first seizure. PLoS One 5, e10839.

Drongelen, W.v. (2007). Signal processing for neuroscientists : introduction to the analysis of physiological signals (Burlington, Mass., Academic Press).

Eckmann, J., Kaphorst, S.O., and Ruelle, D. (1987). Recurrence plots of dynamical systems. Europhysics Letters 5, 973-977.

Elnashaie, S., and Grace, J.R. (2007). Complexity, bifurcation and chaos in natural and man-made lumped and distributed systems. Chemical Engineering Science 62, 3295-3325.

Fine, A.S., Nicholls, D.P., and Mogul, D.J. (2010). Assessing instantaneous synchrony of nonlinear nonstationary oscillators in the brain. J Neurosci Methods 186, 42-51.

Fransson, P., Skiold, B., Horsch, S., Nordell, A., Blennow, M., Lagercrantz, H., and Aden, U. (2007). Resting-state networks in the infant brain. Proc Natl Acad Sci U S A 104, 15531-15536.

Fuchs, E., Ayali, A., Robinson, A., Hulata, E., and Ben-Jacob, E. (2007). Coemergence of regularity and complexity during neural network development. Dev Neurobiol 67, 1802-1814.

Gans, F., Schumann, A.Y., Kantelhardt, J.W., Penzel, T., and Fietze, I. (2009). Cross-modulated amplitudes and frequencies characterize interacting components in complex systems. Phys Rev Lett 102, 098701.

Gao, J. (2007). Multiscale analysis of complex time series : integration of chaos and random fractal theory, and beyond (Hoboken, N.J., Wiley-Interscience).

Gao, J.B., Hu, J., Tung, W.W., and Cao, Y.H. (2006). Distinguishing chaos from noise by scale-dependent Lyapunov exponent. Phys Rev E Stat Nonlin Soft Matter Phys 74, 066204.

Gao, Z., and Jin, N. (2009). Complex network from time series based on phase space reconstruction. Chaos 19, 033137.

Gautama, T., Mandic, D.P., and Van Hulle, M.M. (2003). Indications of nonlinear structures in brain electrical activity. Phys Rev E Stat Nonlin Soft Matter Phys 67, 046204.

Gentili, C., Ricciardi, E., Gobbini, M.I., Santarelli, M.F., Haxby, J.V., Pietrini, P., and Guazzelli, M. (2009). Beyond amygdala: Default Mode Network activity differs between patients with social phobia and healthy controls. Brain Res Bull 79, 409-413.

Greicius, M.D., Kiviniemi, V., Tervonen, O., Vainionpaa, V., Alahuhta, S., Reiss, A.L., and Menon, V. (2008). Persistent default-mode network connectivity during light sedation. Hum Brain Mapp 29, 839-847.

Hironaga, N., and Ioannides, A.A. (2007). Localization of individual area neuronal activity. Neuroimage 34, 1519-1534.

Holland, J. (1995). Hidden Order: How Adaptation Builds Complexity (New York, Addison-Wesley Publishing Co.).

Hornero, R., Abasolo, D., Escudero, J., and Gomez, C. (2009). Nonlinear analysis of electroencephalogram and magnetoencephalogram recordings in patients with Alzheimer's disease. Philos Transact A Math Phys Eng Sci 367, 317-336.

Hu, J., Gao, J., Tung, W.W., and Cao, Y. (2009a). Multiscale analysis of heart rate variability: a comparison of different complexity measures. Ann Biomed Eng 38, 854-864.

Hu, J., Gao, J., Tung, W.W., and Cao, Y. (2009b). Multiscale Analysis of Heart Rate Variability: A Comparison of Different Complexity Measures. Ann Biomed Eng.

Huang, N.E., and Wu, Z. (2005). An Adaptive Data Analysis Method for nonlinear and Nonstationary Time Series:

The Empirical Mode Decomposition and Hilbert Spectral Analysis (Greenbelt, MD, NASA Goddard Space Flight Center).

Huang, N.E., Wu, Z., Long, S.R., and Arnold, K.C. (2009). On Instantaneous Frequency. Advances in Adaptive Data Analysis 1, 177-229.

Hyman, S.E. (2007). Can neuroscience be integrated into the DSM-V? Nat Rev Neurosci 8, 725-732.

Irani, F., Platek, S.M., Bunce, S., Ruocco, A.C., and Chute, D. (2007). Functional near infrared spectroscopy (fNIRS): an emerging neuroimaging technology with important applications for the study of brain disorders. Clin Neuropsychol 21, 9-37.

Johnson, M.H., Griffin, R., Csibra, G., Halit, H., Farroni, T., de Haan, M., Tucker, L.A., Baron-Cohen, S., and Richards, J. (2005). The emergence of the social brain network: evidence from typical and atypical development. Dev Psychopathol 17, 599-619.

Kandel, E.R. (1998). A new intellectual framework for psychiatry. Am J Psychiatry 155, 457-469.

Kandel, E.R., Schwartz, J.H., and Jessell, T.M. (2000). Principles of neural science, 4th edn (New York, McGraw-Hill Health Professions Division).

Kennedy, D.P., and Courchesne, E. (2008). The intrinsic functional organization of the brain is altered in autism. Neuroimage 39, 1877-1885.

Kleinhans, N.M., Richards, T., Sterling, L., Stegbauer, K.C., Mahurin, R., Johnson, L.C., Greenson, J., Dawson, G., and Aylward, E. (2008). Abnormal functional connectivity in autism spectrum disorders during face processing. Brain 131, 1000-1012.

Kreuz, T., Mormann, F., Andrzejak, R., Kraskov, A., and Lehnertz, K. (2007). Measuring synchronization in coupled model systems: A comparison of different approaches. Physica D 225, 29-42.

Kulisek, R., Hrncir, Z., Hrdlicka, M., Faladova, L., Sterbova, K., Krsek, P., Vymlatilova, E., Palus, M., Zumrova, A., and Komarek, V. (2008). Nonlinear analysis of the sleep EEG in children with pervasive developmental disorder. Neuro Endocrinol Lett 29, 512-517.

Kuusela, T.A., Jartti, T.T., Tahvanainen, K.U., and Kaila, T.J. (2002). Nonlinear methods of biosignal analysis in assessing terbutaline-induced heart rate and blood pressure changes. Am J Physiol Heart Circ Physiol 282, H773-783.

Lagioia, A., Van De Ville, D., Debbane, M., Lazeyras, F., and Eliez, S. (2010). Adolescent resting state networks and their associations with schizotypal trait expression. Front Syst Neurosci 4.

Laughlin, R.B. (2005). A different universe : reinventing physics from the bottom down (New York, Basic Books).

Le Van Quyen, M. (2003). Disentangling the dynamic core: a research program for a neurodynamics at the large-scale. Biol Res 36, 67-88.

Li, Y., Tong, S., Liu, D., Gai, Y., Wang, X., Wang, J., Qiu, Y., and Zhu, Y. (2008). Abnormal EEG complexity in patients with schizophrenia and depression. Clin Neurophysiol 119, 1232-1241.

LLorente, A.M., Lopresti, C., and Satz, P. (2009). Neurobehavioral and Neurodevelopmental Sequelae Associated with Pediatric HIV Infection. In Handbook of Clinical Child Neuropsychology, C.R. Reynolds, and E. Fletcher-Janzen, eds. (New York, Springer Science + Business Media), pp. 635-669.

Marwan, N., Romano, M.C., Thiel, M., and Kurths, J. (2007). Recurrence plots for the analysis of complex systems. Physics Reports 438, 237-329.

Marwan, N., Wessel, N., Meyerfeldt, U., Schirdewan, A., and Kurths, J. (2002). Recurrence-plot-based measures of complexity and their application to heart-rate-variability data. Phys Rev E Stat Nonlin Soft Matter Phys 66, 026702.

Mizuhara, H., Wang, L.Q., Kobayashi, K., and Yamaguchi, Y. (2005). Long-range EEG phase synchronization during an arithmetic task indexes a coherent cortical network simultaneously measured by fMRI. Neuroimage *27*, 553-563.

Montez, T., Linkenkaer-Hansen, K., van Dijk, B.W., and Stam, C.J. (2006). Synchronization likelihood with explicit time-frequency priors. Neuroimage *33*, 1117-1125.

Muehlemann, T., Haensse, D., and Wolf, M. (2008). Wireless miniaturized in-vivo near infrared imaging. Opt Express *16*, 10323-10330.

Mulert, C., Leicht, G., Pogarell, O., Mergl, R., Karch, S., Juckel, G., Moller, H.J., and Hegerl, U. (2007). Auditory cortex and anterior cingulate cortex sources of the early evoked gamma-band response: relationship to task difficulty and mental effort. Neuropsychologia *45*, 2294-2306.

Na, S.H., Jin, S.H., Kim, S.Y., and Ham, B.J. (2002). EEG in schizophrenic patients: mutual information analysis. Clin Neurophysiol *113*, 1954-1960.

Niedermeyer, E. (2003). The clinical relevance of EEG interpretation. Clinical EEG (electroencephalography) *34*, 93-98.

Niedermeyer, E., and Lopes da Silva, F.H. (2005). Electroencephalography : basic principles, clinical applications, and related fields, 5th edn (Philadelphia, Lippincott Williams & Wilkins).

Noonan, S.K., Haist, F., and Muller, R.A. (2009). Aberrant functional connectivity in autism: evidence from low-frequency BOLD signal fluctuations. Brain Res *1262*, 48-63.

Norris, P.R., Stein, P.K., and Morris, J.A., Jr. (2008). Reduced heart rate multiscale entropy predicts death in critical illness: a study of physiologic complexity in 285 trauma patients. J Crit Care *23*, 399-405.

Nunez, P.L., and Srinivasan, R. (2006). Electric fields of the brain : the neurophysics of EEG, 2nd edn (New York, Oxford University Press).

Ozonoff, S., Iosif, A.M., Baguio, F., Cook, I.C., Hill, M.M., Hutman, T., Rogers, S.J., Rozga, A., Sangha, S., Sigman, M., *et al.* (2010). A prospective study of the emergence of early behavioral signs of autism. J Am Acad Child Adolesc Psychiatry *49*, 256-266 e251-252.

Pascalis, O., de Haan, M., and Nelson, C.A. (2002). Is face processing species-specific during the first year of life? Science *296*, 1321-1323.

Percha, B., Dzakpasu, R., Zochowski, M., and Parent, J. (2005). Transition from local to global phase synchrony in small world neural network and its possible implications for epilepsy. Phys Rev E Stat Nonlin Soft Matter Phys *72*, 031909.

Pikovsky, A., Rosenblum, M., and Kurths, J. (2001). Synchronization : a universal concept in nonlinear sciences (Cambridge, Cambridge University Press).

Power, J.D., Fair, D.A., Schlaggar, B.L., and Petersen, S.E. (2010). The development of human functional brain networks. Neuron *67*, 735-748.

Puthanakit, T., Aurpibul, L., Louthrenoo, O., Tapanya, P., Nadsasarn, R., Insee-ard, S., and Sirisanthana, V. (2010). Poor cognitive functioning of school-aged children in thailand with perinatally acquired HIV infection taking antiretroviral therapy. AIDS Patient Care STDS *24*, 141-146.

Raghavendra, B.S., Dutt, D.N., Halahalli, H.N., and John, J.P. (2009). Complexity analysis of EEG in patients with schizophrenia using fractal dimension. Physiol Meas *30*, 795-808.

Ramoni, M.F., Sebastiani, P., and Kohane, I.S. (2002). Cluster analysis of gene expression dynamics. Proc Natl Acad Sci U S A *99*, 9121-9126.

Ravasz, E., and Barabasi, A.L. (2003). Hierarchical organization in complex networks. Phys Rev E Stat Nonlin Soft Matter Phys *67*, 026112.

Robertson, K., Liner, J., Hakim, J., Sankale, J.L., Grant, I., Letendre, S., Clifford, D., Diop, A.G., Jaye, A., Kanmogne, G., *et al.* (2010). NeuroAIDS in Africa. J Neurovirol *16*, 189-202.

Sabeti, M., Katebi, S., and Boostani, R. (2009). Entropy and complexity measures for EEG signal classification of schizophrenic and control participants. Artif Intell Med *47*, 263-274.

Saito, N., Kuginuki, T., Yagyu, T., Kinoshita, T., Koenig, T., Pascual-Marqui, R.D., Kochi, K., Wackermann, J., and Lehmann, D. (1998). Global, regional, and local measures of complexity of multichannel electroencephalography in acute, neuroleptic-naive, first-break schizophrenics. Biol Psychiatry *43*, 794-802.

Sakkalis, V., Doru Giurc Neanu, C., Xanthopoulos, P., Zervakis, M.E., Tsiaras, V., Yang, Y., Karakonstantaki, E., and Micheloyannis, S. (2009). Assessment of linear and nonlinear synchronization measures for analyzing EEG in a mild epileptic paradigm. IEEE Trans Inf Technol Biomed *13*, 433-441.

Sakkalis, V., Tsiaras, V., Michalopoulos, K., and Zervakis, M. (2008). Assessment of neural dynamic coupling and causal interactions between independent EEG components from cognitive tasks using linear and nonlinear methods. Conf Proc IEEE Eng Med Biol Soc *2008*, 3767-3770.

Sauseng, P., and Klimesch, W. (2008). What does phase information of oscillatory brain activity tell us about cognitive processes? Neurosci Biobehav Rev *32*, 1001-1013.

Scher, M.S., Ludington-Hoe, S., Kaffashi, F., Johnson, M.W., Holditch-Davis, D., and Loparo, K.A. (2009). Neurophysiologic assessment of brain maturation after an 8-week trial of skin-to-skin contact on preterm infants. Clin Neurophysiol *120*, 1812-1818.

Schinkel, S., Marwan, N., and Kurths, J. (2007). Order patterns recurrence plots in the analysis of ERP data. Cogn Neurodyn *1*, 317-325.

Schreiber, T., and Schmitz, A. (1997). Discrimination power of measures for nonlinearity in a time series. Phys Rev E Stat Nonlin Soft Matter Phys *55*, 5443-5447.

Shannon, C.E., and Weaver, W. (1949). A Mathematical Theory of Communication. Bell System Technical Journal *27*, 379-423.

Shaw, P., Greenstein, D., Lerch, J., Clasen, L., Lenroot, R., Gogtay, N., Evans, A., Rapoport, J., and Giedd, J. (2006). Intellectual ability and cortical development in children and adolescents. Nature *440*, 676-679.

Sheline, Y.I., Barch, D.M., Price, J.L., Rundle, M.M., Vaishnavi, S.N., Snyder, A.Z., Mintun, M.A., Wang, S., Coalson, R.S., and Raichle, M.E. (2009). The default mode network and self-referential processes in depression. Proc Natl Acad Sci U S A *106*, 1942-1947.

Singh, I., and Rose, N. (2009). Biomarkers in psychiatry. Nature *460*, 202-207.

Sörnmo, L., and Laguna, P. (2005). Bioelectrical signal processing in cardiac and neurological applications (Amsterdam ; Boston, Elsevier Academic Press).

Stam, C.J. (2005). Nonlinear dynamical analysis of EEG and MEG: review of an emerging field. Clin Neurophysiol *116*, 2266-2301.

Stauffer, D. (2006). Biology, sociology, geology by computational physicists, 1st edn (Amsterdam ; Boston, Elsevier).

Supekar, K., Musen, M., and Menon, V. (2009). Development of large-scale functional brain networks in children. PLoS Biol *7*, e1000157.

Supekar, K., Uddin, L.Q., Prater, K., Amin, H., Greicius, M.D., and Menon, V. Development of functional and structural connectivity within the default mode network in young children. Neuroimage 52, 290-301.

Supekar, K., Uddin, L.Q., Prater, K., Amin, H., Greicius, M.D., and Menon, V. (2010). Development of functional and structural connectivity within the default mode network in young children. Neuroimage 52, 290-301.

Takahashi, T., Cho, R.Y., Mizuno, T., Kikuchi, M., Murata, T., Takahashi, K., and Wada, Y. (2010). Antipsychotics reverse abnormal EEG complexity in drug-naive schizophrenia: a multiscale entropy analysis. Neuroimage 51, 173-182.

Uddin, L.Q., Kelly, A.M., Biswal, B.B., Margulies, D.S., Shehzad, Z., Shaw, D., Ghaffari, M., Rotrosen, J., Adler, L.A., Castellanos, F.X., et al. (2008). Network homogeneity reveals decreased integrity of default-mode network in ADHD. J Neurosci Methods 169, 249-254.

Uddin, L.Q., Supekar, K., and Menon, V. (2010). Typical and atypical development of functional human brain networks: insights from resting-state FMRI. Front Syst Neurosci 4, 21.

Uhlhaas, P.J., Haenschel, C., Nikolic, D., and Singer, W. (2008). The role of oscillations and synchrony in cortical networks and their putative relevance for the pathophysiology of schizophrenia. Schizophr Bull 34, 927-943.

Van Rie, A., Dow, A., Mupuala, A., and Stewart, P. (2009). Neurodevelopmental trajectory of HIV-infected children accessing care in Kinshasa, Democratic Republic of Congo. J Acquir Immune Defic Syndr 52, 636-642.

Varela, F., Lachaux, J.P., Rodriguez, E., and Martinerie, J. (2001). The brainweb: phase synchronization and large-scale integration. Nat Rev Neurosci 2, 229-239.

Wackermann, J., Lehmann, D., Dvorak, I., and Michel, C.M. (1993). Global dimensional complexity of multi-channel EEG indicates change of human brain functional state after a single dose of a nootropic drug. Electroencephalogr Clin Neurophysiol 86, 193-198.

Ward, L.M. (2003). Synchronous neural oscillations and cognitive processes. Trends Cogn Sci 7, 553-559.

Whittington, M.A. (2008). Can brain rhythms inform on underlying pathology in schizophrenia? Biol Psychiatry 63, 728-729.

Xie, H.-B., He, W.-X., and Liu, H. (2008). Measuring time series regularity using nonlinear similarity-based sample entropy. Physics Letters A 372, 7140-7146.

Zavaglia, M., Astolfi, L., Babiloni, F., and Ursino, M. (2008). The effect of connectivity on EEG rhythms, power spectral density and coherence among coupled neural populations: analysis with a neural mass model. IEEE Trans Biomed Eng 55, 69-77.

Zbilut, J.P., Thomasson, N., and Webber, C.L. (2002). Recurrence quantification analysis as a tool for nonlinear exploration of nonstationary cardiac signals. Med Eng Phys 24, 53-60.

Zhang, D., Ding, H., Liu, Y., Zhou, C., and Ye, D. (2009). Neurodevelopment in newborns: a sample entropy analysis of electroencephalogram. Physiol Meas 30, 491-504.

13

Educational Opportunities in BME Specialization - Tradition, Culture and Perspectives

Wasilewska-Radwanska Marta[1], Augustyniak Ewa[2],
Tadeusiewicz Ryszard[1,3] and Augustyniak Piotr[1,3]
[1]*Multidisciplinary School of Engineering in Biomedicine,
AGH-University of Science and Technology,*
[2]*Faculty of Humanities, AGH-University of Science and Technology,*
[3]*Faculty of Electrical Engineering, Automatics, Computer Science and Electronics,
AGH-University of Science and Technology,
Poland*

1. Introduction

1.1 The traditions of biomedical physics and engineering in Poland

Medical physics and engineering education in Poland started in the 1930s with the foundation of the Radium Institute in Warsaw by Maria Sklodowska-Curie. Prof. Cezary Pawlowski, one of the assistants and then collaborators of Mme Curie (fig. 1), organized the first courses in medical physics and biomedical engineering at the Physics Department of the Radium Institute.

The first course in medical engineering started at the Faculty of Electrical Engineering of the Warsaw University of Technology in the 1950s. Then, at the Faculty of Electrical Engineering, Automatics, Computer Science and Electronics of the AGH University of Science and Technology (former University of Mining and Metallurgy) in Krakow, Prof. Ryszard Tadeusiewicz organized the first courses in biomedical engineering in the 1970s. Fig. 2 shows the first Polish textbooks in Medical Electronics and in Biocybernetics. Note the year of the issue of both books, 1978.

Until the academic year 2005/2006, education in biomedical engineering was offered only as a specialization in other fields of studies, e.g. mechanics, automatics & robotics and electronics. The development of new technologies in medical diagnosis and therapy required a new approach to biomedical engineering education. Therefore, a consortium was set up of six technical universities (in alphabetical order): The AGH University of Science and Technology (Krakow), The Gdansk University of Technology (Gdansk), The Silesian University of Technology (Gliwice), The Technical University of Lodz (Lodz), The Warsaw University of Technology (Warsaw) and The Wroclaw University of Technology (Wroclaw). The consortium developed a new programme of education and then applied to the Ministry of Science and Higher Education for an official permit to create a new field of studies referred to as "Biomedical Engineering" (BME). In June 2006, the Ministry gave its consent to this proposal. The AGH University of Science and Technology was first in Poland to enroll students in BME in the academic year 2006-2007. In 2007-2008, all the members of the consortium had students in BME. In the academic year 2010-2011, BME education is being offered by 16 technical universities in Poland (Table 1).

Fig. 1. Dated 1911. Probably one of the earliest photographs to show Maria Sklodowska Curie (first from the left), a double Nobel Prize laureate and professor at the Sorbonne (Paris), and Walery Goetel (first from the right), professor and future rector of the Mining Academy (now the AGH-UST), on a mountain trip which could have led to Maria Sklodowska Curie's becoming three years later Director of the Red Cross Radiology Service during World War I. She is also known to have converted many ordinary cars into ambulances equipped with mobile radiology units. These cars, called "petite Curie," transported X-ray apparatus to the wounded at the battle front, thus saving the lives of many French soldiers.

Medical physics education in Poland started in 1950 with the technical physics specialization created by Prof. Cezary Pawlowski at the Warsaw University of Technology, and at the AGH University of Science and Technology (former University of Mining and Metallurgy) in Krakow by Prof. Marian Miesowicz. In the 1970s, a medical physics programme was initiated at Warsaw University and at the Jagiellonian University in Krakow. In 1990, a specialization in Radiation Physics and Dosimetry was started at the AGH University of Science and Technology in Krakow, which since 1991-1992 has been run as Medical Physics and Dosimetry in close cooperation with the Collegium Medicum (Faculty of Medicine) of the Jagiellonian University. In the academic year 2010-2011, about 15 universities and technical universities offered courses for students in medical physics (Table 2).

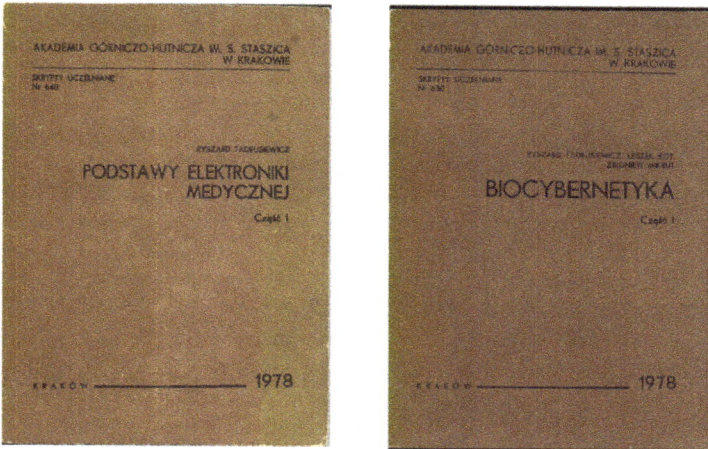

Fig. 2. The first Polish textbooks for biomedical engineering students issued by AGH-UST.

2. A multidisciplinary school – organizational background and curricula

As teaching Biomedical Engineering requires specialists representing many different areas of research and different competences, it was impossible to select a staff on the basis of the individual personnel resources of one particular faculty. Therefore to teach biomedical engineering students, a Multidisciplinary School of Engineering in Biomedicine (MSIB – the acronym used in all Polish documents describing the aims and the scope of the school) was founded (fig. 3). Below you will find the general outline of the structure and main guidelines for the School (Wasilewska-Radwanska & Augustyniak, 2009).

Fig. 3. Rector of the AGH-UST, Professor Ryszard Tadeusiewicz, signs the MSIB foundation act (2005).

2.1 The external situation

The external situation in Poland in 2005, when the Multidisciplinary School of Engineering in Biomedicine was founded, provided several opportunities and challenges (Augustyniak 2008). These initial external conditions can be classified into three groups:

The first group referred to Polish medical technology-related enterprises and/institutions:

- Local industry was rather undeveloped; we estimate the number of local medical technology-related enterprises to be about 100, but most of them (40%) were very small businesses, so-called micro-enterprises, having 1-5 employees, or small enterprises (30%) with 6-50 employees. Bigger enterprises were usually sales- or service representatives of international corporations, without independent human resources management (fig. 4);
- The relation between research and industry was weak; the way from technical innovation to marketing of a final product was very formalized;
- The results of research done by technical universities were financially unattractive and did not match industry needs; industry management preferred independent research rather than cooperation with universities;
- The average technological level of the health care was low, with notable exceptions in some selected centers.

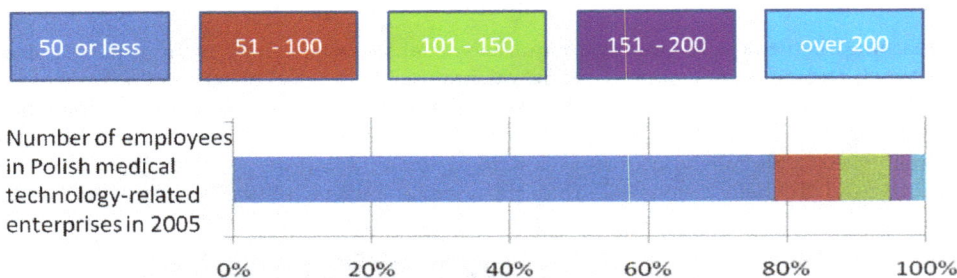

Fig. 4. The employment structure in Polish medical technology-related enterprises in 2005 (ROTMED Consortium 2006).

The second group was related to Polish medical technology needs based on social demands:

- There was an urgent and important need for the development of medical technology because of the poorer quality of social health services in Poland compared with those in highly developed EU countries (fig. 5)

The third group was related to Polish university traditions and traditional models of teaching:

- The experience with two-tier structure of degree courses/university studies (Bachelor's and Master's) was very inadequate; there were no clear guidelines for curricula, syllabuses and examinations, nor for assessment of the teaching quality; the existing government regulations were insufficient;
- There was no experience of teaching in English, and professional bibliography in Polish was very limited;
- The organizational chart of a multidisciplinary school was innovative and rarely implemented by universities, the university funds' distribution mechanisms being inadequate;

- Biomedicine-related research was carried out in several faculties within the framework of other disciplines such as computer or material sciences, electrical or mechanical engineering; there was an adequate number of professors and assistant-professors representing high-performance field-oriented output;
- The principles of school organization enabled quality-based staff selection unlimited by state employment regulations;
- There was a growing interest in medical technology from good candidates; and
- Some recent governmental regulations, aiming at improving the quality of health care, facilitated the employment of clinical engineers.

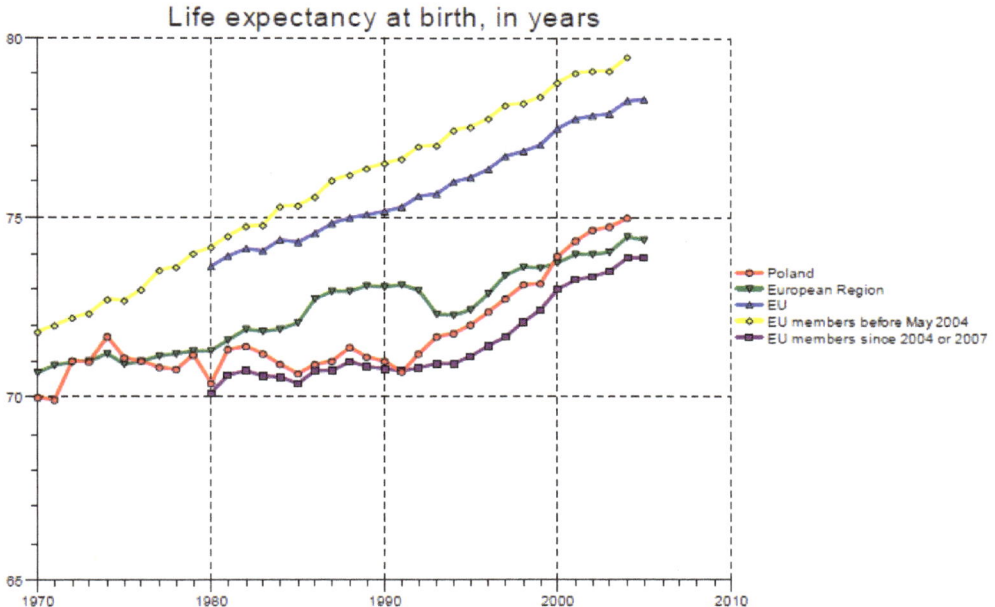

Fig. 5. A comparative plot showing the situation in social health services in Poland vs. the developed EU countries (ROTMED Consortium 2006).

All the above issues and challenges have led to ever stronger inter-university cooperation and integration into society. Representatives of eight Polish universities made every effort to establish educational standards in biomedical engineering as a separate field of study currently offered in Poland for about 1000 candidates each year.

2.2 The organizational scheme and place of MSIB in university structure
The MSIB is located at the AGH University of Science and Technology and has been in operation since the academic year 2005-2006. Although MSIB has been formed on the basis of the staff formally belonging to five faculties, it is treated as a separate part of the AGH University and has its own students. Formally, MSIB's structure is similar to that of other faculties. It is governed by a Board of 18 persons. This Board, approved by the University Senate, is made up of professors with not less than a DSc degree who are teaching at MSIB,

as well as of an adequate representation of students. At present, the professors represent five faculties:

- Faculty of Electrical Engineering, Automatics, Computer Science and Electronics,
- Faculty of Materials Science and Ceramics,
- Faculty of Mechanical Engineering and Robotics,
- Faculty of Metals Engineering and Industrial Computer Science, and
- Faculty of Physics and Applied Computer Science.

One of the Board's tasks is to recommend to the Rector appointments for the Head and the Deputy Head of the School. The appointed Head is also President of the Board. The main responsibility of the Board is to supervise the education process, assure its highest quality, verify and, if necessary, correct academic curricula, prepare staff assignments and implement other objectives of the School. The Head also represents the MSIB in the University Board on par with deans of other faculties.

From the student's viewpoint, there is no organizational difference between the faculty and the Multidisciplinary School. Both have a Dean's Office, a staff of qualified teachers, a social support system and a student board. As far as education is concerned, the rights and responsibilities of the Head of the School are identical to those of a Dean, the only difference being that research is carried out in laboratories in various faculties run by individual professors rather than in the organizational framework of MSIB.

Since medical sciences are not represented in the AGH-University of Science and Technology, six medicine-oriented lectures (e.g. anatomy, physiology, medical deontology, history of medicine) are given by professors of the Collegium Medicum (Medical College) of the Jagiellonian University. The agreement between the universities gives students the opportunity to attend lectures and to participate in laboratory exercises in the Faculty of Medicine. This cooperation is mutually beneficial since it provides an alternative, i.e. medicine-based viewpoint, for our medical colleagues and medicine students. Unfortunately, current medical curricula in Poland do not include engineering aspects in medicine, however some lecturers from AGH-UST are among those who take part in postgraduate studies and technology-oriented teaching projects for medicine students or medical doctors.

2.3 General layout of curricula

The BME teaching programs in the Multidisciplinary School of Engineering in Biomedicine AGH University of Science and Technology follow all the Polish legal regulations, including the national standards for academic teaching set out by the Ministry of Science and Higher Education (Ministry of Science and Higher Education 2007), and the guidelines of the Bologna Process (including the Educational Credits Transfer System and Accumulation System-ECTS). The current program presented in the block diagram in Fig. 6 consists of:

- A single 7-semester track leading to the First (Undergraduate) Degree (Bachelor's/Engineer's);
- Five domain-oriented 3 or 4-semester tracks leading to the Second (Graduate) Degree (Master's);
- A single 8-semester track leading to the Third Degree (Doctor's).

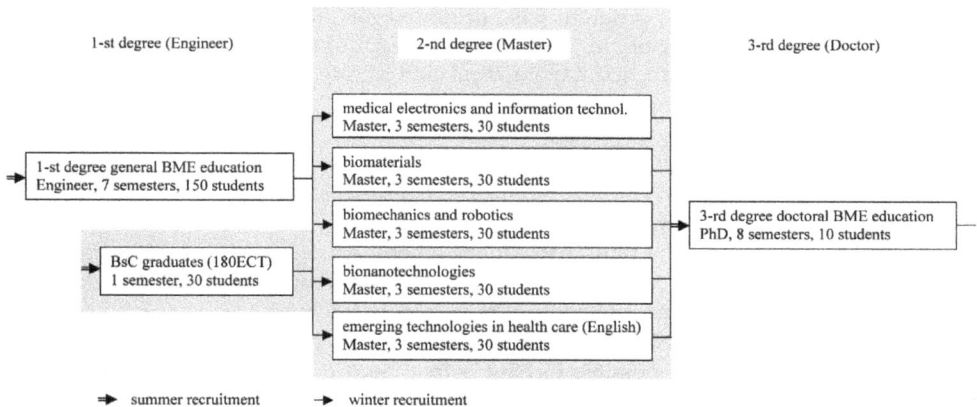

Fig. 6. An education track block diagram for biomedical engineering at MSIB AGH-UST.

After a careful review of the needs from prospective employers' point of view, the availability of existing infrastructure and resources and a detailed study of reports from more experienced colleagues, we have decided to formulate and put into practice several rules and mechanisms that provide a broad basic education in all possible BME domains and fast adaptation of the program to the variability of an unstable, constantly developing local employment market.

A complete description of the offer consisting of 189 lectures [64 for the Bachelor's (First) degree , 104 in five domains for the Master's (Second) degree and 21 for the Doctor's (Third) degree)] is available from the MSIB web service www.biomed.agh.edu.pl (see Appendix, table 3). By way of example, we have selected two curricula for the specialization called "Emerging Health Care Technologies" taught in English for the Master's Degree and for the Doctor's Degree (PhD) studies (see Appendix, table 4) and three course sheets -syllabuses (see Appendix, table 5). Some selected themes for PhD theses may also be worth mentioning:

- 3D Reconstruction of Brain Glial Cells (2010),
- Automatic Facial Action Recognition in Face Images and the Analysis of Images in the
- Human-Machine Interaction Context (2009),
- 3D Segmentation of Medical Data from Computed Tomography and Endoscopic Video Records (2008).

Full texts of these theses are available on-line from the Biomedical Engineering PhD students' web service www.embio.agh.edu.pl.

2.4 Program adaptation mechanisms

In addition to the Bachelor's Degree program of a single track (for 150 students each year), various measures were implemented in order to increase the adaptation range of the proposed tracks. A high degree of flexibility, and a wide offer of elective courses involve students as partners in the educational process. This method of program organization with the essential participation of the students has an additional advantage: it helps to develop the students' responsibility and flexibility that is necessary in the difficult workplaces employing biomedical engineers.

The Second Degree (Master's) program (also proposed for 150 students per year) requires that students select one of five offered parallel tracks. Four of them are taught in Polish and oriented towards the main branches of biomedical engineering:

- Medical electronics and information technologies,
- Biomaterials,
- Biomechanics and robotics,
- Bionanotechnologies,
- Emerging health care technologies.

Fortunately, the formula of the multidisciplinary school that is based on human resources and the infrastructure of five faculties takes advantage of the support thus provided and is sufficient to comply with the requirements of high-quality teaching.

This variety of human and technological resources in the absence of institutional limitations and work contracts makes it possible to have a free hand in creating teaching syllabuses depending on the current market needs and students' interests. The creation of learning programs (new syllabuses and specialties) which comes under the authority of the Head of the faculty is also supplemented by three types of individual lines of study:

- Implementation of some selected additional subjects of study in addition to the standard syllabus;
- Individual schedule of study. i.e. implementation of the standard program following the accelerated of decelerated schedule, or
- Individual course of study making possible the selection of subjects outside the standard syllabus to replace elective subjects.

The creation of the course of study is based on quality indexes systematically collected from students, lecturers and employers[1]. The assessment of the quality of the learning process is regularly made by:

- Estimation of current learning progress made by lecturers;
- Estimation of learning results by the Head of the Faculty based on statistical evaluation;
- Questionnaires filled in by students about their lecturers (carried out by Department Deans and made public in a adapted from to the Head of the MSIB);
- Questionnaires filled out by students on the course of their studies;
- Questionnaires on the professional abilities of students carried out in institutions responsible for the students' field work.

All the quality indexes are analyzed by the Board and are used in making the Head responsible for implementing decisions such as a face-to-face talk with the lecturer, change of the lecturer, modification in the sequence of subjects or a change in the syllabus.

3. Specific characteristics of BME-related corporate culture

Although every profession has its own best business practices, biomedical engineering has some specific characteristics. This relatively new discipline bridges the gap between medicine and technology, and, by applying various novel methods and techniques, directly influences the duration and quality of human life (Schwartz 1988). Consequently, a single medical procedure involves multiple actors, many of whom, mainly engineers, usually remain behind the scene. The cooperation between engineers and members of the

---

[1] In future

medical profession can succeed only if they all share some common understanding about their roles and mutually dependencies. Moreover, a medical procedure requires, among other things, technical excellence and high ethical awareness of the participating engineers. Finally, the clinical engineer frequently works shoulder to shoulder with medics in an emergency and under time pressure, burdened by stress and responsibility, where common feelings, ideas and behavior are a decisive factor in the final result. Therefore three key aspects may be identified as those related to BME professional culture:

- Understanding and good cooperation within a multidisciplinary team;
- Striving for the technical excellence and
- Human-centric ethical background.

As such circumstances are specific to this profession, the need for the organizational culture is fairly high. In case of a biomedical engineer, however, the relations are far more complex than the schemes sketched by corporate culture promoters for a typical customer service (Denison & Mishra, 1995). The above-mentioned elements of the BME-related corporate culture cannot be formed only as spontaneous forms of behavior of clinical engineers. They must be prepared by special methods and techniques of teaching and learning to help young students become highly qualified and well-prepared specialists also in the ethical dimension of their work. This process of ethical formation must be closely combined with professional skill formation, because ethics is not simply an additional competence of the future biomedical engineer, but rather a crucial element determining the use of basic knowledge in all practical applications. Therefore teachers who teach biomedical engineering students must take this additional demand into account (Augustyniak E. & Augustyniak P. 2010).

3.1 Mutual understanding and good cooperation within a multidisciplinary team

Biomedical engineering integrates various technical disciplines, but it also assumes the presence of an engineer in every situation where medicine is being practiced. Therefore, one of the principal requirements is the ability to work in a multidisciplinary team, in which common ideas, values and behavioral patterns have a very practical implication influencing the emergency response, adaptation to conditions of any healthcare mission and a sense of intellectual curiosity about technological development. In such teams, various professional learning-related particular interests are present, and mutual understanding instead of competition is the key to success.

The organizational culture of the School created by all workers from various departments of the School fosters an emphasis on teamwork and intercommunication, especially in the group of specialists from various fields. Common understanding, personal involvement and dedication constitute an essential pattern of a learning organization. In a team which includes a physician, nurse, pharmacist and an engineer, the effective use of knowledge and skill by separate persons is dependent on the mutual understanding within each person's field of competence. In addition, the cooperating specialists motivate each other and thus create novel solutions resulting from the synergy of various experiences.

3.2 Striving for technical excellence

Working in a team with a doctor, a nurse and a pharmacist, the biomedical engineer is responsible for all technology-related ideas, and is perceived as someone capable of solving all problems in his or her area of responsibility; all of this combined may become a source of abnormal stress. Moreover, the support of life, as no other application of technology, has a direct relation to the human being, his or her health and happiness and to fundamental values. Consequently, an assumption of excellence distinguishes a biomedical engineer from engineers of other specialties. Furthermore, since technology is currently responsible for the ever greater efficiency and effectiveness of medicine, it can also be blamed for any adverse effects of medical procedures.

Striving for excellence at all levels of the implementation of detailed goals of achieving cooperation and suggesting technical solutions to problems is a typical characteristic of the organizational culture. In addition to tasks involving the implementation of educational programs (curricula and syllabuses), one of the School's main objectives is to transmit standards. Thus, when the School becomes an archetype or model of the future workplace and the lecturers employed by the School are expected to set an example of how excellence should be achieved, the care taken to provide a proper form of the organizational culture is a *sine qua non* educational requirement (O'Reilly & Chatman, 1996). If the future graduate student acquires adequate competences during the course of his or her studies, their position will be determined in a multidisciplinary team in real situations in health services. Organizational rules and regulations that are in force in the School, which employs lecturers from various faculties, make the presentation of the lecturers' various roles in the learning process easier by taking into account the person's specific knowledge, skills and attitudes. This approach has been used to make an analogy pointing to the technical excellence of an engineer as a basic argument supporting his or her key role in the multidisciplinary team employed in health services.

Fig. 7. Students discussing laboratory projects (2008).

3.3 The ethical background

Engineers who are specialists in other fields of technology than medical or biological engineering are rarely confronted with human life in its various forms and stages, such as birth, adolescence, disablement, disease, pain, death, etc. Therefore, special forms and patterns of education should be dedicated to attitudes about human nature in general. Since the engineering process in biomedicine is always part of a medical procedure, the ethical aspects of medicine should be a mandatory element of lectures and practical work in a BME-teaching institution.

The attitudes, norms and values promoted in the organizational culture of the School have an effect on the way how the future graduates will think and act in their professional life. The sense of responsibility and reliability combined with a sense of humanitarianism, curiosity and creativity are among the most desirable attributes of a biomedical engineer, while also held in high value by employers in other fields. The above values are not easily transferable to traditional university teaching methods such as lectures, problem-solving or experiments, therefore elements of the organizational culture may be of some help in this context.

The most important elements of this culture are:

- Forms of a specific tradition established within a given group to strengthen the feeling of identifying with the group's aims;
- Undertaking novel professional activities such as writing textbooks, promoting the new profession in mass media or encouraging volunteer work by both lecturers and students;
- Permanent evaluation of the School's activity and the teaching and learning programs based on external assessment and the opinions provided by lecturers and students.

4. Opportunities and perspectives of BME-teaching institutions

The MSIB authorities have designed a unique, well-balanced corporate culture program and continue to develop it with success. The MSIB itself as a teaching institution was started from scratch, so from the beginning it was based on the general principles of a learning organization.

4.1 A blueprint for an organizational culture

The establishment of a new institution created an opportunity for devising a profession-specific, independent system of shared values and beliefs implying and reinforcing a collective identity centered on the mission of the biomedical engineer. Our aims exceeded "standard faculty" programs, providing students with government-designed minimum curricula. Our main objective is for MSIB to become a leading research and educational center (Augustyniak et al. 2010). Despite a relatively short history of only five years, the School has undoubtedly earned high marks in Poland and has become well-recognized abroad. This was possible due to:

- Employment of the best professionals in the field as teachers;
- Active creation of cooperation links with other universities in Poland and abroad;
- Publication of a BME-student quarterly - a periodical promoting the profession and reinforcing the students' intellectual confidence and emotional group commitment;

- Leadership and promotion of common activities, exchanges, encouragement of staff and students' mobility, additional lectures etc.;
- Dissemination of knowledge and participation of staff members in international conferences and their activity in commissions and other opinion-forming bodies.

Fig. 8. Professor Zbigniew Kakol, Rector of AGH-UST, confers an Engineering Diploma (2010).

The organizational culture of the MSIB is based on a deeply motivated concept of a high-quality teaching and learning institution, aiming at the education of marketable BME professionals (Gordon & DiTomaso, 1992). The School is student-friendly and supports students' educational initiatives in various fields of activity. What is expected and/or requested from the staff and students are well-defined goals to achieve. In this spirit, thanks to the common sense of responsibility, the internal atmosphere in the School supports cooperation rather than competition.

To make common attitudes and values more popular and generally recognized, the organizational culture of the MSIB offers the following:

- Ceremonies (matriculation, the Dean's address at the New Year, summarizing achievements challenges, and future plans, etc.);
- Corporate design (a well-recognized logo, a precisely defined color, organizational badges for Board members, corporate T-shirts for students);
- Outstanding personalities - undoubtedly this honor goes to Professor Ryszard Tadeusiewicz, the founder of the School and former Rector of AGH-UST, a pioneer in biomedicine and author of several BME-related books since the mid-70s;
- Common recreational activities (mountain trekking, sailing, skiing, student sports teams in various disciplines, volunteer-based events in hospitals and hospices, initiatives on behalf of the handicapped, extramural students' organizations and forums).

The corporate culture is built on selected values and behavioral norms that influence both the employees and students through the internal and external institutional policy. The commonly accepted standards constitute the internal ambience and friendliness of the School, but as a long-term investment become everyday habits. We easily observe teachers transferring these standards to their own departments, which justifies the belief that in the future the graduates will also promote such standards in their workplaces.

Fig. 9. Informal mountain trekking of MSIB students and the Dean (first from the left).

4.2 The employment perspectives of graduates

The high quality of educational outcomes is of vital interest for students, their prospective employers and university staff. All three groups of partners aim for a common goal using different approaches and capabilities. Cultivating the understanding of the School as a common value and sharing the collective identity and commitment reinforces compatibility and multi-professional team building. This training field for interpersonal learning reveals three enhancing feedback loops:

- Attractive employers increase students' motivation and good graduates extend the development perspectives of their employers;
- Innovative employers offer career opportunities for good researchers in applied sciences, and more effective professors provide high-quality knowledge to industry which in turn enhances the university's reputation;
- Good candidates are inclined to look for good lectures and challenging projects, and become a source of professional satisfaction to professors who make efforts to improve their lectures.

These relations, demonstrated above in a simplified form, are fundamental for the team-building spirit, which is based on the commitment of students, tutors and professionals to a quality-focused efficient education. In the MSIB, the healthcare professionals participate in the preparation of curricula, provide opportunities for one-month internships and formulate challenges for young teams of scientists. The university staff cultivates the students' creativity by supervising laboratory exercises and interim projects which constitute a demonstration of the graduates' skills, and can be partly applied in healthcare and industry. The excellence of the university staff in science is also a good example attracting best students who often try to follow or to participate in research themselves.

The most highly appreciated educational initiatives are those focused on education-oriented basics, behavioral norms and values including self-teaching. Equally important to scope and breadth of knowledge are technical skills and human-centric attitudes. The basics of human sciences, ethics and law are indispensable elements of education and practice of the School. Two other important features are:

- Flexibility which makes it possible for the School to adapt itself to changing trends of the employment market and which enables graduates to keep pace with quickly-developing technical sciences; and
- A set of relations between the teacher and/or the dean and the student which focuses on communication and exchange of information. This not only makes relations within the School administration easier but also establishes a pattern for future relations in a multi-professional environment.

Current employment perspectives mostly result from the growing importance of extensive use of technology in health care. Clinical engineer, manufacturing engineer, researcher and sales representative are the four main types of specializations in terms of future professional careers. Nevertheless, local employment markets are weak, and the effort to raise the awareness of a new profession to a higher level is among the principal concerns of the MSIB Management. General statistics are favorable for the AGH-UST graduates: 75% of them find their first job within one month after graduation, and 95% within three months. According to other surveys, in 500 of Poland's biggest enterprises, AGH-UST graduates constitute the second largest percentage among senior management staff.

5. Concluding remarks

The special demands imposed on BME education require that we should seek unusual organizational solutions and non-standard teaching and learning methods, related to the specific and unique educational challenges posed by BME. We hope that these theoretical and practical methods and techniques developed at the MSIB and presented above may be of interest to other educational institutions in BME field.

6. Acknowledgments

This work was supported by the Rector of the AGH University of Science and Technology in Krakow, Poland. The authors want to express their gratitude to the Walery Goetel family for their consent to publish the archival photo of Maria Sklodowska Curie.

7. Appendix

Institutions	Faculty/ Department	Organisation (current state)/specializations
AGH Univ. of Science and Technology (Krakow)	Electrical Eng., Automatics, Comp. Sc.& Electronics	Multidisciplinary School of engineering In Biomedicine (MSIB)/Med. Electronics&Inform.; Biomat.,; Biomech.&Rob.;Bionanot; ; Emerging Techn.Health Car
Bialystok Univ. of Techn.(Bialystok)	Mechanical Engineering	/Biomechanics; Rehabilitation Engineering; Medical Materials
Czestochowa Univ. of Techn. (Czestochowa)	Mechanical Engineering & Comp. Sc.	Rehabiltation Engineering; Medical Informatics.
Gdansk Univ. of Technology (Gdansk)	Electronics, Telecomunications &Informatics	Inter-Faculty full-time studies /Inf.Med.; ElectrM.; ChemM; PhysM
Koszalin Univ. of Techn. (Koszalin)	Institute of Mechatronics, Nanotechnology& Vac. Techn.	/Implants; Medical Instrumentation; Manipulators
Krakow Univ. of Technology (Krakow)	Mechanical Engineering	/Clinical Eng. Dental Biomechanics; Biomechanics of Injuries
Lublin Univ. of Techn. (Lublin)	Mechanical Engineering	In cooperation with Lublin Medical Univ./CT; MR; Aut&Rob; Biomat Art Organs; M Electr ;Infor in Medicine
Lodz Univ. of Techn. (Lodz)	El., Electronic, Comp.& Control Eng.	Inter-Faculty Education Centre with the Fac. of El., Electronic, Comp.& Control Eng./Biomats; Biocorrosion; Biomeasurements
Poznan Univ. of Techn. (Poznan)	Mechanical Engineering&Management	Mechatronics/ Biomedical Engineering
Silesian Univ. of Technology (Gliwice)	Biomedical Engineering	/InformaticsM; M Electr.;Sensors & Biomed;; Implants Reahabilitation
Silesian Univ. (Katowice)	Computing Sc. and Material Sc.	/ Biomed.Inform;; Med. Imaging; Telemed&Clinical Inform. Systems;
Warsaw Univ. of Technology (Warsaw)	Mechatronics/ I. of Metrology&Biomed. Engineering and Electronics and Information Technology	Inter-Faculty studies:e F. of Mechatronics& F. of Electronics&Inf. Techn./Biomech.; Biomat.s; Biomed. Sensors; Biomed.Imaging; Rehab.Eng; ;Prostheses&Art Organs; Clin.Eng. ; Biotechnology; Medical Informatics
Wroclaw Univ. of Technology (Wroclaw)	Fund. Problems of Technology	/Electromed Equip; Smart Transducers; ; Fiber Optics&d Lasers in Med.Equip.; Computer MDiagn.
Zielona Gora University (Zielona Gora)	Mechanics; Electrical Engineering, Comp.Sc.&Telecommunic.; Biological Sc.	Inter-Faculty studies. /Medical Informatics; Medical Electronics; Biomechanics; Biomaterias
Bydgoszcz Univ. of Tech. (Bydgoszcz)	Mechanics	/Technical Medical Consulting; Medical Informatics
West Pomeranian Univ. of Techn. (Szczecin)	Electrical Engineering	/Biomedical and Acoustic Engineering

Table 1. A list of institutions which provide education in Biomedical Engineering (BME) in Poland (since academic year 2010-2011).

Institutions	Faculty/ Department	Discipline	Specialization
Univ. of Bialystok (Bialystok) www.uwb.edu.pl	Physics	P*	MP
Univ.of Gdansk (Gdansk) www.ug.gda.pl	Math, Physics &Informatics in cooperation with Gdansk Medical Univ.	MP*	MP
Univ.of Silesia (Katowice) www.us.edu.pl	Math., Phys. and Chem.	MP	Clinical Dosimetry; Optics in Medicine; Electroradiology
Jan Kochanowski Univ. of Humanities and Sc. (Kielce) www.ujk.edu.pl	Math.&Sc.	P	MP
AGH Univ. of Sc .&Techn. (Krakow) www.agh.edu.pl	Phys.& Appl. Comp. Sc.	MP	Dosimetry and Electronics in Medicine; Imaging and Biometrics
M. Curie-Sklodowska Univ. (Lublin) http://mfi.umcs.lublin.pl	Math., Phys. Comp. Sc.	P	Molecular and medical biophysics
Un. of Lodz (Lodz) www.uni.lodz.pl	Phys& Chem.	P	MP
Techn.Univ.Lodz (Lodz) www.p.lodz.pl	TPhys., Comp Sc.& Appl.Math.	TP*	MP
Univ.of Opole (Opole) www.wmfi.uni.opole.pl	Math., Phys & Chem.	MP	MP
Adam Mickiewicz Univ. (Poznan) http://amu.edu.pl	Physcis	B****	MP
Ignacy Lukasiewicz Techn.Univ. (Rzeszow) http://portal.prz.edu.pl	Math.& Appl. Physics	TP	Physics and Informatics in Medicine
Univ.of Szczecin (Szczecin) www.us.szc.pl	Math.& Physics	P	Biomedical Physics
Nicolaus Copernicus Univ. (Torun) www.umk.pl	Phys., Astr& Informatics	MP	MP and Comp. Appl.
Warsaw Univ. (Warsaw) www.uw.edu.pl	Physics/Inst. of Exp.Phys.	APBM*****	MP
Warsaw Univ. of Techn. (Warsaw) www.pw.edu.pl	Physics	TP	MP
Wroclaw Univ. of Techn. (Wroclaw) www.pwr.wroc.pl	Fund. Problems of Techn.	TP	Nano-engineering
Univ. of Wroclaw (Wroclaw) www.uni.wroc.pl	Physics and Astr.	TP	MP

*)MP=Medical Physics ; **) P=Physics ;***)TP=Technical Physics; ****B=Biophysics;
*****APBM=Applications of Physics in Medicine and Biology

Table 2. A list of institutions which provide education in Medical Physics (MP) in Poland (since academic year 2010-2011).

semester	First Degree(Bachelor's)
1.	• Information technologies • Mathematics • Physics • General chemistry • Biocybernetics • Biology and genetics • Propedeutics of medical sciences
2.	• Mathematics • Physics • Statistics and probability theory • Principles of electrical engineering • Principles of electronics • Organic chemistry
3.	• Sport • Foreign languages • Physics laboratory • Materials sciences • Principles of metrology • Mechanics and strength of materials • Fundamental anatomy • Principles of physiology • Computer programming
4.	• Sport • Foreign language • Computer Aided Design -or- Design with Finite Elements Method • Elements of Biochemistry • Biophysics • Biomaterials • Sensors and non-electrical measurements -or- Integrated measurement systems • Digital signal processing

semester	First Degree (continued)
5.	• Foreign languages • Medical Physics Biomechanics • Computer graphics • Fundamentals of graphical programming languages • Programming languages -or- Object programming • Automatics and Robotics
6.	• Foreign languages • Biomechanics - project • Implants and Artificial Organs • Electronic Medical Instrumentation • Medical Imaging Technology • History of medicine • Elective 1 - Cryptography and data ciphering systems - Chemometry - Ergonomics and occupational medicine • Elective 2 - Principles of management in biotechnical systems - Microcontroller programming in C/C++ - Globalization and modernization problems - Introduction to environmental philosophy - Glass- and glass-ceramic materials in medicine
7.	• Introduction to diagnostics with ionizing radiation • Medical deontology • Introduction to philosophy • Legal and ethic issues in biomedical engineering • Diploma seminar • Engineering project and examination for Bachelor's degree • Elective 3 - Biomineral science - Practical electronics - Programming of control and measurement systems

Table 3. (a) Syllabus of the First Degree (Bachelor's) studies in biomedical engineering

semester	Second Degree (Master's)		
	medical electronics and information technologies	biomaterials	biomechanics and robotics
1.	• Identification and modelling of biological structures and processes • Tissue and genetic engineering • Fundamentals of telemedicine • Neural networks • Electronics Systems for Clinical Applications • Information systems in health care • Picture archiving and communication systems • Elective 1 - Design of VLSI circuits - Multimedia systems in medicine - Advanced methods for programming of multithreaded applications	• Clinical trials • Ceramic Biomaterials • Polymer Biomaterials • Identification and modeling of biological structures and processes • Tissue and genetic engineering • Information systems in health care • Implantation technologies • Elective 1 - Electron microscopy in biomedical engineering - Neurochemistry and neuropharmacology	• Biomedical signals processing • Identification and modelling of biological structures and processes • Tissue and genetic engineering • Rehabilitation Technology • Biomechanical designs • Servomechanisms and advanced control systems • Control systems in medical devices • Visual surgery support techniques
2.	• Research of biomaterials and tissues • Dedicated medical diagnostics algorithms • Multimodal interfaces • Rehabilitation Technology • Medical imaging systems • Telesurgery and medical robots • Individual project • Elective 2 - Algorithms for medical image analysis and processing - Cognitive informatics	• Research of biomaterials and tissues • Composite biomaterials • Metallic biomaterials Telesurgery and medical robots • Rehabilitation Technology • Fundamentals of applied crystalography • Individual project	• Research of biomaterials and tissues • Ergonomics • Intelligent materials and structures • Acoustical diagnosis • Information systems in health care • Telesurgery and medical robots • Image processing for surgery support • Individual project
3.	• Voice computer communication • Computer support for acoustic diagnostics • Fundamentals of embedded systems design • Diploma seminar • **Master thesis and examination**	• Fundamentals of regenerative medicine • Diploma seminar • **Master thesis and examination** • Elective 3 - Selected problems of neurobiology - Surface engineering	• Bionics • Pharmaceutical industry equipment • Diploma seminar • **Master thesis and examination** • Elective 3 - Computer aided of engineering - EPLAN - Pharmaceutical industry materials and designs - Nanotechnology

Table 3. (b) Syllabus of the Second Degree (Master's) studies in biomedical engineering.

semester	Second Degree (Master's)	
	bionanotechnologies	emerging health care technologies
1.	• Symmetries and structures, solid body and molecules • Soft tissue physics • Polymers • Physics of thin film surfaces • Identification and modeling of biological structures and processes • Tissue and genetic engineering • Physical methods in biology and medicine • Information systems in health care • Elective 1 - Biotechnological challenges in biophysics - Instrumental analysis methods - Fundamentals of cell and tissue engineering - Structural backgrounds of cell biology	• Biomaterials and artificial organs • Electronics Systems for Clinical Applications • Information systems in health care • Physical methods in biology and medicine • Telemedicine and e-health • Tissue and genetic engineering • Assisted Living Technologies
2.	• Research of biomaterials and tissues • Magnetic nanomaterials • X-ray applications in biomedicine • Applications of magnetic resonance in biomedical research • Telesurgery and medical robots • Rehabilitation Technology • Introduction to radiobiology • Individual project	• Design of biomechatronical systems • Identification and modeling of biological structures and processes • Medical imaging systems • Research of biomaterials and tissues • Rehabilitation Technology • Telesurgery and medical robots • X-ray applications in biomedicine • Development of VLSI systems • Individual project
3.	• Optical method of matter investigation • Neuroelectronics • Diploma seminar • **Master thesis and examination** • Elective 3 - General and molecular genetics - Protein engineering - Leukocyte and cancer cells transportation - Fluorescent and confocal - microscopy - Molecular modeling of bioparticles - Photobiology and photomedicine	• Implantation Techniques • Introduction to Biometrics • Neurochemistry and neuropharmacology • Diploma seminar • **Master thesis and examination**

Table 3. (c) Syllabus of the Second Degree (Master's) studies in biomedical engineering.

semester	Third Degree (Doctor's)
1.	• Graph theory • Methods of systems optimization • Biocybernetics • Medical sensors and measurements
2.	• Graph theory • Methods of systems optimization • Medical imaging in clinical practice • Biometry and medical statistics
3.	• Information systems in telemedicine • Biomechanics i acoustics • Biomedical digital signal processing • Biomaterials and artificial organs
4.	• Electronic medical instrumentation • Medical physics
5.	• Medical image analysis • Modeling of biological systems
6.	• Electives (2 of 6) - Dedicated algorithms for biosignal interpretation - Integrated systems SoC in medical diagnostics and therapy - Intelligent sensor arrays - Biophysics - Biological interfaces - Advanced equipment in medicine and rehabilitation
7.	• Philosophy / Economy
8.	• individual research

Table 3. (d) Syllabus of the Third Degree (Doctor's) studies in biomedical engineering.

Faculty: Biomedical Engineering, degree: II (Master of Science)
Track: Emerging Health Care Technologies

ID		exa	scr	HΣ	lec	cla	lab	prj	lec	cla	lab	prj	E	ECTS	lec	cla	lab	prj	E	ECTS	lec	cla	lab	prj	E	ECTS
						15 weeks HOURS SUMMARY					**1 spring**						**2 fall**						**3 spring**			
Common compulsory courses																										
s5	Research on biomaterials and tissues S	0	3	30	15	0	15	0						3												3
s3	Identification and modeling of biological structures and processes	1	1	60	30	0	15	15	*2*		1	1	5	5												5
s6	Rehabilitation technology S	0	2	30	15	0	15	0	1				3	3												3
s1	Information systems in medicine S	0	2	30	15	0	15	0						3	1		1		3							3
s4	Tissue and genetic engineering S	0	2	45	15	15	15	0	1	1	1		4	4												4
s2	Telesurgery and medical robotics S	1	2	60	30	0	15	15						5	*2*		1	1	5							5
Recommended courses																										
1	Electronic systems for clinical applications	1	1	60	30	0	15	15	*2*		1	1	4	4												4
2	Medical imaging systems	1	1	60	30	0	15	15	*2*		1	1	4	3	*2*	1	1	1	3							3
3	Telemedicine and e-health	0	2	60	30	0	15	15	2		1	1	3	3												3
4	Design of biomechatronical systems	1	3	60	30	15	15	0					4	4	*2*	1	1		4							4
5	Implantation techniques	0	2	45	15	30	0	0		2				4							1	2			3	4
6	Biomaterials and artificial organs	1	2	60	30	0	30	0	*2*		2		4	4												4
7	Introduction to bionetrics	0	2	60	30	0	30	0						4	1		1	1								4
8	Physical methods in biology and medicine	1	3	75	30	15	15	15	*2*	1	1	1	4	4	*2*	1	1	1	3							4
9	X-ray applications in biomedicine	1	2	60	30	0	15	15	*2*		1	1	4	3							2		2		4	3
Elective																										
10	Elective course 1	0	3	60	30	0	30	0						3	2		2		3							3
11	Elective course 2	0	2	60	30	0	30	0						3	2		2		3							3
	Elective course 3	0	2	60	30	0	30	0						3							2		2		3	3
12	Individual project	0	1	60	0	0	0	60						6				4	6							6
13	Master diploma	1		300	300	0	0	0						20							*20*				20	20
SUMMARY		9	37	1335	765	75	330	165	14	2	10	4		30	12	1	8	7		30	25	2	4	0		90
															28						31					
	hours weekly				30		30																			
									exa 2	scr 10					exa 4	scr 8					exa 1	scr 4				
	scores summary	exa 4	scr 4																							
Obligatory courses (hours)				255																						90
Practical classes (percentage)				55.1																						
Elective courses (percentage)				30.8																						

Legend

2 Lecture, two hours weekly with examination

Elective
Development of VLSI systems
Neurochemistry and neuropharmacology
Assisted living technology

Table 4. (a) Syllabus of the Second Degree (Master's) studies in biomedical engineering, specialization Emerging health care technologies.

Postgraduate BME studies AGH – TRACK TABLE

discipline "biocybernetics and biomedical engineering"

Id	Couse name	Professor (confidential)	year I Sem.1	year I Sem.2	year II Sem.3	year II Sem.4	year III Sem.5	year III Sem.6	year IV Sem. 7	year IV Sem. 8	remarks
				Common courses							
1	Graph theory		30	30							
3	Methods of systems optimization		30	30							
5	Philosophy / Economy								30		
				Discipline-oriented courses							
6	Biocybernetics		30								
7	Medical imaging in clinical practice			30							
8	Medical sensors and measurements		30								
9	Biometry and medical statistics			30							
3bi1	Information systems in telemedicine				30						4 obligatory courses
3bi2	Biomechanics and acoustics				30						
3bi3	Biomedical digital signal processing				30						
3bi4	Biomaterials and artificial organs				30						
4bi1	Electronic medical instrumentation					30					2 obligatory courses
4bi2	Medical physics					30					
5bi1	Medical image analysis						30				2 obligatory courses
5bi2	Modeling of biological systems						30				
6bi1	Dedicated algorithms for biosignal interpretation										2 obligatory courses to select out of 6 total duration 60 h
6bi2	Integrated systems SoC in medical diagnostics and therapy							30			
6bi3	Intelligent sensor arrays							30			
6bi4	Biophysics										
6bi5	Biological interfaces										
6bi6	Advanced equipment in medicine and rehabilitation										

Table 4. (b) Syllabus of Third Degree (Doctor's) studies in biomedical engineering.

msib	Medical imaging systems	Master (MSc) Level Recommended course Biomedical Engineering

Coordinating person:	**Prof. dr hab. inż. Ryszard Tadeusiewicz**	Teaching team:	Dr inż. Przemysław Korohoda Dr inż. Zbigniew Bubliński

Department **Automatics and Control**   Year/Semester: 5/9

Lecture language: **English**

Lecture hours: **30 h**   Laboratory/Exercise hours: **1/1**   Hours per week: **2/-/1/1**

Expected knowledge

Elements of signal and image processing, elements of computer science, basic physics, fundamentals mathematical analysis and algebra of vectors and matrices

Principal aims

Theoretical and practical presentation of principal aspects of medical images acquisition (different methods!), filtering, storing, transmission, processing, segmentation, analysis, recognition, and also automatic understanding

Teaching rules

Lectures are given in auditory and supported by additional e-learning material,
Students participate in 7 laboratory exercises presenting various aspects of telemedicine, students are expected to solve a particular measurement or methodological problem,
Students working in supervised teams are expected to develop a custom project or to solve a complex practical problem under given conditions.

Progress evaluation rules

Report-based score after each laboratory exercise. Project evaluation based on report and practical presentation.

Lecture description

1. Introduction to medical images acquisition methods
2. X-ray images and properties
3. CT – computer tomography
4. MRI – magnetic resonance imaging
5. SPECT, PET – nuclear imaging methods
6. USG – ultrasonography
7. TG – thermography
8. Endoscopic visualization
9. Medical images registration and transmission. DICOM standard
10. Medical images filtration and denoising
11. Image segmentation
12. Image analysis and diagnosis aid
13. Image recognition
14. Medical images automatic understanding
15. Conclusions

Bibliography

1. Burger W., Burge M.J. eds.: Digital Image Processing - An Algorithmic Introduction using Java. Texts in Computer Science series. Springer Science + Business Media, New York, 2008
2. Baert A. L., ed.: Encyclopedia of Diagnostic Imaging. Springer-Verlag, Berlin - New York, 2008
3. Ogiela M. R., Tadeusiewicz R.: Modern Computational Intelligence Methods for the Interpretation of Medical Images, Studies in Computational Intelligence, Volume 84, Springer-Verlag, Berlin – Heidelberg – New York, 2008
4. Tadeusiewicz R., Ogiela M. R.: Medical Image Understanding Technology, Series: Studies in Fuzziness and Soft Computing, Vol. 156, Springer-Verlag, Berlin – Heidelberg – New York, 2004
5. http://www.medical-image-processing.info/
6. http://mipav.cit.nih.gov/

Table 5. (a) Syllabus of the course "Medical imaging systems" (Second Degree-Master's).

![msib logo] msib	Research of biomaterials and tissues	Master (MSC) Level Compulsory course **Biomedical Engineering**

Coordinating person:	**Dr hab. inż. Anna M. Ryniewicz, prof. AGH**	Teaching team:	**Dr hab. inż. Anna M. Ryniewicz, prof. AGH** **Dr inż. Tomasz Madej**
Department	**Department of Machine Design and Technology**		Year/semester: 4/8

Lecture language: **English**

Lecture hours: **15 h**   Exercise/Laboratory: **15h**   Hour per week: 1 L, 1 lab.

Expected knowledge

The subject requires knowledge from the fields of the materials science, technology and use of biomaterials, the analysis of biomechanical extortion, histology and physiology of tissues.

Principal aim

The aim of the subject is the study the research methods and the estimation of biomaterials in the aspect of application of carrying construction, stabilizing and biodegradation.

Teaching rules

The lectures in the range of two hours per week are carried out in the form of multimedia presentations and precede in its matter laboratory study, which is also realized in two hours per week after the completion of laboratory cycle.

Progress evaluation rules

The credit is obtained on the basis of positieve note from test and the positive particular notes from laboratory jobs.

Subject description

Lecture:
Biomechanics of human locomotive and stomatognathic systems. The Doppler methods of testing the flow. Criterions of biomaterials selection.
The methods of testing biomaterials resulting from the function in human body. The mechanical tests of biomaterials and tissues: testing of resistance by static and dynamic loads and the tribological tests. Ultrasonic tests.
The methods of testing the surface layer structure of biomaterials and tissues in micro, macro, nano scale using new research techniques (optical microscopy, scanning electron microscopy SEM and scanning transmission electron microscope STEM, roentgen diffraction, photoelectron spectroscopy, atom force microscopy AFM, tunnel microscopy, infrared spectroscopy
The analysis of contact: bone tissue – implant. The analysis of stresses and displacements distributions in the numerical models with the purpose to obtain optimization of selection of implant and estimation the state of tissues effort. Biocompatibility tests.
The correlation between in vitro tests, tests on animals and application tests. The standards that regulate the estimation of biocompatibility.
Laboratory:
The reconstruction of three-dimensional geometry of regular hip joint and implanted endoprosthesis in pathological hip joint on the basis of imaging diagnostics (MR CT) using Amira software.
The estimation of working and kinematics parameters of movement of hip joint using HIP98 software. The development of strategy and testing of resistance on compression and bending of bone tissues using testing machine INSTRON.
The testing of resistance on compression and bending of selected biomaterials on testing machine INSTRON.
The tribological research of selected biomaterials and cartilaginous tissue on testing machine ROXANA
The testing of adhesion of multilayer biomaterials on testing machine INSTRON.
The comparison of conditions of transfer of loads in endoprosthesis of hip joint made from different biomaterials using finite element method FEM.

Bibliography

Powers J.M., Sakaguchi R.L.: Craig's restorative dental materials, cop. Elsevier Inc., USA, 2006,
Temenoff, J. S.: Biomaterials the intersection of biology and materials science, Upper Saddle River : Pearson Prentice Hall, 2008.
Prendergast, P J.: Biomechanical techniques for pre-clinical testing of prostheses and implants; Institute of Fundamental Technological Research, Polish Academy of Sciences. Warsaw : Centre of Excellence for Advanced Materials and Structures, 2001.
Fung Y. C.: Biomechanics : mechanical properties of living tissues, New York : Springer-Verlag, cop. 1981.

Table 5. (b) Syllabus of the course "Research of biomaterials and tissues" (Second Degree-Master's).

msib	Tissue and genetic engineering	Master (MSc) Level Compulsory course **Biomedical Engineering**

Coordinating person:	**Dr hab. inż. Elżbieta Pamuła**	Teaching team:	Dr hab. inż. Elżbieta Pamuła Dr Elżbieta Menaszek Mgr inż. Małgorzata Krok
Department	**Department of Biomaterials**		Year/Semester: 4/8

Lecture language: English

Lecture hours: 15 h	Laboratory/Exercise hours: 15/15	Hours per week: 2/2/4(4 weeks)

Expected knowledge

Basic knowledge of biology, genetics, biochemistry, materials science, biomaterials, implants and artificial organs

Principal aims

Introduction to key issues of tissue engineering and genetic engineering

Teaching rules

2-h lectures are given every second week
Students participate in 2-h seminars every second week
Laboratories are blocked: students participate in four 4-h exercises

Progress evaluation rules

Test (lecture), oral presentation (seminars) and reports (laboratories)

Lecture description

Scope of tissue engineering. Cell and tissue cultures, in vitro techniques, bioreactors. Cell types and sources. Growth factors. Extracellular matrix as a biological scaffold for tissue engineering. Interactions at the interfaces: materials/biomacromolecules/cells/tissues. Methods to investigate material/biological environment interface. Scaffolds design and fabrication. Tailoring structure, microstructure and surface properties of scaffolds for cell and tissue culture. Tissue engineering in vivo – guided tissue regeneration. Tissue engineering products (skin, cartilage). Genetic engineering and gene therapies. Enzymes (restriction endonucleases, ligases), expression vectors, gene cloning. Construction and analysis of recombinant DNA. Molecular probes. Expression of cloned genes. DNA amplification by PCR. DNA sequencing. Genetic engineering products (synthetic human insulin, synthetic human erythropoietin, insects or herbicide resistant plants).

Seminars

Based on the most recent publications the students elaborate and present a particular topic related to tissue and/or genetic engineering.

Laboratories

The students are introduced to elementary rules of cell cultures. The learn how to count mammalian cells in suspension and analyze their viability. They seed cells on different biomaterials, and following a short incubation and staining cell adhesion, morphology and secretion are evaluated. The students are also introduced to in vitro biomaterials evaluation methods and cytotoxicity tests.

Bibliography

1. Tissue engineering, C. Van Blitterswijk (Senior Editor), Amsterdam, Elsevier, 2008
2. Y. Ikada, Tissue engineering. Fundamentals and applications, Amsterdam, Elsevier, 2006
3. Methods of Tissue Engineering A. Atala, R. P. Lanza (Editors), Elsevier, 2002
4. D.S.T. Nicholl, An Introduction to Genetic Engineering, 3rd edition, Cambridge University Press, 2008

Table 5. (c) Syllabus of the course "Tissue and genetic engineering" (Second Degree-Master's).

8. References

Augustyniak P. (2008) Proceedings of the First National Conference on Biomedical Engineering Education (OKIBEdu). *Acta Bio-optica et Informatica Medica - Biomedical Engineering*. Vol 3?

Augustyniak E, Augustyniak P. (2010) From the Foundation Act to the Corporate Culture of a BME Teaching Institute, *Proceedings of the 32nd annual international conference of the IEEE Engineering in Medicine and Biology Society*, 2010, pp. 319–322

Augustyniak P, Tadeusiewicz R & Wasilewska-Radwańska M. (2010) BME Education Program Following the Expectations from the Industry, Health Care and Science, In: *Medicon 2010, XII Mediterranean Conference on Medical and Biological Engineering and Computing*, Bamidis PD, Pallikarakis N. (Eds.) 2010 Springer, (IFMBE Proceedings vol. 29), pp. 945–948.

Denison D. & Mishra A. (1995) Toward a theory of organisational culture and effectiveness, *Organisation Science*, Vol. 6 No.2, 1995, pp. 204-23.

Gordon G. & DiTomaso N. (1992) Predicting corporate performance from organization culture, *Journal of Management Studies*, Vol. 29 No.6, 1992 pp. 783-98

Ministry of Science and Higher Education (2007) Educational Standards for Higher Education, No 49 Biomedical Engineering (in Polish)

Monzon E. (2005) The Challenges of Biomedical Engineering Education in Latin America. *Proceedings of the 27th annual international conference of the IEEE Engineering in Medicine and Biology Society*, 2005: 2403-2405

O'Reilly C. & Chatman J. (1996) Culture as social control: corporations, cults and commitment, *Research in Organisational Behaviour*, Vol. 18, 1996 pp. 157-200.

ROTMED Consortium (2006) Rapport from foresight examinations entitled: Analysis of the state of Polish medical technologies sector, IBIB PAN, Warsaw 2006 (in Polish)

Schwartz MD. (1988) Biomedical Engineering Education, In: *Encyclopedia of medical Devices and Instrumentation*, Webster JG. (Ed.), Wiley, New York, 1988: 392- 403

Wasilewska-Radwanska M. & Augustyniak P. (2009) Multidisciplinary School as a BME Teaching Option. *IFMBE Proceedings*, Vol. 25, 2009, pp. 200–203

14

Life Support System Virtual Simulators for Mars-500 Ground-Based Experiment

Eduard Kurmazenko[1], Nikolay Khabarovskiy[1], Guzel Kamaletdinova[1],
Evgeniy Demin[2] and Boris Morukov[2]

[1]*Joint-Stock Co. 'NIIchimmash'*,
[2]*SSC RF-IMBP RAS,*
Russia

1. Introduction

A Mars manned mission is practically impossible without resolving some problems on the ground with test subjects involved related to crew life-support psychological stability, fitness to work during a long-duration, self-sustained space mission. One of the problems to be resolved in spaceflight is the crew's health and fitness work. These factors are possible to investigate in a ground experiment to make more effective preparation for interplanetary missions including a Mars mission. Another problem lies in failure-free functioning of on-board systems and first of all the Integrated Regenerative Life-Support System (IRLSS) in the Mars-500 project. The crew plays a key role in maintaining system operability and reliability the entire mission.

In order to make long-duration, self-sustained interplanetary missions a reality it is necessary to provide crew support and its activities under conditions essentially different from those of earth orbital flights.

Specifics of interplanetary flights include:

- long duration (over 960 day) missions with the crew being in confined space that demands:
  - expansion of functions of IRLSS functions related to crew personal;
  - prompt parameters of the crew's environment under spaceflight conditions;
  - the necessity of carrying out the crew's medical control and strain relief on-board the spacecraft.
- Self-sustained manned flight is characterized by:
  - lack of renewal expendables units;
  - the systems incorporated in the IRLSS architecture shell ensure trouble-free performance over the entire flight with minimal spare parts and expendables required;
  - the necessity of decision-making by crew as identification and localization of possible off-normal situations related to the IRLSS due to limited intervention on the part of the ground mission control center to control the crew's actions.

Investigations into prolonged influence of the conditions of self-sustained interplanetary flight on crewmembers' intellectual faculties in operation of the IRLSS are of prime importance to make a mission a success and cannot be predicted on the basis of theoretical and experimental data, ground experiments conducted shall be aimed first of all at evaluating the operator's effective activities and his psychological condition (attention, vigilance, perception, memory, thinking training) in combination with physiological parameters of the central and vegetative nervous systems.

Actual participation of test subjects in serving operational standard system at first phases of ground simulation of spaceflight to Mars is extremely complicated and economically unprofitable. A more rational approach is the application of standard system virtual simulators interacting with simulation models for both environment and crew as a load component and integrated in a single Hardware/Software Complex for Serving Operational Systems by crew (HSCSOS) intended for system functioning in normal, off-normal and emergency situations resulted in failure of some systems and deviation of environment controllable parameters from specified values. Those situations may include human factor (crew members' decline in fitness to work, activities, etc.).

The purpose of this charter is analysis of all possible approaches to development of similar complexes based on simulation taking into account a long-duration stay of Man in confined space. The analysis results may be used in development of similar hardware/software complexes to analyze complicated human-machine interaction and specialist training for various-purpose Man-Made Ecosystems (MMES).

2. The simulating object: Its engineering architecture and properties

The crews, environment, IRLSS placed in Pressurized Manned Modules (PMM) define a MMES of Interplanetary Spacecraft intended for crewmembers' support and activity as well as other biological object under conditions essentially different from those on Earth (Kurmazenko E.A. at al., 2009).

The distinguished features of the MMES in comparison with natural ecosystem include:

- the necessity of creating system architecture based on processes with intensity significantly exceeding the intensity of natural transformation processes;
- the system architecture should incorporate finite number of engineering devices and units with built-in or embedded man-made technologies;
- the capability of stable functioning is governed by the value of a substance reserve fund of the substances disposed in the PMM confined volume.

The IRLSS as an abiotic component of the MMES intended for long-duration manned missions may be defined as a sophisticated engineering system with devices interacting in time and space to provide crew support based on metabolic product recovery, a minimum of spare parts and expendables to create the conditions which ensure that the crew will be provided with physical and mental stability to a specified degree of reliability (Pravetskiy V.N. et al., 1981).

In generalized state space the MMES as a complicated integral system may be given by

$$C \equiv C\left(|C|, \bar{C}, \underline{C}\right). \tag{1}$$

The initial system decomposition allows two interrelated and interacted components to be signed out the IRLSS and external environment:

$$S \equiv S\left(|S|, \bar{S}\right); \tag{2}$$

$$E \equiv E\left(|E|, \overline{E}, \underline{C}\right), \tag{3}$$

$$C = S \cap E ; \tag{4}$$

$$\overline{C} \equiv \overline{S}\left(|S|, |E|, \overline{E}\right) \cap \overline{E}\left(|E|, \overline{E}, |S|\right), \tag{5}$$

in accordance with the purpose of the **C** system over each time of functioning.

In the expressions (1)÷(5) the following notation is adopted: **C, S, E**=the state functions of the MMES, IRLSS and an environment systems respectively; $|C|, |S|, |E|$ =the technological structure its systems; $\overline{C}, \overline{S}, \overline{E}$ =the system functioning regularities; $\underline{C}$ =the purpose of the MMES.

An IRLSS is selected as a baseline for HSCSOS development consists of (Figure 1):

- a System for Water Recovery from Humidity Condensate (WRS-AC) based on sorption/catalytic process to remove of organic and inorganic contaminants from condensate;
- a System for Water Reclamation from Urine (WRS-U) based on the low-tem-premature distillation and sorption/catalytic process for the removal of organic and inorganic contaminants from urine condensate;
- a System for Oxygen Generation (OGS) by water electrolysis from electrolyte solution (the OGS-1 is a virtual simulator and the OGS-2 is an Electrical Trainer (ET) of standard system enabling a number of manual operations);
- a Oxygen Solid Fuel Generator (OSFG) based on sodium perchlorates;
- a Bottle Filling System (BFS) for extra-vehicular activity based on post-purification and compression of oxygen produced by OGS;
- a Trace Contaminant Control System (TCCS) based on sorption/catalytic process for the removal of trace contaminants from the pressurized manned module cabin atmosphere;
- a Carbon Dioxide Removal System (CDRS) based on regenerable absorbent vacuum desorption;
- a Carbon Dioxide Concentration System (CDCS) based on regenerable absorbent vapor desorption;
- a Carbon Dioxide Reduction System (CRS) based on carbon dioxide conversion to methane by hydrogen;
- an Atmosphere Leakage Make-up System (ALMS) by means of nitrogen supply with supply correction based on the oxygen partial pressure value;
- Atmosphere Leakage Tracing System (ALTS).

The individual systems as part of the IRLSS interact via interfaces between the systems and environment. The analysis results of the operational interfaces shows the HSCSOS architecture shall incorporated the environment components such as the atmosphere of the PMM and the crew taking into account specifics of its functioning in interplanetary flight conditions as well as on-board Power Supply System (PSS) and a convective/radiation Thermal Control System (TCS).

The operator's skills which crewmembers shall posses in the critical phases of long self-sustained mission owing to localization of the off-nominal situations (ONS) related the IRLSS functioning imply a high degree of sophistication and may be disturbed under influence of many kinds of stresses. Therefore, when off-normal situation localization simulation is carried out it is necessary to approach it as closely as possible.

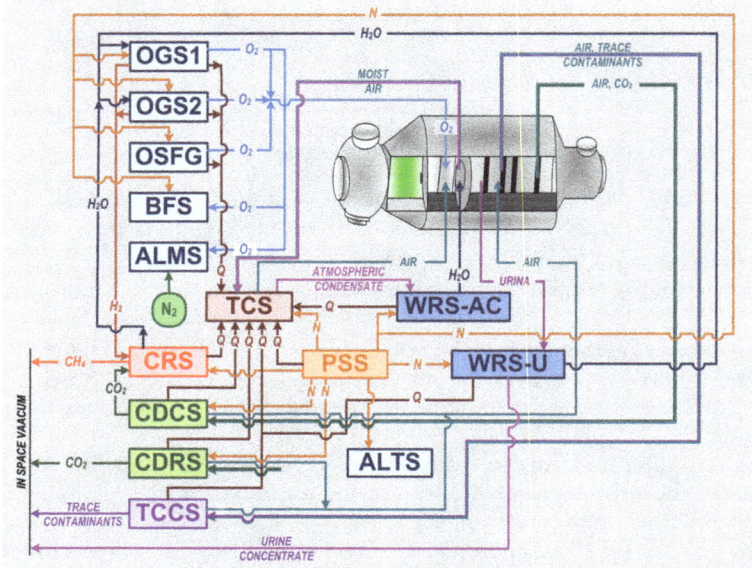

Fig. 1. Integrated Regenerative Life-Support System baseline selected for HSCSOS.

3. Approach to the HSCOCS software development

Approach to the HSCSOS software development is based on the application of the simulation modeling for analysis for the IRLSS system performance in the normal, off-normal and emergency situations, the monitoring of the environmental controllable parameter values and efficient assessment of the crew's actions when off-normal situations are being localized. The simulation model is a logic-mathematical description of the object which may be applied to staging computational experiments in designing, analyzing and assessing object functioning.

Simulation modeling as a specific type of modeling is applied when:

- an experiment upon an actual object is difficult or impossible;
- an analytical model cannot be built (there is cause/effect relationships, how-linearity, stochastic variables);
- time dependent system mode of operation.

The analysis results of the IRLSS performance show that the HSCSOS software architecture shall include the following interacting units (Figure 2):

- IRLSS individual system functioning Virtual Simulators (VS);
- a simulation model of the PMM atmosphere integrating the VSs as a whole and providing the monitoring of crew's environment controllable parameters;
- a crew simulation model as component of environment governing the loads on individual systems;
- procedures for generating probable off-normal situation (ONS) in operation of the IRLSS system, ONS identification and preparation of guidelines and rules for crew in localization of given ONS;

- procedure for crewmember's action efficiency assessment in localization of ONS;
- a specially-created database of monitoring and measurement of crewmember's physiological parameters.

In complex development such as the HSCSOS it is most important for formation of the closed formalized description to select the approaches for generation a closed formalized description of the above specific components of the complex architecture in order to obtain required and sufficient information on system performance analysis in normal and off-normal situations.

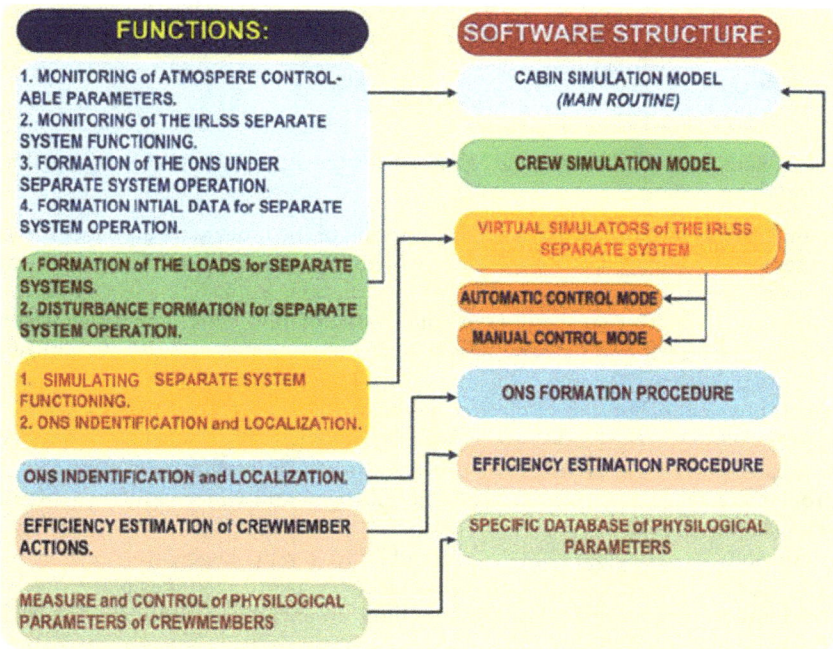

Fig. 2. The architecture and basic functions of the HSCSOS software some components.

3.1 An approach applied to generating formalized descriptions of virtual simulators

Formalized descriptions of individual system virtual simulators based on the detailed level as the aggregate are applied (Kurmazenko E.A. at al., 1997; Kurmazenko E.A. at al., 2008).

In this case an individual system is presented as an aggregate which implies generation of a closed mathematical description including (Figure 3 a):

- the set {X}=alphabet of state in parameters;
- the set {Y}=alphabet of state output parameters;
- the set {Z}=alphabet of state inner parameters;
- the set {U}=alphabet of state controlling parameters;
- the set {W}=alphabet of outer and inner perturbation actions.

When applying the simulation models the alphabet of in parameters, out parameters and controlling parameters shall be correspond to controllable parameters of the system being simulated. The alphabet of perturbation actions is governed by time-varying controllable parameter values of the environment with a specific system in operation and crew present in the PMM atmosphere. The inner perturbation actions are mainly governed by controllable parameter values of the system being simulated. In order to formulate the alphabet of inner states an approach based on a functional description of this description is applied.

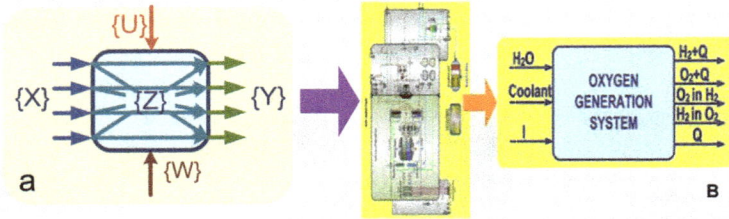

Fig. 3. The system presented as the aggregate: a = design aggregate schematic; b = the OGS presentation as aggregate.

As an example generation of a formalized description of the inner state alphabet for the OGS VS is presented on Figure 3 b. In order to formulate the alphabet of inner states the following basic assumptions are made:
- A formalized description shall take into consideration only values of controllable parameters of inflow and outflow which govern the regulatory and behavior of system functioning;
- thermal/physical properties of electrolyte–produced gas mixtures and coolant over the temperature range investigated are assumed constant;
- owing to a short time of transient process for a current the mass flows of oxygen, hydrogen in oxygen, hydrogen, oxygen in hydrogen and water vapor are described by algebraic equations;
- the main sources of oxygen in hydrogen and hydrogen in oxygen are electrolyzer water supply headers in which an uncontrollable electrolysis process takes place.

With consideration for the given assumptions the formalized description of the inner state alphabet may be presented as:

$$I_{el}^{\tau} = I\frac{R_{col}^{\tau}}{R_{col}^{\tau} + R_{el}^{\tau}} ; \tag{6}$$

$$U^{\tau} = e_a^0 + e_c^0 + a_a + a_c + 0{,}001\left(b_a + b_c\right)\frac{I_{el}^{\tau}}{S_{ed}} + I_{el}^{\tau}R_{el}^{\tau} ; \tag{7}$$

$$I_{col} = I - I_{el} ; \tag{8}$$

$$G_{O_2(H_2)}^{\tau} = \frac{n}{2F}\left[2A_{O_2(H_2)}I_{el}^{\tau} + \left(A_{O_2} + A_{H_2}\right)I_{col}^{\tau}\right] ; \tag{9}$$

$$G_{H_2 \rightarrow O_2(O_2 \rightarrow H_2)}^{\tau} = \frac{nA_{O_2}\left(A_{H_2}\right)}{2F} , \tag{10}$$

where: A_{O_2}, A_{H_2} =chemical equivalents of oxygen and hydrogen, kg/mol, respectively; a_a, a_c, b_a, b_c =Tafel's constants for the anode and cathode; e_a^0, e_c^0 = theoretical potentials of the anode and cathode, V for a_a and a_c and Vm²/A, for b_a and b_c, respectively; F=Faraday constant, Kl/mol; $G_{O_2(H_2)}^{\tau}$ =the mass flow-rate of gas being produced in the electrolyzer oxygen (hydrogen) compartment, kg/s; $G_{H_2 \to O_2(O_2 \to H_2)}^{\tau}$ = the mass flow-rate of oxygen (hydrogen) produced in the hydrogen (oxygen) compartment; I, I_{el}, I_{col} = the total electrolyzer current, current through the electrolytic cell and current in the header, A; n= the quantity of an electrolytic cells; R_{col}^{τ}, R_{el}^{τ} = the electric resistance of the header and electrolyzer, Ohm; S_{ed} =the electrolytic cell surface area, m²; U^{τ} = the electrolyzer voltage, V. The average electrolyzer temperature T_{el}, K, as a function of the supply current I_{el} is determined from regressive dependence

$$T_{el} = f_1(T_{cool})\left(-0.3305 + 0.0034 I_{el}^2 + 2.085 I_{el}^{0.5} + \frac{21.29}{I_{el}^{0.5}}\right), \tag{11}$$

obtained as a result of processing of the data of a computer experiments conducted by using the OGS detailed simulation model.

The hydrogen (oxygen) moisture content downstream the separator, kg/kg

$$d_{H_2(O_2)}^{\tau} = \frac{\left[27.6 + 0.23\left(T_{sep}^{\tau} - 273\right)^{1.5}\right]}{m_{sep}} \cdot \frac{\mu_{H_2O}}{\mu_{H_2(O_2)}}, \tag{12}$$

where $\mu_{H_2(O_2)}$, μ_{H_2O} =the molar masses of hydrogen (oxygen) and water, kg/mol.

The temperature of the mass flows of hydrogen and oxygen downstream from the separator T_{sep}, K, is determined by regressive dependence as

$$T_{sep} = f_2(T_{cool})\left(-15.387 + 0.249 I_{el} + \frac{173.184}{I_{el}^{0.5}} + \frac{319.898}{I_{el}}\right). \tag{13}$$

The temperature functions f_1 (T_{cool}) in the relationship (11) and f_2 (T_{cool}) in the relationship (13) are determined as

$$f_1(T_{cool}) = \left[1 + 0.82(T_{el} - 273) + 0.03(T_{el} - 273)^2\right]T_{el}^{-1}; \tag{14}$$

$$f_2(T_{cool}) = \left[1 + 0.9(T_{el} - 273)\right]T_{sep}^{-1}, \tag{15}$$

where T_{cool}=the coolant temperature, K.
The similar approach to generation of formalized descriptions of system functioning is adopted for other virtual simulators.

3.2 Approach used for formation of the PMM atmosphere formalized description

When generation the formalized PMM atmosphere description the following basic assumptions are made (Kurmazenko E.A. at al., 1998):

- the PMM atmosphere is considered as an open thermodynamic system;
- man-made atmosphere is considered as a mixture of ideal gases the heat capacity of which is governed by its chemical composition and temperature–independent;
- trace contaminants due to their low content do not affect the generation of total pressure in the PMM and thermal/physical properties of man-made gaseous atmosphere.

Considering the assumptions made the nonlinear equations of mass balances for the basic components (oxygen, carbon dioxide, nitrogen and water vapor) and trace contaminants as well as the non-linear equation of internal energy balance for the PMM atmosphere reference volume may be written as the equations in deviations:

$$M_{PMMa}(\tau) = \sum_{i=1}^{i=4} M_{PMMi}(\tau) ; \tag{16}$$

$$M_{PMMi}(\tau) = M_{PMMi}(\tau - \Delta\tau) + \sum_{j=1}^{j=n} (\pm G_{ij}\Delta\tau) ; \tag{17}$$

$$M_{TC}(\tau) = M_{TC}(\tau - \Delta\tau) + \sum_{k=1}^{k=s} (\pm G_{TCk}\Delta\tau) \tag{18}$$

$$U(\tau) = U(\tau - \Delta\tau) \pm \sum_{l=1}^{l=p} c_{p_i} G_{il}\Delta\tau \pm \sum_{m=1}^{m=t} q_m\Delta\tau \tag{19}$$

In the equations (16)÷(19) the following notation is adopted: $M_{PMMa}, M_{PMMi}, M_{TC}, U$=the value of atmosphere total mass, i-basic component mass, k-trace contaminant mass, kg, and atmosphere internal energy, J, respectively; G_{ij}, G_{TCk}=i- basic component mass flow-rate and k=trace contaminant mass flow-rate entering and leaving the reference volume, kg/s; q_m=heat flows due to heat conduction entering and leaving the volume under consideration, W; c_{p_i} = the i- basic component specific heat capacity, J/kg °C; $\tau - \Delta\tau, \tau, \Delta\tau$ = previous time, current time and integration step in time, respectively, s.

The current values of mass flow-rates of the atmosphere basic components and trace contaminants upstream and downstream the reference volume as well as heat flows entering and leaving together with mass flows of atmosphere components and heat conduction are determined at each integration step by the current values of the ingoing and outgoing flows with the system performance virtual simulator values.

3.3 Approach used for formation of the 'crew' unit formalized description

When generating a formalized description of the 'crew' unit the following assumptions are made:

- a single crewmember is considered as the structure of interrelated functioning systems in which incoming mass and energy flows are converted into outgoing mass and heat flows, and activity;

- the cosmonaut's energy expenditure when doing various kinds of activity is balanced by caloric value of food ration and total value of energy expenditure;
- potable water is consumed with food;
- the main factor that governs the basic point is the crew's activity defined by the spaceflight program;
- in order to describe mass and heat flows in the 'crew' unit an international model of a conventional human where the mass and heat flows are proportional to energy expenditures (Adamovich BA., and Gorshenin V.A., 1997). In doing so, the coefficients of this model are corrected based on the results of the computational experiments used the detailed simulation model on the basis which the human organism main functional systems (Figure 4), governing the mass/exchange with the environment are simulated (Kurmazenko E.A. at al., 2000).

The initial data used for generating a formalized description of the 'crew' unit also include an activity/rest cycloramas for every crewmember.

Oxygen consumption G_{O_2}, g/h, is a function of energy expenditure N

$$G_{O_2} = a_0 N ,$$ (20)

in which coefficient a_0 varies in the range from 0.28 to 0.31, g/kcal.

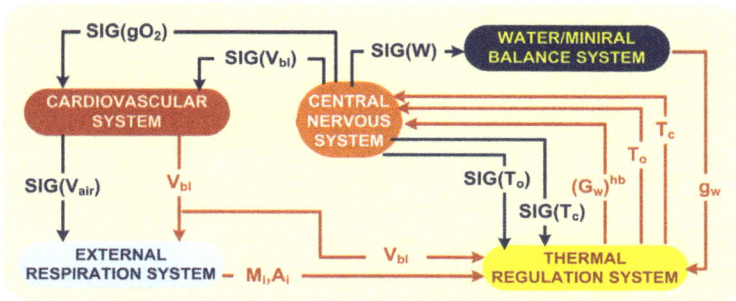

Fig. 4. Enlarged flowchart of human body main functional systems, where: $SIG(gO_2)$, $SIG(V_{bl})$, $SIG(V_{air})$, $SIG(W)$, $SIG(T_0)$, $SIG(T_c)$ =signal functions of the Central Nervous System controlled by the oxygen consumption, blood flow, breath volume, water consumption and Thermal Regulation System in dependence from values of ambient temperature and human's body core temperature, respectively; $V_{bl}, M_i, A_i, (G_w)^{hb}, g_w, T_0, T_c$ =the controllable parameters for signal functions.

Carbon dioxide released G_{CO_2} is a function of energy expenditure N is determined as

$$G_{CO_2} = a_0 K_r N ,$$ (21)

with the value of the respiratory coefficient K_r is determined from empirical dependence

$$K_r = -0.801 + 0.142 \exp\left(\frac{-96.432}{A}\right) .$$ (22)

The moisture losses as a result of perspiration and respiration J, g/h, are also of energy expenditure N is determined as

$$J = 91.7 - 0.19N .$$ (23)

The quantity of urine donated U, g/h, is a function of energy expenditure N

$$U = 0.19N .$$ (24)

Trace contaminant realized TC_i to be considered as a first order approximation are proportional energy expenditure N

$$TC_i = A_i N .$$ (25)

In calculation of trace contaminants realized the data presented in work (Savina, V.P., & Kuznetsova, T.I., 1980) are used. The data processed for the mixed ration (50 %-natural food products and 50 % sublimated food products) as a function of the ambient temperature are given in Table 1.

Trace contaminant	Temperature-dependency specific secretion intensity A_i, mg kcal /h for:	
	t=20	20< t ≤40
Ammonia and its compounds	0.0144	0.0144+0.0031t
Ketones	0.0577	0.0577+0.0003t
Carbon monoxide	0.195	0.195+0.0186t
Aldehydes	0.005	0.005+0.0005t
Inorganic acids	0.018	0.018+0.0017t
Total Alcohols	0.012	0.012+0.0005t
Hydrocarbons	0.008	0.008+0.0008t
Methane	0.033	0.033+0.0003t
Acetaldehyde	0.003	0.003+0.0001t
Methanol	0.008	0.008+0.0001t
Ethanol	0.011	0.011+0.001t
Dimethylamine	0.007	0.007+0.0006t
Acetone	0.046	0.046+0.0043t

Table 1. Initial data for simulation of trace contaminant secretion.

4. HSCSOS software implementation

The HSCSOS architecture includes the hardware and software components integrated in a single hardware/software complex.

A general view of the Mars-500 project Ground Experimental Stand (GES) and HSCSOS hardware arrangement are given on Figure 5 and Figure 6, respectively.

The HSCSOS hardware features two work places located in the PMM: Operator's Terminal in PMM-150 (OT1) and Operator's Terminal in PMM-50 (OT2), and the Instructor's Work Place (IWP) located in the Control Experimental Center (CEC).

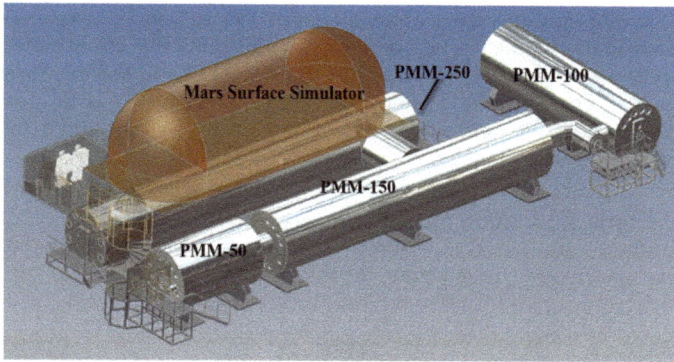

Fig. 5. The general view of the Mars-500 project Ground Experimental Stand.

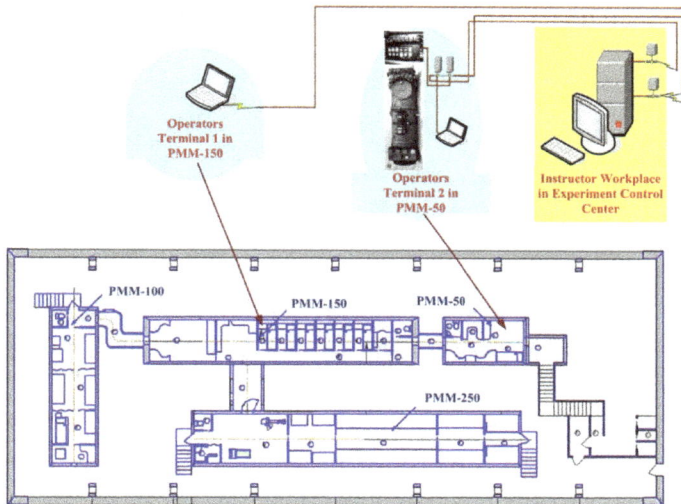

Fig. 6. HSCSOS hardware arrangement at the Mars-500 project Ground Experimental Stand.

According to the HSCSOS architecture the complex software based on 'client –server' technology of data processing and transfer has been developed. The software includes a server and a client sections. The server and client sections are interrelated via specialized technology of data transfer DataSocet integrated in the programming environment LabView, v. 8.5.

The server section contains various calculation modules of 'PMM atmosphere', and 'crew' simulation models, system virtual simulators, off-normal situation generation module, a crew audio warning module, a crew action efficiency assessment module and a protocol module.

The server section provides computations and interrelation with the client section for displaying the parameter values in order to classify the environment and individual system performance and translate control command from client section for correction of system production rates or ONS localization.

The client section displays the data obtained from the server section, generates control signals to transfer them to the server section. In addition, the client section allows the operator to identify and trace ONS develops in the HSCSOS and the time spent to conduct some operations in servicing the systems and localizing ONS. It is not necessary to permanently interconnect the server and client sections. The operator's terminal can be connected in case if the performance requires intervention.

4.1 'PMM atmosphere' simulation model software implementation

The 'PMM atmosphere' simulation model routine is a main program. This simulation model is the basis one to ensure integration of system virtual simulators in the HSCSOS architecture and monitoring of the crew environment parameters (Figure 7).

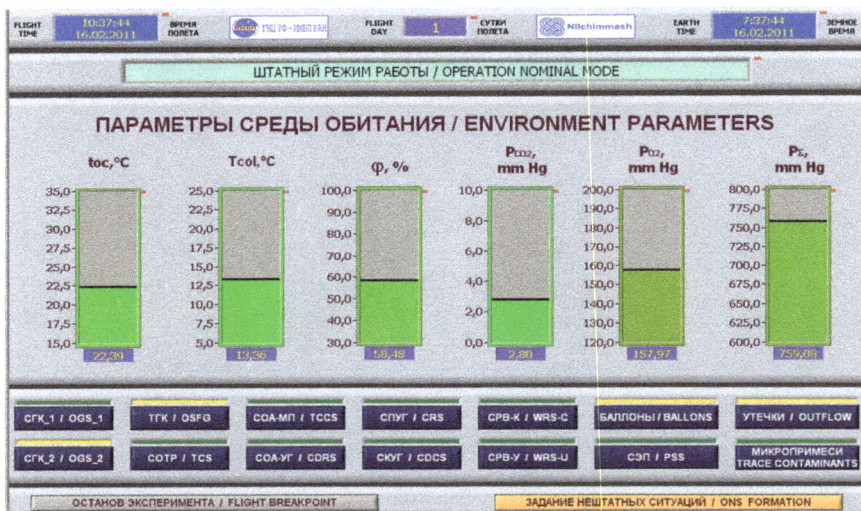

Fig. 7. Control panel of the 'PMM atmosphere' simulation model, where: t_{oc}, °C, T_{col}, °C, φ, %, p_{CO_2}, mm of Hg, p_{O_2}, mm of Hg, p_{Σ}, mm of Hg = controllable parameter values.

The crew's environment parameter values are displayed on the main routine control panel both as, values and in color when the parameter value is changed. When the 'TRACE CONTAMNANTS' key is pressed there appears an extra window on the routine control panel to display current trace contaminant concentrations (Figure 8).

In the lower part of the program front panel there are pushbuttons enabling call-in the front panel the corresponding program simulating the functioning of the system when the line appears.

At the bottom of the routine control panel there are call-in keys for control panels of inputs systems simulating system performance. In doing so, color indication 'SYSTEM STATUS CONDITION' appears. Two keys, located below, fully assessable from the instructor's work place and disconnected at the operator's terminal have inscriptions 'ASSIGNMENT OF OFF-NORMAL SITUATIONS' and 'EXPERIMENT SHUTDOWN'.

Experimental verification of the model is carried out on the basis of algorithm assessment for inconsistency, analysis results of computational experiment related to calculation of the balance relationships and ergonomic requirements.

Fig. 8. 'Trace Contaminant Concentration' extra windows on control panel 'PMM atmosphere' simulation model.

4.2 'Crew' simulation model software implementation

The 'Crew' simulation model as part of the HSCSOS software simulates mass/energy exchange between the crew and the environment as the results of which the loads required for functioning of virtual simulators are generated. This model is also as a loading component which is the source of disturbance.

The given subroutine operates in the background mode and is not directly displayed when the main 'PMM atmosphere' routine is in operation although being its subroutine (Figure 9).

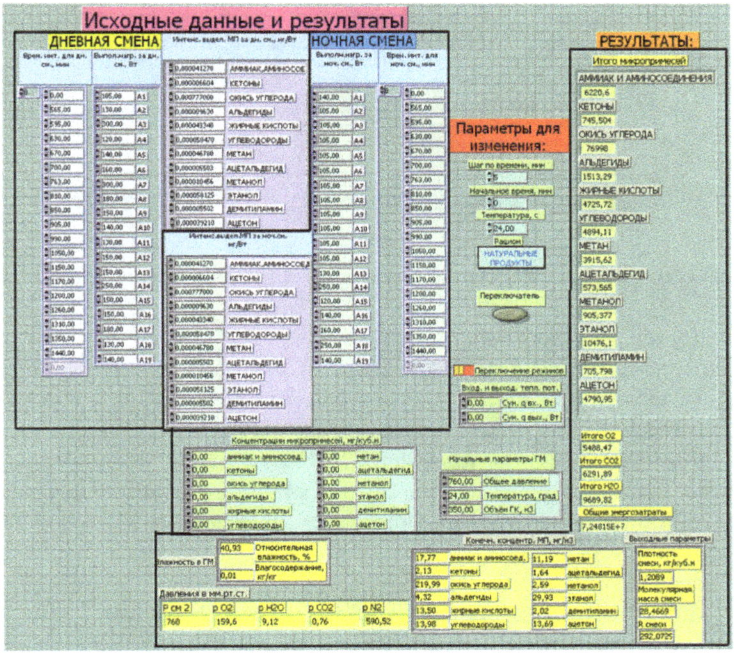

Fig. 9. Control panel of 'Crew' simulation model.

The main routine is interrelated via 'DAY/NIGHT' subroutine which simulates the crew's energy expenditure in the day and night shifts depending on the activity/rest regimens and according to 'Mass Balance' subroutine operation.

The 'Crew' simulation model is experimentally verified based on a correlation between computational experiment resulted (Figure 10) and published specification.

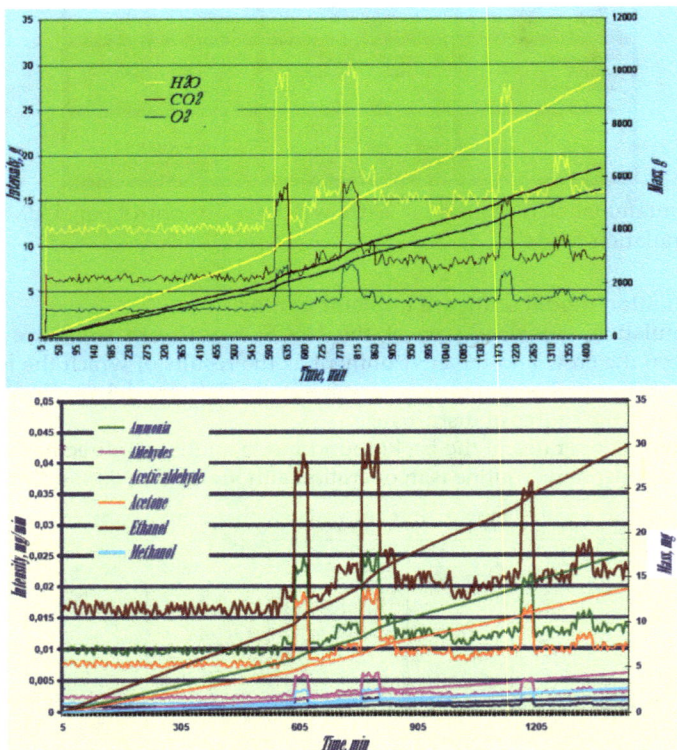

Fig. 10. Typical computational experiment results on 'Crew' simulation model.

4.3 Virtual simulator program implementation

When program implementing virtual simulators of some IRLSS systems its performance data governed by technologies applied they are based on and principles of design execution are adopted as a baseline.

The Air Revitalization and Monitoring Systems (ARMS) are designed to obviate the need to replace units and/or components in prolonged operation. If a unit or a component fails 'cold' or 'hot' redundancy is used to ensure system functioning. Thus, the ONS may be localized without unit replacement.

The systems such as WRS-AC based on sorption/catalytic processes and modular construction require the replacement of some units run out of their lives.

As an example, OGS program implementation as part of the ARMS is considered below (Figure 11).

Fig. 11. Front panel of the 'OGS' virtual simulator subroutine in 'OPERATION NOMINAL MODE' (НБП/SUP=supply unit pump; КОВ/PWC=pre-purification water container; КЭI/EVI=electromagnetic valve; МНО/MMP and МНР/RMP=main micro-pump and reserve micro-pump; БХ/CU=cooling unit; БР/SU=separator unit; GA=gas analyzer).

A possible off-normal situation generated at the IWP is illustrated in Figure 12.

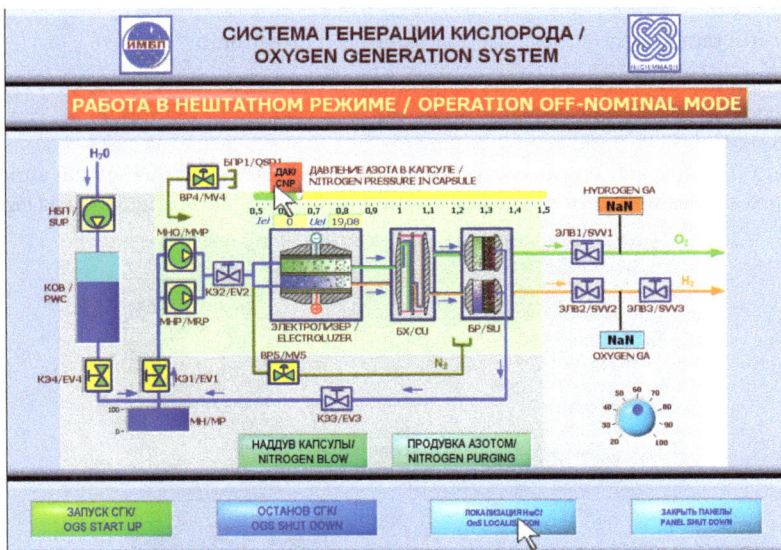

Fig. 12. Front panel of the 'OGS' virtual simulator subroutine in 'OPERATION OFF-NOMINAL MODE' (Pressure in the canister is below norm.).

The line 'CURRENT STATUS' displays 'NORMAL OPERATION' or 'ONS' inscriptions in the upper part of the control panel. The subroutine generates the following signals:

- current system status;
- a combined signal indicating the necessity of maintenance or ONS localization.

When the 'OFF-NORMAL SITUATION' signal is displayed on the control panel of the 'PMM atmosphere' main routine the operator shall switch to the OGS subroutine control panel by pressing the key and jump to ONS localization operations.

As an example of OGS shutdown the ONS by the signal 'PRESSURE IN THE CANISTER BELOW NORM (CNP)' is considered. The 'OFF-NORMAL OPERATION' inscription and 'CNP' inscription on the OGS control panel light up. To localize the ONS the operator shall press the 'ONS LOCALIZATION' key on the OGS control panel.

The canister pressurization panel (Figure 13) opens and the operator carries out all the required operations to pressurize the canister.

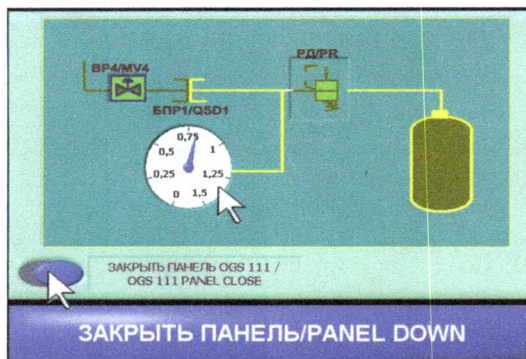

Fig. 13. Control panel of the subroutine 'Canister pressurization'.

Then the operator shall return to the subroutine by pressing 'PANEL DOWN' key. After ONS has been localized the operator shall put the system in operation by pressing the 'SYSTEM START-UP' key and check the startup.

The program implementation of on-board the PMM and TCS syste ms virtual simulators is accomplished similar to that of the IRLSS system virtual simulators (Figure 14 and Figure 15).

Fig. 14. Control panel of the PSS VS.

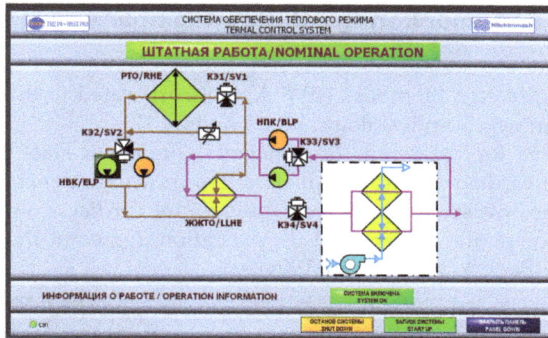

Fig. 15. Control panel of the TCS VS.

OGS-2 electrical trainer (ET) program implementation is for inculcating in crewmembers the practical skills in start-up, normal functioning, and shutdown, and in case of off-normal situations.

The ET (Figure 16) consists: an electrical operational breadboard of the liquid unit (LU); a post-purification unit; a signal and command synchronization unit (SCSU); a commutation unit; an Electron-VM monitoring and control unit implemented by an individual subroutine integrated in the HSCSOS software architecture.

Fig. 16. Electrical trainer of the OGS-2 system based on the Electron-VM standard system.

The ET architecture is based on a combination of the simulation model of functioning realized at the IWP, and standard system hardware. A set of existing units incorporated in the ET architecture is used due to availability, and necessity of carrying out manual operations on the standard system hardware. The signals generated by the sensors of the LU electrical breadboard, as well as the signals and commands are simulated on the IWP computer according to a control algorithm, then converted in the communication unit and enter the SCSU unit to be executed by the LU components.

5. Measurement and monitoring of crewmember's psycho-physiological parameters

A medico-engineering system Biomouse (BMEA) is incorporated in the HSCSOS hardware architecture to perform psycho-physiological tests (Figure 17).

The applied procedure for assessing the functional organism state is based on use of calculation analysis of cardiac rhythm parameters. This procedure enables rapid assessment of the influence of cardiovascular system on the basis of cardiac rhythm parameters. The functionality of cardiovascular system and the excitation degree of the vegetative nervous system are calculated (Baevsky, R.M. at al., 1998).

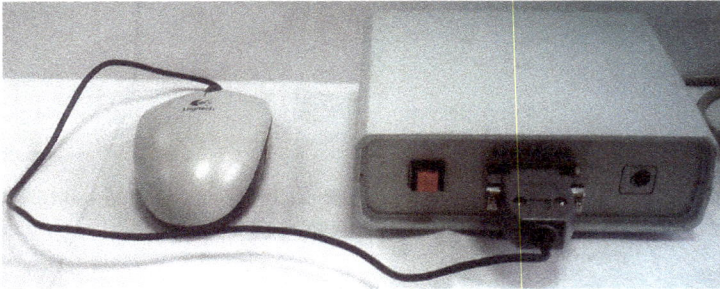

Fig. 17. A general view of the Biomouse medico-engineering system.

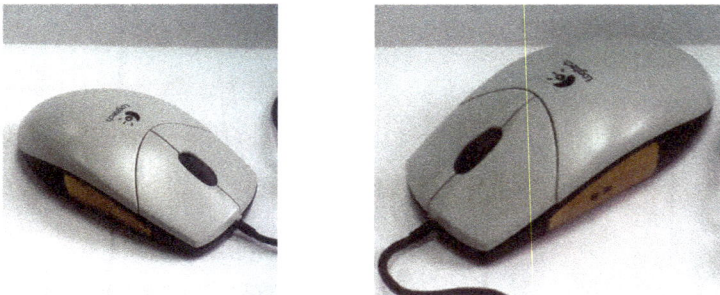

Fig. 18. Manipulator of mouse type with the built-in combined sensors: a view from the PPG sensor is shown in the left and a view from GSR and ECR sensors is shown in the right.

The BMEA system consists of a measurement unit and manipulators of mouse type installed on the OT1 and OT2 with the built-in combined sensors allowing simultaneously and continuously to register three physiological parameters (Figure 18): photo-plethysmogram (PPG), galvanic skin reflex (GSR) and electro-coetaneous resistance (ECR).

The estimation of a crewmember's psycho-physiological condition prior to the beginning of activities and on the termination of activities on localization ONS was conducted in tests mentioned below (ZAO "Neurolab", 2008):

- *Variation hronokardiometriya (WRC)* is method of rapid assessment of the cardio-interval-grams cardiovascular system regulatory mechanisms. Calculated level of the cardiovascular system functionality and autonomic homeostasis, in addition, recorded: maximum, minimum and average values, as well as fashion, mode amplitude, standard deviation and magnitude of the sequence of cardio. Primary information is photo-plethysmogram (a signal from an optical infrared sensor in a digital form). From this signal is allocated an array of cardio intervals, which is subjected to statistical processing.

- *Complex visual-motor reaction (CVMR)* is designed to study the functional state of central nervous system and elements of the operator's attention to human efficiency. The test is based on a study of the statistical characteristics of distribution of the set reaction time. On the screen appear consistently distinguished by the color of light stimuli - circles red or green, the test subject must quickly put out by pressing the right or left mouse button. Recorded response time and response error (omission, premature depression, abnormal response). The following parameters are calculated: the average response time, standard deviation, the number of errors of each type. Based on the statistical parameters of the algorithm on the attached class state of the operator's central nervous system for two-dimensional scale is calculated.

- *Reaction to a moving object (MOR)* is the test to evaluate balance of excitation and inhibition in the nervous system, as well as functional changes under the influence of the load. In this test, the test subject must stop the moving hour hand as close to 12 o'clock by using the Space key.

- *Mirror coordinograph (MC)* is designed to determine the level of stress stability of the test subject. In this test, the operator must use mouse to quickly pass a curved path on the screen without touching its edges. Time and the fact of the contour, the number of touches and time are recorded. These parameters determine the quality of the operator's actions. In addition, before and after the passage of the contour levels are recorded and the mean pacing heart rate. Changes of these parameters are interpreted as the 'value' of the operator. Performance assessment is based on two criteria: 'value' and 'quality'. 'Quality' is composed of indicators such as time of the circuit, the number of touches, while touches.

The BHEA software integrated in the HSCSOS server section software at the IWP is presented by specific database, which executes the following functions:

- storage of a database surveyed crewmembers;
- formation of a set of tests for examination;
- processing results of examination and storage of results in an archival file.

6. HSCSOS operational use in 105-day experiment under Mars-500 project

The main purpose of research is an estimation of efficiency of servicing by the crew of the IRLSS systems. The following problems should be solved for achievement of the given purpose:

- estimation of sufficiency of the controllable parameters list for the analysis of functioning and servicing on the basis of use of the HSCSOS and an electrical trainer of the Electron-VM integrated in complex in conditions of long autonomous mission;
- estimation of efficiency of activity of the operator on localization of the off-nominal situations arising at functioning of systems and/or deviations of the environment controllable parameter values from prescribed values;
- estimation of efficiency of acceptance of independent solutions by crewmembers in the ONS localization;

- estimation of efficiency of ways of display of the information on values of the environment controllable parameters in analysis of functioning and servicing with the use of virtual simulators of systems;
- estimation of the ONS localization influence on the mental and physiological state of crewmembers in conditions of long-term autonomous mission.

6.1 Technique of an experiment

In realization of initial phase, being the final stage of the crew training makes tentative estimation of the crewmember action efficiency in the ONS localization. Formation of the particular situation arising in the IRLSS specific system operation is made on the basis of random sample by the Instructor.

Localization of the arisen ONS is made by each crewmember with the use of the on-board instruction from the operator's workplace disposed in the Main Control Board. In this case both a rigid copy of the on-board instruction, and its electronic version which is available in a format *.PDF on Operator's Terminal 1 and Operator's Terminal 2, can be used without Mission Control Center recommendations.

Generated off-normal situations are characterized by different degrees of complexity in their localization:

- simple (service of a complex, localization of some ONS, not demanding replacements of units, etc.);
- average complexity (the most part of ONS entered demanded the replacement of units);
- complicated (actuate crewmember's activity with the Electron-VM electrical trainer, and also ONS, demanding a long-period operation on elimination or monitoring (imitation of a fire or leakages).

For assessment of the crewmember gained efficiency the following is considered:

- complexity of necessary camera skills depending on the solved problem of ONS localization;
- time of reaction on ONS being in parameter time-dependent day of the ONS occurrence and from congestion of crew other problems;
- attentiveness of crew during ONS localization;
- time spent for ONS localization;
- total amount of the activities considering total of solved tasks including monitoring of system operation, activity with an electrical simulator and realization of maintenance;
- dynamics of activity formation.

The increased duration of localization of failure indicates absence of attention concentration in some crewmembers that is connected or with realization of some additional activities, either with fatigue, or with presence of distracting factors.

6.2 Results of experiment

During experiment 52 tasks in total (including activity with the Electron-VM electrical trainer) are generated. 51 tasks are successfully solved. Views generated of the ONS and times expended on its localization are given in Table 2. Results of the crewmember action efficiency estimation in the ONS localization are presented in Table 3.

The typical results of experiments are shown on Figure 19 and Figure 20.

ONS localized	Time spent by a crewmember on ONS localization, min						Total time, min	Average time, min
	1001	1002	1003	1004	1005	1006		
TCCS ONS:								
Failure of fan		2		12		12	26	6.5
Failure of electroheater		5	14		10	2	33	6.6
Failure of valve	2			5			7	3.5
CDRS ONS:								
Failure of fan	2	4	2	7		3	18	2.8
Failure of valve			6	1		5	12	4
WRS-AC ONS:								
Failure of pump		4	11	5	5	8	33	4.7
OGS-1 ONS:								
ONS 'CURRENT < 2 A'	5	11					16	8
ONS 'PRESSURE in CAPSULE < NORM'			3				3	3
Failure of pump						2	2	2
ONS 'ELECROLYZER TEMPERATURE < NORM'					4	8	12	6
TCS ONS:								
Failure of pump		11	8				19	9.5
Failure of fan		2		4	5	3	14	3.5
Failure of valve		6		3		9	18	6

Table 2. ONS types generated and times expended on its localization.

CREWMEMBER	ESTIMATION of CREWMEMBER'S ACTION EFFICIENCY															
	IN TOTAL ONS LOCALIZATION	TOTAL of DIFFICULT TASKS	FROM ITS SOLVED	TOTAL of AVERAGE COMPLIXITY	FROM ITS SOLVED	TOTAL of SIMPLE TASKS	FROM ITS SOLVED	ANALYSIS of SYSTEM OPERATION	USE of ONBORD INSTRUCTION under INDEPENT ACTIONS	ONBOARD INSTRUCTION CORRECTION	USE of EXTERNAL HELP under INDEPENT ACTIONS	ATTENTIVENESS under ACTIONS	REACTION under ACTIONS	AVERAGE TIME for ONS LOCALIZATION of AVERAGE COMPLEXITY, min	AVERAGE TIME for ONS LOCALIZATION of LOW COMPLEXITY, min
1001	6	1	0	1	1	1	1	0	2	0	2	2.5	1	4	1
1002	11	2	1	8	8	1	1	1	2	1	2	2	1	5.5	3
1003	10	1	0	7	7	2	2	0	1	0	1	1.5	1	5.6	5
1004	11	1	0	7	7	3	3	0	1	0	1	2.5	1	4.3	2.3
1005	5	0	0	5	5	0	0	0	1	0	1	1.5	1	4.6	0
1006	11	0	0	7	7	1	1	0	1	0	1	2	1	5.1	3.8

Table 3. Results of the crewmember action efficiency estimation at ONS localization.

With a task of elimination of leakages in the PMM illuminators conducted from 06.29.09 to 07.08.09, the crew has failed. The reason is easing of attentiveness and fatigue of crew at the final stage of the experiment.

Commissioning of Electron-VM electrical trainer has been successfully conducted by the 06.25.09, operator 1002.

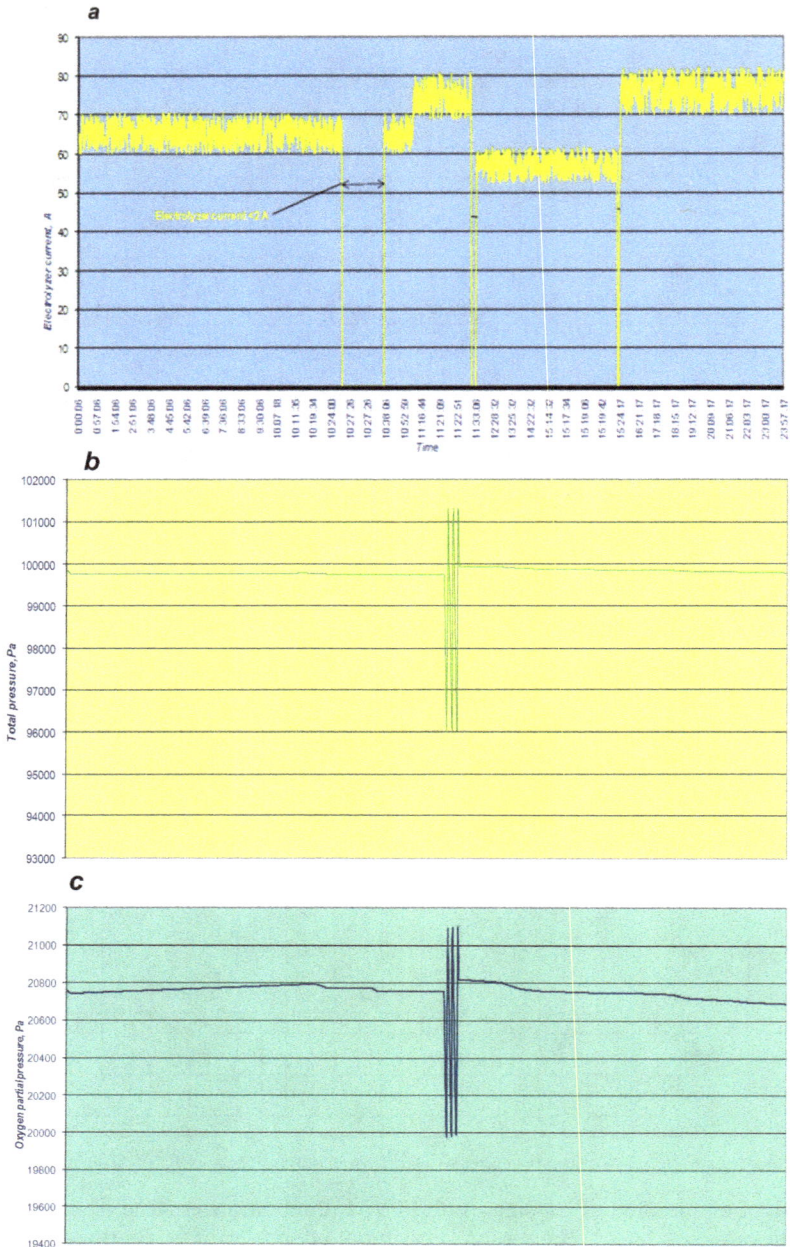

Fig. 19. Localization of the ONS *'ELECTROLYZER CURRENT < 2 A'* (Crewmember 1002):
a - electrolyzer current change; b - total pressure change; c - oxygen partial pressure change.

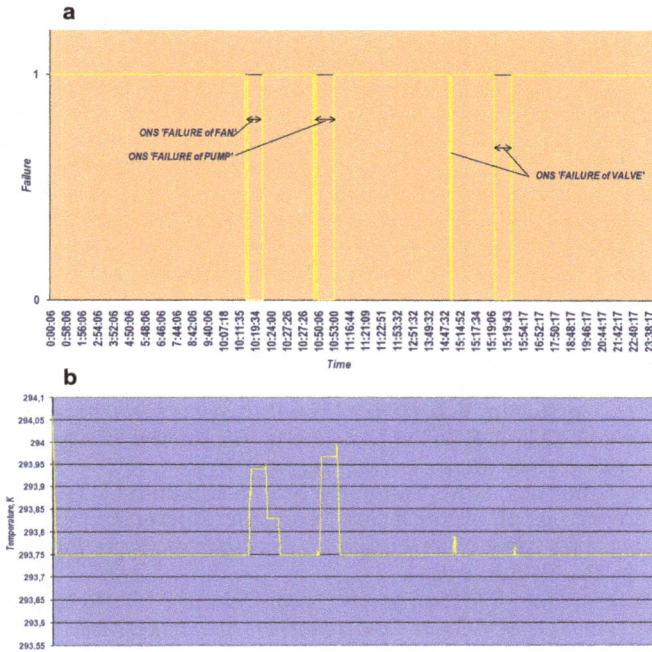

Fig. 20. Localization of the ONS in TCS (Crewmembers: 1005, 1006, 1001): a – failures; b – change of temperature in the PPM.

7. Conclusion

In the chapter the approach to analysis of interaction in the Man-IRLSS system relating to a complex man-machine system, in which human-operator interacts with a technical device during production of environment components, management, processing information, etc is considered.

The offered approach is advanced for the ground medical/engineering experiment imitating interplanetary flight to Mars at which use of standard aboard regeneration life support systems is complicated and is economically unprofitable.

The HSCSOs application has allowed solving the following primary problems:

- to conduct an estimation of crewmember action efficiency at service of the IRLSS and at ONS localization caused by probable failures in its functioning and deviations of the environment controllable parameters from preset values in view of a degree of readiness of crewmembers and conditions of long isolation;

- to research interactions in the IRLSS-Crew system in real time with the purpose of the medical/engineering and ergonomic requirements to IRLSS systems.

Including of standard systems in the HSCSOS architecture is effected by replacement of corresponding virtual simulators with standard systems. As this takes place information channels to and from systems are locked in corresponding logic devices controlling directly actual hardware. Data exchange procedure between specific virtual systems corresponds to a logic structure of flow exchange between actual systems therefore replacement of virtual systems with actual ones will not be problematic.

The considered approaches to research of the IRLSS virtual simulators can be used at development of 'man – machine' for other particular mission.

8. Acknowledgment

Authors express thanks to the colleagues from Joint-Stock Company 'NIIchimmash' Lev Gavrilov, Aleksey Kochetkov, Victor Andreev and Roman Sachkov, as well as the our colleagues from SRC RF – IMBP RAN Juriy Sinyak and Vladimir Trikolkin in many respects promoting fulfillment of the given charter.
Authors are grateful to the reviewer of this chapter Aleksandar Lazinica from CEO for valuable comments made at its preparation.

9. References

Adamovich, B. A., & Gorshenin, V. B. (1997). *Life outside of the Earth.* Moscow, Science.
Baevsky, R.M., Kirillov, O.I., & Kletskin, S.Z. (1984). *The mathematical analysis of changes of a cordial rhythm at stress.* Moscow, Science.
Kurmazenko, E.A., Samsonov, N.M., Farafonov N.S., & Dokunin I.V (1997). *Simulation Subsystem Models for Analysis of the Integrated Regenerative Life Support Systems Functioning.* Proceeding of 6th European Symposium on Space Environmental Systems, Noordwijk, the Netherlands.
Kurmazenko, E.A., Gavrilov, L.I., Samsonov, N.M. Dokunin, I.V., Romanov S.Ju., & Ryabkin, A.M. (1998) *A Man-Made Gas Atmosphere Simulation Model of International Space Station's Russian Segment.* SAE Technical Paper Series #981718.
Kurmazenko, E.A., Samsonov, N.M., Farafonov, N.S., Gavrilov, L.I., Dokunin I.V., & Kotelnikova, M.A. (2000). *Crew software simulation for Integrated Life Support System Operation Study.* SAE Technical Paper Series #2000-2120.
Kurmazenko, E.A., Gavrilov, L.I., Kochetkov, A.A., Khabarovskiy, N.N. Demin, E.P., Grigorjev A.I, & Baranov V.N. (2008). *Life Support System Virtual Simulators for Mars-500 Ground-Based Experiment.* Proceeding of 59th International Astronautical Congress, Glasgow, Scotland.
Kurmazenko, E.A., Gavrilov, L.I., Kochetkov, A.A., & Khabarovskiy, N.N. (2009). *Space Ecological/Engineering System for the Manned Interplanetary Vehicles Crew: Status and Key Technologies for its Development.* Proceeding of the 60th International Austronautical Congress, Daejeon, Republic of Korea.
Kurmazenko, E.A., Kochetkov, A.A., Gavrilov, L.I., Khabarovskiy, N.N., Kamaletdinova, G.R., Morukov,B.V., Demin, E.P., & Trikolkin, V.I (2010). *Crew's Service of the Virtual Integrated Regenerative Life-Support System Operation: 105-Day Experiment Results under Mars-500 Project.* AIAA Technical Paper Series #2010-6168.
Pravetskiy, V.N., Samsonov, V.M., Utyamyshev, R.I., & Kurmazenko, E.A. (1981). *Separate Problems of the Life-Support and Safety System Development for a Spaceflight Vehicle Crew. Scientific Reading on Aircraft and Astronautics 1980.* Moscow, Science.
Savina, V.P., & Kuznetsova, T.I. (1980) *Trace contaminant sources and their toxicological evaluation.* Space Biology, *v. 42,* Moscow, Science.
User guide complex BioMouse CPP and CPP-01-01b (options "Professional" and "Research "), introducing the principles of operation of the product with the functions of the software, and implemented methods (2008). Moscow, ZAO "Neurolab", www.neurolab.ru

Permissions

The contributors of this book come from diverse backgrounds, making this book a truly international effort. This book will bring forth new frontiers with its revolutionizing research information and detailed analysis of the nascent developments around the world.

We would like to thank Prof. Dhanjoo N. Ghista, for lending his expertise to make the book truly unique. He has played a crucial role in the development of this book. Without his invaluable contribution this book wouldn't have been possible. He has made vital efforts to compile up to date information on the varied aspects of this subject to make this book a valuable addition to the collection of many professionals and students.

This book was conceptualized with the vision of imparting up-to-date information and advanced data in this field. To ensure the same, a matchless editorial board was set up. Every individual on the board went through rigorous rounds of assessment to prove their worth. After which they invested a large part of their time researching and compiling the most relevant data for our readers. Conferences and sessions were held from time to time between the editorial board and the contributing authors to present the data in the most comprehensible form. The editorial team has worked tirelessly to provide valuable and valid information to help people across the globe.

Every chapter published in this book has been scrutinized by our experts. Their significance has been extensively debated. The topics covered herein carry significant findings which will fuel the growth of the discipline. They may even be implemented as practical applications or may be referred to as a beginning point for another development. Chapters in this book were first published by InTech; hereby published with permission under the Creative Commons Attribution License or equivalent.

The editorial board has been involved in producing this book since its inception. They have spent rigorous hours researching and exploring the diverse topics which have resulted in the successful publishing of this book. They have passed on their knowledge of decades through this book. To expedite this challenging task, the publisher supported the team at every step. A small team of assistant editors was also appointed to further simplify the editing procedure and attain best results for the readers.

Our editorial team has been hand-picked from every corner of the world. Their multi-ethnicity adds dynamic inputs to the discussions which result in innovative outcomes. These outcomes are then further discussed with the researchers and contributors who give their valuable feedback and opinion regarding the same. The feedback is then collaborated with the researches and they are edited in a comprehensive manner to aid the understanding of the subject.

Apart from the editorial board, the designing team has also invested a significant amount of their time in understanding the subject and creating the most relevant covers. They scrutinized every image to scout for the most suitable representation of the subject and create an appropriate cover for the book.

The publishing team has been involved in this book since its early stages. They were actively engaged in every process, be it collecting the data, connecting with the contributors or procuring relevant information. The team has been an ardent support to the editorial, designing and production team. Their endless efforts to recruit the best for this project, has resulted in the accomplishment of this book. They are a veteran in the field of academics and their pool of knowledge is as vast as their experience in printing. Their expertise and guidance has proved useful at every step. Their uncompromising quality standards have made this book an exceptional effort. Their encouragement from time to time has been an inspiration for everyone.

The publisher and the editorial board hope that this book will prove to be a valuable piece of knowledge for researchers, students, practitioners and scholars across the globe.

List of Contributors

Husheng Yan and Miao Guo
Key Laboratory of Functional Polymer Materials, Ministry of Education, Institute of Polymer Chemistry, Nankai University, Tianjin, P. R. China

Keliang Liu
Beijing Institute of Pharmacology and Toxicology, Beijing, P. R. China

N. De Geyter and R. Morent
Research Unit Plasma Technology – Department of Applied Physics, Faculty of Engineering and Architecture – Ghent University, Belgium

Lin Xiao, Bo Wang and Guang Yang
College of Life Science and Technology, Huazhong University of Science and Technology, China
National Engineering Research Center for Nano-Medicine, Huazhong University of Science and Technology, China

Mario Gauthier
Department of Chemistry, University of Waterloo, Canada

Barry M. O'Connell and Michael T. Walsh
Centre for Applied Biomedical Engineering Research (CABER), Department of Mechanical Aeronautical and Biomedical Engineering and Materials and Surface Science Institute, University of Limerick, Limerick, Ireland

Letizia Fracchia, Massimo Cavallo and Maria Giovanna Martinotti
Department of Chemical, Food, Pharmaceutical and Pharmacological Sciences, Drug and Food Biotechnology Center, Università del Piemonte Orientale "Amedeo Avogadro", Novara, Italy

Ibrahim M. Banat
School of Biomedical Sciences, Faculty of Life and Health Sciences, University of Ulster, Coleraine, N. Ireland, UK

Dušan Koji´c, Božica Bojovi´c, Dragomir Stamenkovi´c, Nikola Jagodi´c and Ðuro Koruga
University of Belgrade, Faculty of Mechanical Engineering, Department of Biomedical Engineering, Serbia

Catauro Michelina and Bollino Flavia
Department of Aerospace and Mechanical Engineering, Second University of Naples, Aversa, Italy

Hua Cao, Deyin Lu and Bahram Khoobehi
Louisiana State University, University of Mississippi Medical Center, LSU Eye Center, USA

Dhanjoo N. Ghista
Department of Graduate and Continuing Education, Framingham State University, Framingham, Massachusetts, USA

U. Rajendra Acharya
School of Engineering, Division of ECE, Ngee Ann Polytechnic, Singapore

Kamlakar D. Desai
Mukesh Patel School of Technology Management & Engineering, India

Sarma Dittakavi
Biomedical Engineering department, Osmania University, India

Adejuwon A. Adeneye
Department of Pharmacology, Faculty of Basic Medical Sciences, Lagos State University College of Medicine, Ikeja, Department of Pharmaceutical Sciences, College of Pharmacy, University of Kentucky, Kentucky, Lagos state, Nigeria

Loh Kah Meng
VicWell Biomedical Private Limited, Singapore

Mario F. Letelier and Juan S. Stockle
University of Santiago of Chile, Chile

Sepp Porta
Institute of Applied Stress Research, Judendorf – Strassengel, Austria
Institute of Pathophysiology, Medical University of Graz, Austria
Institute of Mathematics and Scientific Computing, KFU Graz, Austria
Rehabilitation Clinique of the AUVA, Tobelbad, Austria
Theresianische Militärakademie, Wiener Neustadt, Austria
St. Anna Hospital, Herne, Germany

William J. Bosl
Harvard Medical School, Children's Hospital Boston Informatics Program, Boston University School of Medicine Behavioral Neuroscience Program, USA

Wasilewska-Radwanska Marta
Multidisciplinary School of Engineering in Biomedicine, AGH-University of Science and Technology, Poland

Augustyniak Ewa
Faculty of Humanities, AGH-University of Science and Technology, Poland

Tadeusiewicz Ryszard and Augustyniak Piotr
Multidisciplinary School of Engineering in Biomedicine, AGH-University of Science and Technology, Poland
Faculty of Electrical Engineering, Automatics, Computer Science and Electronics, AGH-University of Science and Technology, Poland

Eduard Kurmazenko, Nikolay Khabarovskiy and Guzel Kamaletdinova
Joint-Stock Co. 'NIIchimmash', Russia

Evgeniy Demin and Boris Morukov
SSC RF-IMBP RAS, Russia